STUDIES IN MICROBIOLOGY

Principles of Gene Manipulation

AN INTRODUCTION TO GENETIC ENGINEERING

R. W. OLD MA, PhD
Department of Biological Sciences
University of Warwick
Coventry

S. B. PRIMROSE BSc, PhD
Amersham International plc
Amersham, Buckinghamshire

FIFTH EDITION

**Blackwell
Science**

STUDIES IN MICROBIOLOGY

SERIES EDITOR

N. G. CARR
Department of Biological Sciences
University of Warwick

First published in 1980
Second edition 1981
Reprinted 1982, 1983 (twice)
Japanese translation 1983
Third edition 1985
Reprinted 1986 (twice), 1987, 1988
Japanese translation 1986
Fourth edition 1989
Reprinted 1990, 1991, 1992, 1993 (twice)
German translation 1992
Japanese translation 1992
Fifth edition 1994
Reprinted 1995 (twice), 1996

Set by Excel Typesetters Co., Hong Kong
Printed and bound in Great Britain at
the Alden Press Limited, Oxford and
Northampton

DISTRIBUTORS

Marston Book Services Ltd
PO Box 87
Oxford OX2 0DT
(*Orders*: Tel: 01865 791155
        Fax: 01865 791927
        Telex: 837515)

USA
Blackwell Science, Inc.
238 Main Street
Cambridge, MA 02142
(*Orders*: Tel: 800 215-1000
            617 876-7000
        Fax: 617 492-5263)

Canada
Copp Clark, Ltd
2775 Matheson Blvd East
Mississauga, Ontario
Canada, L4W 4P7
(*Orders*: Tel: 800 263-4374
            905 238-6074)

Australia
Blackwell Science Pty Ltd
54 University Street
Carlton, Victoria 3053
(*Orders*: Tel: 03 9347-0300
        Fax: 03 9349 3016)

A catalogue record for this title
is available from the British Library

ISBN 0-632-03712-1

Library of Congress
Cataloging-in-Publication Data

Old, R.W.
Principles of gene manipulation:
    an introduction to genetic
    engineering/R.W. Old,
    S. B. Primrose. – 5th ed.
        p.   cm. – (Studies in microbiology)
    Includes bibliographical references
    and index.
    ISBN 0-632-03712-1
1. Genetic engineering.
I. Primrose, S.B. II. Title.   III. Series.
[DNLM: 1. Genetic Engineering.
2. DNA, Recombinant.
W1 ST924 1994/QH 442 044p 1994]
QH442.042   1994
575.1′0 0724 – dc20
DNLM/DLC
for Library of Congress

Contents

Preface

It is 15 years since the first edition of *Principles of Gene Manipulation* was published. In writing the first edition, our aim was to explain a new and rapidly growing technology. Our basic philosophy was to present the principles of gene manipulation, and its associated techniques, in sufficient detail to enable the non-specialist reader to understand them. We also intended that the scope of this technology, and its potential impact on virtually all areas of biology, would be evident. It was assumed that the reader would have a reasonable working knowledge of molecular biology.

The second, third and fourth editions were enlarged, to cope with the advances in technology which were quickly broadening the field. It had become apparent that, around a core of fundamental techniques concerning the manipulation of DNA *in vitro*, there was developing an ever-expanding repertoire of transformation techniques, library construction and screening methods, expression systems, and host–vector systems. In this fifth edition the biggest change has been the need to devote an entire chapter to the polymerase chain reaction because now it can facilitate every step in gene manipulation.

In this new edition we have stuck to the basic philosophy of the previous ones, which we feel has been justified with the passage of time: we think it is valuable to identify and explain the basic principles. As before, the book is intended to be an introduction to the subject. But with the field having grown as much as it now has, we hope that the book will provide a useful overview of the state-of-the-art for researchers who already have experience of recombinant DNA work. We recognize that previous editions have been used widely as textbooks for undergraduates, and we want to continue to serve this readership. Therefore we have not changed the level at which the book is written, nor the general style. We also have retained many of the early experiments on gene manipulation and some of the earlier but classical work on bacterial and viral genetics. Although older readers may consider this work 'dated' the current generation of students often have never been exposed to it. Without this core knowledge they cannot appreciate fully what has been achieved. Inevitably the book is larger than before but we hope that it still communicates the essential simplicity of gene manipulation.

Once again we should like to acknowledge the assistance of our colleagues, at the University of Warwick, at Amersham International plc, and elsewhere, in the impossible task of keeping up-to-date in this exciting field. In particular, we would like to thank Sue Dale, Andrew Gooday, Alan Hamilton and Colin Harwood who took the time to review certain chapters and suggest ways in which they could be improved. Thanks also are due to Margaret Courtney for typing large sections of the book and providing much needed secretarial support and to Vera Butterworth and Lynne Goodman for assistance in finding and checking many of the references.

R.W. Old
S.B. Primrose

Abbreviations and conversion scale

amber (mutation)	*am*
covalently closed circles	CCC
dihydrofolate reductase	DHFR
gene for DNA ligase	*lig*
kilobases	kb
megadaltons	MDa
molecular weight	mol. wt
plaque-forming unit	p.f.u.
purine	Pu
pyrimidine	Py
resistance (to an antibiotic)	R
sensitivity (to an antibiotic)	S
temperature-sensitive (mutation)	*ts*
thymidine kinase	TK

Abbreviations of antibiotic names

ampicillin	Ap
chloramphenicol	Cm
kanamycin	Km
neomycin	Nm
streptomycin	Sm
sulphonamide	Su
tetracycline	Tc
trimethoprim	Tp

Scale for conversion between kilobase pairs of duplex DNA and molecular weight.

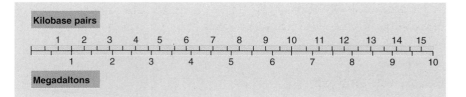

1 Gene manipulation: an all-embracing technique

Introduction

Occasionally technical developments in science occur that enable leaps forward in our knowledge and increase the potential for innovation. Molecular biology and biomedical research experienced such a revolutionary change 20 years ago with the development of gene manipulation. Although the initial experiments generated much excitement it is unlikely that any of the early workers in the field could have predicted the breadth of applications to which the technique has been put. Nor could they have envisaged that the methods they developed would spawn an entire industry comprising several hundred companies, of varying sizes, in the US alone.

The term gene manipulation can be applied to a variety of sophisticated *in vivo* genetics as well as to *in vitro* techniques. In fact, in most Western countries there is a precise *legal* definition of gene manipulation as a result of government legislation to control it. In the United Kingdom, gene manipulation is defined as

> the formation of new combinations of heritable material by the insertion of nucleic acid molecules, produced by whatever means outside the cell, into any virus, bacterial plasmid or other vector system so as to allow their incorporation into a host organism in which they do not naturally occur but in which they are capable of continued propagation.

The definitions adopted by other countries are similar and all adequately describe the subject matter of this book. Simply put, gene manipulation permits stretches of DNA to be isolated from their host organism and propagated in the same or a different host, a technique known as *cloning*. The ability to clone DNA has far-reaching consequences as will be shown below.

Sequence analysis

Cloning permits the isolation of discrete pieces of a genome and their amplification. This in turn enables the DNA to be sequenced. Analysis of the sequences of some genetically well-characterized genes led to the identification of the sequences and structures which characterize the

[1]

principal control elements of gene expression, e.g. promoters, ribosome binding sites, etc. As this information built up it became possible to scan new DNA sequences and identify potential new genes, or *open reading frames*, because they were bounded by characteristic motifs. Initially this sequence analysis was done manually but to the eye long runs of nucleotides have little meaning and patterns evade recognition. Fortunately such analyses have been facilitated by rapid increases in the power of computers and improvements in software which have taken place contemporaneously with advances in gene cloning. Now sequences can be scanned quickly for a whole series of structural features, e.g. restriction enzyme recognition sites, start and stop signals for transcription, inverted palindromes, sequence repeats, Z-DNA, etc. A good example of the power of computers in analysing sequence data comes from the work of Koonin *et al.* (1994). They analysed the information contained in the complete nucleotide sequence of yeast chromosome III (see below) and found that 61% of the probable gene products have significant similarities to sequences in the current databases. As many as 54% have known functions or are related to functionally characterized proteins and 19% are similar to proteins of known three-dimensional structure. New proteins identified include a sugar kinase related to ribokinases, a phosphatidyl serine synthetase, a putative transcription regulator, a flavodoxin-like protein and a zinc finger protein belonging to a distinct subfamily.

From the nucleotide sequence of a gene it is easy to deduce the protein sequence which it encodes. Unfortunately, we are unable to formulate a set of general rules that allows us to predict a protein's three-dimensional structure from the amino acid sequence of its polypeptide chain. However, based on crystallographic data from over 300 proteins certain structural motifs can be predicted. Nor does an amino acid sequence on its own give any clue as to function. The solution is to compare the amino acid sequence with that of other better-characterized proteins: a high degree of homology suggests similarity in function. Again, computers are of great value since algorithms exist for comparing two sequences or for comparing one sequence with a group of other sequences simultaneously. This comparison of sequence information operates best when the user has access to all the sequence data available in the world. This is possible due to the existence of central data storage banks in Germany and the USA (Rice *et al.* 1993, Benson *et al.* 1993) from which sequences can be retrieved and transferred to local systems by subscribers.

The original goal of sequencing was to determine the precise order of nucleotides in a gene. Then the goal became the sequence of a small genome. First it was that of a small virus (ΦX174). This was followed by a number of plasmid genomes, e.g. pBR322, larger viruses such as bacteriophage λ, Epstein–Barr virus and Vaccinia virus and most recently a complete chromosome from yeast (315 316 nucleotides; Oliver *et al.* 1992) and the nematode *Caenorhabditis elegans* (2181 032 nucleotides; Wilson *et al.* 1994). The current goals are the entire DNA sequences of *E. coli*, yeast, *C. elegans*, various plant genomes and, most ambitious of all, the entire human genome. At the time of going to press, almost eight of the yeast chromosomes have been entirely sequenced. The biggest surprise from the sequences obtained from *C. elegans* (Sulston *et al.* 1992) and yeast (Oliver *et al.* 1992) is that there are more genes in the regions sequenced than had been expected, e.g. yeast chromosome III has a gene content five times greater than previously believed. Once the sequencing of large genomes becomes easier and more routine, one can envisage that it will be used to dissect the process of evolution at a molecular level. For example, how different at the nucleotide level are the finches of the Galapagos islands? Such analyses will tell us what DNA changes occurred, and perhaps how, but not why.

Sequence information can be used in a different way for the DNA fragment sequenced provides a molecular hybridization probe of absolute purity. Such probes can be used diagnostically to determine or confirm the identity of a microorganism. They also can be used forensically as a molecular fingerprint either of a criminal or for pedigree analysis of plants, animals or even humans, e.g. the recent identification of the remains of the last Russian Tsar and his family (Gill *et al.* 1994). Such probes also can be used to diagnose genetic disease. So far, most have been limited to simple single gene disorders, e.g. thalassaemia, but in time will be extended to common multifactorial disorders such as cancer, heart disease and mental illness.

In vivo biochemistry

Any living cell, regardless of its origin, carries out a plethora of biochemical reactions. To analyse these different reactions, biochemists break open cells, isolate the key components of interest and measure their levels. They purify these components and try to determine their performance characteristics. For example, in the case of an enzyme they might determine its substrate specificity, kinetic parameters such as K_m and V_{max}, and identify inhibitors and their mode of action. From these data they try to build up a picture of what happens inside the cell. However, the properties of a purified enzyme in a test tube may bear little resemblance to its behaviour when it shares the cell cytoplasm or a cell compartment with thousands of other enzymes and chemical compounds. Understanding what happens inside cells has been facilitated by the use of mutants. These permit the determination of the consequences of altered regulation or loss of a particular component or activity. Mutants also have been useful in elucidating macromolecule structure and function. However, the use of mutants is limited by the fact that with classical technologies one usually has little control over the type of mutant isolated and/or location of the mutation.

Gene cloning provides elegant solutions to the above problems. Once isolated entire genes or groups of genes can be introduced back into the cell type from whence they came, or different cell types or completely new organisms, e.g. bacterial genes in plants or animals. The levels of gene expression can be measured directly or through the use of reporter mol-

Box 1.1 The birth of an industry

Biotechnology is not new. Cheese, bread and yoghurt are products of biotechnology and have been known for centuries. However, the stock market excitement about biotechnology stems from the potential of gene manipulation, which is the subject of this book. The birth of this modern version of biotechnology can be traced to the founding of the company Genentech.

In 1976, a 27-year-old venture capitalist called Robert Swanson had a discussion over a few beers with a University of California professor, Herb Boyer. The discussion centred on the commercial potential of gene manipulation. Swanson's enthusiasm for the technology and his faith in it was contagious. By the close of the meeting the decision was taken to found Genentech (Genetic Engineering Technology). Though Swanson and Boyer faced scepticism from both the academic and business communities they forged ahead with their idea. Successes came thick and fast (see Table B1.1) and within a few years they had proved their detractors wrong. Over 600 biotechnology companies have been set up in the USA alone since the founding of Genentech but very, very few have been as successful.

Table B1.1 Key events at Genentech

1976	Genentech founded
1977	Genentech produced first human protein (somatostatin) in a microorganism
1978	Human insulin cloned by Genentech scientists
1979	Human growth hormone cloned by Genentech scientists
1980	Genentech went public, raising $35 million
1982	First recombinant DNA drug (human insulin) marketed (Genentech product licensed to Eli Lilly & Co.)
1984	First laboratory production of Factor VIII for therapy of haemophilia. Licence granted to Cutter Biological
1985	Genentech launched its first product, Protropin (human growth hormone) for growth hormone deficiency in children
1987	Genentech launched Activase (tissue plasminogen activator) for dissolving blood clots in heart attack patients
1990	Genentech launched Actimmune (interferon-$\gamma_{1\beta}$) for treatment of chronic granulomatous disease
1990	Genentech and the Swiss pharmaceutical company Roche complete a $2.1 billion merger

ecules and can be modulated up or down at the whim of the experimenter. Also, specific mutations ranging from a single base-pair to large deletions or additions can be built into the gene at any position to permit all kinds of structural and functional analyses. Function in different cell types also can be analysed, e.g. do those structural features of a protein which result in its secretion from a yeast cell enable it to be exported from bacteria or higher eukaryotes? Experiments like these permit comparative studies of macromolecular processes and in some cases gene cloning and sequencing provides the only way to begin to understand such events as mitosis, cell division, telomere structure, intron splicing, etc.

Over-production of cellular components

The determination of the structure, function or utility of a protein demands that adequate amounts of purified material are available for study. This is not always an easy task, particularly when the protein is normally present at very low levels in the cell. Genetic engineering provides a means of generating sufficient material: the relevant gene is cloned and over-expressed at whatever scale is required. For example, 5 mg of somatostatin was first isolated from half a million sheep brains and a smaller amount of epidermal growth factor from 40 000 gallons of human urine. After gene cloning the same amount of material was obtained from a few litres of bacterial culture. This principle has been applied to a wide range of mammalian cellular proteins and was the basis for many of the biotechnology start-up companies in the US (see Box 1.1). Many of these companies are hoping that data from the Human Genome Sequencing Project will result in the identification of a large number of new proteins with potential for human therapy. Others are using gene manipulation to understand the regulation of transcription of particular genes arguing that it would make better therapeutic sense to modulate the process with low-molecular-weight, orally-active drugs.

Over-production need not be restricted to proteins. It is possible to raise the levels of most intracellular components provided they are not toxic to the producing organism. This can be done by cloning all the genes for a particular biosynthetic pathway and over-expressing them. Alternatively, it is possible to shut down particular metabolic pathways and thus re-direct particular intermediates towards the desired end-product.

2   Basic techniques

Introduction

The initial impetus for gene manipulation *in vitro* came about in the early 1970s with the simultaneous development of techniques for:
- transformation* of *Escherichia coli*;
- cutting and joining DNA molecules;
- monitoring the cutting and joining reactions.

In order to explain the significance of these developments we must first consider the essential requirements of a successful gene-manipulation procedure.

The basic problems

Before the advent of modern gene-manipulation methods there had been many early attempts at transforming pro- and eukaryotic cells with foreign DNA. But, in general, little progress could be made. The reasons for this are as follows. Let us assume that the exogenous DNA is taken up by the recipient cells. There are then two basic difficulties. First, where detection of uptake is dependent on gene expression, failure could be due to lack of accurate transcription or translation. Second, and more importantly, the exogenous DNA may not be maintained in the transformed cells. If the exogenous DNA is integrated into the host genome, there is no problem. The exact mechanism whereby this integration occurs is not clear and it is usually a rare event. However this occurs, the result is that the foreign DNA sequence becomes incorporated into the host cell's genetic material and subsequently will be propagated as part of that genome. If, however, the exogenous DNA fails to be integrated, it will probably be lost during subsequent multiplication of the host cells. The reason for this is simple. In

*The sudden change of an animal cell possessing normal growth properties into one with many of the growth properties of the cancer cell is called *growth transformation*. Growth transformation is mentioned in Chapter 15 and should not be confused with bacterial transformation which is described here.

[6]

order to be replicated, DNA molecules must contain an *origin of replication*, and in bacteria and viruses there is usually only one per genome. Such molecules are called *replicons*. Fragments of DNA are not replicons and in the absence of replication will be diluted out of their host cells. It should be noted that even if a DNA molecule contains an origin of replication, this may not function in a foreign host cell.

There is an additional, subsequent problem. If the early experiments were to proceed, a method was required for assessing the fate of the donor DNA. In particular, in circumstances where the foreign DNA was maintained because it had become integrated in the host DNA, a method was required for mapping the foreign DNA and the surrounding host sequences.

The solutions: basic techniques

If fragments of DNA are not replicated, the obvious solution is to attach them to a suitable replicon. Such replicons are known as *vectors* or *cloning vehicles*. Small plasmids and bacteriophages are the most suitable vectors for they are replicons in their own right, their maintenance does not necessarily require integration into the host genome and their DNA can be isolated readily in an intact form. The different plasmids and phages which are used as vectors are described in detail in Chapters 4 and 5. Suffice it to say at this point that initially plasmids and phages suitable as vectors were only found in *Escherichia coli*. An important consequence follows from the use of a vector to carry the foreign DNA: simple methods become available for purifying the vector molecule, complete with its foreign DNA insert, from transformed host cells. Thus not only does the vector provide the replicon function, but it also permits the easy bulk preparation of the foreign DNA sequence, free from host-cell DNA.

Composite molecules in which foreign DNA has been inserted into a vector molecule are sometimes called DNA *chimaeras* because of their analogy with the Chimaera of mythology – a creature with the head of a lion, body of a goat and the tail of a serpent. The construction of such composite or *artificial recombinant* molecules has also been termed *genetic engineering* or *gene manipulation* because of the potential for creating novel genetic combinations by biochemical means. The process has also been termed *molecular cloning* or *gene cloning* because a line of genetically identical organisms, all of which contain the composite molecule, can be propagated and grown in bulk, hence *amplifying* the composite molecule and *any gene product whose synthesis it directs*.

Although conceptually very simple, cloning of a fragment of foreign, or *passenger*, or *target* DNA in a vector demands that the following can be accomplished.
- The vector DNA must be purified and cut open.
- The passenger DNA must be inserted into the vector molecule to create the artificial recombinant. DNA joining reactions must therefore be performed. Methods for cutting and joining DNA molecules are now so sophisticated that they warrant a chapter of their own (Chapter 3).

- The cutting and joining reactions must be readily monitored. This is achieved by the use of gel electrophoresis.
- Finally, the artificial recombinant must be transformed into *E. coli*, or other host cell. Further details on the use of gel electrophoresis and transformation of *E. coli* are given in the next section. As we have noted, the necessary techniques became available at about the same time and quickly led to many cloning experiments, the first of which were reported in 1972 (Jackson *et al.* 1972, Lobban & Kaiser 1973).

Agarose gel electrophoresis

The progress of the first experiments on cutting and joining of DNA molecules was monitored by velocity sedimentation in sucrose gradients. However, this has been entirely superseded by gel electrophoresis. Gel electrophoresis is not only used as an analytical method, it is routinely used preparatively for the purification of specific DNA fragments. The gel is composed of polyacrylamide or agarose. Agarose is convenient for separating DNA fragments ranging in size from a few hundred to about 20 kb (Fig. 2.1). Polyacrylamide is preferred for smaller DNA fragments.

The mechanism responsible for the separation of DNA molecules by molecular weight during gel electrophoresis is not well understood

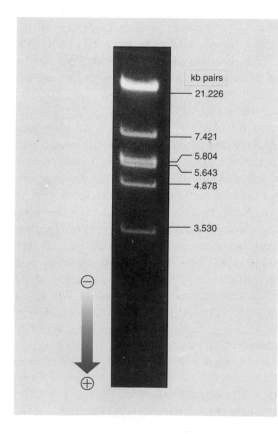

Fig. 2.1 Electrophoresis of DNA in agarose gels. The direction of migration is indicated by the arrow. DNA bands have been visualized by soaking the gel in a solution of ethidium bromide (which complexes with DNA by intercalating between stacked base-pairs) and photographing the orange fluorescence which results upon ultraviolet irradiation.

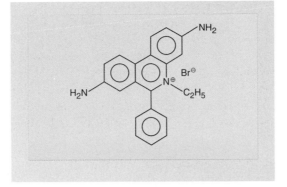

Fig. 2.2 Ethidium bromide.

(Holmes & Stellwagen 1990). The migration of the DNA molecules through the pores of the matrix must play an important role in molecular weight separations since the electrophoretic mobility of DNA in free solution is independent of molecular weight. An agarose gel is a complex network of polymeric molecules whose average pore size depends on the buffer composition and the type and concentration of agarose used. DNA movement through the gel was originally thought to resemble the motion of a snake (reptation). However, real-time fluorescence microscopy of stained molecules undergoing electrophoresis has revealed more subtle dynamics (Smith *et al.* 1989, Schwartz & Koval 1989). DNA molecules display elastic behaviour by stretching in the direction of the applied field and then contracting into dense balls. The larger the pore size of the gel, the greater the ball of DNA which can pass through and hence the larger the molecules which can be separated. Once the globular volume of the DNA molecule exceeds the pore size, the DNA molecule can only pass through by reptation. This leads to size-independent mobility and loss of separation. If very large DNA molecules (1000–2000 kb) are to be separated by electrophoresis, it is necessary to use pulsed electrical fields (see Chapter 6).

Aaij & Borst (1972) showed that the migration rates of the DNA molecules were inversely proportional to the logarithms of the molecular weights. Subsequently, Southern (1979a,b) showed that plotting fragment length or molecular weight against the reciprocal of mobility gives a straight line over a wider range than the semi-logarithmic plot. In any event, gel electrophoresis is frequently performed with marker DNA fragments of known size which allow accurate size determination of an unknown DNA molecule by interpolation. A particular advantage of gel electrophoresis is that the DNA bands can be readily detected at high sensitivity. The bands of DNA in the gel are stained with the intercalating dye ethidium bromide (Fig. 2.2), and as little as 0.05 µg of DNA in one band can be detected as visible fluorescence when the gel is illuminated with ultraviolet light.

In addition to resolving DNA fragments of different lengths, gel electrophoresis can be used to separate different molecular configurations of a DNA

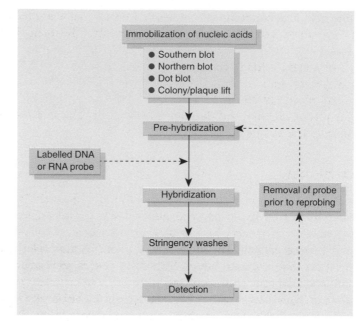

Fig. 2.3 Overview of nucleic acid blotting and hybridization (reproduced courtesy of Amersham International).

molecule. Examples of this are given in Chapter 4 (see p. 48). Gel electrophoresis also can be used for investigating protein–nucleic acid interactions in the so-called *gel retardation* or *band shift* assay. It is based on the observation that binding of a protein to DNA fragments usually leads to a reduction in electrophoretic mobility. The assay typically involves the addition of protein to linear double-stranded DNA fragments, separation of complex and naked DNA by gel electrophoresis and visualization. A review of the physical basis of electrophoretic mobility shifts and their application is provided by Lane *et al.* (1992).

Nucleic acid blotting

Nucleic acid labelling and hybridization on membranes have formed the basis for a range of experimental techniques central to recent advances in our understanding of the organization and expression of the genetic material. These techniques may be applied in the isolation and quantification of specific nucleic acid sequences and in the study of their organization, intracellular localization, expression and regulation. A variety of specific applications includes the diagnosis of infectious and inherited disease. Each of these topics is covered in depth in subsequent chapters.

An overview of the steps involved in nucleic acid blotting and membrane hybridization procedures is shown in Fig. 2.3. *Blotting* describes the immobilization of sample nucleic acids onto a solid support, generally nylon or nitrocellulose membranes. The blotted nucleic acids are then used as 'targets' in subsequent hybridization experiments. The main blotting procedures are:

- blotting of nucleic acids from gels;
- dot and slot blotting;
- colony and plaque blotting.

Colony and plaque blotting are described in detail in Chapter 7 and dot and slot blotting in Chapter 17.

Southern blotting

The original method of blotting was developed by Southern (1975, 1979b) for detecting fragments in an agarose gel that are complementary to a given RNA or DNA sequence. In this procedure, referred to as Southern blotting, the agarose gel is mounted on a filter paper wick which dips into a reservoir containing transfer buffer (Fig. 2.4). The hybridization membrane is sandwiched between the gel and a stack of paper towels (or other absorbent material) which serves to draw the transfer buffer through the gel by capillary action. The DNA molecules are carried out of the gel by the buffer flow and immobilized on the membrane. Initially, the membrane material used was nitrocellulose. The main drawback with this membrane is its fragile nature. Supported nylon membranes have since been developed which have greater binding capacity for nucleic acids in addition to high tensile strength.

For efficient Southern blotting, gel pre-treatment is important. Large DNA fragments (>10 kb) require a longer transfer time than short fragments. To allow uniform transfer of a wide range of DNA fragment sizes, the electrophoresed DNA is exposed to a short depurination treatment

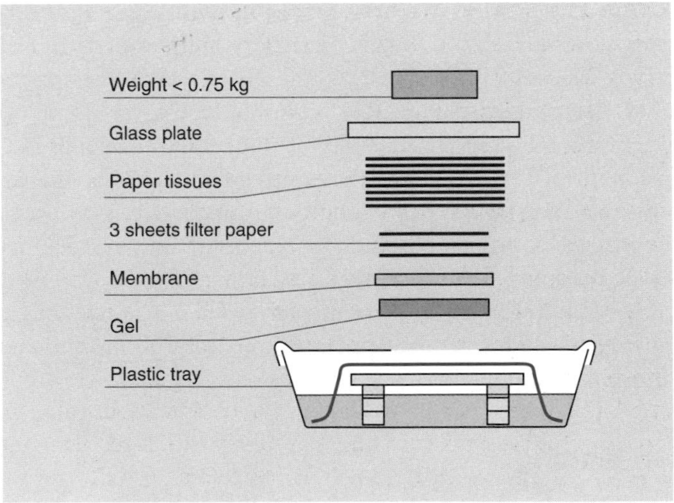

Fig. 2.4 A typical capillary blotting apparatus.

(0.25 M HCl) followed by alkali. This shortens the DNA fragments by alkaline hydrolysis at depurinated sites. It also denatures the fragments prior to transfer ensuring that they are in the single-stranded state and accessible for probing. Finally, the gel is equilibrated in neutralizing solution prior to blotting. An alternative method uses positively-charged nylon membranes which remove the need for extended gel pre-treatment. With them the DNA is transferred in native (non-denatured) form and then alkali denatured *in situ* on the membrane.

After transfer, the nucleic acid needs to be fixed to the membrane and a number of methods are available. Oven baking at 80°C is the recommended method for nitrocellulose membranes and this also can be used with nylon membranes. Due to the flammable nature of nitrocellulose, it is important that it is baked in a vacuum oven. An alternative fixation method utilizes ultraviolet cross-linking. It is based on the formation of crosslinks between a small fraction of the thymine residues in the DNA and positively-charged amino groups on the surface of nylon membranes. A calibration experiment must be performed to determine the optimal fixation period.

Following the fixation step the membrane is placed in a solution of labelled (radioactive or non-radioactive) RNA, single-stranded DNA, or oligodeoxynucleotide which is complementary in sequence to the blot-transferred DNA band or bands to be detected. Conditions are chosen so that the labelled nucleic acid hybridizes with the DNA on the membrane. Since this labelled nucleic acid is used to detect and locate the complementary sequence it is called the *probe*. Conditions are chosen which maximize the rate of hybridization, compatible with a low background of non-specific binding on the membrane (see Box 2.1). After the hybridization reaction has been carried out, the membrane is washed to remove unbound radioactivity and regions of hybridization are detected autoradiographically by placing the membrane in contact with X-ray film (see Box 2.2). A common approach is to carry out the hybridization under conditions of relatively low stringency which permit a high rate of hybridization, followed by a series of post-hybridization washes of increasing stringency (i.e. higher temperature or, more commonly, lower ionic strength). Autoradiography following each washing stage will reveal any DNA bands that are related to, but not perfectly complementary with, the probe and will also permit an estimate of the degree of mismatching to be made.

The Southern blotting methodology can be extremely sensitive. It can be applied to mapping restriction sites around a single-copy gene sequence in a complex genome such as that of man (Fig. 2.5), and when a 'mini-satellite' probe is used it can be applied forensically to minute amounts of DNA (see Chapter 17).

Northern blotting

Southern's technique has been of enormous value, but it was thought that it could not be applied directly to the blot-transfer of RNAs separated by

Box 2.1 Hybridization of nucleic acids on membranes

The hybridization of nucleic acids on membranes is a widely-used technique in gene manipulation and analysis. Unlike solution hybridizations, membrane hybridizations tend not to proceed to completion. One reason for this is that some of the bound nucleic acid is embedded in the membrane and is inaccessible to the probe. Prolonged incubations may not generate any significant increase in detection sensitivity.

The composition of the hybridization buffer can greatly affect the speed of the reaction and the sensitivity of detection. The key components of these buffers are shown below:

Rate enhancers	Dextran sulphate and other polymers act as volume excluders to increase both the rate and extent of hybridization
Detergents and blocking agents	Dried milk, heparin and detergents such as SDS have been used to depress non-specific binding of the probe to the membrane. Denhardt's solution (Denhardt 1966) uses Ficoll, polyvinylpyrrolidone and bovine serum albumin
Denaturants	Urea or formamide can be used to depress the melting temperature of the hybrid so that reduced temperatures of hybridization can be used
Heterologous DNA	This can reduce non-specific binding of probes to non-homologous DNA on the blot

Stringency control

Stringency can be regarded as the specificity with which a particular target sequence is detected by hybridization to a probe. Thus, at high stringency, only completely complementary sequences will be bound, whereas low-stringency conditions will allow hybridization to partially matched sequences. Stringency is most commonly controlled by the temperature and salt concentration in the post-hybridization washes, although these parameters can also be utilized in the hybridization step. In practice, the stringency washes are performed under successively more stringent conditions (lower salt or higher temperature) until the desired result is obtained.

The melting temperature (T_m) of a probe-target hybrid can be calculated to provide a starting point for the determination of correct stringency. The T_m is the temperature at which the probe and target are 50% dissociated. For probes longer than 100 base-pairs:

$$T_m = 81.5°C + 16.6\log M + 0.41 \, (\%G + C)$$

where M = ionic strength of buffer in moles/litre. With long probes, the hybridization is usually carried out at $T_m - 25°C$. When the probe is used to detect partially matched sequences, the hybridization temperature is reduced by 1°C for every 1% sequence divergence between probe and target.

continued

Box 2.1 *continued*

Oligonucleotides can give a more rapid hybridization rate than long probes as they can be used at a higher molarity. Also, in situations where target is in excess to the probe, for example dot blots, the hybridization rate is diffusion limited and longer probes diffuse more slowly than oligonucleotides. It is standard practice to use oligonucleotides to analyse putative mutants following a site-directed mutagenesis experiment where the difference between parental and mutant progeny is often only a single base-pair change (see Chapter 11).

The availability of the exact sequence of oligonucleotides allows conditions for hybridization and stringency washing to be tightly controlled so that the probe will only remain hybridized when it is 100% homologous to the target. Stringency is commonly controlled by adjusting the temperature of the wash buffer. The 'Wallace rule' (Lay Thein & Wallace 1986) is used to determine the appropriate stringency wash temperature:

$$T_m = 4 \times (\text{number of G:C base-pairs}) + 2 \times (\text{number of A:T base-pairs})$$

In filter hybridizations with oligonucleotide probes, the hybridization step is usually performed at 5°C below T_m for perfectly matched sequences. For every mismatched base-pair, a further 5°C reduction is necessary to maintain hybrid stability.

The design of oligonucleotides for hybridization experiments is critical to maximize hybridization specificity. Consideration should be given to:
• probe length—the longer the oligonucleotide, the less chance there is of it binding to sequences other than the desired target sequence under conditions of high stringency;
• oligonucleotide composition—the GC content will influence the stability of the resultant hybrid and hence the determination of the appropriate stringency washing conditions. Also the presence of any non-complementary bases will have an effect on the hybridization conditions.

gel electrophoresis, since RNA was found not to bind to nitrocellulose. Alwine *et al.* (1979), therefore, devised a procedure in which RNA bands are blot-transferred from the gel onto chemically reactive paper, where they are bound covalently. The reactive paper is prepared by diazotization of aminobenzyloxymethyl paper (creating diazobenzyloxymethyl, DBM, paper), which itself can be prepared from Whatman 540 paper by a series of uncomplicated reactions. Once covalently bound, the RNA is available for hybridization with radiolabelled DNA probes. As before, hybridizing bands are located by autoradiography. Alwine's method thus extends that of Southern and for this reason it has acquired the jargon term *northern* blotting.

Subsequently it has been found that RNA bands can indeed be blotted onto nitrocellulose membranes under appropriate conditions (Thomas 1980) and suitable nylon membranes have been developed. Because of the convenience of these more recent methods which do not require freshly activated paper, the use of DBM paper has been superseded.

Box 2.2 The principles of autoradiography

The localization and recording of a radiolabel within a solid specimen is known as autoradiography and involves the production of an image in a photographic emulsion. Such emulsions consist of silver halide crystals suspended in a clear phase composed mainly of gelatin. When a β-particle or γ-ray from a radionuclide passes through the emulsion the silver ions are converted to silver atoms. This results in a latent image being produced which is converted to a visible image when the image is developed. Development is a system of amplification in which the silver atoms cause the entire silver halide crystal to be reduced to metallic silver. Unexposed crystals are removed by dissolution in fixer giving an autoradiographic image which represents the distribution of radiolabel in the original sample.

In direct autoradiography, the sample is placed in intimate contact with the film and the radioactive emissions produce black areas on the developed autoradiograph. It is best suited to detection of weak to medium strength β-emitting radionuclides (^{3}H, ^{14}C, ^{35}S). Direct autoradiography is not suited to the detection of highly energetic β-particles such as those from ^{32}P or for γ-rays emitted from isotopes like ^{125}I. These emissions pass

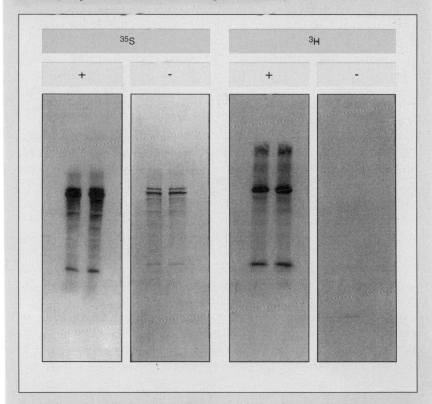

Fig. B2.1 Autoradiographs showing the detection of ^{35}S- and ^{3}H-labelled proteins in acrylamide gels with (+) and without (−) fluorography. (Photo courtesy of Amersham International.)

continued

Box 2.2 *continued*

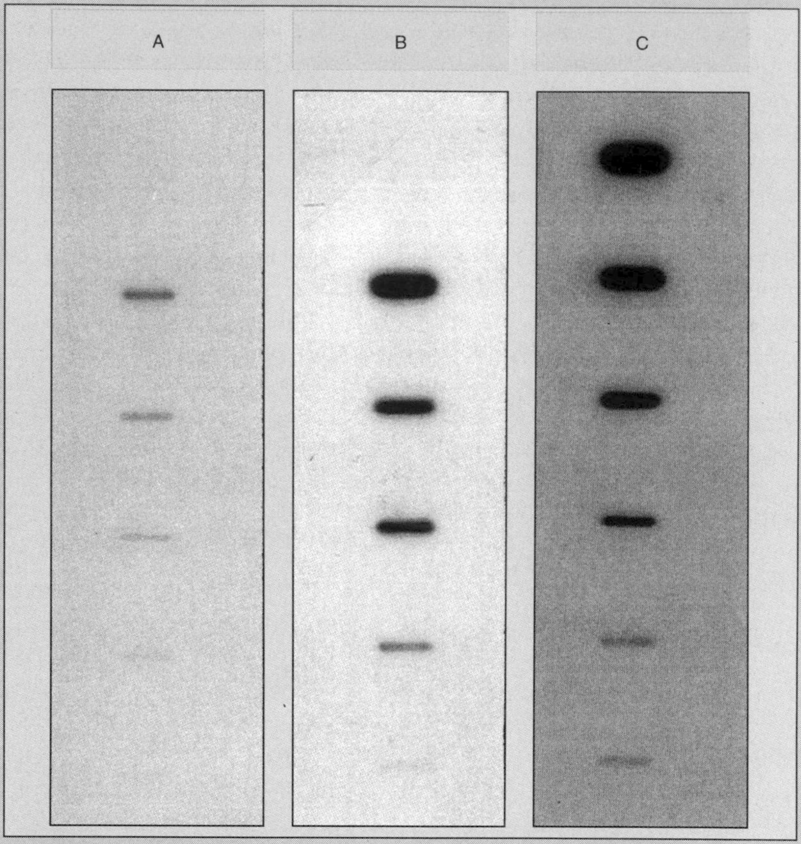

Fig. B2.2 The improvement in sensitivity of detection of ^{125}I-labelled IgG by autoradiography obtained by using an intensifying screen and pre-flashed film. A, no screen and no pre-flashing; B, screen present but film not pre-flashed; C, use of screen and pre-flashed film. (Photo courtesy of Amersham International.)

through and beyond the film with the majority of the energy being wasted. Both ^{32}P and ^{125}I are best detected by indirect autoradiography.

Indirect autoradiography describes the technique by which emitted energy is converted to light by means of a scintillator, using fluorography or intensifying screens. In fluorography the sample is impregnated with a liquid scintillator. The radioactive emissions transfer their energy to the scintillator molecules which then emit photons which expose the photographic emulsion. Fluorography is mostly used to improve the detection of weak β-emitters (Fig. B2.1). Intensifying screens are sheets of a solid inorganic scintillator which are placed behind the film. Any emissions passing through the photographic emulsion are absorbed by the screen and converted to light, effectively superimposing a photographic image upon the direct autoradiographic image.

continued

Box 2.2 *continued*

The gain in sensitivity which is achieved by use of indirect autoradiography is offset by non-linearity of film response. A single hit by a β-particle or γ-ray can produce hundreds of silver atoms, but a single hit by a photon of light produces only a single silver atom. Although two or more silver atoms in a silver halide crystal are stable, a single silver atom is unstable and reverts to a silver ion very rapidly. This means that the probability of a second photon being captured before the first silver atom has reverted is greater for large amounts of radioactivity than for small amounts. Hence small amounts of radioactivity are under-represented with the use of fluorography and intensifying screens. This problem can be overcome by a combination of pre-exposing a film to an instantaneous flash of light (pre-flashing) and exposing the autoradiograph at −70°C. Pre-flashing provides many of the silver halide crystals of the film with a stable pair of silver atoms. Lowering the temperature to −70°C increases the stability of a single silver atom increasing the time available to capture a second photon (Fig. B2.2).

Western blotting

The term 'western' blotting (Burnette 1981) refers to a procedure which does not directly involve nucleic acids, but which is of importance in gene manipulation. It involves the transfer of electrophoresed protein bands from a polyacrylamide gel onto a membrane of nitrocellulose or nylon, to which they bind strongly (Gershoni & Palade 1982, Renart & Sandoval 1984). The bound proteins are then available for analysis by a variety of specific protein–ligand interactions. Most commonly, antibodies are used to detect specific antigens. Lectins have been used to identify glycoproteins. In these cases the probe may itself be labelled with radioactivity, or some other 'tag' may be employed. Often, however, the probe is, unlabelled and is itself detected in a 'sandwich' reaction using a second molecule which is labelled, for instance a species-specific second antibody, or protein A of *Staphylococcus aureus* (which binds to certain subclasses of IgG antibodies), or streptavidin (which binds to antibody probes that have been biotinylated). These second molecules may be labelled in a variety of ways with radioactive, enzyme or fluorescent tags. An advantage of the sandwich approach is that a single preparation of labelled second molecule can be employed as a general detector for different probes. For example, an antiserum may be raised in rabbits which reacts with a range of mouse immunoglobins. Such a rabbit anti-mouse (RAM) antiserum may be radio-labelled and used in a number of different applications to identify polypeptide bands probed with different, specific, monoclonal antibodies, each monoclonal antibody being of mouse origin. The sandwich method may also give a substantial increase in sensitivity, owing to the multivalent binding of antibody molecules.

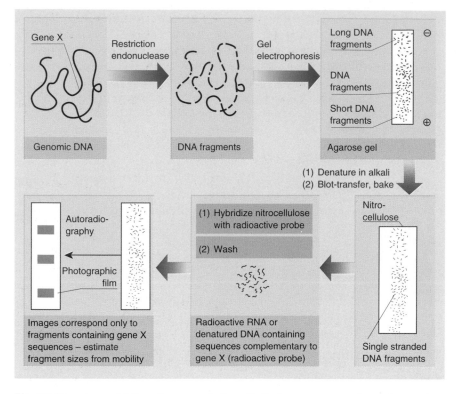

Fig. 2.5 Mapping restriction sites around a hypothetical gene sequence in total genomic DNA by the Southern blot method.

Genomic DNA is cleaved with a restriction endonuclease into hundreds of thousands of fragments of various sizes. The fragments are separated according to size by gel electrophoresis and blot-transferred on to nitrocellulose paper. Highly radioactive RNA or denatured DNA complementary in sequence to gene X is applied to the nitrocellulose paper bearing the blotted DNA. The radiolabelled RNA or DNA will hybridize with gene X sequences and can be detected subsequently by autoradiography, so enabling the sizes of restriction fragments containing gene X sequences to be estimated from their electrophoretic mobility. By using several restriction endonucleases singly and in combination, a map of restriction sites in and around gene X can be built up.

Alternative blotting techniques

The original blotting technique employed capillary blotting but nowadays the blotting is usually accomplished by electrophoretic transfer of polypeptides from an SDS-polyacrylamide gel onto the membrane (Towbin *et al.* 1979). Electrophoretic transfer also is the method of choice for transferring DNA or RNA from low pore-size polyacrylamide gels. It can also be used with agarose gels. However, in this case, the rapid electrophoretic transfer process requires high currents which can lead to extensive heating effects resulting in distortion of agarose gels. The use of an external cooling system is necessary to prevent this.

Another alternative to capillary blotting is vacuum-driven blotting (Olszewska & Jones 1988) for which several devices are commercially available. Vacuum blotting has several advantages over capillary or electrophoretic transfer methods: transfer is very rapid and gel treatment can be performed *in situ* on the vacuum apparatus. This ensures minimal gel handling and, together with the rapid transfer, prevents significant DNA diffusion.

Transformation of *E. coli*

Early attempts to achieve transformation of *E. coli* were unsuccessful and it was generally believed that *E. coli* was refractory to transformation. However, Mandel and Higa (1970) found that treatment with $CaCl_2$ allowed *E. coli* cells to take up DNA from bacteriophage λ. A few years later Cohen *et al.* (1972) showed that $CaCl_2$-treated *E. coli* cells are also effective recipients for plasmid DNA. Almost any strain of *E. coli* can be transformed with plasmid DNA, albeit with varying efficiency, whereas it was thought that only $recBC^-$ mutants could be transformed with linear bacterial DNA (Cosloy & Oishi 1973). Later, Hoekstra *et al.* (1980) showed that $recBC^+$ cells can be transformed with linear DNA, but the efficiency is only 10% of that in otherwise isogenic $recBC^-$ cells. Transformation of $recBC^-$ cells with linear DNA is only possible if the cells are rendered recombination proficient by the addition of a *sbc*A or *sbc*B mutation. The fact that the *recBC* gene product is an exonuclease explains the difference in transformation efficiency of circular and linear DNA in $recBC^+$ cells.

As will be seen from the next chapter, many bacteria contain restriction systems which can influence the efficiency of transformation. Although the complete function of these restriction systems is not known yet, one role they do play is the recognition and degradation of foreign DNA. For this reason it is usual to use a restriction-deficient strain of *E. coli* as a transformable host.

Since transformation of *E. coli* is an essential step in many cloning experiments, it is desirable that it be as efficient as possible. Several groups of workers have examined the factors affecting the efficiency of transformation. It has been found that *E. coli* cells and plasmid DNA interact productively in an environment of calcium ions and low temperature (0–5°C), and that a subsequent heat shock (37–45°C) is important, but not strictly required. Several other factors, especially the inclusion of metal ions in addition to calcium, have been shown to stimulate the process.

A very simple, moderately efficient transformation procedure for use with *E. coli* involves resuspending log-phase cells in ice-cold 50 mM calcium chloride at about 10^{10} cells/ml and keeping them on ice for about 30 min. Plasmid DNA (0.1 mg) is then added to a small aliquot (0.2 ml) of these now *competent* (i.e. competent for transformation) cells, and the incubation on ice continued for a further 30 min, followed by a heat shock of 2 min at 42°C. The cells are then usually transferred to nutrient medium and incubated for some time (30 min to 1 hour) to allow phenotypic properties conferred by the plasmid to be expressed, e.g. antibiotic resistance commonly used as a selectable marker for plasmid-containing cells. (This so-called *phenotypic lag* may not need to be taken into consideration with high-level ampicillin resistance. With this marker, significant resistance builds up very rapidly, and ampicillin exerts its effect on cell-wall biosynthesis only in cells which have progressed into active growth.) Finally the cells are plated out on selective medium. Just why such a transformation procedure is effective is not fully understood. The calcium chloride affects the cell wall and may also be responsible for binding DNA to the cell surface. The actual uptake of DNA is stimulated by the brief heat shock.

Hanahan (1983) has re-examined factors that affect the efficiency of transformation, and has devised a set of conditions for optimal efficiency (expressed as transformants per μg plasmid DNA) applicable to most *E. coli* K12 strains. Typically, efficiencies of 10^7 to 10^9 transformants/μg can be achieved depending on the strain of *E. coli* and the method used (Liu & Rashidbaigi 1990). There are many enzymic activities in *E. coli* which can destroy incoming DNA from non-homologous sources (see Chapter 3) and reduce the transformation efficiency. Large DNAs transform less efficiently, on a molar basis, than small DNAs. Even with such improved transformation procedures, certain potential gene-cloning experiments requiring large numbers of clones are not reliable. One approach which can be used to circumvent the problem of low transformation efficiencies is to package recombinant DNA into virus particles *in vitro*. A particular form of this approach, the use of cosmids, is described in detail in Chapter 5. Another approach is electroporation which is described below.

Transformation of other organisms

Although *E. coli* often remains the host organism of choice for cloning experiments, many other hosts are now used, and with them transformation may still be a critical step. In the case of Gram-positive bacteria, the two most important groups of organisms are *Bacillus* spp. and actinomycetes. That *B. subtilis* is naturally competent for transformation has been known for a long time and hence the genetics of this organism are fairly advanced. For this reason *B. subtilis* is a particularly attractive alternative prokaryotic cloning host. The significant features of transformation with this organism are detailed in Chapter 12. Of particular relevance here is that it is possible to transform protoplasts of *B. subtilis*, a technique which leads to improved transformation frequencies. A similar technique is used to transform actinomycetes, and recently it has been shown that the frequency can be increased considerably by first entrapping the DNA in liposomes that then fuse with the host cell membrane.

In later chapters we discuss ways in which cloned DNA can be introduced into eukaryotic cells. With animal cells there is no great problem as only the membrane has to be crossed. In the case of yeast, protoplasts are required (Hinnen *et al.* 1978). With higher plants one strategy that has been adopted is either to package the DNA in a plant virus or to use a bacterial plant pathogen as the donor. It has also been shown that proto-plasts prepared from plant cells are competent for transformation. A further remarkable approach that has been demonstrated with plants and animals (Klein & Fitzpatrick-McElligott 1993) is the use of microprojectiles shot from a gun (Chapter 14).

A rapid and simple method for introducing cloned genes into a wide variety of microbial, plant and animal cells is *electroporation* (for review, see Potter 1988). This technique depends upon the original observation by Zimmerman *et al.* (1983) that high-voltage electric pulses can induce cell plasma membranes to fuse. Subsequently, it was found that when sub-jected to electric shock (typically a brief exposure to a voltage gradient of 4000–8000 V/cm), the cells take up exogenous DNA from the sus-pending solution, apparently through holes momentarily created in the plasma membrane. A proportion of these cells become stably transformed (Newman *et al.* 1982, Potter *et al.* 1984), and can be selected if a suitable marker gene is carried on the transforming DNA. Potter *et al.* (1984) also found that treating eukaryotic cells with colcemid before electroporation increases the transformation efficiency. This is probably because the cells, which are arrested in metaphase by the drug, lack a nuclear membrane or have an unusually permeable membrane. These workers also found that linear DNA transforms more efficiently than supercoiled DNA. This tech-nique appears to be a general one. It is successful with animal cell types that have not been amenable to other approaches (Chu *et al.* 1987). With *E. coli*, electroporation has been found to give plasmid transformation effici-encies (10^9/mg DNA) comparable with the $CaCl_2$ method (Dower *et al.* 1988). For the generation of plasmid cDNA libraries, electroporation was 100-fold more efficient than $CaCl_2$-mediated transformation (Bottger 1988).

Animal cells, and protoplasts of yeast, plant and bacterial cells are susceptible to transformation by liposomes (Deshayes *et al.* 1985). A simple transformation system has been developed which makes use of liposomes prepared from a cationic lipid (Felgner *et al.* 1987). Small unilamellar (single bilayer) vesicles are produced. DNA in solution spontaneously and ef-ficiently complexes with these liposomes (in contrast to previously em-ployed liposome encapsidation procedures involving non-ionic lipids). The positively-charged liposomes not only complex with DNA, but also bind to cultured animal cells and are efficient in transforming them, probably by fusion with the plasma membrane. The use of liposomes as a transforma-tion or transfection system is called *lipofection*.

3 Cutting and joining DNA molecules

Cutting DNA molecules

Before 1970 there was simply no method available for cutting a duplex DNA molecule into discrete fragments. It became apparent that the related phenomena of host-controlled restriction and modification might lead towards a solution to the problem when it was discovered that restriction involves specific endonucleases. The favourite organisms of molecular biologists, *E. coli* K12, was the first to be studied in this regard, but turned out to be an unfortunate choice. Its endonuclease is perverse in the complexity of its behaviour. The breakthrough in 1970 came with the discovery in *Haemophilus influenzae* of an enzyme that behaves more simply. Present-day DNA technology is totally dependent upon our ability to cut DNA molecules at specific sites with restriction endonucleases. An account of host-controlled restriction and modification therefore forms the first part of this chapter.

Host-controlled restriction and modification

Restriction systems allow bacteria to monitor the origin of incoming DNA and to destroy it if it is recognized as foreign. Restriction endonucleases recognize specific sequences in the incoming DNA and cleave the DNA into fragments, either at specific sites or more randomly. When the incoming DNA is a bacteriophage genome, the effect is to reduce the efficiency of plating, i.e. to reduce the number of plaques formed, in plating tests. The phenomena of restriction and modification were well illustrated and studied by the behaviour of phage lambda on two *E. coli* host strains.

If a stock preparation of phage λ, for example, is made by growth upon *E. coli* strain C and this stock is then titred upon *E. coli* C and *E. coli* K, the titres observed on these two strains will differ by several orders of magnitude, the titre on *E. coli* K being the lower. The phage are said to be *restricted* by the second host strain (*E. coli* K). When those phage that do

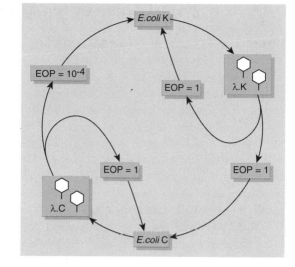

Fig. 3.1 Host-controlled
restriction and modification
of phage λ in *E. coli* strain K,
analysed by efficiency of
plating (EOP). Phage
propagated by growth on
strains K or C (i.e. λK or λC)
have EOPs on the two strains
as indicated by arrows.

result from the infection of *E. coli* K are now re-plated on *E. coli* K they are
no longer restricted; but if they are first cycled through *E. coli* C they are
once again restricted when plated upon *E. coli* K (Fig. 3.1). Thus the
efficiency with which phage λ plates upon a particular host strain depends
upon the strain on which it was last propagated. This non-heritable change
conferred upon the phage by the second host strain (*E. coli* K) that allows it
to be re-plated on that strain without further restriction is called *modification*.

The restricted phages adsorb to restrictive hosts and inject their DNA
normally. When the phage are labelled with ^{32}P it is apparent that their
DNA is degraded soon after injection (Dussoix & Arber 1962) and the
endonuclease that is primarily responsible for this degradation is called a
restriction endonuclease or restriction enzyme (Lederberg & Meselson 1964).
The restrictive host must, of course, protect its own DNA from the potentially lethal effects of the restriction endonuclease and so its DNA must be
appropriately modified. Modification involves methylation of certain bases
at a very limited number of sequences within DNA which constitute the
recognition sequences for the restriction endonuclease. This explains why
phage that survive one cycle of growth upon the restrictive host can
subsequently re-infect that host efficiently; their DNA has been replicated
in the presence of the modifying methylase and so it, like the host DNA,
becomes methylated and protected from the restriction system. Although
phage infection has been chosen as our example to illustrate restriction and
modification, these processes can occur whenever DNA is transferred from
one bacterial strain to another.

The restriction endonuclease of *E. coli* K was the first to be isolated and
studied in detail. Meselson and Yuan (1968) achieved this by devising an
ingenious assay in which a fractionated cell extract was incubated with a
mixture of unmodified and modified phage λ DNAs which were differentially radiolabelled – one with ^{3}H, the other with ^{32}P – so that they could be

distinguished. After incubation, the DNA mixture was analysed by sedimentation through a sucrose gradient, where the appearance of degraded unmodified DNA in the presence of undegraded modified DNA indicated the activity of restriction endonuclease.

The enzyme from *E. coli* K, and the similar one from *E. coli* B, have unusual properties. In addition to magnesium ions, they require the cofactors ATP and S-adenosylmethionine (SAM). The enzymes are composed of three subunits, a specificity subunit which determines the recognition specificity, a modification subunit, and a restriction subunit (Wilson & Murray 1991). In the presence of SAM, the enzyme binds to a bipartite recognition sequence (AACN$_6$GTGC in the case of the enzyme from *E. coli* K) irrespective of its state of methylation. If the recognition sequence is methylated in both strands (at the second A in the strand shown, and at the A in the complementary strand opposite the T shown) the ATP stimulates the dissociation of enzyme from DNA. If the site is hemimethylated (i.e. methylated in only one strand) the ATP stimulates the methylation of the other strand, with SAM being the methyl donor. If the site is unmethylated, cleavage occurs (Yuan *et al.* 1975, Bickle *et al.* 1978). Cleavage takes place in a surprising way: the enzyme first translocates along the DNA by looping it past the enzyme which remains bound to the recognition site (Yuan *et al.* 1980). Cleavage then occurs at a 'random' site several kilobases from the recognition site. Cleavage occurs by nicking the two strands, and it is likely that two separate restriction events involving two molecules of enzyme are required for the double strand break (Studier & Bandyopadhyay 1988). Translocation and nicking require ATP hydrolysis (Rosamond *et al.* 1979). Enzymes with these properties are now known as type I restriction endonucleases, and their cognate restriction and modification systems are type I systems. Like all restriction endonucleases they recognize specific nucleotide sequences, but they are not particularly useful for gene manipulation since their cleavage sites are essentially random.

While these bizarre properties of type I restriction enzymes were being unravelled, a restriction endonuclease from *H. influenzae* Rd was discovered (Kelly & Smith 1970, Smith & Wilcox 1970) that was to become the prototype of a large number of restriction endonucleases — now known as type II enzymes — that have none of the unusual properties displayed by type I enzymes and which are fundamentally important in the manipulation of DNA. The type II enzymes recognize a particular target sequence in a duplex DNA molecule and break the polynucleotide chains within, or near to, that sequence to give rise to discrete DNA fragments of defined length and sequence. In fact, the activity of these enzymes is often assayed and studied by gel electrophoresis of the DNA fragments which they generate (see Fig. 2.1). As expected, digests of small plasmid or viral DNAs give characteristic simple DNA band patterns.

Type II recognition sequences are symmetric. Some sequences are continuous (e.g. GATC), some are interrupted (e.g. GANTC). Unlike type I enzymes, type II endonucleases consist of a single polypeptide, usually as a homodimer. The type II endonucleases require no cofactor other than

[handwritten margin note: Type I rest. endos. -cut bases away from recognition site]

[handwritten margin note: Type II r. endo's -cut at recognition site.]

**Box 3.1 Restriction: from a phenomenon in bacterial genetics
to a biological revolution**

In the two related phenomena of host-controlled restriction and modification, a single cycle of phage growth, in a particular host bacterium, alters the host range of the progeny virus. The virus may fail to plate efficiently on a second host; its host range is restricted. This modification of the virus differs fundamentally from mutation because it is imposed by the host cell on which the virus has been grown but it is not inherited; when the phage is grown in some other host, the modification may be lost. In the 1950s restriction and modification were recognized as common phenomena, affecting many virulent and temperate (i.e. capable of forming lysogens) phages, and involving various bacterial species (Luria 1953, Lederberg 1957).

Arber and Dussoix clarified the molecular basis of the phenomena. They showed that restriction of phage λ is associated with rapid breakdown of the phage DNA in the host bacterium. They also showed that modification results from an alteration of the phage DNA which renders the DNA insensitive to restriction. They deduced that a single modified strand in the DNA duplex is sufficient to prevent restriction (Arber & Dussoix 1962, Dussoix & Arber 1962). Subsequent experiments implicated methylation of the DNA in the modification process (Arber 1965).

Detailed genetic analysis, in the 1960s, of the bacterial genes (in *E. coli* K and *E. coli* B) responsible for restriction and modification supported the duality of the two phenomena. Mutants of the bacteria could be isolated that were both restriction-deficient and modification-deficient (R^-M^-), or that were R^-M^+. The failure to recover R^+M^- mutants was correctly ascribed to the suicidal failure to confer protective modification upon the host's own DNA.

The biochemistry of restriction advanced with the isolation of the restriction endonuclease from E. coli K (Meselson & Yuan 1968). It was evident that the restriction endonucleases from *E. coli* K and *E. coli* B were important examples of proteins that recognize specific structures in DNA, but the properties of these type I enzymes as they are now known, were complex. Although the recognition sites in the phage could be mapped genetically (Murray *et al.* 1973a), determined efforts to define the DNA sequences cleaved were unsuccessful (Eskin & Linn 1972, Murray *et al.* 1973b).

The breakthrough came with Hamilton Smith's discovery of a restriction endonuclease from *Haemophilus influenzae* strain Rd (Smith & Wilcox 1970), and the elucidation of the nucleotide sequence at its cleavage sites in phage T7 DNA (Kelly & Smith 1970). This enzyme is now known as *Hind*II. The choice of T7 DNA as the substrate for cleavage was a good one, because the bacterium also contains another type II restriction enzyme, *Hind*III, in abundance. Fortunately, *Hind*III does not cleave T7 DNA, and so any contaminating *Hind*III in the *Hind*II preparation could not be problematical (Old *et al.* 1975). Shortly after the discovery of *Hind*II, several other type II

continued

Box 3.1 *continued*

restriction endonucleases were isolated and characterized. *Eco*RI was foremost amongst these (Hedgepeth *et al.* 1972). They were rapidly exploited in the first recombinant DNA experiments.

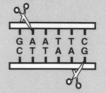

By the mid-1960s, restriction and modification had been recognized as important and interesting phenomena within the field of bacterial genetics (see, for example, Hayes 1968), but who could have foreseen the astonishing impact of restriction enzymes upon biology?

magnesium ions. The cognate modification methylases act independently of the endonucleases and only require SAM.

Type IIs (shifted cleavage) systems, e.g. *Mbo*II, differ from standard type II systems in having asymmetric recognition sequences. Cleavage occurs only on one side of the recognition sequence, at a point some distance away (fewer than 20 nucleotides). Cleavage usually occurs by creating staggered nicks a few nucleotides apart on opposite strands.

Type II endonucleases are much the most common type. Very many type II endonucleases have been isolated from a wide variety of bacteria. About 150 type II endonucleases have been identified and at least partially characterized (Roberts 1990). Because of their biotechnological importance, the number continues to grow as more bacterial species are surveyed. Some recognition specificities have been found many times, thus there are many examples of enzymes recognizing the GGCC site. It is clearly common for bacteria to have restriction–modification systems, and it is not unusual for a single bacterium to have more than one system. The genes that encode restriction–modification systems may be found in the bacterial chromosome, or may reside on a plasmid (as is the case for *Eco*RI), or may be encoded by a temperate virus (e.g. phage P1). It is worth noting that many so-called restriction endonucleases have not formally been shown to correspond with any genetically identified restriction and modification system of the bacteria from which they have been prepared: it is usually assumed that a site-specific endonuclease which is inactive upon host DNA and active upon exogenous DNA is, in fact, a restriction endonuclease.

Type III restriction systems (e.g. *Eco*P1) are relatively rare and do not provide endonucleases for gene manipulation. Type III endonucleases act as complexes of two subunits, one subunit (M subunit) responsible for site recognition and modification, the other (R subunit) responsible for nuclease action. DNA cleavage requires magnesium ions, ATP, and is stimulated by SAM. The recognition sites are asymmetric and cleavage occurs by nicking one strand at a measured distance to one side of the recognition sequence.

Table 3.1 Characteristics of restriction endonucleases (Hadi *et al.* 1982, Iida *et al.* 1982, Kessler *et al.* 1985, Wilson & Murray 1991)

Characteristic	Type I	Type II	Type III
Restriction and modification activities	Single multifunctional enzyme	Separate endonuclease and methylase	Separate enzymes with a subunit in common
Protein structure of restriction endonuclease	Three different subunits	Simple	Two different subunits
Requirements for restriction	ATP, Mg^{2+} S-adenosylmethionine	Mg^{2+}	ATP, Mg^{2+} (S-adenosylmethionine)
Sequence of host specificity sites	*Eco*B: $TGAN_8TGCT$ *Eco*K: $AACN_6GTGC$	Rotational symmetry (not in type IIs)	*Eco*P1: AGACC *Eco*P15: CAGCAG
Cleavage sites	Possibly random, at least 1000 bp from host specificity site	At or near host specificity site	24–26 bp to 3' of host specificity site
Enzymatic turnover	No	Yes	Yes
DNA translocation	Yes	No	No
Site of methylation	Host specificity site	Host specificity site	Host specificity site

N = any nucleotide.

Two sites in opposite orientations are necessary to break the DNA duplex (Kruger *et al.* 1990, Meisel *et al.* 1992).

The classification of restriction–modification systems into types I to III (Table 3.1) is convenient, but may require alteration as new discoveries are made. For example, the *Eco*57I system comprises a single polypeptide which has both modification and nuclease activities, with the nuclease stimulated by SAM. It has been proposed that this system should be reclassified from type IIs to become the first type IV (Petrusyte *et al.* 1988). The biochemical and evolutionary relationships of restriction–modification systems have been reviewed by Wilson and Murray (1991). There is evidence for selection favouring divergence of restriction specificities (Murray *et al.* 1993).

Other restriction systems are known which fall outside the type I–IV classification. Among these are the *Mcr* systems (Raleigh *et al.* 1988). In the 1980s it became apparent that DNA from various bacterial and eukaryotic sources could be cloned only with low efficiency in certain commonly-used *E. coli* host strains (Woodcock *et al.* 1989). The DNA was restricted in the host strains. The phenomenon is caused by methylcytosine in the DNA, and was called *modified cytosine restriction (Mcr)*. There are two *Mcr* systems: one (*Mcr*A) is encoded by a prophage-like element, the other (*Mcr*BC) is encoded by two genes, *mcrB* and *mcrC*, that are actually located very close to the genes encoding the *E. coli* K type I system. The recognition site for the *Mcr*BC system has been recently determined as R mC (N_{40-80}) R mC, where R = A or G. Cleavage occurs at multiple sites in both strands between the methylcytosines. The cleavage requires GTP (Sutherland *et al.* 1992). The *McrBC* system is now known to be identical to a system, formerly called *RglB*, discovered in 1952 (Luria & Human 1952). The early investigations were of T-even (T2, T4 and T6) phages. Wild-type T-even phages have glucosylated 5-hydroxymethylcytosine in their DNA instead of cytosine. It was found that if glucosylation of the 5-hydroxymethylcytosine was prevented by mutation of phage or host genes, the phage was restricted (Revel 1967). The name *RglB* stands for *restricts glucoseless phage*. It was not originally clear why *E. coli* should possess a system for restricting DNA containing 5-hydroxymethylcytosine, since this is an uncommon form of DNA. But it is now apparent that the *RglB* (= *McrBC*) system restricts DNA containing 5-hydroxymethylcytosine, 5-methylcytosine, or 4-methylcytosine.

The type II restriction systems are important for providing reagents for gene manipulation, and they are discussed further in that context below. However, the other restriction systems are important for recombinant DNA technology because they must be eliminated from host bacteria where they could otherwise prevent efficient transformation by recombinant DNA (see p. 19).

Nomenclature

The discovery of a large number of restriction enzymes called for a uniform nomenclature. A system based upon the proposals of Smith and Nathans (1973) has been followed for the most part. The proposals were as follows.

- The species name of the host organism is identified by the first letter of the genus name and the first two letters of the specific epithet to form a three-letter abbreviation in italic: for example, *Escherichia coli* = *Eco* and *Haemophilus influenzae* = *Hin*.
- Strain or type identification is written as a subscript, e.g. Eco_K. In cases where the restriction and modification system is genetically specified by a virus or plasmid, the abbreviated species name of the host is given and the extrachromosomal element is identified by a subscript, e.g. Eco_{PI}, Eco_{RI}.
- When a particular host strain has several different restriction and modification systems, these are identified by roman numerals, thus the systems from *H. influenzae* strain Rd would be Hin_d I, Hin_d II, Hin_d III, etc. These roman numerals should not be confused with those in the classification of restriction enzymes into type I, etc.
- All restriction enzymes have the general name endonuclease R, but, in addition, carry the system name, e.g. endonuclease R.Hin_d III. Similarly, modification enzymes are named methylase M followed by the system name. The modification enzyme from *H. influenzae* Rd corresponding to endonuclease R.Hin_d III is designated methylase M.Hin_d III.

In practice this system of nomenclature has been simplified further: (a) subscripts are typographically inconvenient: the whole abbreviation is now usually written on the line, e.g. HindIII, and (b) where the context makes it clear that restriction enzymes only are involved, the designation

Table 3.2 Some restriction endonucleases and their cleavage sites. The cleavage pattern is shown by the lines. The original bacterial source may not be the actual source for enzymes produced commercially; many enzymes are now cloned and over-expressed in *E. coli*

Bacterial source	Enzyme abbreviation	Sequence 5′ → 3′ 3′ ← 5′	Note
Haemophilus aegyptius	*Hae*III	G G | C C C C | G G	1
Staphylococcus aureus 3A	*Sau*3AI	| G A T C C T A G |	2
Bacillus amyloliquefaciens H	*Bam*HI	G | G A T C C C C T A G | G	2
Escherichia coli RY13	*Eco*RI	G | A A T T C C T T A A | G	2, 6, 7
Haemophilus influenzae Rd	*Hin*dII	G T Py | Pu A C C T Pu | Py T G	1, 5
	*Hin*dIII	A | A G C T T T T C G A | A	2
Providencia stuartii	*Pst*I	C T G C A | G G | A C G T C	3
Serratia marcescens	*Sma*I	C C C | G G G G G G | C C C	1
Xanthomonas malvacearum	*Xma*I	C | C C G G G G G G C C | C	2
Moraxella bovis	*Mbo*II	G A A G A N$_8$ | C T T C T N$_7$ |	4

Notes:
1. Produces blunt ends.
2. Produces cohesive end, with 5′ single-stranded overhang.
3. Produces cohesive end, with 3′ single-stranded overhang.
4. This is a type IIs enzyme. It does not cut within the recognition sequence, but at whatever sequence lies to the right as shown. N is any nucleotide.
5. Pu is any purine (A or G), Py is any pyrimidine (C or T). All possible structures, consistent with base-pairing, are cleaved.
6. Under certain conditions (low ionic strength, alkaline pH or 50% glycerol) the *Eco*RI specificity is reduced, so that only the internal tetranucleotide sequence of the canonical hexanucleotide sequence is necessary for cleavage. This reduced specificity of *Eco*RI is called *Eco*RI• or *Eco*RI-star activity. The star activity is inhibited by *p*-chloromercuribenzoate, whereas the normal activity is insensitive (Tikchonenko *et al*. 1978). Many other restriction enzymes exhibit star activity, i.e. reduced specificity, under suboptimal conditions.
7. The name of *Eco*RI is anomalous. The names *Eco*RI and *Eco*RII (which cleaves ↓ CC(A/T)GG) were derived from two *different* bacterial strains carrying different antibiotic resistance transfer factors (plasmids) which encoded the restriction enzymes. The two resistance transfer factors were called RI and RII.

endonuclease R is omitted. This is the system used in Table 3.2, which lists some of the more commonly used restriction endonucleases. A more extensive list is given in Appendix 1.

Target sites

The vast majority of, but not all, type II restriction endonucleases recognize and break DNA within particular sequences of tetra-, penta-, hexa- or hepta-nucleotides, which have a twofold axis of *rotational symmetry*; for example, *Eco*RI cuts at the positions indicated by arrows in the sequence

Axis of symmetry

$$5' - G \overset{\downarrow}{A} \overset{*}{A} \mid T T C -$$
$$3' - C T T \mid A A \uparrow G -$$

giving rise to termini bearing 5'-phosphate and 3'-hydroxyl groups. Such sequences are sometimes said to be *palindromic* by analogy with words that read alike backwards and forwards.

The structure of the enzyme–DNA complex has been determined by X-ray crystallography (McClarin *et al.* 1986). It is evident that the endonuclease acts as a dimer of identical subunits, and that the palindromic nature of the target sequence reflects the twofold rotational symmetry of the dimeric protein. If the target sequence is modified by methylation so that 6-methyladenine residues are found at *one* or *both* of the positions indicated by asterisks, then the sequence is resistant to endonuclease R. *Eco*RI. The resistance of the hemimethylated site protects the bacterial host's own duplex DNA from attack immediately after semiconservative replication of the fully methylated site until the modification methylase can once again restore the daughter duplexes to the fully methylated state.

We can see that *Eco*RI makes single-strand breaks four nucleotide pairs apart in the opposite strands of its target sequence, and so generates fragments with protruding 5'-termini. These DNA fragments can associate by hydrogen bonding between overlapping 5'-termini, or the fragments can circularize by intramolecular reaction, and for this reason the fragments are said to have *sticky* or *cohesive* ends (Fig. 3.2). In principle, DNA fragments from diverse sources can be joined by means of the cohesive ends, and it is possible, as we shall see later, to seal the remaining nicks in the two strands to form an intact *artificially recombinant* duplex DNA molecule.

It is clear from Table 3.2 that not all type II enzymes cleave their target sites like *Eco*RI. Some enzymes (e.g. *Pst*I) produce fragments bearing 3'-cohesive ends. Others (e.g. *Hae*III) make even cuts giving rise to flush- or blunt-ended fragments with no cohesive end at all. Some enzymes recognize tetranucleotide sequences, others recognize longer sequences, and this of course determines the average fragment length produced. We would expect any particular tetranucleotide target to occur about once every 4^4 (i.e. 256) nucleotide pairs in a long random DNA sequence, assuming all bases are equally frequent. Any particular hexanucleotide target would be

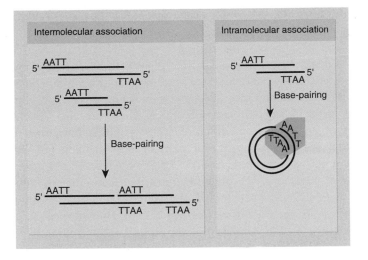

Fig. 3.2 Cohesive ends of DNA fragments produced by digestion with *Eco*RI.

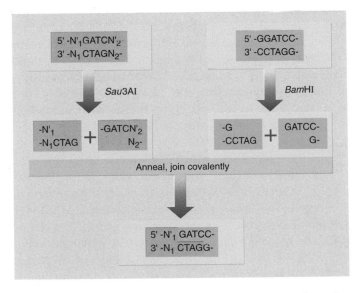

Fig. 3.3 Production of a hybrid site by cohesion of complementary sticky ends generated by *Sau*3A I and *Bam*HI.

expected to occur once in every 4^6 (i.e. 4096) nucleotide pairs. Some enzymes (e.g. *Sau*3AI) recognize a tetranucleotide sequence that is included within the hexanucleotide sequence recognized by a different enzyme (e.g. *Bam*HI). The cohesive termini produced by these enzymes are such that fragments produced by *Sau*3AI will cohere with those produced by *Bam*HI. If the fragments are then covalently joined, the 'hybrid site' so produced will be once again sensitive to *Sau*3AI, but may not constitute a target for

*Bam*HI; this will depend upon the nucleotides adjacent to the original *Sau*3AI site (Fig. 3.3). Several other combinations of enzymes have this property.

From Table 3.2 we can also see that *Hind*II, the first type II enzyme to be discovered, is an example of an enzyme recognizing a sequence with some ambiguity; in this case all possible sequences corresponding to the structure given in Table 3.2 are substrates. There are also several known examples of enzymes from different sources which recognize the same target. They are *isoschizomers*. Some pairs of isoschizomers cut their target at different places (e.g. *Sma*I, *Xma*I).

In our discussion of the phenomena of restriction and modification of phage λ by *E. coli* K, we saw that methylation was the basis of modification in that system. In the wide variety of type II restriction enzymes now known there are some curious and useful examples of the influence of methyl groups at restriction sites. The enzymes *Hpa*II and *Msp*I are isoschizomers with the target sequence CCGG. *Hpa*II will not cut the target when it contains 5-methylcytosine as indicated by the asterisk CC̊GG (indeed this is the product of M.*Hpa*II). However, *Msp*I is known to be indifferent to methylation at this nucleotide: it cleaves whether or not this C residue is methylated. Now it has been found that over 90% of the methyl groups in genomic DNA of many animals, including vertebrates and echinoderms, occur as 5-methylcytosine in the sequence CG. Many of these methyl groups occur as *Msp*I sites, and their presence can be detected by comparing digests of the DNA generated by *Hpa*II and *Msp*I. Indeed methylation at *Msp*I sites around a single gene can be investigated in detail by combining the use of these two enzymes with the Southern-blot technique (Bird & Southern 1978, Razin & Riggs 1980).

A potentially troublesome effect of DNA methylation concerns the *dam* and *dcm* methylation activities of many *E. coli* strains. The *dam* (DNA *a*denine *m*ethylase) activity methylates adenine, creating 6-methyladenine, in the sequence GATC. The *dcm* (DNA *c*ytosine *m*ethylase) activity methylates the internal cytosine, creating 5-methylcytosine, in the sequence CC(A/T)GG. The *dam* methylation appears to play a role in strand discrimination for mismatch repair (Lu *et al.* 1983, Radman & Wagner 1984). Further postulated functions for the adenine and cytosine methylation in the initiation of DNA replication or recombination are less well understood. From the point of view of the gene manipulator, the main importance of these methylation functions is their effect of limiting the ability of certain restriction endonucleases to act on DNAs from *dam*$^+$ and *dcm*$^+$ *E. coli* strains: two examples follow. *Dam* methylation of the sequence GATC renders it resistant to restriction by *Mbo*I. This problem is overcome by the availability of the *Sau*3AI isoschizomer, which cuts the sequence whether or not the adenine residue is methylated. It is mainly for this reason that *Sau*3AI has replaced *Mbo*I in the catalogues of commercially available restriction endonucleases. *Dcm* methylation of the CC(A/T)GG sequence prevents restriction by *Eco*RII. In this example also there is isoschizomer available, *Bst*NI, which is indifferent to the *dcm* methylation. An alternative solution to the problem in this case, and in several other examples where no methylation–indifferent isoschizomer is available, is to produce the DNA which is to be restricted in a *dam*⁻ or *dcm*⁻ *E. coli* host. In fact, *dcm*⁻ *dam*⁻ double-mutant strains are available for this application (Marinus *et al.* 1983). (Refer to Chapter 6 for the involvement of *dam* methylation in cleavage by *Dpn*I.)

The wide variety of properties exhibited by restriction endonucleases described in the preceding paragraphs has provided great scope for ingenious and resourceful gene manipulators. This will be apparent from examples in following chapters.

What is the function of restriction endonucleases *in vivo*? Clearly host-controlled restriction acts as a mechanism by which bacteria distinguish self from non-self. It is analogous to an immunity system. Restriction is moderately effective in preventing infection by some bacteriophages. It may be for this reason that the T-even phages (T2, T4 and T6) have evolved with glucosylated hydroxymethylcytosine residues replacing cytosine in their DNA, so rendering it resistant to many restriction endonucleases. A mutant strain of T4 is available which does have cytosine residues in its DNA and is therefore amenable to conventional restriction methodology (Velten *et al.* 1976, Murray *et al.* 1979, Krisch & Selzer 1981). As an alternative to the unusual DNA structure of the T-even phages, other mechanisms appear to have evolved in T3 and T7 for overcoming restriction *in vivo* (Spoerel *et al.* 1979, Bandyopadhyay *et al.* 1985, Kruger *et al.* 1985).

In fact bacteriophages commonly have anti-restriction mechanisms of one sort or another (Kruger & Bickle 1983, Sharp 1986). Mechanisms include:

• inhibition of restriction endonucleases by phage-encoded proteins, some of which are injected along with phage DNA;
• virus-encoded DNA modification;
• phage stimulation of host-encoded modification;
• unusual bases, or glucosylation of bases;
• phage-encoded proteins that destroy cofactors of restriction endonucleases;
• absence of restriction sites in phage DNA.

Korona *et al.* (1993) has surveyed the sensitivity of a number of coliphages, newly isolated from natural sources, to a range of type I and II restriction–modification systems. A substantial fraction of the coliphages were not restricted. But some were sensitive, supporting the notion that selection for immunity from phage infection is responsible for the maintenance of restriction–modification in *E. coli*. In spite of this evidence we may be mistaken in concluding that immunity to phage infection is the sole or main function of all restriction endonucleases in nature (Roberts 1978, Price & Bickle 1986).

The importance of eliminating restriction systems in *E. coli* strains used as hosts for recombinant molecules

If foreign DNA is introduced into an *E. coli* host it may be attacked by restriction systems active in the host cell. An important feature of these systems is that the fate of the incoming DNA in the restrictive host depends not only on the sequence of the DNA but also upon its history: the DNA sequence may or may not be restricted, depending upon its source immediately prior to transforming the *E. coli* host strain. As we have seen, post-replication modifications of the DNA, usually in the form of methylation of particular adenine or cytosine residues in the target sequence, protects against cognate restriction systems but not, in general, against different restriction systems.

Because restriction provides a natural defence against invasion by foreign DNA, it is usual to employ a K restriction-deficient *E. coli* K12 strain as a host in transformation with newly-created recombinant molecules. Thus where, for example, mammalian DNA has been ligated into a plasmid vector, transformation of the *Eco*K restriction-deficient host eliminates the possibility that the incoming sequence will be restricted, even if the mammalian sequence contains an unmodified *Eco*K target site. If the host happens to be *Eco*K restriction-deficient but *Eco*K modification-*proficient*, propagation on the host will confer modification methylation and hence allow subsequent propagation of the recombinant in *Eco*K restriction-proficient strains, if desired.

*Eco*K is not the only restriction-like activity in *E. coli* K12 strains. Two other activities have been found in many laboratory strains of *E. coli* K12. By contrast with the examples above, the targets for restriction by these activities are *methylated* DNAs. In particular, DNA containing 5-methylcytosine is restricted by these strains. The restriction is due to two genetically distinct systems, that differ in their sequence specificities and are named *Mcr*A and *Mcr*BC (for *m*odified *c*ytosine *r*estriction) (Raleigh & Wilson 1986, Raleigh *et al.* 1988, Sutherland *et al.* 1992). If DNA is methylated at the cytosine of 5'-GC 3' dinucleotide sequences it will be subject to restriction by the *Mcr*B system, and the degree of restriction is a function of the number of methylated sites. Many bacterial species contain methylases that produce DNA that will be a substrate for *Mcr*BC if that DNA is transformed into an $McrB^+C^+$ *E. coli* strain. Modification methylases M. *Pvu*II (C-A-G-^{m}C-T-G) and M.*Hae*III (G-G-^{m}C-C) are examples of such methylases. Of course bacteria are not the only organisms that contain 5-methylcytosine in their DNA. In plants, as much as 40–50% of all the C may be methylated, including the majority of C-G, C-A-G and C-T-G sequences. In vertebrates, the overall amount of methylation is less (about 10% of all C residues in mammals), the great majority is in the C-G dinucleotide sequence, and the cell-type specific methylation appears to be involved in gene regulation. The C methylation in DNA from such sources may give rise to a methylated G-C dinucleotide (e.g. in mammals where the sequence G-^{m}C-G occurs) and therefore render the DNA susceptible to

Table 3.3 Restriction properties of some commonly used *E. coli* host strains

E. coli strain	Notes	Restriction properties			
		*Eco*K or *Eco*B	*Eco*P1	*Mcr*A	*Mcr*BC
HB101	K strain carrying hsd−*mcr*B region from *E. coli* B. Is *hsd* 20 and thus phenotypically R−M− for *Eco*K and *Eco*B		−	+	−
DH1	*recA gyrA*	$R_k^- M_k^+$	−	+	+
DH5	derived from DH1	$R_k^- M_k^+$	−	+	+
JM101	M13 host	$R_k^+ M_k^+$	−	+	?
JM107	M13 host	$R_k^- M_k^+$	−	−	+
Y1090	λgt11 host	$R_k^+ M_k^+$	−	−	+
LE392	*sup E sup F*	$R_k^- M_k^+$	−	−	+
MC1061		$R_k^- M_k^+$	?	-	−
K802		$R_k^- M_k^+$	−	−	−
K803		$R_k^- M_k^-$	−	−	−
GM2163	*dam− dcm−*	$R_k^- M_k^+$	−	−	−

Sources: Raleigh & Wilson (1986), Raleigh *et al.* (1988).

restriction by the *Mcr*BC activity. The *Mcr*A activity is less well understood at present, but is known to restrict DNA with the following methylated sequence, C ^{m}C-G-G, such as that produced by M.*Hpa*II. The main importance of these systems, for the gene manipulator, is their possible influence in leading to the under-representation of certain genomic sequences during the creation of genomic libraries. It is therefore wise to use for such cloning an *E. coli* host strain that not only lacks the three familiar restriction systems (*Eco*K, *Eco*B and the prophage P1-specific *Eco*P1 system) found in many laboratory strains of *E. coli*, but which also lacks the Mcr systems. Table 3.3 shows the restriction properties of some commonly used host strains. Note that the foreign methylation pattern will be lost upon replication of the cloned DNA in *E. coli*, so that after the critical transformation the cloned DNAs can be freely moved into Mcr$^+$ strains.

Mechanical shearing of DNA

In addition to digesting DNA with restriction endonucleases to produce discrete fragments, there are a variety of treatments which result in non-specific breakage. Non-specific endonucleases and chemical degradation can be used but the only method that has been much applied to gene manipulation involves mechanical shearing.

The long, thin threads which constitute duplex DNA molecules are sufficiently rigid to be very easily broken by shear forces in solution. Intense sonication with ultrasound can reduce the length to about 300 nucleotide pairs. More controlled shearing can be achieved by high-speed stirring in a blender. Typically, high-molecular-weight DNA is sheared to a population of molecules with a mean size of about 8 kb pairs by stirring at 1500 rev/min for 30 min (Wensink *et al*. 1974). Breakage occurs essentially at random with respect to DNA sequence. The termini consist of short, single-stranded regions which may have to be taken into account in subsequent joining procedures.

Joining DNA molecules

Having described the methods available for cutting DNA molecules we must consider the ways in which DNA fragments can be joined to create artificially recombinant molecules. There are currently three methods for joining DNA fragments *in vitro*. The first of these capitalizes on the ability of DNA ligase to join covalently the annealed cohesive ends produced by certain restriction enzymes. The second depends upon the ability of DNA ligase from phage T4-infected *E. coli* to catalyse the formation of phosphodiester bonds between blunt-ended fragments. The third utilizes the enzyme terminal deoxynucleotidyltransferase to synthesize homopolymeric 3'-single-stranded tails at the ends of fragments. We can now look at these three methods a little more deeply.

DNA ligase

E. coli and phage T4 encode an enzyme, DNA ligase, which seals single-stranded nicks between adjacent nucleotides in a duplex DNA chain (Olivera *et al*. 1968, Gumport & Lehman 1971). Although the reactions catalysed by the enzymes of *E. coli* and T4-infected *E. coli* are very similar, they differ in their cofactor requirements. The T4 enzyme requires ATP, whilst the *E. coli* enzyme requires NAD^+. In each case the cofactor is split and forms an enzyme–AMP complex. The complex binds to the nick, which must expose a 5'-phosphate and 3'-OH group, and makes a covalent bond in the phosphodiester chain as shown in Fig. 3.4.

When termini created by a restriction endonuclease that creates cohesive ends associate, the joint has nicks a few base-pairs apart in opposite strands. DNA ligase can then repair these nicks to form an intact duplex. This reaction, performed *in vitro* with purified DNA ligase, is fundamental to many gene manipulation procedures such as that shown in Fig. 3.5.

The optimum temperature for ligation of nicked DNA is 37°C, but at this temperature the hydrogen-bonded joint between the sticky ends is unstable. *Eco*RI-generated termini associate through only four AT base-pairs and these are not sufficient to resist thermal disruption at such a high temperature. The optimum temperature for ligating the cohesive termini is therefore a compromise between the rate of enzyme action and association of the termini, and has been found by experiments to be in the range 4–15°C (Dugaicyzk *et al*. 1975, Ferretti & Sgaramella 1981).

The ligation reaction can be performed so as to favour the formation of recombinants. First, the population of recombinants can be increased by performing the reaction at a high DNA concentration; in dilute solutions *circularization* of linear fragments is relatively favoured because of the

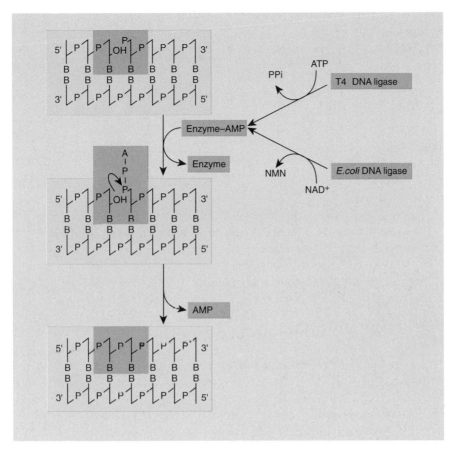

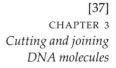

Fig. 3.4 Action of DNA ligase. An enzyme–AMP complex binds to a nick bearing 3'-OH 5'-P groups. The AMP reacts with the phosphate group. Attack by the 3'-OH group on this moiety generates a new phosphodiester bond which seals the nick.

reduced frequency of intermolecular reactions. Second, by treating linearized plasmid vector DNA with alkaline phosphatase to remove 5'-terminal phosphate groups, both recircularization and plasmid dimer formation are prevented (Fig. 3.6). In this case, circularization of the vector can occur only by insertion of non-phosphatase-treated foreign DNA which provides one 5'-terminal phosphate at each join. One nick at each join remains unligated, but after transformation of host bacteria, cellular repair mechanisms reconstitute the intact duplex.

Joining DNA fragments with cohesive ends by DNA ligase is a relatively efficient process which has been used extensively to create artificial recombinants. A modification of this procedure depends upon the ability of T4 DNA ligase to join blunt-ended DNA molecules (Sgaramella 1972). The *E. coli* DNA ligase will not catalyse blunt ligation except under special reaction conditions of macromolecular crowding (Zimmerman & Pheiffer 1983). Blunt ligation is most usefully applied to joining blunt-ended fragments via *linker* molecules; in an early example of this, Scheller *et al.* (1977) synthesized self-complementary decameric oligonucleotides, which contain sites for one or more restriction endonucleases. One such molecule is shown in Fig. 3.7. The molecule can be ligated to both ends of the foreign

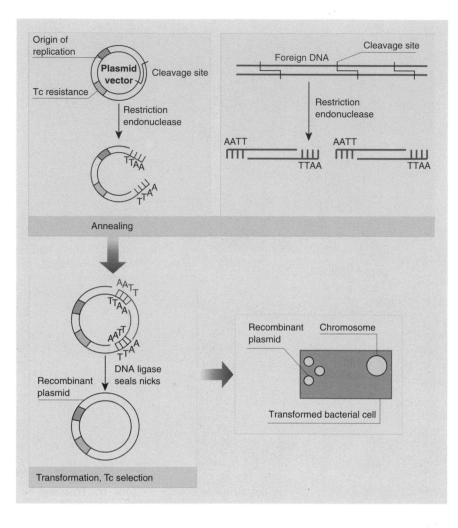

Fig. 3.5 Use of DNA ligase to create a covalent DNA recombinant joined through association of termini generated by *Eco*RI.

DNA to be cloned, and then treated with restriction endonuclease to produce a sticky-ended fragment which can be incorporated into a vector molecule that has been cut with the same restriction endonuclease. Insertion by means of the linker creates restriction enzyme target sites at each end of the foreign DNA and so enables the foreign DNA to be excised and recovered after cloning and amplification in the host bacterium.

Double-linkers

Plasmid vectors have been derived which contain a set of closely clustered cloning sites. An example of such a vector is pUC8, which is described in more detail on page 67. This vector has been used to clone duplex cDNA molecules by the double-linker approach (Kurtz & Nicodemus 1981, Helfman *et al.* 1983), in which *different* linker molecules are added to the opposite ends of the cDNA (Fig. 3.8). This has the following advantages.

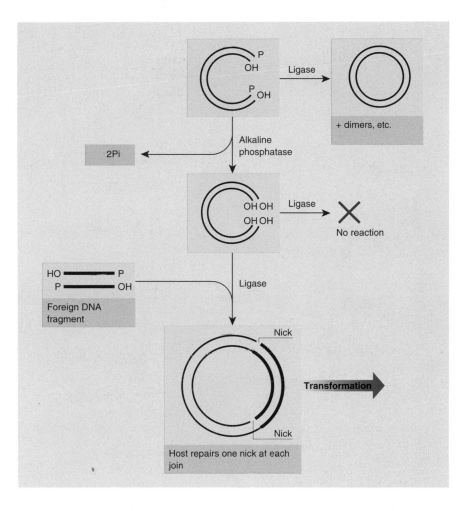

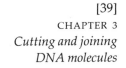

Fig. 3.6 Application of alkaline phosphatase treatment to prevent recircularization of vector plasmid without insertion of foreign DNA.

- The problem of vector reclosure without insertion of foreign DNA is overcome. Partly for this reason the method is efficient, i.e. cDNAs have been cloned which were derived from rare mRNA molecules in the starting population.
- The use of linkers, rather than homopolymers, is desirable when expression from a vector-borne promoter is sought (see Chapter 8).
- The orientation of the inserted DNA is fixed.

Adaptors

It may be the case that the restriction enzyme used to generate the cohesive ends in the linker will also cut the foreign DNA at internal sites. In this situation the foreign DNA will be cloned as two or more subfragments. One solution to this problem is to choose another restriction enzyme, but there may not be a suitable choice if the foreign DNA is large and has sites for several restriction enzymes. Another solution is to methylate internal

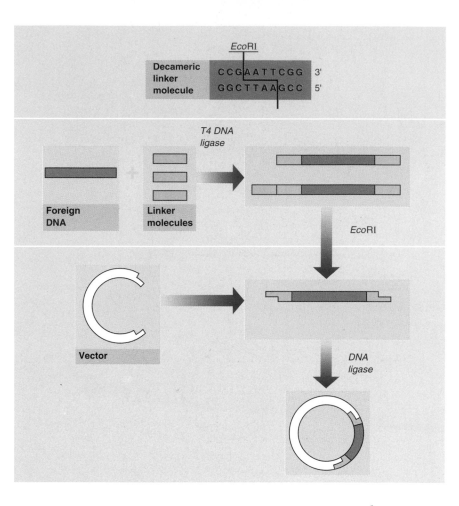

Fig. 3.7 A decameric linker molecule containing an *Eco*RI target site is joined by T4 DNA ligase to both ends of flush-ended foreign DNA. Cohesive ends are then generated by *Eco*RI. This DNA can then be incorporated into a vector that has been treated with the same restriction endonuclease.

restriction sites with the appropriate modification methylase. An example of this is described in Chapter 5. Alternatively, a general solution to the problem is provided by chemically synthesized adaptor molecules which have a *preformed* cohesive end (Wu *et al.* 1978). Consider a blunt-ended foreign DNA containing an internal *Bam*HI site (Fig. 3.9), which is to be cloned in a *Bam*HI-cut vector. The *Bam* adaptor molecule has one blunt end, bearing a 5′-phosphate group, and a *Bam* cohesive end which is not phosphorylated. The adaptor can be ligated to the foreign DNA ends. The foreign DNA plus added adaptors is then phosphorylated at the 5′-termini and ligated into the *Bam*HI site of the vector. If the foreign DNA were to be recovered from the recombinant with *Bam*HI, it would be obtained in two fragments. However, the adaptor is designed to contain two other restriction sites (*Sma*I, *Hpa*II) which may enable the foreign DNA to be recovered intact.

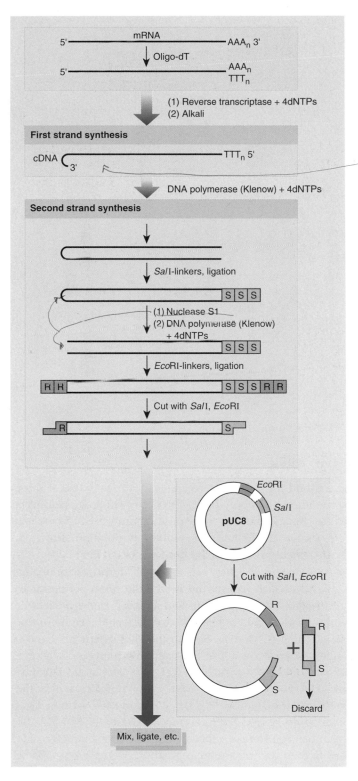

DNA pol Klenow will fill this in by using hairpin as a primer — giving dsDNA.

Fig. 3.8 Double linkers. mRNA is copied into double-stranded cDNA and *Sal*I-linkers (S) are added. The hairpin loop formed by self-priming of second strand synthesis is removed by nuclease S1. In this procedure, any raggedness left by nuclease S1 (i.e. short, single-strand projections at the terminus) is removed by polishing with Klenow polymerase. *Eco*RI-linkers are then ligated to the duplex molecule, and cohesive termini revealed by restriction with *Sal*I and *Eco*RI. The cDNA plus linkers is then ligated into a vector cut with the same two enzymes.

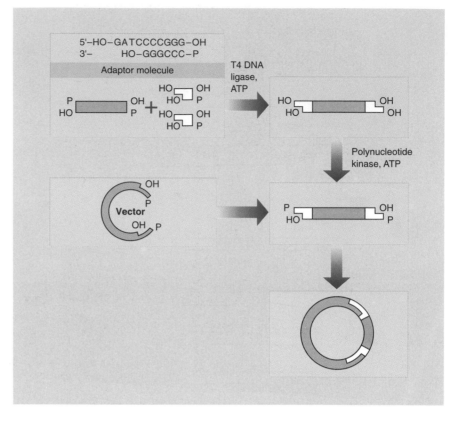

Fig. 3.9 Use of a *Bam*HI adaptor molecule. A synthetic adaptor molecule is ligated to the foreign DNA. The adaptor is used in the 5'-hydroxyl form to prevent self-polymerization. The foreign DNA plus ligated adaptors is phosphorylated at the 5'-termini and ligated into the vector previously cut with *Bam*HI.

Homopolymer tailing

A general method for joining DNA molecules makes use of the annealing of complementary homopolymer sequences. Thus, by adding oligo(dA) sequences to the 3'-ends of one population of DNA molecules, and oligo(dT) blocks to the 3'-ends of another population, the two types of molecule can anneal to form mixed dimeric circles (Fig. 3.10).

An enzyme purified from calf thymus, terminal deoxynucleotidyl-transferase, provides the means by which the homopolymeric extensions can be synthesized, for if presented with a single deoxynucleotide triphosphate it will repeatedly add nucleotides to the 3'-OH termini of a population of DNA molecules (Chang & Bollum 1971). DNA with exposed 3'-OH groups, such as arise from pre-treatment with phage λ exonuclease or restriction with an enzyme such as *Pst*I, is a very good substrate for the transferase. However, conditions have been found in which the enzyme will extend even the shielded 3'-OH of 5'-cohesive termini generated by *Eco*RI (Roychoudhury *et al.* 1976, Humphries *et al.* 1978).

The terminal transferase reactions have been characterized in detail with regard to their use in gene manipulation (Deng & Wu 1981, Michelson & Orkin 1982). Typically, 10–40 homopolymeric residues are added to each end.

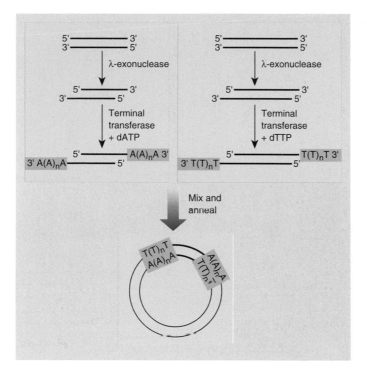

Fig. 3.10 Use of calf-thymus terminal deoxynucleotidyl-transferase to add complementary homopolymer tails to two DNA molecules.

In 1972, Jackson *et al.* were among the first to apply the homopolymer method when they constructed a recombinant in which a fragment of phage λ DNA was inserted into SV40 DNA. In their experiments, the single-stranded gaps which remained in the two strands at each join were repaired *in vitro* with DNA polymerase and DNA ligase so as to produce covalently closed circular molecules, which were then used to transfect susceptible mammalian cells (see Chapter 15).

Subsequently, the homopolymer method, employing either dA.dT or dG.dC homopolymers, has been applied extensively in constructing recombinant plasmids for cloning in *E. coli*. Commonly, the annealed circles are used directly for transformation with repair of the gaps occurring *in vivo*. In the example which follows we shall see how homopolymer tailing can be applied to cloning DNA copies of eukaryotic messenger RNA and how a careful choice of which homopolymers are used can be important.

Cloning cDNA by homopolymer tailing

If we wish to construct a clone containing sequences derived from eukaryotic mRNA, we must first obtain the sequence in DNA form. We can do this by making a complementary (cDNA) copy of the mRNA, using the enzyme reverse transcriptase, which is a type of DNA polymerase

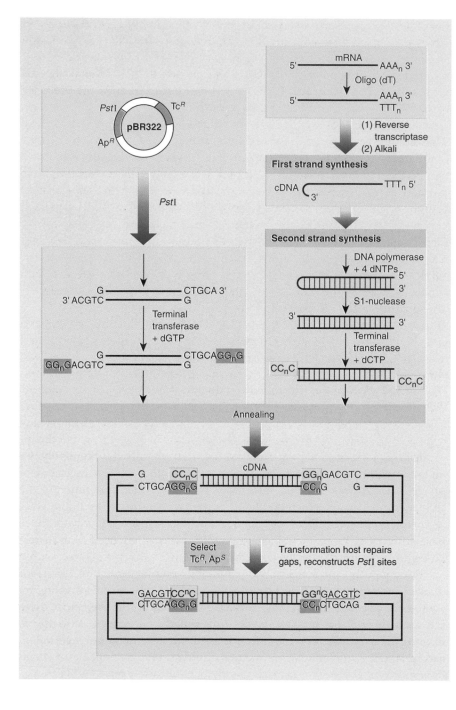

Fig. 3.11 Synthesis of a cDNA copy of a polyadenylated mRNA and insertion into a vector molecule by homopolymer tailing. See text for explanation.

found in retroviruses, and whose function is to synthesize DNA upon an RNA template.

Like other true DNA polymerases, reverse transcriptase can only synthesize a new DNA strand if provided with a growing point in the form of a pre-existing primer which is base-paired with the template and bears a free 3'-OH group. Fortunately, most eukaryotic mRNAs occur naturally in a polyadenylated form with up to 200 adenylate residues at their 3'-termini and hence we can provide a primer simply by hybridizing a short oligo(dT) molecule with this poly(A) sequence. The primer is then suitably located for synthesis of a complete cDNA by reverse transcriptase in the presence of all four deoxynucleoside triphosphates (Fig. 3.11).

The immediate product of the reaction is an RNA–DNA hybrid. The RNA strand can then be destroyed by alkaline hydrolysis, to which DNA is resistant, leaving a single-stranded cDNA which can be converted into the double-stranded form in a second DNA polymerase reaction. This reaction depends upon the observation that cDNAs can form a transient self-priming structure in which a hairpin loop at the 3'-terminus is stabilized by enough base-pairing to allow initiation of second-strand synthesis. Once initiated, subsequent synthesis of the second strand stabilizes the hairpin (Efstratiadis *et al.* 1976, Higuchi *et al.* 1976). The hairpin and any single-stranded DNA at the other end of the cDNA molecule are then trimmed away by treatment with the single-strand-specific nuclease S1, giving rise to a fully duplex molecule.

In our example (Fig. 3.11) the duplex cDNA is tailed with oligo(dC) and annealed with the pBR322 vector which has been cut open with *Pst*I and tailed with oligo(dG). It will be seen that these homopolymers have been chosen so that *Pst*I target sites are reconstructed in the recombinant molecule, thus providing a simple means for excising the inserted sequences after amplification (Smith *et al.* 1979). This can be accomplished in another way, by constructing dA.dT joins. In that case the homopolymeric regions will have a lower melting temperature than the rest of the recombinant molecule, and so under partially denaturing conditions can be cleaved by nuclease S1 to release the inserted sequence.

Full-length cDNA cloning

In the two cDNA schemes illustrated in this chapter, second-strand synthesis is self-primed, resulting in the formation of a duplex cDNA with a hairpin loop that is subsequently removed by nuclease S1. This step necessarily leads to the loss of a certain amount of sequence corresponding to the 5' end of the mRNA, and unless the nuclease S1 is very pure, there can be adventitious damage to the duplex cDNA. For this reason, self-priming of second-strand synthesis is now rarely carried out; improved methods have been developed and are discussed in Chapters 6 and 10.

4 Plasmids as cloning vehicles for use in *E. coli*

Basic properties of plasmids

Plasmids are widely used as cloning vehicles, but before discussing their use in this context it is appropriate to review some of their basic properties. Plasmids are replicons which are stably inherited in an extrachromosomal state. It should be emphasized that extrachromosomal nucleic acid molecules are not necessarily plasmids, for the definition given above implies genetic homogeneity, constant monomeric unit size, and the ability to replicate independently of the chromosome. Thus the heterogeneous circular DNA molecules which are found in *Bacillus megaterium* (Carlton & Helinski 1969) are not necessarily plasmids. The definition given above, however, does include the prophages of those temperate phages, e.g. P1, which are maintained in an extrachromosomal state, as opposed to those such as λ (see Chapter 5) which are maintained by integration into the host chromosome. Also included are the replicative forms of the filamentous coliphages which specify the continued production and release of phage particles without concomitant cell lysis.

Most plasmids exist as double-stranded circular DNA molecules. If both strands of DNA are intact circles the molecules are described as *c*ovalently *c*losed *c*ircles or CCC DNA (Fig. 4.1). If only one strand is intact, then the molecules are described as *o*pen *c*ircles and OC DNA. When isolated from cells, covalently closed circles often have a deficiency of turns in the double helix such that they have a supercoiled configuration. The enzymatic interconversion of supercoiled, relaxed CCC DNA* and OC DNA is shown in Fig. 4.1. Because of their different structural configurations, supercoiled and OC DNA separate upon electrophoresis in agarose

---

*The reader should not be confused by the terms *relaxed circle* and *relaxed plasmid*. Relaxed circles are CCC DNA that does not have a supercoiled configuration. Relaxed plasmids are plasmids with multiple copies per cell.

[46]

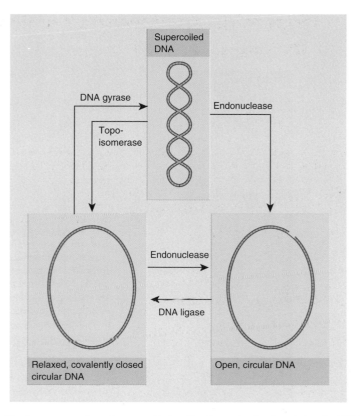

Fig. 4.1 The interconversion of supercoiled, relaxed covalently closed circular DNA, and open circular DNA.

gels (Fig. 4.2). Addition of an intercalating agent, such as ethidium bromide, to supercoiled DNA causes the plasmid to unwind. If excess ethidium bromide is added the plasmid will rewind in the opposite direction (Fig. 4.3). Use of this fact is made in the isolation of plasmid DNA (see p. 51).

In recent years, a number of unusual plasmids have been identified in a variety of bacteria. For example, double-stranded linear plasmids are found in *Borrelia* sp. (Barbour 1988, 1993) and *Streptomyces* (Kinashi *et al.* 1987, 1992). Both groups of plasmids have molecular structures more in common with animal viruses, e.g. terminal hairpin loops, inverted terminal repeats, covalently attached protein. Multicopy single-stranded DNA (msDNA) with RNA attached at the 5′ end has been found in bacteria as diverse as *E. coli* and myxobacteria (Inouye & Inouye 1991). Since special techniques sometimes have to be used to detect these plasmids, e.g. orthogonal-field-alteration gel electrophoresis (Carle & Olson 1984) their frequency and distribution in bacteria as a whole is not known.

Plasmids are widely distributed throughout the prokaryotes, vary in size from less than 1×10^6 daltons to greater than 200×10^6, and are generally dispensable. Some of the phenotypes which these plasmids confer

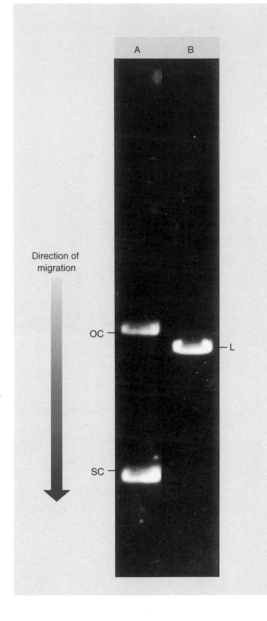

Direction of migration

OC —

SC —

A   B

— L

Fig. 4.2 Electrophoresis of DNA in agarose gels. The direction of migration is indicated by the arrow. DNA bands have been visualized by soaking the gel in a solution of ethidium bromide (which complexes with DNA by intercalating between stacked base-pairs) and photographing the orange fluorescence which results upon ultraviolet irradiation. (A) Open circular (OC) and super-coiled (SC) forms of a plasmid of 6.4 kb pairs. Note that the compact super-coils migrate considerably faster than open circles (B). Linear plasmid (L) DNA produced by treatment of the preparation shown in lane (A) with *Eco*RI for which there is a single target site. Under the conditions of electrophoresis employed here, the linear form migrates just ahead of the open circular form.

on their host cells are listed in Table 4.1. Plasmids to which phenotypic traits have not yet been ascribed are called *cryptic* plasmids.

Plasmids can be categorized into one of two major types — conjugative or non-conjugative — depending upon whether or not they carry a set of transfer genes, called the *tra* genes, that promotes bacterial conjugation. Plasmids can also be categorized on the basis of their being maintained as multiple copies per cell (*relaxed* plasmids) or as a limited number of copies per cell (*stringent* plasmids). Generally, conjugative plasmids are of relatively high molecular weight and are present as one to three copies

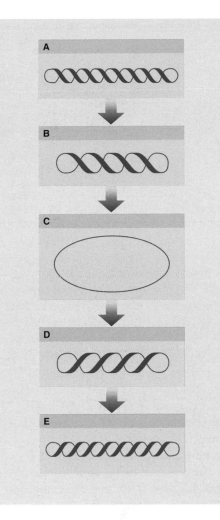

Fig. 4.3 Effect of intercalation of ethidium bromide on supercoiling of DNA. As the amount of intercalated ethidium bromide increases, the double helix untwists with the result that the supercoiling decreases until the open form of the circular molecule is produced. Further intercalation introduces excess turns in the double helix resulting in supercoiling in the opposite sense (note the direction of coiling at B and D). For clarity, only a single line represents the double helix.

Table 4.1 Some phenotypic traits exhibited by plasmid-carried genes

Antibiotic resistance
Antibiotic production
Degradation of aromatic compounds
Haemolysin production
Sugar fermentation
Enterotoxin production
Heavy-metal resistance
Bacteriocin production
Induction of plant tumours
Hydrogen sulphide production
Host-controlled restriction and modification

Table 4.2 Properties of some conjugative and non-conjugative plasmids of Gram-negative organisms

Plasmid	Size (MDa)	Conjugative	No. of plasmid copies/chromosome equivalent	Phenotype
Col E1	4.2	No	10–15	Col E1 production
RSF 1030	5.6	No	20–40	Ampicillin resistance
clo DF13	6	No	10	Cloacin production
R6K	25	Yes	13–38	Ampicillin and streptomycin resistance
F	62	Yes	1–2	–
RI	62.5	Yes	3–6	Multiple drug resistance
Ent P 307	65	Yes	1–3	Enterotoxin production

per chromosome whereas non-conjugative plasmids are of low molecular weight and present as multiple copies per chromosome (Table 4.2). An exception is the conjugative plasmid R6K which has a molecular weight of 25×10^6 daltons and is maintained as a relaxed plasmid.

Often it is important to measure plasmid copy number. Although a number of techniques are available they are not without their limitations. A commonly used method (Twigg & Sherratt 1980) is to extract total DNA from a cell paste and to separate the chromosomal DNA from the plasmid DNA by agarose gel electrophoresis. After photographing the gel, the photographic negative is scanned with a densitometer. The density of the plasmid DNA relative to the chromosomal DNA is a measure of plasmid copy number. An alternative method is to quantify the plasmid DNA by high-performance liquid chromatography (Coppella *et al*. 1987). Both these methods suffer from the disadvantage that errors can arise if there is differential or incomplete release of plasmid DNA from cell extracts.

An alternative procedure for estimating plasmid copy number is to measure the amount of a plasmid-encoded gene product. For example, β-lactamase activity can be measured if the plasmid specifies ampicillin resistance. Before this method can be used it is essential to show that the selected gene is expressed constitutively under all physiological conditions. Klotsky and Schwartz (1987) have shown that this is indeed the case for the ampicillin-resistance gene present on many cloning vectors used with *E. coli*.

Plasmid *incompatibility* is the inability of two different plasmids to coexist in the same host cell in the absence of selection pressure. The term incompatibility can only be used when it is certain that entry of the second plasmid has taken place and that DNA restriction is not involved. Groups of plasmids which are mutually incompatible are considered to belong to the same incompatibility group. Currently, over 30 incompatibility groups have been defined among plasmids of *E. coli* and 13 for plasmids of *Staphylococcus aureus*. Plasmids belonging to incompatibility classes P, Q and W are termed *promiscuous* for they are capable of promoting their own transfer to a wide range of Gram-negative bacteria (group P and W) and of being stably maintained in these diverse hosts. Such promiscuous plasmids thus offer the potential of readily transferring cloned DNA molecules into a wide range of genetic environments (see p. 210).

An extremely useful article explaining the terminology used in plasmid genetics is that of Novick *et al.* (1976). A much fuller discussion of the topics outlined above is provided by Falkow (1975) and Broda (1979).

The purification of plasmid DNA

An obvious prerequisite for cloning in plasmids is the purification of the plasmid DNA. Although a wide range of plasmid DNAs are now routinely purified, the methods used are not without their problems. Undoubtedly the trickiest stage is the lysis of the host cells; both incomplete lysis and total dissolution of the cells result in greatly reduced recoveries of plasmid DNA. The ideal situation occurs when each cell is just sufficiently broken to permit the plasmid DNA to escape without too much contaminating chromosomal DNA. Provided the lysis is done gently, most of the chromosomal DNA released will be of high molecular weight and can be removed, along with cell debris, by high-speed centrifugation to yield a *cleared lysate*. The production of satisfactory cleared lysates from bacteria other than *E. coli*, particularly if large plasmids are to be isolated, is frequently a combination of skill, luck and patience.

Many methods are available for isolating pure plasmid DNA from cleared lysates but only two will be described here. The first of these is the 'classical' method and is due to Vinograd (Radloff *et al.* 1967). This method involves isopycnic centrifugation of cleared lysates in a solution of CsCl containing ethidium bromide (EtBr). EtBr binds by intercalating between the DNA base-pairs, and in so doing causes the DNA to unwind. A CCC DNA molecule such as a plasmid has no free ends and can only unwind to a limited extent, thus limiting the amount of EtBr bound. A linear DNA molecule, such as fragmented chromosomal DNA, has no such topological constraints and can therefore bind more of the EtBr molecules. Because the density of the DNA–EtBr complex decreases as more EtBr is bound, and because more EtBr can be bound to a linear molecule than a covalent circle, the covalent circle has a higher density at saturating concentrations of EtBr. Thus covalent circles (i.e. plasmids) can be separated from linear chromosomal DNA (Fig. 4.4).

Currently the most popular method of extracting and purifying plasmid DNA is that of Birnboim and Doly (1979). This method makes use of the observation that there is a narrow range of pH (12.0–12.5) within which denaturation of linear DNA, but not covalently closed circular DNA, occurs. Plasmid-containing cells are treated with lysozyme to weaken the cell wall and then lysed with sodium hydroxide and sodium dodecyl sulphate. Chromosomal DNA remains in a high molecular weight form but is denatured. Upon neutralization with acidic sodium acetate, the chromosomal DNA renatures and aggregates to form an insoluble network.

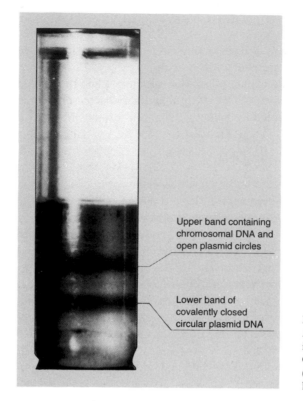

Upper band containing
chromosomal DNA and
open plasmid circles

Lower band of
covalently closed
circular plasmid DNA

Fig. 4.4 Purification of Col E1
Kan^R plasmid DNA by
isopycnic centrifugation in a
CsCl–EtBr gradient.
(Photograph by courtesy of
Dr G. Birnie.)

Simultaneously, the high concentration of sodium acetate causes precipitation of protein–SDS complexes and of high molecular weight RNA. Provided the pH of the alkaline denaturation step has been carefully controlled the CCC plasmid DNA molecules will remain in a native state and in solution while the contaminating macromolecules co-precipitate. The precipitate can be removed by centrifugation and the plasmid concentrated by ethanol precipitation. If necessary, the plasmid DNA can be purified further by gel filtration.

Although most cloning vehicles are of low molecular weight (see next section) it sometimes is necessary to use the much larger conjugative plasmids. Although these high molecular weight plasmids can be isolated by the methods just described, the yields often are very low. Either there is inefficient release of the plasmids from the cells as a consequence of their size or there is physical destruction caused by shear forces during the various manipulative steps. A number of alternative procedures have been described (Gowland & Hardmann 1986) many of which are a variation on that of Eckhardt (1978). Bacteria are suspended in a mixture of Ficoll and lysozyme and this results in a weakening of the cell walls. The samples then are placed in the slots of an agarose gel where the cells are lysed by the addition of detergent. The plasmids subsequently are extracted from the gel following electrophoresis. The use of agarose which melts at low temperature facilitates extraction of the plasmid from the gel.

Desirable properties of plasmid cloning vehicles

An ideal cloning vehicle would have the following three properties:
- low molecular weight;
- ability to confer readily selectable phenotypic traits on host cells;
- single sites for a large number of restriction endonucleases, preferably in genes with a readily scorable phenotype.

The advantages of a low molecular weight are several. First, the plasmid is much easier to handle, i.e. it is more resistant to damage by shearing, and is readily isolated from host cells. Second, low molecular weight plasmids are usually present as multiple copies (see Table 4.2), and this not only facilitates their isolation but leads to gene dosage effects for all cloned genes. Finally, with a low molecular weight there is less chance that the vector will have multiple substrate sites for any restriction endonuclease (see below).

After a piece of foreign DNA is inserted into a vector the resulting chimaeric molecules have to be transformed into a suitable recipient. Since the efficiency of transformation is so low it is essential that the chimaeras have some readily scorable phenotype. Usually this results from some gene, e.g. antibiotic resistance, carried on the vector, but could also be produced by a gene carried on the inserted DNA.

One of the first steps in cloning is to cut the vector DNA and the DNA to be inserted with either the same endonuclease or ones producing the same ends. If the vector has more than one site for the endonuclease, more than one fragment will be produced. When the two samples of cleaved DNA are subsequently mixed and ligated, the resulting chimaeras will, in all probability, lack one of the vector fragments. It is advantageous if insertion of foreign DNA at endonuclease-sensitive sites inactivates a gene whose phenotype is readily scorable, for in this way it is possible to distinguish chimaeras from cleaved plasmid molecules which have self-annealed. Of course, readily detectable insertional inactivation is not essential if the vector and insert are to be joined by the homopolymer tailing method (see p. 42) or if the insert confers a new phenotype on host cells.

Some examples will be presented which illustrate the points raised above, but first we shall consider how some of the common plasmids rate as cloning vehicles.

Usefulness of 'natural' plasmids as cloning vehicles

The term 'natural' is used loosely in this context to describe plasmids which were not constructed *in vitro* for the sole purpose of cloning. Col E1 is a naturally occurring plasmid which specifies the production of a bacteriocin, colicin E1. By necessity this plasmid also carries a gene which confers on host cells immunity to colicin E1. RSF 2124 is a derivative of Col E1 which carries a transposon specifying ampicillin resistance. For long enough the origin of pSC101 was not clear. It is now known to be a natural

Table 4.3 Properties of some 'natural' plasmids used for cloning DNA

Plasmid	Size (MDa)	Single sites for endonucleases	Marker for selecting transformants	Insertional inactivation of
pSC101	5.8	*Xho*I, *Eco*RI *Pvu*II, *Hinc*II *Hpa*I	Tetracycline resistance	–
		*Hind*III, *Bam*HI *Sal*I	–	Tetracycline resistance
Col E1	4.2	*Eco*RI	Immunity to colicin E1	Colicin E1 production
RSF 2124	7.4	*Eco*RI, *Bam*HI	Ampicillin resistance	Colicin E1 production

plasmid of *Salmonella panama* previously named SP-219 (Manen & Caro 1991). Details of these plasmids are shown in Table 4.3.

To clone DNA in pSC101, the plasmid DNA and the DNA to be inserted are digested with *Eco*RI, mixed, and treated with DNA ligase. The ligated molecules are then used to transform a suitable recipient to tetracycline resistance. Unfortunately, there is no easy genetical method of distinguishing chimaeras from reconstituted vector DNA unless the insert confers a new phenotype on the transformants. Two examples of the use of pSC101 for cloning DNA are presented in the next section. When using *Eco*RI, cloning with Col E1 as the vector is a little simpler. Transformants are selected on the basis of immunity to colicin E1 and chimaeras recognized by their inability to produce colicin E1. Unfortunately, screening for immunity to colicin E1 is not technically simple, and plasmid RSF 2124 is more useful in this respect since transformants are selected by virtue of their ampicillin resistance.

Col E1 and plasmids derived from it (see later) have two distinct advantages over pSC101. They have a higher copy number and they can be enriched with chloramphenicol. When chloramphenicol is added to a late log-phase culture of a Col E1-containing strain of *E. coli*, chromosome replication ceases because of the need for continued protein synthesis. However, the cessation of protein synthesis has no effect on Col E1 replication, such that after 10–12 hours over 50% of the DNA in the cells is plasmid DNA (Hershfield *et al.* 1974). Since there may be 1000–3000 copies of the plasmid in each cell it is easy to see why chloramphenicol enrichment is a useful step in plasmid isolation.

Examples of the use of pSC101 for cloning

1 Expression of Staphylococcus *plasmid genes in* E. coli

For this experiment Chang and Cohen (1974) considered *S. aureus* plasmid p1258 (molecular weight 20×10^6) as being particularly appropriate for experiments involving interspecies genome construction, since it carries several different genetic determinants that were potentially detectable in *E.*

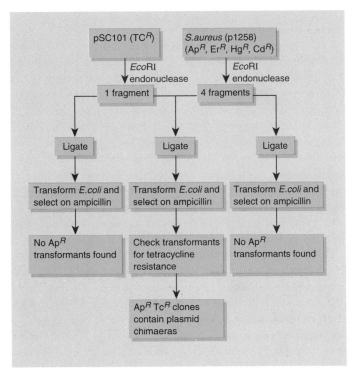

Fig. 4.5 Insertion of an *S. aureus* gene specifying ampicillin resistance into plasmid pSC101.

coli. Moreover, agarose gel electrophoresis indicated that this plasmid is cleaved by the *Eco*RI restriction endonuclease into four easily identifiable fragments. Molecular chimaeras containing DNA derived from both *Staphylococcus* and *E. coli* were constructed by ligation of a mixture of *Eco*RI-cleaved pSC101 and p1258 DNA and then were used to transform a restrictionless strain of *E. coli* (Fig. 4.5). *E. coli* transformants that expressed the ampicillin resistance determinant carried by the *Staphylococcus* plasmid were selected and checked for tetracycline resistance.

Caesium chloride gradient analysis of one ampicillin-resistant, tetracycline-resistant chimaera showed that its buoyant density was intermediate to the buoyant densities of the parental plasmids. In addition, treatment of this chimaera with *Eco*RI produced two fragments, one the size of *Eco*RI cleaved pSC101 and the other the size of one of the *Eco*RI fragments of p1258.

2 *Cloning of* Xenopus *DNA in* E. coli

This experiment by Morrow *et al.* (1974) involved the construction *in vitro* of plasmid chimaeras composed of both prokaryotic and eukaryotic DNA, and the recovery of recombinant DNA molecules from transformed *E. coli* in the absence of selection for genetic determinants carried by the eukaryotic

Box 4.1 The Cohen and Boyer patents

The examples of cloning in pSC101 shown in Figs 4.5 and 4.6 have a special significance for they form the basis of two well-known patents assigned to Stanford University and known universally as the Cohen and Boyer patents. On 2 December 1980, the US Patent Office issued patent number 4 237 224 entitled Process for Producing Biologically Functional Chimeras. This patent covers a series of steps involved in replicating exogenous genes in microorganisms. Specifically, a plasmid or viral DNA is cleaved, a new gene inserted, the replicon is placed in a microorganism and the transformants isolated. In issuing the patent the US Patent Office showed that it considered gene-splicing methods to be patentable. An important facet of this patent is that foreign applications were not filed. This is a result of a prior publication by the inventors which destroyed the non-US patent rights. This meant that recombinant DNA technology could be used for profit in other countries and the products of the technology could be imported into the USA. Consequently, a second patent was filed in the USA which laid claim to the products of the technology regardless of country of origin. This patent, number 4 468 464, was issued on 28 August 1984.

The two patents are infamous because of their scope. Because they lay claim to the fundamental technology of gene splicing on which much of the biotechnology industry is based, they dominate most other patents in the field. Thus all commercial users of the technology in the USA are required to pay royalties to Stanford University. Although these royalties are not unreasonable, many companies challenged the validity of the patents to such an extent that they were a *cause célèbre* for much of the 1980s. There were two major challenges. The first of these was that the original patent was non-enabling. This patent covers a key plasmid, pSC101, which was used as the vehicle for inserting new genes in *E. coli* and described the method for producing this plasmid. However, in 1977, Cohen published a paper that admits an error in the original procedure. Deposition of the plasmid in a culture collection would have avoided this problem except that deposition was made too late—after the patent had been filed. The patent assignees claimed that Cohen and Boyer made the plasmid available to other scientists but there were restrictions on access to it. Furthermore, they argued that the patent was enabling since pSC101 is not vital as the instructions provided in the patent explain how to select, isolate and use other suitable plasmids. However, it should be noted that in a submission to the Patent Office in June 1977, Stanford University took the opposite view; namely, that pSC101 was indeed vital to the patent claims and that this new patent found by the inventors was what distinguished their claims from potentially competing work by others! The second objection to the patent was prior disclosure. In an article in *New Scientist* on 25 October 1973, more than a year before the first patent application was filed, a detailed report was given of the work of Cohen and Boyer. Both these objections were dismissed by the Patent Office. Although it was generally expected at the

continued

Box 4.1 *continued*

time, no major court challenge to the validity of the patents has occurred. In retrospect, this is not surprising since Stanford University set a very reasonable royalty rate which minimized the financial burden on companies and so encouraged the widespread use of the technology covered by the patent.

Two other points about the Cohen and Boyer patents are worth noting. First, these patents are unusual because they dominate almost all other patents in the field of molecular biotechnology. In no other industry are there patents with such an all-embracing impact. Second, the fact that the inventors of such an important patent as US 4 237 224 were academics and that the patent assignee was a university, has had an unprecedented effect in making other universities around the world aware of the potential commercial value of scientific discoveries made by their staff.

DNA. The amplified ribosomal DNA from *Xenopus laevis* oocytes was used as the source of eukaryotic DNA for these experiments, since this DNA can be purified readily and had been well characterized. In addition, the repeat unit of *X. laevis* rDNA is susceptible to cleavage by *Eco*RI, resulting in the production of discrete fragments that can be linked to the pSC101 vector.

In this experiment (Fig. 4.6) a mixture of *Eco*RI-cleaved pSC101 DNA and *X. laevis* rDNA was ligated and was used to transform *E. coli* to tetracycline resistance. Fifty-five separate transformants were selected and their plasmid DNA extracted and analysed. All 55 plasmids gave a fragment of molecular weight 5.8×10^6 on *Eco*RI digestion and 13 of them yielded additional fragments corresponding in size to *Eco*RI-produced fragments of *X. laevis* rDNA. Since there was no direct selection method for the *Xenopus*-derived DNA it was indeed fortuitous that such a high percentage (23.6%) of clones contained chimaeric molecules.

Construction and characterization of a new cloning vehicle: pBR322

Although pSC101, Col E1 and RSF 2124 can be used to clone DNA they suffer from a number of disadvantages as outlined above. For this reason considerable effort has been expended on constructing, *in vitro*, superior cloning vehicles. Undoubtedly the most versatile and widely used of these artificial plasmid vectors is pBR322. Plasmid pBR322 contains the Ap^R and Tc^R genes of RSF 2124 and pSC101 respectively, combined with replication elements of pMB1, a Col E1-like plasmid (Fig. 4.7a). The origins of pBR322, and its progenitor pBR313, are shown in Fig. 4.7b, and details of its construction can be found in the papers of Bolivar *et al.* (1977a,b).

Plasmid pBR322 has been completely sequenced. The original published sequence (Sutcliffe 1979) was 4362 base-pairs long. Position O of the sequence was arbitrarily set between the A and T residues of the *Eco*RI recognition sequence (GAATTC). The sequence was revised by the inclusion

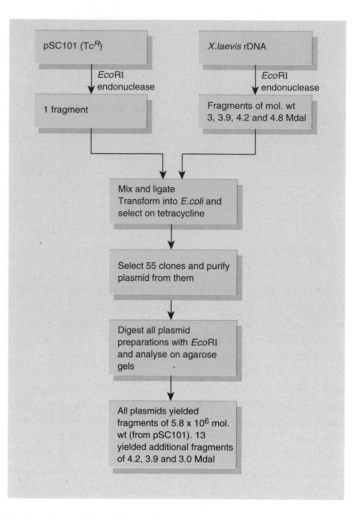

Fig. 4.6 Cloning of genes from *Xenopus laevis* in *Escherichia coli* with the aid of pSC101.

of an additional CG base-pair at position 526, thus increasing the size of the plasmid to 4363 base-pairs (Backman & Boyer 1983, Peden 1983). The most useful aspect of the DNA sequence is that it totally characterizes pBR322 in terms of its restriction sites such that the exact length of every fragment can be calculated. These fragments can serve as DNA markers for sizing any other DNA fragment in the range of several base-pairs up to the entire length of the plasmid.

There are over 20 enzymes with unique cleavage sites on the pBR322 genome (Fig. 4.8). The target sites of seven of these enzymes (*Eco*RV, *Nhe*I, *Bam*HI, *Sph*I, *Sal*I, *Xma*III and *Nru*I) lie within the Tc^R gene, and there are sites for a further two (*Cla*I and *Hin*dIII) within the promoter of that gene. There are unique sites for three enzymes (*Pst*I, *Pvu*I and *Sca*I) within the Ap^R gene. Thus, cloning in pBR322 with the aid of any one of those 12 enzymes will result in insertional inactivation of either the Ap^R or the Tc^R

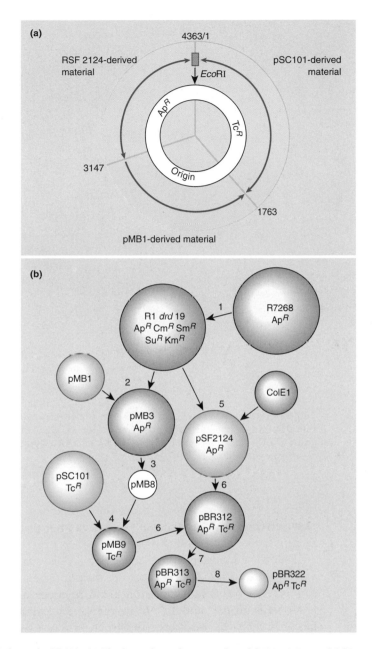

Fig. 4.7 The origins of plasmid pBR322. (a) The boundaries between the pSC101, pMB1 and RSF 2124-derived material. The numbers indicate the positions of the junctions in base-pairs from the unique *Eco*RI site. (b) The molecular origins of plasmid pBR322. R7268 was isolated in London in 1963 and later renamed R1. 1, A variant, R1*drd*19, which was de-repressed for mating transfer, was isolated. 2, The *Ap*^R transposon, Tn*3*, from this plasmid was transposed onto pMB1 to form pMB3. 3, This plasmid was reduced in size by *Eco*RI* rearrangement to form a tiny plasmid, pMB8, which carries only colicin immunity. 4, *Eco*RI* fragments from pSC101 were combined with pMB8 opened at its unique *Eco*RI site and the resulting chimaeric molecule rearranged by *Eco*RI* activity to generate pMB9. 5, In a separate event, the Tn*3* of R1*drd*19 was transposed to Col E1 to form pSF2124. 6, The Tn*3* element was then transposed to pMB9 to form pBR312. 7, *Eco*RI* rearrangement of pBR312 led to the formation of pBR313, from which (8) two separate fragments were isolated and ligated together to form pBR322. During this series of constructions, R1 and Col E1 served only as carries for Tn*3*. (Reproduced by courtesy of Dr G. Sutcliffe and Cold Spring Harbor Laboratory.)

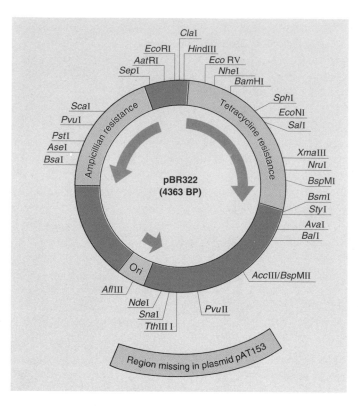

Fig. 4.8 The structure of pBR322 showing the unique cleavage sites. The grey arrows inside the circle show the direction of transcription of the Ap^R and Tc^R genes. The short arrow shows the direction of DNA replication. Also shown is the region of pBR322 missing in plasmid pAT153 (see p. 67).

markers. However, cloning in the other unique sites does not permit the easy selection of recombinants because neither of the antibiotic resistance determinants is inactivated.

Following manipulation *in vitro*, *E. coli* cells transformed with plasmids with inserts in the Tc^R gene can be distinguished from those cells transformed with recircularized vector. The former are Ap^R and Tc^S whereas the latter are both Ap^R and Tc^R. In practice, transformants are selected on the basis of their Ap resistance and then replica-plated onto Tc-containing media to identify those that are Tc^S. Cells transformed with pBR322 derivatives carrying inserts in the Ap^R gene can be identified more readily (Boyko & Ganschow 1982). Detection is based upon the ability of β-lactamase produced by Ap^R cells to convert penicillin to penicilloic acid, which in turn binds iodine. Transformants are selected on rich medium containing soluble starch and Tc. When colonized plates are flooded with an indicator solution of iodine and penicillin β-lactamase-producing (Ap^R), colonies clear the indicator solution whereas Ap^S colonies do not.

The *Pst*I site in the Ap^R gene is particularly useful because the 3′-tetranucleotide extensions formed on digestion are ideal substrates for terminal transferase. Thus this site is excellent for cloning by the homopolymer tailing method described on the previous chapter (see p. 42). If oligo(dG.dC) tailing is used, the *Pst*I site is regenerated (see Fig. 3.11) and the insert may be cut out with that enzyme. In addition, the *Pst*I site is particularly useful for obtaining expression of cloned genes and this aspect is covered in detail in Chapter 8 (p. 143).

Plasmid pBR322 is the most widely used cloning vehicle. In addition, it has been widely used as a model system for the study of prokaryotic transcription and translation as well as investigation of the effects of topological changes on DNA conformation. The popularity of pBR322 is a direct result of the availability of an extensive body of information on its structure and function. This in turn is increased with each new study. The reader wishing more detail on the structural features, transcriptional signals, replication, amplification, stability and conjugal mobility of pBR322 should consult the review of Balbás *et al.* (1986).

Examples of the use of plasmid pBR322 as a vector

1 *Isolation of DNA fragments which carry promoters*

Cloning into the *Hin*dIII site of pBR322 generally results in loss of tetracycline resistance. However, in some recombinants Tc^R is retained or even increased. This is because the *Hin*dIII site lies within the promoter rather than the coding sequence. Thus whether or not insertional inactivation occurs depends on whether the cloned DNA carries a promoter-like sequence able to initiate transcription of the *Tc^R* gene. Widera *et al.* (1978) have used this technique to search for promoter-containing fragments.

Four structural domains can be recognized within *E. coli* promoters (see Chapter 8). These are:
● position 1, the purine initiation nucleotide from which RNA synthesis begins;
● position −6 to −12, the Pribnow box;
● the region around base-pair −35;
● the sequence between base-pairs −12 and −35.

Although the *Hin*dIII site lies within the Pribnow box (Rodriguez *et al.* 1979) the box is re-created on insertion of a foreign DNA fragment (Fig. 4.9). Thus when insertional inactivation occurs it must be the region from −13 to −40 which is modified.

2 *Expression in* E. coli *of a chemically synthesized gene for the hormone somatostatin*

There are two reasons for selecting this particular early experiment to illustrate the use of pBR322 as a vector. First, the fact that gene manipulation had been successfully used to obtain a functional gene product from a chemically synthesized gene indicated that the potential of genetic engineering was not just a dream but had already been realized. Second,

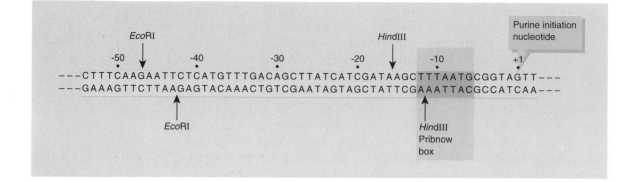

Fig. 4.9 DNA base sequence of the promoter of the tetracycline resistance gene. The bases are numbered on the basis of the purine initiation nucleotide being position +1. (In the conventional pBR322 map the bases are numbered from the *Eco*RI site.) The arrows indicate the positions of cleavage by restriction endonucleases *Eco*RI and *Hin*dIII.

success was dependent on the use of a number of elegant 'tricks' and some of these warrant detailed examination. Although it does not make easy reading for the novice, the original paper is worth studying in detail for it is one of the classics of gene manipulation.

The rationale of the experiment was in fact to show that recombinant DNA technology can be used to fuse chemically synthesized genes to plasmid elements for expression in *E. coli* or other bacteria. As a model, Itakura *et al.* (1977) designed and synthesized a gene for the hormone somatostatin. The somatostatin 'gene' was chosen because somatostatin is a small polypeptide of known amino acid composition, there were sensitive radioimmune and biological assays, and being a hormone it was of intrinsic biological interest.

The somatostatin gene was synthesized chemically such that there was an *Eco*RI site at one end and a *Bam*HI site at the other (Fig. 4.10). Also, a methionine codon preceded the normal NH₂-terminal amino acid of somatostatin and the COOH-terminal amino acid was followed by two stop codons.

In the first part of the experiment three new plasmids were created from pBR322. The control region of the *lac* operon, comprising the *lac* promoter, catabolite gene activator-protein binding site, the operator, the ribosome binding site, and the first seven triplets of the β-galactosidase structural gene, were inserted into the *Eco*RI site of pBR322 to create pBH10. Plasmid pBH10 has two *Eco*RI sites and one of these was removed to generate pBH20. Finally, the synthetic somatostatin gene was inserted next to the *lac* control gene to yield pSOM I (Fig. 4.11).

The DNA sequence of pSom I indicated that the clone carrying this plasmid should produce a peptide containing somatostatin, but no somatostatin was found. However, in reconstruction experiments it was observed that exogenous somatostatin was degraded rapidly in *E. coli* extracts. Thus the failure to find somatostatin activity could be accounted for by intracellular degradation by endogenous proteolytic enzymes. Such proteolytic degradation might be prevented by attachment of the somatostatin to a large protein, e.g. β-galactosidase. The β-galactosidase structural gene has an *Eco*RI site near the COOH-terminus and the available data on the amino acid sequence of this protein suggested that it would be possible to insert

```
           Met Ala Gly Cys Lys Asn Phe Phe Trp Lys Thr Phe Thr Ser Cys Stop Stop
     5' AATTC ATG GCT GGT TGT AAG AAC TTC TTT TGG AAG ACT TTC ACT TCG TGT TGA TAG
           TAC CGA CCA ACA TTC TTG AAG AAA ACC TTC TGA AAG TGA AGC ACA ACT ATCCTAG 5'
```

the synthetic gene into this site and still maintain the proper reading frame. In order to do this, two new plasmids pSom II and pSom II-3 were created.

Fig. 4.10 Sequence of the chemically synthesized somatostatin gene.

The formation of pSom II (Fig. 4.11) Plasmid pSom I was digested with *Eco*RI and *Pst*I and the larger fragment, which contains the synthetic somatostatin gene, purified by gel electrophoresis. This fragment has lost the *lac* control region and part of the *Ap*^R gene. The *Ap*^R gene was restored by ligating this large fragment to the small fragment produced by digesting pBR322 with *Eco*RI and *Pst*I. The ligated mixture was used to transform *E. coli* to ampicillin resistance and the plasmid from one *Ap*^R clone selected and called pSom II.

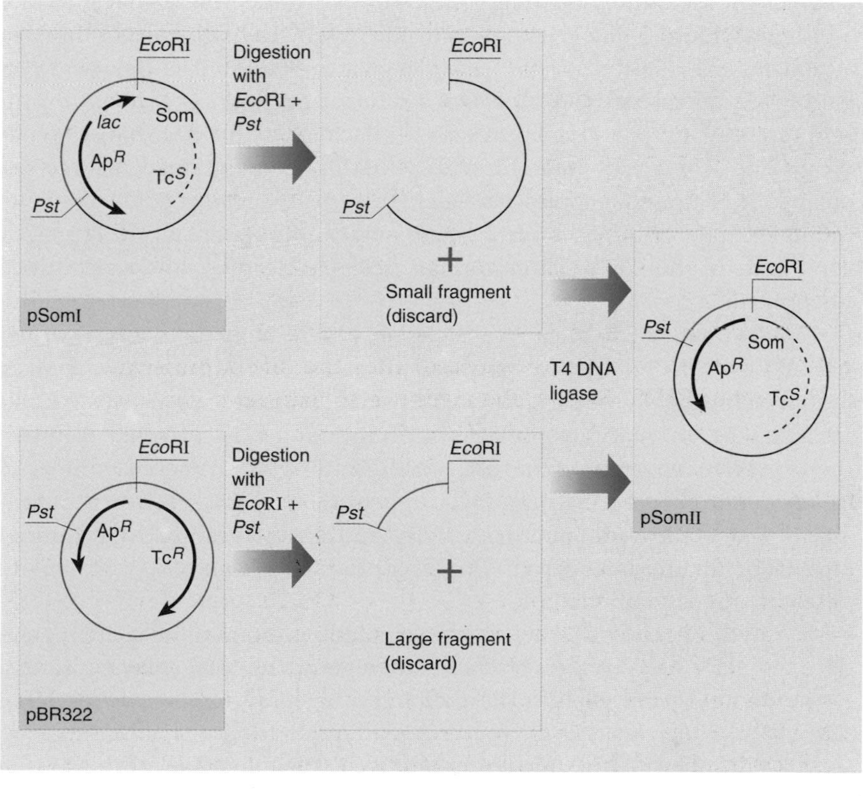

Fig. 4.11 Formation of pSom II. For clarity, only one strand of the plasmid DNA is shown.

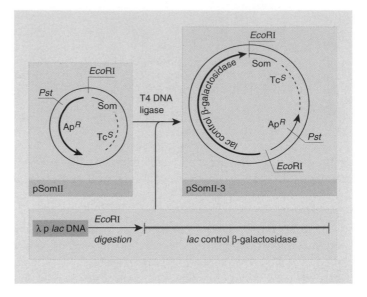

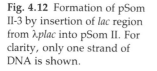

Fig. 4.12 Formation of pSom II-3 by insertion of *lac* region from λ*plac* into pSom II. For clarity, only one strand of DNA is shown.

The formation of pSom II-3 (Fig. 4.12) Bacteriophage λ*plac*5 which carries the *lac* control region and the entire β-galactosidase gene was digested with *Eco*RI and the mixture ligated with *Eco*RI-treated pSom II. The ligated mixture was used to transform *E. coli* and selection was made for blue colonies on medium containing ampicillin and the chromogenic substrate 5-bromo-4-chloro-3-indolyl-β-D-galactoside (Xgal). The rationale for this was as follows. Xgal is not an inducer of β-galactosidase but is cleaved by β-galactosidase, releasing a blue indolyl derivative. Since Xgal is not an inducer, only mutants constitutive for β-galactosidase produce blue colonies on medium containing Xgal. Plasmid pSom II-3 is maintained as a relaxed plasmid, i.e. multiple copies per cell. Thus, cells carrying pSom II-3 have multiple copies of the *lac* control region and can titrate out all the repressor produced by the single chromosomal *lac*I gene leading to a constitutive phenotype.

Approximately 2% of the transformants were blue and analysis of the plasmid from them showed the presence of a 4.4 MDa fragment identical to that found by digesting λ*plac*5 with *Eco*RI. Two orientations of the *Eco*RI *lac* fragment of λ*plac*5 are possible but only one of these would maintain the proper reading frame into the somatostatin gene. When a number of independent clones were examined, approximately 50% produced detectable somatostatin radioimmune activity and all of these had the desired orientation of the *lac* operon. The non-producing clones were found to have the opposite orientation.

It should be noted that no somatostatin radioimmune activity was detected prior to cyanogen bromide cleavage of the total cellular protein. Since the antiserum used in the radioimmune assay requires a free NH$_2$-terminal alanine, no activity was expected prior to cleavage. Methionine residues are the site of cyanogen bromide cleavage and it is for this reason

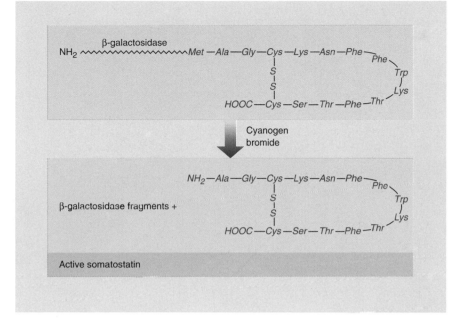

Fig. 4.13 Cleavage of the chimaeric protein by cyanogen bromide to yield active somatostatin. The somatostatin can readily be purified from cyanogen bromide and the fragments of β-galactosidase.

that a methionine codon was included in the synthetic somatostatin gene (Figs 4.10 and 4.13). Cyanogen bromide can only be used to cleave fusion proteins when the wanted portion has no internal methionine residues. Alternative methods for releasing the desired protein include the use of trypsin (Smith *et al*. 1984) and citraconic anhydride (Shine *et al*. 1980).

Improved vectors derived from pBR322

Over the years numerous different derivatives of pBR322 have been constructed, many to fulfil special-purpose cloning needs (Table 4.4). A compilation of the properties of some of these plasmids has been provided by

Table 4.4 Summary of modified vectors derived from pBR322

Function of modified vector	Reference
Vectors to facilitate expression of proteins	Chapter 8
Vectors for the identification of regulatory signals	This chapter
Vectors for the direct selection of recombinants	This chapter
Vectors with additional restriction sites	This chapter
Vectors with different, additional or improved selective markers	This chapter
Vectors with increased stability	Chapter 8
Vectors with altered copy numbers	This chapter, chapter 8
Plasmids for DNA sequencing	Chapter 9
Plasmids to permit secretion of proteins	Chapter 8
Gene-fusion vectors to facilitate protein purification	Chapter 8
Vectors for use in *E. coli* and unrelated organisms (shuttle vectors)	Chapters 12, 13, 14 and 15

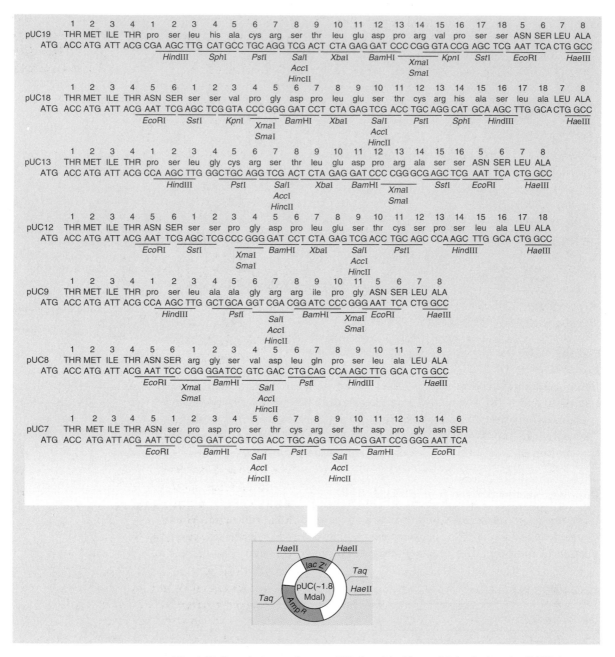

Fig. 4.14 Genetic maps of some pUC plasmids. The multiple cloning site (MCS) is inserted into the *lacZ* gene but does not interfere with gene function. The additional codons present in the *lacZ* gene as a result of the polylinker are labelled with lower-case letters. These polylinker regions (MCS) are identical to those of the M13 mp series of vectors (see p. 91).

Balbás *et al.* (1986). The design principles behind the improved general-purpose vectors are outlined below. Details of the more specialist vectors will be presented in later chapters.

Much of the early work on the improvement of pBR322 centred on the insertion of additional unique restriction sites and selectable markers, e.g. pBR325 encodes chloramphenicol resistance in addition to ampicillin and tetracycline resistance and has a unique *Eco*RI site in the *Cm*^{R} gene. Each new vector was constructed in a series of steps analogous to those used in the generation of pBR322 itself (Fig. 4.7). More recently, the construction of improved vectors has been simplified (Vieira & Messing 1982, 1987, Yanisch-Perron *et al.* 1985) by the use of *polylinkers* or *multiple cloning sites* (MCS) as exemplified by the pUC vectors (Fig. 4.14). A MCS is a short DNA sequence, 2.8 kb in the case of pUC19, carrying sites for many different restriction endonucleases. An MCS increases the number of potential cloning strategies available by extending the range of enzymes that can be used to generate a restriction fragment suitable for cloning. By combining them within an MCS, the sites are made contiguous, so that any two sites within it can be cleaved simultaneously without excising vector sequences.

The pUC vectors also incorporate a DNA sequence which permits rapid visual detection of an insert. The MCS is inserted into the 5' end of the *lac* operon of *E. coli*. It contains the promoter and a small segment of the β-galactosidase (β-*gal*) gene. This encodes a sequence known as the α peptide. The α peptide will combine with a mutant β-gal protein which lacks the normal *N*-terminus, to generate a functional enzyme molecule. When an *E. coli* cell containing such a mutant β-gal within its genome is transformed with a pUC vector, enzymatically active β-gal accumulates. Such transformed clones can be recognized by including in the agar a β-gal substrate (Xgal) which produces a blue precipitate when hydrolysed. Normally, any fragment of DNA inserted into the MCS will at some point along its length happen to contain a termination codon in the same reading frame as the α peptide. This will prevent synthesis of the α peptide. Hence bacterial colonies containing recombinant DNA molecules are colourless and can be easily identified from the 'background' of blue, non-recombinant clones.

A derivative of pBR322 which has been widely used as a cloning vehicle is pAT153 (Twigg & Sherratt 1980). In terms of biological containment pAT153 has a great advantage over pBR322. Although pBR322 is not self-transmissible, it can be mobilized at a frequency of 10^{-1} from cells containing a conjugative plasmid plus plasmid Col K. However, the fragment removed from pBR322 during the formation of pAT153 (see Fig. 4.8) contains a DNA sequence (called *nic* or *bom*) essential for conjugal transfer. As a consequence pAT153 cannot be mobilized and this provides a means of biological containment. Another reason for its widespread use is that it has a copy number three- to fivefold higher than that of pBR322.

Another widely used series of vectors are pBluescript II and its derivatives pBC and pBS which are commercially available. These vectors have a number of useful features which facilitate gene manipulation. Like the pUC vectors described above they have a multiple cloning site inserted into the β-galactosidase α peptide sequence so that blue/white colour selection can be used to detect inserts. A feature not found in the pUC vectors is that the multiple cloning site has a phage T7 promoter (see p. 94) at one end and a phage T3 promoter at the other enabling expression of the cloned insert to be obtained regardless of its orientation. The vectors also have two origins of DNA replication, one derived from Col E1 and the other from the single-stranded DNA phage f1 (see p. 86). Under normal circumstances the Col E1 origin is used for plasmid replication but following phage f1 infection the other origin is used and single-stranded DNA is produced. This is particularly useful for sequencing (see p. 170). Because these vectors have origins of replication derived from both a plasmid and a phage they are known as *phagemids*.

Direct selection vectors

Most plasmid vectors contain at least two selectable markers. One marker is kept intact and is used to select transformants. The transformants are then screened to detect inactivation of the second marker, indicating the insertion of foreign DNA. It would be more convenient if recombinant plasmids could be selected directly after transformation, and consequently a number of workers have constructed direct selection vectors. (Roberts *et al.* 1980, Dean 1981, Hennecke *et al.* 1982, Hashimoto-Gotoh *et al.* 1986). Many of these vectors suffer from some or all of the following limitations:
- requirement for specialized strains or media for selection
- limited availability of cloning sites
- inability to achieve regulated expression of the inserted fragments
- high level of false positives

These problems have been overcome by Stevis and Ho (1987) who made use of the fact that high levels of xylose isomerase activity in wild-type *E. coli* strains results in a Xyl − phenotype. They constructed plasmid pLX100, in which a gene for xylose isomerase, containing contiguous unique sites for *Hind*III, *Pst*I, *Bam*HI and *Xho*I, is placed under the control of the *lac* promoter. *E. coli* transformants containing pLX100 cannot grow in minimal medium with xylose unless a DNA fragment is inserted into one of the unique restriction sites.

Low-copy-number plasmid vectors

High-copy-number plasmid vectors such as pBR322 are widely used because they are easily purified and, via a gene dosage effect, they can direct the synthesis of high levels of cloned gene products. However there are some genes which cannot be cloned on high-copy-number vectors because their presence seriously disturbs the normal physiology of the cell. Examples include genes which encode surface structural proteins, e.g. *ompA* (Beck & Bremer 1980), proteins that regulate basic cellular metabolism, e.g. polA (Murray & Kelley 1979), the cystic fibrosis transmembrane conductance regulator (Gregory *et al.* 1990) and lentivirus envelope sequences (Cunningham *et al.* 1993). One strategy for cloning such genes is to use low-copy-number vectors. Two suitable low-copy-number vectors are phage λ in its lysogenic state (see Chapter 5) and plasmid pSC101. Although pSC101 was the first plasmid to be used for cloning *in vitro* (p. 54), it is seldom used because it carries only a single marker, and insertional inactivation cannot be used to screen for recombinant clones.

Two sets of vectors have been constructed which retain the low-copy-number of pSC101 but have two more antibiotic-resistance markers and unique cleavage sites for many restriction endonucleases (Hashimoto-Gotoh *et al.* 1981, Stoker *et al.* 1982). Sankar *et al.* (1993) have developed a low-copy vector based on pSC101 in which expression of cloned genes is very tightly controlled to minimize toxic effects. In this case expression is independent of normal physiological controls (induction and regression) because it is dependent on the presence of phage T7 RNA polymerase.

Runaway plasmid vectors

One reason for cloning a gene on a multicopy plasmid is to increase greatly its expression and hence facilitate purification of the protein it encodes. However, as indicated above, some genes cannot be cloned on high-copy-number vectors because excess gene product is lethal to the cell. The use of low-copy-number vectors avoids cell killing, but this may be self-defeating since expression of the cloned gene will be reduced. A solution to this problem is to use runaway plasmid vectors.

The first runaway plasmid vectors were described by Uhlin *et al.* (1979). At 30°C the plasmid vector is present in a moderate number of copies per cell. Above 35°C all control of plasmid replication is lost and the number of plasmid copies per cell increases continuously. Cell growth and protein synthesis continue at the normal rates for 2–3 hours at the higher temperature. During this period, products from genes on the plasmid are overproduced. Eventually, inhibition of cell growth occurs and the cells lose viability, but at this stage plasmid DNA may account for 50% of the DNA in the cell. Since the original description by Uhlin *et al.* (1979) of runaway vectors a number of groups have constructed improved derivatives. The biology of these vectors and their uses have been reviewed by Nordstrom and Uhlin (1992).

5 Bacteriophage and cosmid vectors for *E. coli*

Bacteriophage λ

Essential features

Bacteriophage λ is a genetically complex but very extensively studied virus of *E. coli*. Because it has been the object of so much molecular genetical research it was natural that, right from the beginnings of gene manipulation, it should have been investigated and developed as a vector. The DNA of phage λ, in the form in which it is isolated from the phage particle, is a linear duplex molecule of about 48.5 kbp. The entire DNA sequence has been determined (Sanger *et al.* 1982). At each end are short single-stranded 5′-projections of 12 nucleotides, which are complementary in sequence and by which the DNA adopts a circular structure when it is injected into its host cell, i.e. λ DNA naturally has cohesive termini which associate to form the *cos* site.

Functionally related genes of phage λ are clustered together on the map, except for the two positive regulatory genes *N* and *Q*. Genes on the left of the conventional linear map (Fig. 5.1) code for head and tail proteins of the phage particle. Genes of the central region are concerned with recombination (e.g. *red*) and the process of lysogenization in which the circularized chromosome is inserted into its host chromosome and stably replicated along with it as a prophage. Much of this central region, including these genes, is not essential for phage growth and can be deleted or replaced without seriously impairing the infectious growth cycle. Its dispensability is crucially important, as will become apparent later, in the construction of vector derivatives of the phage. To the right of the central

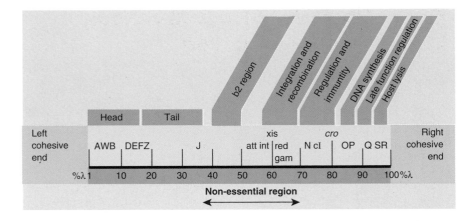

Fig. 5.1 Map of the λ chromosome, showing the physical position of some genes on the full-length DNA of wild-type bacteriophage λ. Clusters of functionally related genes are indicated.

region are genes concerned with gene regulation and prophage immunity to superinfection (*N, cro, cI*), followed by DNA synthesis (*O, P*), late function regulation (*Q*) and host cell lysis (*S, R*). Figure 5.2 illustrates the λ life cycle.

Promoters and control circuits

As we shall see, it is possible to insert foreign DNA into the chromosome of phage λ derivatives, and in some cases foreign genes can be expressed efficiently via λ promoters. We must therefore briefly consider the promoters and control circuits affecting λ gene expression (see Ptashne 1992 for an excellent monograph on phage λ control circuits).

In the lytic cycle, λ transcription occurs in three temporal stages: early, middle and late. Basically, early gene transcription establishes the lytic cycle (in competition with lysogeny); middle gene products replicate and recombine the DNA, and late gene products package this DNA into mature phage particles. Following infection of a sensitive host, early transcription proceeds from major promoters situated immediately to the left (P_L) and right (P_R) of the repressor gene (*cI*) (Fig. 5.3). This transcription is subject to repression by the product of the *cI* gene and in a lysogen this repression is the basis of immunity to superinfecting λ. Early in infection transcripts from P_L and P_R stop at termination sites t_L and t_{R_1}. The site t_{R_2} stops any transcripts that escape beyond t_{R_1}. Lambda switches from early- to middle-stage transcription by anti-termination. The *N* gene product, expressed from P_L, directs this switch. It interacts with RNA polymerase and, antagonizing the action of host termination protein ρ, permits it to ignore the stop signals so that P_L and P_R transcripts extend into genes such as *red* and *O* and *P* necessary for the middle stage. The early and middle transcripts and patterns of expression therefore overlap. The *cro* product, when sufficient has accumulated, prevents transcription from P_L and P_R. The gene *Q* is expressed from the distal portion of the extended P_R transcript and is responsible for the middle-to-late switch. This also operates by anti-

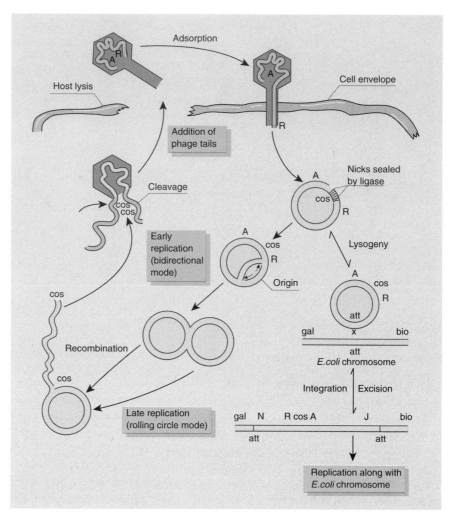

Fig. 5.2 Replication of phage
λ DNA in lytic and lysogenic
cycles.

termination. The *Q* product specifically anti-terminates the short P_R transcript, extending it into the late genes, across the cohered *cos* region, so that many mature phage particles are ultimately produced.

Both *N* and *Q* play positive regulatory roles essential for phage growth and plaque formation; but an N^- phage *can produce* a small plaque if the termination site t_{R_2} is removed by a small deletion termed *nin* (*N* independent) as in $\lambda N^- \; nin$.

Vector DNA

Wild-type λ DNA contains several target sites for most of the commonly used restriction endonucleases and so is not itself suitable as a vector. Derivatives of the wild-type phage have therefore been produced that either have a single target site at which foreign DNA can be inserted

Box 5.1 Bacteriophage λ: its important place in molecular biology and recombinant DNA technology

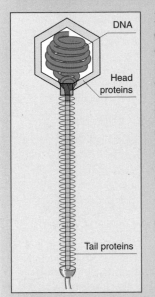

In the early 1950s, following some initial studies on *Bacillus megaterium*, André Lwoff and his colleagues at the Institut Pasteur, described the phenomenon of lysogeny in *E. coli*. It became clear that certain strains of *E. coli* were lysogenized by phage, that is to say, these bacteria harboured phage λ in a dormant form called a prophage. The lysogenic bacteria grew normally and might easily not have been recognized as lysogenic. However when Lwoff exposed the bacteria to a moderate dose of ultraviolet light, the bacteria stopped growing, and after about 90 minutes of incubation the bacteria lysed, releasing a crop of phage into the medium.

The released phage were incapable of infecting more *E. coli* that had been lysogenized by phage λ (this is called immunity to superinfection), but non-lysogenic bacteria could be infected to yield another crop of virus. Not every non-lysogenic bacterium yielded virus; some bacteria were converted into lysogens because the phage switched to the dormant lifestyle—becoming prophage—rather than causing a lytic infection.

By the mid-1950s it was realized that the prophage consisted of a phage λ genome that had become integrated into the *E. coli* chromosome. It was also apparent to Lwoff's colleagues, Jacob and Monod, that the switching between the two states of the virus—the lytic and lysogenic lifestyles—was an example of a fundamental aspect of genetics that was gaining increasing attention, gene regulation.

Intensive genetic and molecular biological analysis of the phage, mainly in the 1960s and 1970s, led to a good understanding of the virus. The key molecule in maintaining the dormancy of the prophage, and in conferring immunity to superinfection, is the phage repressor, which is the product of the phage *cI* gene. In 1967 the phage repressor was isolated by Mark Ptashne (Ptashne 1967a,b). The advanced molecular genetics of the phage made it a good candidate for development as a vector, beginning in the 1970s and continuing to the present day, as described in the text. The development of vectors exploited the fact that a considerable portion of the phage genome encodes functions that are not needed for the infectious cycle. The ability to package recombinant phage DNA into virus particles *in vitro* was an important development for library construction (Hohn & Murray 1977).

A landmark in molecular biology was reached when the entire sequence of the phage λ genome, 48 502 nucleotide pairs, was determined by Fred Sanger and his colleagues (Sanger 1982).

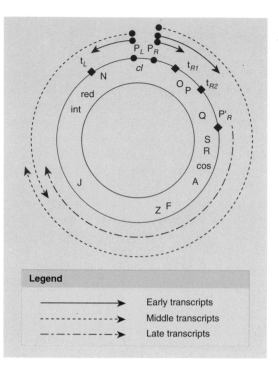

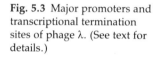

Fig. 5.3 Major promoters and transcriptional termination sites of phage λ. (See text for details.)

(*insertional* vectors), or have a pair of sites defining a fragment that can be removed and replaced by foreign DNA (*replacement* vectors). Since phage λ can accommodate only about 5% more than its normal complement of DNA, vector derivatives are constructed with deletions to increase the space within the genome. The shortest λ DNA molecules that produce plaques of nearly normal size are 25% deleted. Apparently, if too much non-essential DNA is deleted from the genome it cannot be packaged into phage particles efficiently. This can be turned to advantage, for if the replaceable fragment of a replacement-type vector either is removed by physical separation, or is effectively destroyed by treatment with a second restriction endonuclease that cuts it alone, then the deleted vector genome can give rise to plaques only if a new DNA segment is inserted into it. This amounts to positive selection for recombinant phage carrying foreign DNA.

Many vector derivatives of both the insertional and replacement types were produced by several groups of researchers early in the development of recombinant DNA technology (e.g. Thomas *et al.* 1974, Murray & Murray 1975, Blattner *et al.* 1977, Leder *et al.* 1977). Most of these vectors were constructed for use with *Eco*RI, *Bam*HI or *Hind*III, but their application could be extended to other endonucleases by the use of linker molecules. These early vectors have been largely superseded by improved vectors for rapid and efficient genomic library or cDNA library construction (see below and Chapter 6). However, we shall discuss them briefly here because they illustrate some basic principles.

The λWES.λB′ phage was an early vector, widely used at one time,

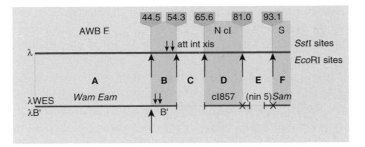

Fig. 5.4 Physical map of λDNA and the vector derivative, λWES.λB'. The λ regions are aligned. Parentheses indicate deletions. Downward and upward arrows are *Sst*I and *Eco*RI restriction sites, respectively. Numbers under *Eco*RI sites indicate the positions of the sites as percentages of the wild-type genome length.

which illustrates several important points. For details of the construction of this phage the reader is referred to papers of Thomas *et al.* (1974) and Leder *et al.* (1977). The DNA map of this vector is shown in Fig. 5.4. We can see that the phage has been constructed with three amber mutations in genes *W*, *E* and *S*. These reduce the likelihood of recombinants escaping from the laboratory environment, since appropriate amber suppressor strains are very uncommon in nature. The fragment designated C in wild-type λ has been deleted by restriction and re-ligation *in vitro*. In addition, the two most righthand *Eco*RI sites have been eliminated and a *nin* deletion introduced. The deletions create space for insertion of foreign DNA. The B' fragment is the replaceable fragment. (The B fragment has inadvertently been inverted during construction of the vector and is designated B'.) In use, the vector DNA is digested with *Eco*RI, then the B' fragment may be removed by preparative gel electrophoresis or other physical methods. Alternatively, this fragment can be destroyed by treatment of the *Eco*RI digest with *Sst*I. The *Eco*RI treated foreign DNA is then added to a mixture of vector arms, the mixture is ligated and used to transfect an appropriate amber suppressor strain of *E. coli* so that viable recombinant phage are recovered. Joining of the two DNA arms without insertion of foreign DNA results in a molecule that is too short (9.8% (B' fragment) + 11.3% (C fragment) + 6.1% (*nin* deletion) = 27.2% less than λ$^+$) to produce viable phage, even though it contains all of the genes necessary for lytic growth.

Other early phage λ vectors were constructed by Blattner *et al.* (1977). These were called Charon phages by their originators, after the old ferryman of Greek mythology who conveyed the spirits of the dead across the River Styx. Charon 16A is an insertional vector with a single *Eco*RI site located in the gene for β-galactosidase (*lacZ*) which is included in the *lac*5 DNA substitution. This is useful because there is a convenient colour test for the production of β-galactosidase. When the chromogenic substrate Xgal (see p. 67) is included in the plating medium, phage carrying *lac*5 give dark blue plaques on Lac⁻ indicator bacteria. Potential success with

Charon 16A cloning is detected by insertional inactivation of the Lac function which results in colourless plaques.

Another screening method employing insertional inactivation was exploited by Murray *et al.* (1977). Insertion of foreign DNA at the single target within the immunity region of one of their vector molecules destroys the ability of the phage to produce a functional repressor so that recombinants give clear plaques which are readily distinguished from the turbid plaques of parental phages formed by simple rejoining of the two fragments of the vector DNA. This insertional inactivation of the *cI* gene also forms the basis of a powerful system for *selecting* recombinant formation in such phages as λgt10 (Nathans & Hogness 1983, Young & Davis 1983) and λNM1149 (Murray 1983). When the non-recombinant phages are plated on the *hfl*A (*h*igh *f*requency of *l*ysogeny) mutant of *E. coli* no plaques are formed. This is because lysogens, which are of course immune to superinfection, are created at such a high frequency in this host that a visible plaque does not appear. However, recombinant phage have an inactive *cI* repressor gene, cannot form lysogens, and therefore do form plaques.

Improved phage λ vectors

As with plasmid vectors, improved phage vector derivatives have been developed. There have been several aims, among which are the following.
● To increase the capacity for foreign DNA fragments, preferably for fragments generated by any one of several restriction enzymes (reviewed by Murray 1983).
● To devise methods for positively selecting recombinant formation.
● To allow RNA probes to be conveniently prepared by transcription of the foreign DNA insert; this facilitates the screening of libraries in chromosome walking procedures. An example of a vector with this property is λZAP (see p. 95).
● To develop vectors for the insertion of eukaryotic cDNA such that expression of the cDNA, in the form of a fusion polypeptide with β-galactosidase, is driven in *E. coli*; this form of expression vector is useful in antibody screening; an example of such a vector is λgt11.

The first two points will be discussed here. The discussion of improved vectors in library construction and screening is deferred until p. 126 and Chapter 6.

The maximum capacity of phage λ derivatives can only be attained with vectors of the replacement type, so that there has also been an accompanying incentive to devise methods for positively selecting recombinant formation without the need for prior removal of the stuffer fragment. Even when steps are taken to remove the stuffer fragment by physical purification of vector arms, small contaminating amounts may remain, so that genetic selection for recombinant formation remains desirable. Two innovative means of achieving this are exploitation of the *A3* gene of phage T5 and the Spi⁻ phenotype. The first of these strategies is interesting but has not been widely used. The second, Spi⁻, strategy is applicable to the widely used λEMBL3 and λEMBL4 vectors (Frischauf *et al.* 1983).

Davison *et al.* (1979) have devised a vector in which the λB stuffer fragment of λWES.λB' has been replaced by two identical 1.8 kb fragments of phage T5. These fragments carry the *A3* gene, which is known to prevent growth of T5 itself, or this adapted λ vector on *E. coli* carrying the plasmid Col Ib. Two fragments were found to be necessary in this new vector, called λgtWES.T5622, because a single 1.8 kb replacement of the λB fragment was too short to give viable phage. Positive selection for recombinant formation is imposed by plating on an *E. coli* host carrying plasmid Col Ib.

Wild-type λ cannot grow on *E. coli* strains lysogenic for phage P2; in other words the λ phage is Spi⁺ (sensitive to P2 inhibition). It has been shown that the products of λ genes *red* and *gam*, which lie in the region 64–69% on the physical map, are responsible for the inhibition of growth in a P2 lysogen (Herskowitz 1974, Sprague *et al.* 1978, Murray 1983). Hence vectors have been derived (e.g. λL47 and λ1059) in which the stuffer fragment includes the region 64–69%, so that recombinants in which this has been replaced by foreign DNA are phenotypically Spi⁻ and can be positively selected by plating on a P2 lysogen (Karn *et al.* 1980, Loenen & Brammar 1980).

Deletion of the *gam* gene has other consequences. The *gam* product is necessary for the normal switch in λDNA replication from the bidirectional mode to the rolling circle mode (see Fig. 5.2). Gam phage cannot generate the concatemeric linear DNA which is normally the substrate for packaging into phage heads. However, *gam*⁻ phage do form plaques because the *rec* and *red* recombination systems act on circular DNA molecules to form multimers which can be packaged. *Gam*⁻ *red*⁻ phage are totally dependent upon *rec*-mediated exchange for plaque formation on *rec*⁺ bacteria. λ DNA is a poor substrate for this *rec*-mediated exchange. Therefore, such phage make vanishingly small plaques unless they contain one or more short DNA sequences called *chi* (*c*ross-over *h*ot-spot *i*nstigator) sites, which stimulate *rec*-mediated exchange. The early replacement vector λWES.λB' generates *red*⁻ *gam*⁺ clones. But many of the new replacement vectors with a large capacity (e.g. λL47, λ1059) generate *red*⁻ *gam*⁻ clones. These vectors have, therefore, been constructed so as to include a *chi* site within the non-replaceable part of the phage.

The most recent generation of λ vectors combine a large capacity for foreign DNA, close to the theoretical limit of 23 kb, together with features that allow simple and efficient library construction (see Chapter 6). The replacement vectors EMBL3 and EMBL4 (Frischauf *et al.* 1983) have convenient polylinker sequences flanking the replaceable fragment. Phages with inserts can be selected by their Spi⁻ phenotype, and are *chi*⁺. Derivatives of EMBL3 containing amber mutations are available (EMBL3 *Sam*, EMBL3 *Aam Bam*, EMBL3 *Aam Sam*). The inclusion of amber mutations in phage λ vectors not only increases biological containment, but can be used in a selective system for isolating DNA sequences linked to suppressor genes (see Chapter 15), and in recombinational screening (see Chapter 7).

High-level expression of genes cloned in λ vectors

It is sometimes the aim of a gene manipulator to promote the expression of a gene which has been cloned so as to amplify the synthesis of a desirable gene product. This is usually achieved with plasmid expression systems (see Chapter 8) but it is also possible with recombinant phage systems, where high-gene-copy numbers are the natural result of virus multiplication. Panasenko *et al.* (1977) described a recombinant phage, constructed *in vitro*, carrying the *E. coli* DNA ligase gene which, after induction of the recombinant lysogen, results in a 500-fold over-production of the enzyme so that it represents 5% of the total cellular protein of *E. coli*. Dramatic amplification depends *inter alia* upon both increasing the gene dosage and ensuring efficient transcription. The gene dosage is increased as a result of phage DNA replication within the host; the level of transcription may be improved by suitable choice of vector and subsequent manipulation of the recombinant phage.

A great deal of our knowledge about expression of genes cloned in phage λ comes from the studies of N.E. Murray, W.J. Brammar and their colleagues on a model system in which genes from the *trp* operon of *E. coli* are inserted in the phage genome either by manipulation *in vitro* or by genetic methods *in vivo* (Hopkins *et al.* 1976, Moir & Brammar 1976). The following discussion is based on their work.

First, we must distinguish between cases where the inserted DNA does or does not include the bacterial *trp* promoter. If the insert *does* include its own promoter, the yield of *trp* enzymes can be enhanced simply by delaying cell lysis so that the number of gene copies is increased and the time available for expression is extended. This was originally achieved by making the vector S^-. Moir and Brammar obtained better amplification of gene products by including mutations in gene Q or N. In Q^- phage all the late functions, including that of S, are blocked, and in addition, packaging of the replicated DNA is prevented which even further extends the availability of the DNA. An N^- phage is also defective in late functions and although it replicates more slowly than $N^+ Q^-$ phage, yields of enzyme achieved were at least as great. In such infected cells anthranilate synthetase, the product of the *trpE* gene, comprised more than 25% of the total soluble protein.

λ *trp* phage *lacking* the *trp* promoter have been constructed so that *trp* expression is initiated at the promoter P_L of the leftward operon of phage λ. This operon has two useful features:
• P_L is a powerful promoter;
• the anti-termination effect of gene N expression permits transcription through sequences which might otherwise prevent expression of a distant inserted gene.

Once again, cell lysis and DNA packaging were prevented by mutations in Q and S and additionally the cro^- mutation was introduced so as to de-repress transcription from P_L. Cells infected with such a phage may contain as much as 10% of the soluble protein as anthranilate synthetase. However, these phage were difficult to construct and propagate so an alternative approach was adopted. The *cro* gene lies within the immunity region and its product is immune-specific. The *cro* product of the hetero-immune phage 434 will not interact with P_L of λ. Hybrid phage containing P_L from λ but *cro* and P_R from 434 are therefore phenotypically Cro$^-$ as far as leftward transcription is concerned. Infection with such a λ*trp* derivative, which also carried the S$^-$ mutation, gave cells in which 25% of the soluble protein was anthranilate synthetase. Derivatives of this type which are *Nam* can be grown on a non-suppressing host providing they also carry the *nin* deletion of t_{R_2}. Thus amplification can be modulated by controlling the suppression of the *Nam* gene. This is a useful property since extreme over-production of a product relaxes the selection that can be imposed on a recombinant clone and may lead to problems of instability.

This elegant exploitation of the genetics of phage λ demonstrates the advantages of a well-characterized genetic system and shows that useful amplification of a variety of gene products may be achieved with λ vectors, even in cases where a strong promoter recognized by the host RNA polymerase does not accompany the inserted gene. Lessons learnt from this model system have been applied to the amplification of *E. coli* DNA polymerase I and T4 DNA ligase in induced lysogens of λ recombinants carrying these genes (Kelley *et al.* 1977, Murray *et al.* 1979, Wilson & Murray 1979).

Packaging phage λ DNA *in vitro*

So far, we have considered only one way of introducing manipulated phage DNA into the host bacterium, i.e. by transfection of competent bacteria (see Chapter 2). Using freshly prepared λ DNA that has not been subjected to any gene manipulation procedures, transfection will result in typically about 10^5 plaques per µg of DNA. In a gene-manipulation experiment in which the vector DNA is restricted, etc. and then ligated with foreign DNA, this figure is reduced to about 10^4–10^3 plaques per µg of vector DNA. Even with perfectly efficient nucleic acid biochemistry, some of this reduction is inevitable. It is a consequence of the random association of fragments in the ligation reaction, which produces molecules with a variety of fragment combinations, many of which are inviable. Yet, in some contexts, 10^6 or more recombinants are required. The scale of such experiments can be kept within a reasonable limit by packaging the recombinant DNA into mature phage particles *in vitro*.

Placing the recombinant DNA in a phage coat allows it to be introduced into the host bacteria by the normal processes of phage infection, i.e. phage adsorption followed by DNA injection. Depending upon the details of the experimental design, packaging *in vitro* yields about 10^7 plaques per µg of vector DNA after the ligation reaction.

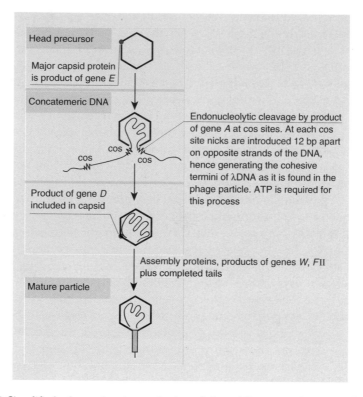

Fig. 5.5 Simplified scheme showing packaging of phage λDNA into phage particles.

Figure 5.5 shows some of the events occurring during the packaging process that take place within the host during normal phage growth and which we now require to perform *in vitro*. Phage DNA in concatemeric form, produced by a rolling circle replication mechanism (see Fig. 5.2), is the substrate for the packaging reaction. In the presence of phage head precursor (the product of gene *E* is the major capsid protein) and the product of gene *A*, the concatemeric DNA is cleaved into monomers and encapsidated. Nicks are introduced in opposite strands of the DNA, 12 nucleotide pairs apart at each *cos* site, to produce the linear monomer with its cohesive termini. The product of gene *D* is then incorporated into what now becomes a completed phage head. The products of genes *W* and *FII*, among others, then unite the head with a separately assembled tail structure to form the mature particle.

The principle of packaging *in vitro* is to supply the ligated recombinant DNA with high concentrations of phage head precursor, packaging proteins and phage tails. Practically this is most efficiently performed in a very concentrated mixed lysate of two induced lysogens, one of which is blocked at the pre-head stage by an amber mutation in gene *D* and therefore accumulates this precursor while the other is prevented from forming any head structure by an amber mutation in gene *E* (Hohn & Murray 1977). In the mixed lysate, genetic complementation occurs and exogenous DNA is packaged (Fig. 5.6). Although concatemeric DNA is the substrate for

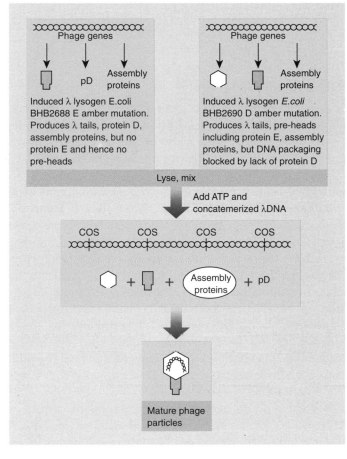

Fig. 5.6 *In vitro* packaging of concatemerized phage λDNA in a mixed lysate.

packaging (covalently joined concatemers are of course produced in the ligation reaction by association of the natural cohesive ends of λ), the *in vitro* system *will* package added monomeric DNA, which presumably first concatemerizes non-covalently.

There are two potential problems associated with packaging *in vitro*. First, endogenous DNA derived from the induced prophages of the lysogens used to prepare the packaging lysate can itself be packaged. This can be overcome by choosing the appropriate genotype for these prophages, i.e. excision upon induction is inhibited by the *b2* deletion (Gottesman & Yarmolinsky 1968) and *imm 434* immunity will prevent plaque formation if an *imm 434* lysogenic bacterium is used for plating the complex reaction mixture. Additionally, if the vector does not contain any amber mutation a non-suppressing host bacterium can be used so that endogenous DNA will not give rise to plaques. The second potential problem arises from recombination in the lysate between exogenous DNA and induced prophage markers. If troublesome, this can be overcome by using recombination-deficient (i.e. *red$^-$ rec$^-$*) lysogens and by UV-irradiating the cells used to prepare the lysate, so eliminating the biological activity of the endogenous DNA (Hohn & Murray 1977).

Cosmid vectors

As we have seen, concatemers of unit-length λDNA molecules can be efficiently packaged if the *cos* sites — substrates for the packaging-dependent cleavage — are 37–52 kb apart (75–105% the size of λ$^+$DNA). In fact, only a small region in the proximity of the *cos* site is required for recognition by the packaging system (Hohn 1975).

Plasmids have been constructed which contain a fragment of λDNA including the *cos* site (Collins & Brüning 1978, Collins & Hohn 1979, Wahl *et al.* 1987, Evans *et al.* 1989). These plasmids have been termed *cosmids* and can be used as gene cloning vectors in conjunction with the *in vitro* packaging system. Figure 5.7 shows a gene-cloning scheme employing a cosmid. Packaging the cosmid recombinants into phage coats imposes a desirable selection upon their size. With a cosmid vector of 5 kb we demand the insertion of 32–47 kb of foreign DNA — much more than a phage λ vector can accommodate. Note that after packaging *in vitro*, the particle is used to infect a suitable host. The recombinant cosmid DNA is injected and circularizes like phage DNA but replicates as a normal plasmid without the expression of any phage functions. Transformed cells are selected on the basis of a vector drug-resistance marker.

Cosmids provide an efficient means of cloning large pieces of foreign DNA. Because of their capacity for large fragments of DNA, cosmids are particularly attractive vectors for constructing libraries of eukaryotic genome fragments. Partial digestion with restriction endonuclease provides suitably large fragments. However, there is a potential problem associated with the use of partial digests in this way. This is due to the possibility of two or more genome fragments joining together in the ligation reaction, hence creating a clone containing fragments that were not initially adjacent in the genome. This would give an incorrect picture of their chromosomal organization. The problem can be overcome by size-fractionation of the partial digest.

Even with sized foreign DNA, in practice cosmid clones may be produced that contain non-contiguous DNA fragments ligated to form a single insert. The problem can be solved by dephosphorylating the foreign DNA fragments so as to prevent their ligation together. This method is very sensitive to the exact ratio of target-to-vector DNAs (Collins & Brüning 1978) because vector-to-vector ligation can occur. Furthermore, recombinants with a duplicated vector are unstable and break down in the host by recombination, resulting in the propagation of non-recombinant cosmid vector.

Such difficulties have been overcome in a cosmid-cloning procedure devised by Ish-Horowicz and Burke (1981). By appropriate treatment of the cosmid vector pJB8 (Fig. 5.8) left-hand and right-hand vector ends are

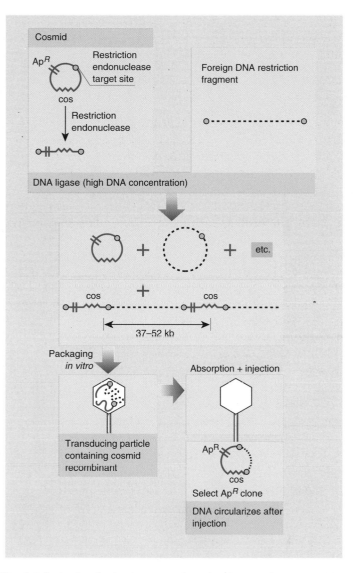

Fig. 5.7 Simple scheme for cloning in a cosmid vector. (See text for details.)

purified which are incapable of self-ligation but which accept dephosphorylated foreign DNA. Thus the method eliminates the need to size the
foreign DNA fragments, and prevents formation of clones containing short
foreign DNA or multiple vector sequences.

An alternative solution to these problems has been devised by Bates
and Swift (1983), who have constructed cosmid c2XB. This plasmid carries
a *Bam*HI insertion site and two *cos* sites separated by a blunt-end restriction
site (Fig. 5.9). The creation of these blunt ends, which ligate only very
inefficiently under the conditions used, effectively prevents vector self-
ligation in the ligation reaction.

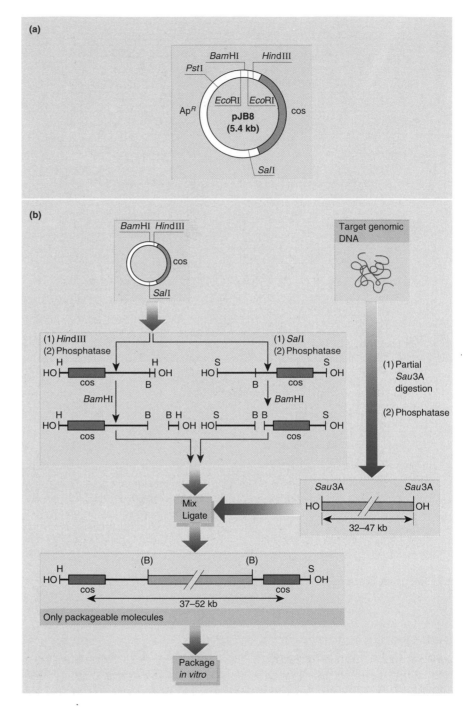

Fig. 5.8 Cosmid cloning scheme of Ish-Horowicz and Burke. (a) Map of cosmid pJB8. (b) Application to the construction of a genomic library of fragments obtained by partial digestion with *Sau*3A. This restriction endonuclease has a tetranucleotide recognition site and generates fragments with the same cohesive termini as *Bam*HI (see p. 31).

Modern cosmids of the pWE and sCos series (Wahl *et al.* 1987, Evans *et al.* 1989) contain features such as: (1) multiple cloning sites (Bates & Swift 1983, Breter *et al.* 1987, Pirrotta *et al.* 1983) for simple cloning using non size-selected DNA; (2) phage promoters flanking the cloning site; (3) unique *Not*I, *Sac*II or *Sfi*I sites (rare cutters, see Chapter 6) flanking the

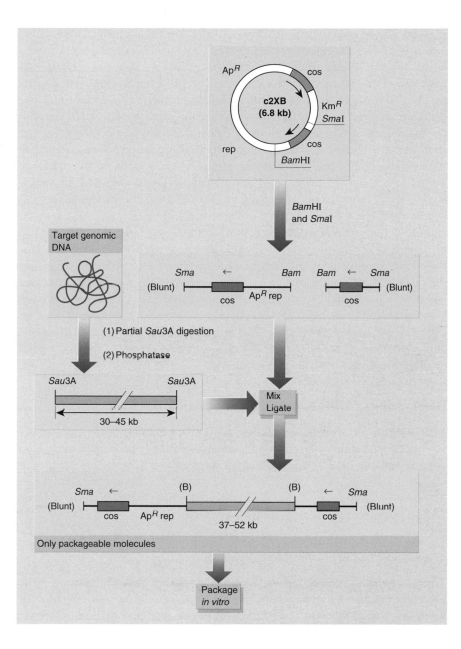

Fig. 5.9 Cosmid cloning scheme of Bates and Swift. The cosmid c2XB contains two *cos* sites, separated by a site for the restriction endonuclease *Sma*I which creates blunt ends. These blunt ends ligate only very inefficiently under the conditions used and effectively prevent the formation of recombinants containing multiple copies of the vector.

cloning site to permit removal of the insert from the vector as single fragments. Mammalian expression modules encoding dominant selectable markers (Chapter 15) may also be present, for gene transfer to mammalian cells if required.

Phasmid vectors

A second combination of plasmid and phage λ sequences has been devised to exploit the virtues of each type of vector. This combination consists of a plasmid vector carrying a λ attachment (λ*att*) site. The plasmid may insert into a phage λ genome by means of the site-specific recombination mechanism of the phage that is normally responsible for recombinational insertion of the phage into the bacterial chromosome during lysogen formation. This reversible recombinational insertion of plasmid into the phage is referred to as 'lifting' the plasmid and generates a phage genome containing one or more plasmid molecules (depending upon the length of the plasmid). These novel genetic combinations are called *phasmids* (Brenner *et al.* 1982). They contain functional origins of replication of the plasmids and of λ, and may be propagated as a plasmid or as a phage in appropriate *E. coli* strains. Reversal of the lifting process releases the plasmid vector.

Phasmids may be used in a variety of ways; for instance, DNA may be cloned in the plasmid vector in a conventional way and then the recombinant plasmid can be lifted onto the phage. Phage particles are easy to store, they have an effectively infinite shelf-life, and screening phage plaques by molecular hybridization often gives cleaner results than screening bacterial colonies (see Chapter 7). Alternatively, a phasmid may be used as a phage-cloning vector, from which subsequently a recombinant plasmid may be released. A highly developed and novel phasmid vector, λZAP, with components of λ, M13, T3, and T7 phages, is described at the end of the next section, where the development of M13 vector derivatives is explained.

DNA cloning with single-stranded DNA vectors

M13, f1 and fd are filamentous coliphages containing a circular single-stranded DNA molecule. These coliphages have been developed as cloning vectors for they have a number of advantages over other vectors, including the other two classes of vector for *E. coli*, plasmids and phage λ. However, in order to appreciate their advantages, it is essential to have a basic understanding of the biology of filamentous phages.

The biology of the filamentous coliphages

The phage particles have dimensions $900 \times 9\,\mathrm{nm}$ and contain a single-stranded circular DNA molecule which is 6407 (M13) or 6408 (fd) nucleotides long. The complete nucleotide sequences of fd and M13 are available and they are 97% homologous. The differences consist mainly of isolated nucleotides here and there, mostly affecting the redundant bases of codons, with no blocks of sequence divergence. Sequencing of f1 DNA indicates that it is very similar to M13 DNA.

The filamentous phages only infect strains of enteric bacteria harbouring F pili. The adsorption site appears to be the end of the F pilus, but exactly how the phage genome gets from the end of F pilus to the inside of the cell is not known. Replication of phage DNA does not result in host cell lysis. Rather, infected cells continue to grow and divide, albeit at a rate slower than uninfected cells, and extrude virus particles. Up to 1000 phage particles may be released into the medium per cell per generation (Fig. 5.10).

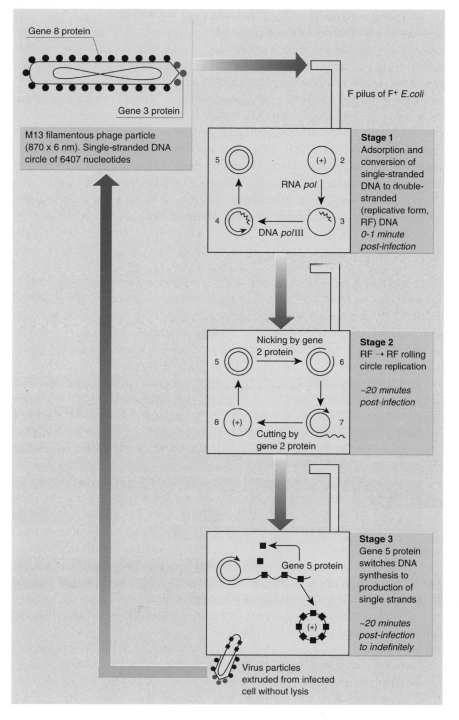

Fig. 5.10 Life cycle and DNA replication of phage M13.

The single-stranded phage DNA enters the cell by a process in which decapsidation and replication are tightly coupled. The capsid proteins enter the cytoplasmic membrane as the viral DNA passes into the cell while being converted to a double-stranded replicative form (RF). The RF multiplies rapidly until about 100 RF molecules are formed inside the cell. Replication of the RF then becomes asymmetric, due to the accumulation of a viral-encoded single-stranded specific DNA-binding protein. This protein binds to the viral strand and prevents synthesis of the complementary strand. From this point on, only viral single strands are synthesized. These progeny single strands are released from the cell as filamentous particles following morphogenesis at the cell membrane. As the DNA passes through the membrane the DNA binding protein is stripped off and replaced with capsid protein.

Why use single-stranded vectors?

For several applications of cloned DNA, single-stranded DNA is required. Sequencing by the original dideoxy method required single-stranded DNA, as do techniques for oligonucleotide-directed mutagenesis, and certain methods of probe preparation. The use of vectors that occur in single-stranded form is an attractive means of combining cloning, amplification, and strand separation of an originally double-stranded DNA fragment.

As single-stranded vectors the filamentous phages have a number of advantages. First, the phage DNA is replicated via a double-stranded circular DNA (RF) intermediate. This RF can be purified and manipulated *in vitro* just like a plasmid. Second, both RF and single-stranded DNA will transfect competent *E. coli* cells to yield either plaques or infected colonies, depending on the assay method. Third, the size of the phage particle is governed by the size of the viral DNA and therefore there are no packaging constraints. Indeed, viral DNA up to six times the length of M13 DNA has been packaged (Messing *et al.* 1981). Finally, with these phages it is very easy to determine the orientation of an insert. Although the relative orientation can be determined from restriction analysis of RF, there is an easier method (Barnes 1980). If two clones carry the insert in opposite directions, the single-stranded DNA from them will hybridize and this can be detected by agarose gel electrophoresis. Phage from as little as 0.1 ml of culture can be used in assays of this sort, making mass screening of cultures very easy.

In summary, as vectors, filamentous phages possess all the advantages of plasmids while producing particles containing single-stranded DNA in an easily obtainable form.

Development of filamentous phage vectors

Unlike λ, the filamentous coliphages do not have any non-essential genes which can be used as cloning sites. However, in M13 there is a 507 base-pair intergenic region, from position 5498 to 6005 of the DNA sequence, which contains the origins of DNA replication for both the viral and the complementary strands. In most of the vectors developed so far, foreign DNA has been inserted at this site although it is possible to clone at the carboxy-terminal end of gene IV (Boeke *et al.* 1979). The wild-type phages are not very promising as vectors because they contain very few unique sites within the intergenic region: *Asu*I in the case of fd, and *Asu*I and *Ava*I in the case of M13. However, a site does not have to be unique to be useful, as the example below shows.

The first example of M13 cloning made use of one of ten *Bsu*I sites in the genome, two of which are in the intergenic region (Messing *et al.* 1977). For cloning, M13 RF was partially digested with *Bsu*I and linear full-length molecules isolated by agarose gel electrophoresis. These linear monomers were blunt-end ligated to a *Hind*II restriction fragment comprising the *E. coli lac* regulatory region and the genetic information for the α-peptide of β-galactosidase. The complete ligation mixture was used to transform a strain of *E. coli* with a deletion of the β-galactosidase α-fragment and recombinant phage detected by intragenic complementation on media containing IPTG and Xgal. The IPTG is a gratuitous inducer of β-galactosidase, and Xgal a chromogenic substrate, where complementation occurs, a blue colour is produced. One of the blue plaques was selected and the virus in it designated M13 mp1.

Insertion of DNA fragments into the *lac* region of M13 mp1 destroys its ability to form blue plaques, making detection of recombinants easy. However, the *lac* region only contains unique sites for *Ava*II, *Bgl*I and *Pvu*I and three sites for *Pvu*II, and there are no sites anywhere on the complete genome for the commonly used enzymes such as *Eco*RI of *Hind*III. To remedy this defect, Gronenborn and Messing (1978) introduced an *Eco*RI site into the *lac* region of mp1 and the way they did so is particularly interesting. From DNA sequence data and restriction mapping (Fig. 5.11) it was known that a single base change (guanine residue 13 → adenine) in the codon for the 5th amino acid residue of the β-galactosidase α-fragment would create an *Eco*RI site. This in turn would lead to an aspartate residue being replaced by an asparagine residue that would have no significant effect on the complementation properties of the α-peptide.

Methylation of guanine has been shown to cause it to mispair with uracil. Therefore single-stranded DNA from M13 mp1 particles was treated with the methylating agent *N*-methyl-*N*-nitrosourea and then transformed into cells and allowed to undergo several cycles of replication. The CCC RF DNA was isolated from these cells and digested with *Eco*RI. Linear molecules of genome length were separated from undigested molecules by agarose gel electrophoresis, excised from the gel, and recircularized by way of their cohesive ends. The ligated molecules were transformed into cells and the resulting virus particles isolated. Following this procedure three individual clones with unique *Eco*RI restriction sites at different positions in the phage genome were isolated. Two of these, M13 mp2 and

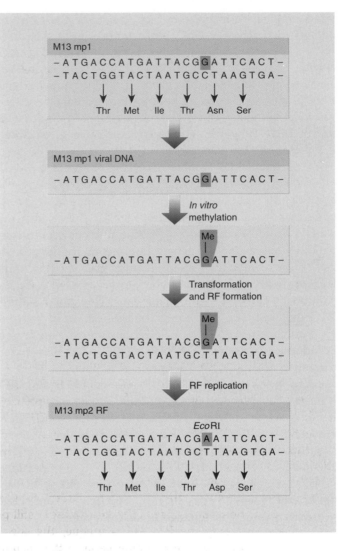

Fig. 5.11 *In vitro* mutagenesis of the *lac* region of M13 mp1 to produce M13 mp2. Note that the methylguanine pairs with thymine but during replication this thymine will pair with adenine resulting in a G:C base-pair being replaced by an A:T base-pair.

M13 mp3, were the result of the conversion of *Eco* RI* sequences in the *lac* fragment of M13 mp1 corresponding to the positions of amino acids 5 and 119, respectively. The *Eco*RI site of the third mutant was elsewhere in the genome.

The introduction of an *Eco*RI site into the *lac* region of M13 mp1, which is itself resistant to cleavage by *Eco*RI, creates a unique site to clone DNA fragments. Gronenborn and Messing (1978) have shown that insertion of *Eco*RI-derived fragments into M13 mp2 leads to inactivation or reduction of the β-galactosidase activity. Furthermore, expression of functions coded by the inserted DNA can be controlled by the *lac* regulatory region. However,

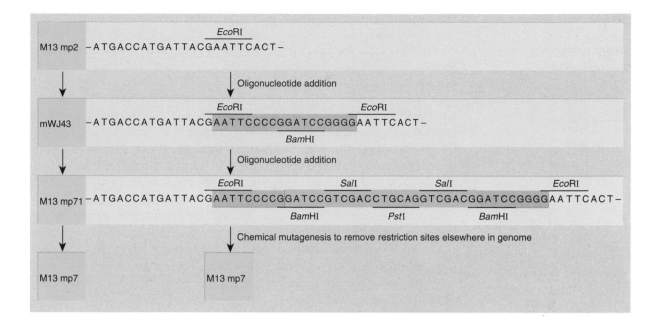

Fig. 5.12 The derivation of M13 mp7. Only the base sequence at the beginning of the *lac* region is shown.

to improve the versatility of M13 as a vector, Messing *et al.* (1981) constructed M13 mp7 which has a multipurpose cloning site in the *lac* region.

The actual construction of M13 mp7 is too complex to describe in detail here but a summary is provided in Fig. 5.12. Starting with M13 mp2, Rothstein *et al.* (1979) removed the single *Bam*HI site in gene III. Then a synthetic linker containing a *Bam*HI site was inserted at the *Eco*RI site to generate mWJ43. This phage still gives blue plaques on media containing IPTG and Xgal. Into this *Bam*HI site Messing *et al.* (1981) introduced yet another oligonucleotide linker, this time one containing *Pst*I and *Sal*I sites, to create M13 mp71. Since the reading frame of the *lac* region of M13 mp71 is still unaltered, a functional α-peptide of β-galactosidase is still produced. Although in this phage only the *Pst*I site is unique, the sites for *Sal*I, *Bam*HI and *Eco*RI are also usable for cloning because the ensuing loss of small inserts still results in a functional *lac* sequence. The sequence 5'-GTCGAC-3' is cleaved by endonucleases *Sal*I, *Acc*I and *Hinc*II. Unfortunately, due to ambiguities in their recognition sequence the latter two endonucleases each cleave one additional site, both located in gene II. Consequently, these two sites were removed by chemical mutagenesis to generate M13 mp7.

The M13 mp7 vector has a symmetrical multiple restriction site, or *polylinker* region. This has the limitation that DNA fragments with dissimilar ends cannot be inserted because treatment of the polylinker site with a pair of restriction enzymes (e.g. *Sal*I and *Eco*RI) will generate a vector with ends derived from the outer pair of restriction sites (*Eco*RI). In order to overcome this problem Messing and his co-workers have constructed new derivatives, M13 mp8, mp10, mp11 (Messing & Vieira 1982), mp18 and mp19 (Norrander *et al.* 1983), which have unpaired restriction sites in non-symmetrical polylinker regions (see p. 67). These vectors have the advantage that DNA fragments with dissimilar ends can be cloned, and the orientation of the insert is fixed. M13 mp8 and mp9 have similar polylinker regions in opposite orientations, so that a foreign DNA fragment can be inserted either way round. The M13 mp10/mp11 pair, and the mp18/mp19 pair, also have their polylinker sites in opposite orientations. In practice, after such vectors have been digested with a pair of restriction enzymes it is often convenient to isolate the large vector band from an agarose gel, hence discarding the small restriction fragment. When ligation reactions are performed in the presence of the foreign DNA fragment, simple reclosure of the non-recombinant vector cannot occur, so that only white, recombinant plaques are obtained.

In this section we have concentrated on the vectors developed by Messing and his co-workers. There are several reasons for this. First, the clever use of mutagenesis to insert and remove restriction sites. Second, the construction of vectors with polylinkers. The convenience afforded by such polylinkers in gene manipulation experiments has been widely appreciated. Plasmid vectors have been constructed which also incorporate such polylinker sites. These are the pUC plasmids (Vieira & Messing 1982, Norrander *et al.* 1983). This principle has also been extended to the specialized vector πVX (see Chapter 7), to the phage λ vectors EMBL3 and EMBL4 and to many other recently developed vectors such as λZAP. Finally, the M13 mp series of vectors have revolutionized site-directed mutagenesis and large-scale DNA sequencing. These topics are discussed in Chapters 9 and 11.

Exploitation of f1 (M13) biology to create a new family of single-stranded/double-stranded DNA vectors: pEMBL

The pUC series (see p. 67) of plasmid vectors was developed after the demonstration of the versatility of polylinker, or multiple, cloning sites in the M13 mp series of phage vectors, combined with blue/white screening of insertional inactivation of the α-peptide of β-galactosidase using Xgal plates. The pUC series have the additional advantage of being very small vectors. The important property of the M13 mp series of vectors is to provide cloned DNA in single-stranded form. However, one problem encountered with the single-stranded phage vectors is the instability of inserts when the insert exceeds a few kilobases in size (Zinder & Boeke 1982). The pEMBL series of vectors has been constructed so as to add to the features of single-stranded vectors the further advantages of a small vector size and stability of large inserts (Dente *et al.* 1983).

The basic principle that permitted this development was discovered by Dotto *et al.* (1981) and Dotto and Horiuchi (1981). They inserted into the *Eco*RI site of pBR322, a region of the f1 genome that contains all the *cis*-acting elements required for DNA replication and phage morphogenesis. They showed that, when F⁺ *E. coli* containing the recombinant pBR322 were superinfected with f1 phage ('helper' phage), virion capsids were

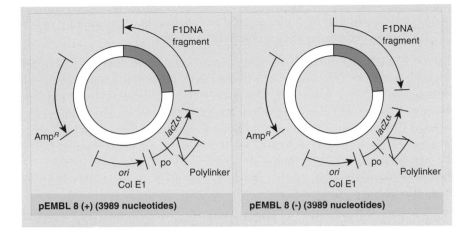

Fig. 5.13 Structures of pEMBL8(+) and pEMBL8(−).

secreted that contained either f1 single-stranded DNA or the recombinant single-stranded pBR322 DNA, in about equal frequency.

In constructing pEMBL8, Dente *et al.* (1983) inserted a 1.3 kb fragment of the f1 genome containing all *cis*-acting elements required for DNA replication and phage morphogenesis into the unique *Nar*I site of pUC8 in both orientations, hence generating pEMBL8(+) and pEMBL8(−). The resulting plasmids (Fig. 5.13) conserved all the pUC8 features, including origin of plasmid replication, ampicillin resistance and blue colony phenotype on Xgal plates. Upon superinfection with f1 helper phage the intact origin of f1 replication is activated so that single-stranded plasmid DNA is produced, packaged, and secreted into the culture medium in virion-like particles. The orientation of the f1 DNA determines which of the two strands is found in the virion: pEMBL8(+) and pEMBL8(−) contain the non-coding and coding strands of the β-galactosidase gene, respectively. The plasmids pEMBL9 were constructed by replacing the polylinker of pEMBL8 with the polylinker of pUC9.

In practice, recombinant single-stranded DNA can be isolated from the mixture of particles in the culture medium and will contain a proportion of the helper phage DNA. For most purposes, such as DNA sequencing and oligonucleotide-directed mutagenesis, it is not necessary to perform purification steps to remove the helper DNA; its presence need not interfere with those applications. Thus, for example, the primer conventionally used in dideoxy sequencing is complementary to a region of the β-galactosidase gene which is present in the M13 mp series of vectors and also present in the pUC and pEMBL plasmid vectors. The 'wild-type' helper phage does not contain β-galactosidase sequences and so does not interfere with sequencing reactions because it does not hybridize with sequencing primer.

Double-stranded recombinant DNA can be prepared from cells harbouring the plasmid, in the absence of helper phage, where it replicates as a pUC-derived plasmid, and typically gives higher yields than RF preparations of the M13 mp series.

λZAP: exploitation of M13 biology and the specificity of phage T3, T7 RNA polymerases for their promoters

λZAP is a sophisticated insertional vector that is particularly suitable for cloning cDNAs. This vector, developed and marketed commercially by Stratagene Cloning Systems (Short *et al.* 1988), illustrates several modern features, one of which is the exploitation of M13 biology so as to allow the cloned DNA insert to be automatically excised from the phage vector into a plasmid vector. In this regard λZAP is a form of phasmid vector. λZAP incorporates the following features:
• Multiple unique cloning sites which can hold inserts up to 10 kb in length.
• Insertional inactivation of β-galactosidase, giving blue/white screening on Xgal plates.
• Expression of hybrid or fusion polypeptides in a manner analogous to λgt11 (see p. 126).
• Automatic excision *in vivo* of the cloned DNA from the phage vector into a plasmid vector, Bluescript SK(−). This excision is brought about by M13 or f1 helper phage, and places the cloned DNA in a small plasmid vector convenient for restriction mapping and sequencing. This process eliminates the need to subclone DNA inserts from the λ phage into a plasmid by restriction and ligation.
• The ability to prepare RNA transcripts of the inserted foreign DNA. Such transcripts are synthesized from either strand by using either the T3 or T7 phage RNA polymerases.

The structure of the λZAP genome is shown in Fig. 5.14. Basically it is a λ insertional vector. The foreign DNA − the figure illustrates an application with cDNA − is inserted at the multiple cloning site region, thus inactivating *lacZ* (and incidentally providing the opportunity for fusion polypeptide synthesis if the reading frame is correct). When an F⁺ (or more commonly an F′) strain is infected with the recombinant phage and then superinfected with M13 helper phage, the helper phage supplies *trans*-acting proteins which recognize two DNA sequences incorporated in the λZAP arms. These two DNA sequences were derived from the f1 (M13) origin of replication, and signal the initiation and termination of DNA synthesis.

In normal f1 (M13) DNA replication these two distinct but overlapping DNA sequences act as follows. One site − the initiator − is recognized by the gene 2 protein which nicks the DNA(+) strand in RF DNA. This nick is then the site at which unidirectional, rolling circle DNA synthesis is initiated, with displacement of the (+) strand. This (+) strand is then cleaved at the terminator site, which contains the same sequence that was nicked for initiation. Following cutting at the terminator by gene 2 protein, the two ends of the (+) strand are ligated to form a circular single (+) strand genome. This is subsequently converted to double-stranded RF DNA (refer to Fig. 5.10).

Short *et al.* (1988) exploited this knowledge by positioning f1 initiator and terminator sequences at separate places in the λ vector. The DNA

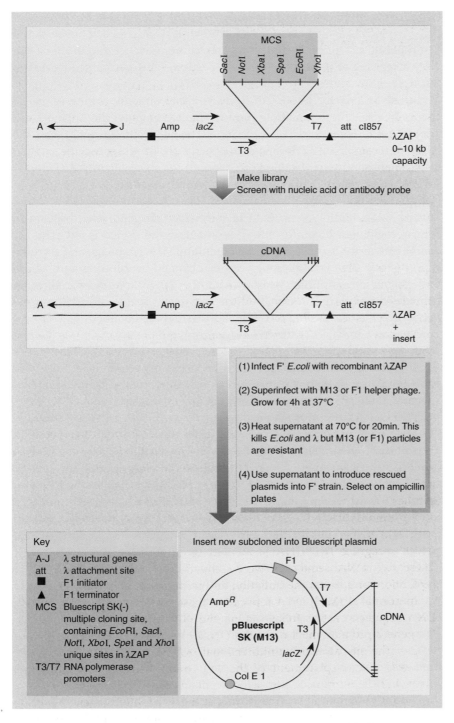

Fig. 5.14 The phage λ insertional vector, λZAP. Automatic excision. (λZAP carries an amber mutation in gene *S*, not shown. λZAPII is *S*⁺).

lying between these two sequences is that of a phagemid, Bluescript SK(−), which is analogous to the pEMBL vectors described above in that it can replicate as a plasmid conferring ampicillin resistance and in that it can be packaged into an M13 virion-like particle. This vector sequence has multiple cloning sites.

In use as a vector, the λZAP is cleaved into arms by cutting at one of the unique sites in the multiple cloning site. Commonly, the unique *Eco*RI site is used. Foreign cDNA is inserted using *Eco*RI linkers. Once the required recombinant λZAP phage has been isolated, the automatic excision is accomplished by coinfecting F′ *E. coli* with the recombinant λZAP and f1 (M13) helper phage. This leads to the synthesis of a DNA strand from between the two signals, with strand displacement. The displaced strand is then automatically circularized in the helper phage-infected bacteria to form recombinant Bluescript SK(−), containing the cloned insert. This recombinant vector is packaged as a filamentous M13-like phage and secreted from the cell. Bluescript plasmids can be obtained by infecting an F′ strain and plating on ampicillin plates where the Col E1 origin of replication provides the replicon function and the *amp* gene provides ampicillin resistance. Colonies containing recombinant plasmid are obtained.

As shown in Fig. 5.14 the DNA insert of the Bluescript SK(−) is flanked by promoters for the RNA polymerases of phages T3 and T7. This allows RNA copies of the inserted DNA to be made conveniently *in vitro*. For example, if the recombinant plasmid is first linearized at the unique *Not*I site and then placed in a suitable reaction mixture containing T7 RNA polymerase and the four ribonucleoside triphosphates, a single-stranded RNA molecule is synthesized initiating from the T7 promoter and terminating at the *Not*I-cleaved end of the DNA. This facility is very useful for a number of applications. The RNA can be synthesized at high specific radioactivity by including an α-^{32}P-nucleoside triphosphate in the reaction. Such a radioactive RNA copy can be used effectively to probe northern blots (RNA–RNA hybrids have a melting temperature at about 15°C higher than DNA–RNA hybrids and give 'clean' northern blot results). These probes can also be used for RNase mapping (Zinn *et al.* 1983).

Clearly, in such applications it is important that the correct strand of the DNA insert is transcribed into RNA. The positioning of T7 and T3 promoters at opposite ends of the inserts allows either strand to be transcribed, as desired, by appropriate choice of RNA polymerase, combined with linearization at a distal site with a suitable restriction endonuclease.

The purified RNA polymerases from coliphages T7, T3, T5 and the *Salmonella* phage SP6 are available commercially for applications such as these.

These polymerases are single-subunit enzymes, much simpler than the *E. coli* RNA polymerase. They exhibit high specificity for their individual promoter sequences. The T3 RNA polymerase, for example, is highly specific for a 23 base-pair promoter sequence which differs by three base-pairs from the T7 RNA polymerase promoter. Several simple plasmid vectors have been produced that incorporate promoters for phage RNA

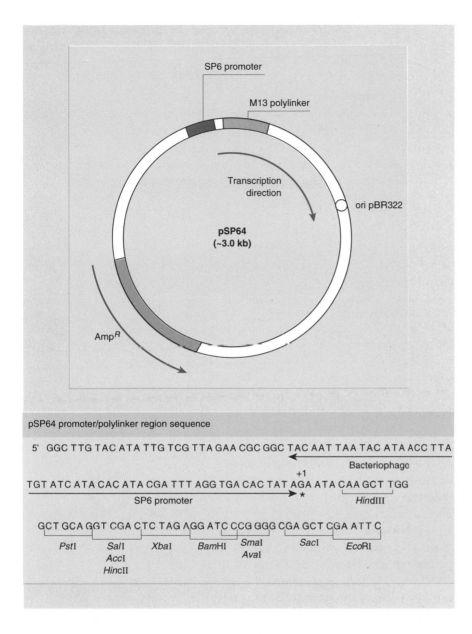

SP6 promoter

M13 polylinker

Transcription direction

pSP64
(~3.0 kb)

ori pBR322

Amp^R

pSP64 promoter/polylinker region sequence

5' GGC TTG TAC ATA TTG TCG TTA GAA CGC GGC TAC AAT TAA TAC ATA ACC TTA

Bacteriophage

+1

TGT ATC ATA CAC ATA CGA TTT AGG TGA CAC TAT AGA ATA CAA GCT TGG

SP6 promoter   *   HindIII

GCT GCA GGT CGA CTC TAG AGG ATC CCG GGG CGA GCT CGA ATT C

PstI   SalI   XbaI   BamHI   SmaI   SacI   EcoRI
        AccI   AvaI
        HincII

Fig. 5.15 Structure of pSP64, a vector for transcription of inserted DNA by phage SP6 RNA polymerase *in vitro*. The related plasmid pSP65 is similar, with an inverted polylinker. Transcription initiation occurs at the G marked with an asterisk (+1).

polymerase, situated adjacent to multiple cloning sites. The prototype on which these are based is the pSP64 and pSP65 pair (Fig. 5.15). This pair of vectors first established the usefulness of the phage polymerases, employing the phage SP6 RNA polymerase. In addition to the applications mentioned above, Krieg and Melton (1984) showed that transcripts of suitable inserts could function as synthetic mRNAs for translation in rabbit reticulocyte lysate or wheat-germ cell-free systems. For efficient translation in these systems the synthetic mRNAs must bear a 'cap' at the 5' end. Fortunately this can conveniently be incorporated at the initiation of strand synthesis by the RNA polymerase if the cap dinucleotide m^{7}GpppG is included

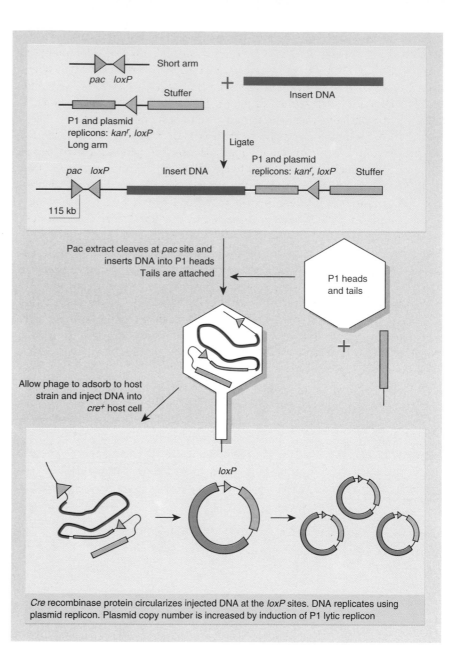

Fig. 5.16 The phage P1
vector system. The P1 vector
Ad10 (Sternberg 1990) is
digested to generate short
and long vector arms. These
are dephosphorylated to
prevent self-ligation. Size-
selected insert DNA (85–
100 kb) is ligated with vector
arms, ready for a two-stage
processing by packaging
extracts. First, the
recombinant DNA is cleaved
at the *pac* site by pacase in
the packaging extract. Then
the pacase works in concert
with head/tail extract to
insert DNA into phage
heads, *pac* site first, cleaving
off a headful of DNA at
115 kb. Heads and tails then
unite. The resulting phage
particle can inject
recombinant DNA into host
E. coli. The host is *cre*⁺. The
cre recombinase acts on *loxP*
sites to produce a circular
plasmid. The plasmid is
maintained at low copy
number, but can be amplified
by inducing the P1 lytic
operon.

in the RNA polymerase reaction mix. Capped RNAs have also been
injected into the cytoplasm of *Xenopus* oocytes for translation or, after
injection into the oocyte nucleus, for studies of RNA processing (Green *et
al.* 1983).

Phage P1 derivatives

Phage P1 is a temperate bacteriophage, widely used for the genetic analysis
of *E. coli* (and even, remarkably, *Myxococcus*) because it is famously able to

act as a generalized transducing phage. Stenberg and co-workers have developed a P1 vector system which has a capacity for DNA fragments as large as 95 kb (Sternberg 1990, Pierce *et al.* 1992). Thus the capacity is about twice that of cosmid clones, but less than that of YAC clones (see Chapter 13). The P1 vector derivatives contain a P1 packaging site (*pac*), which is necessary to package the recombinant vector into phage particles. Packaging can be performed *in vitro*. Vectors contain two *loxP* sites which are the sites recognized by the phage recombinase (the product of the phage *cre* gene) and which therefore lead to circularization of the packaged DNA after it has been injected into a host *E. coli* expressing the recombinase (Fig. 5.16). Clones are maintained in *E. coli* as low-copy-number plasmids by selection for a vector kanamycin-resistance marker. A high copy number can be induced by exploitation of the P1 lytic replicon (Sternberg 1990).

An improved vector has a system for positive selection of vectors with inserts (Pierce *et al.* 1992). This system is based on the properties of the *sacB* gene from *Bacillus amyloliquefaciens*. The *sacB* gene encodes the enzyme levansucrase, which catalyses the hydrolysis of sucrose to products lethal to *E. coli*. By inserting foreign DNA into a unique *Bam*HI cloning site, insertional inactivation of the vector-borne *sacB* gene permits growth of plasmid-containing *E. coli* in medium containing sucrose. The *Bam*HI site is flanked by the rare-cutter restriction sites *Not*I, *Sal*I and *Sfi*I, and by phage T7 and SP6 promoters, to facilitate characterization of inserts.

6 Cloning strategies, gene libraries and cDNA cloning

Cloning strategies

Any DNA cloning procedure has four essential parts: a method for generating DNA fragments, reactions which join foreign DNA to the vector, a means of introducing the artificial recombinant into a host cell in which it can replicate, and a method of selecting or screening for a clone of recipient cells that has acquired the recombinant (Fig. 6.1). In previous chapters DNA cutting and joining reactions have been described, and the properties of several phage and plasmid vectors have been discussed together with the factors governing the choice between the various cutting and joining methods and different vector molecules. These choices will depend upon what type of clones are wanted, e.g. cDNA or genomic DNA clones.

Genomic DNA libraries

As an example, let us suppose that we wish to clone a single-copy gene from the human genome. We might simply digest total human DNA with a restriction endonuclease such as *Eco*RI, insert the fragments into a suitable phage λ vector and then attempt to isolate the desired clone. How many recombinants would we have to screen in order to isolate the right one? Assuming *Eco*RI gives, on average, fragments about 4 kb long, and given that the human haploid genome is 2.8×10^6 kb, we can see that over 7×10^5 independent recombinants must be prepared and screened in order to have a reasonable chance of including the desired sequence. In other words we have to obtain a very large number of recombinants, which together contain a complete collection of all (or nearly all) of the DNA sequences in the entire human genome. Such a collection from which we withdraw the desired clone is called a *gene library* or *gene bank*.

There are two problems with the above approach. First, the gene may be cut internally one or more times by *Eco*RI so that it is not obtained as a

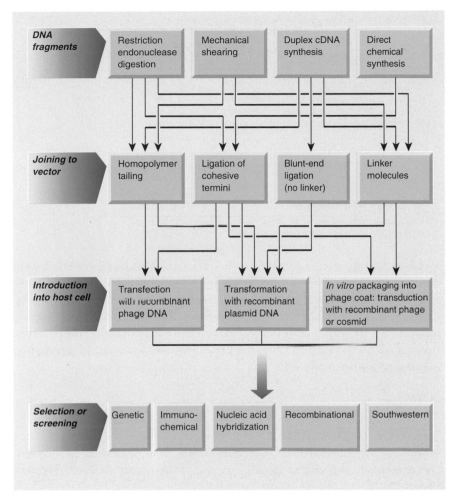

Fig. 6.1 Generalized scheme for DNA cloning in *E. coli*. Favoured routes are shown by arrows.

single fragment. This is likely if the gene is large. Also, it may be desired to obtain extensive regions flanking the gene or whole gene clusters. Fragments averaging about 4 kb are likely to be inconveniently short. Alternatively, the gene may be contained on an *Eco*RI fragment that is larger than the vector can accept. In this case the appropriate gene would not be cloned at all.

These problems can be overcome by cloning *random* DNA fragments of a large size (~20 kb). Since the DNA is randomly fragmented, there will be no systematic exclusion of any sequence. Furthermore, clones will overlap one another, giving an opportunity to 'walk' from one clone to an adjacent one (see p. 107). Because of the larger size of each cloned DNA, fewer clones are required for a complete or nearly complete library. How many clones are required? Let n be the size of the genome relative to a single cloned fragment. Thus for the human genome, 2.8×10^6 kb* and for

*See Table 6.1 for the genome size of various organisms.

Table 6.1 Genome sizes of some organisms

Organism	Genome size (kb) (haploid where appropriate)
Escherichia coli	4.0×10^3
Yeast (*Saccharomyces cerevisiae*)	1.35×10^4
Arabidopsis thaliana (higher plant)	7.0×10^4
Tobacco	1.6×10^6
Wheat	5.9×10^6
Zea mays	15×10^6
Drosophila melanogaster	1.8×10^5
Mouse	2.3×10^6
Human	2.8×10^6
Xenopus laevis	3.0×10^6

a cloned fragment size of 20 kb, $n = 1.4 \times 10^5$. The number of independent recombinants required in the library must be greater than n, because sampling variation will lead to the inclusion of some sequences several times, and the exclusion of other sequences in a library of just n recombinants. Clarke and Carbon (1976) have derived a formula which relates the probability (P) of including any DNA sequence in a random library of N independent recombinants:

$$N = \frac{\ln(1 - P)}{\ln\left(1 - \dfrac{1}{n}\right)}.$$

Therefore, to achieve a 95% probability ($P = 0.95$) of including any particular sequence in a random human genomic DNA library of 20 kb fragment size:

$$N = \frac{\ln(1 - 0.95)}{\ln\left(1 - \dfrac{1}{1.4 \times 10^5}\right)}$$

$$= 4.2 \times 10^5.$$

Notice that a considerably higher number of recombinants is required to achieve a 99% probability, for here $N = 6.5 \times 10^5$.

How can appropriately-sized random fragments be produced? Various methods are available. Random breakage by mechanical shearing is appropriate, but a much more commonly used procedure involves restriction endonucleases. In the strategy devised by Maniatis *et al.* (1978) (Fig. 6.2) the target DNA is restricted with a mixture of *two* restriction enzymes. These enzymes have tetranucleotide recognition sites, which therefore occur frequently in the target DNA and in a *limit* double-digest would produce fragments averaging less than 1 kb. The restriction is carried out only to a partial extent, so that the bulk of fragments are relatively large, in

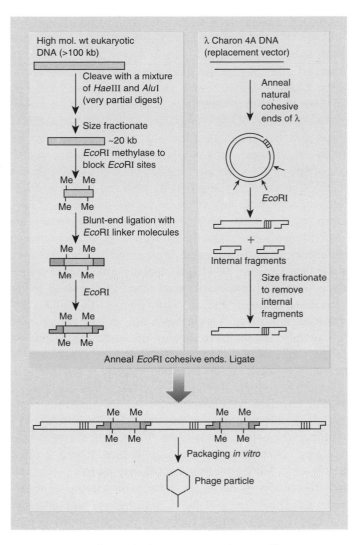

Fig. 6.2 Maniatis' strategy for producing a representative gene library.

the range 10–30 kb. These are effectively a random set of overlapping fragments. These can be fractionated by velocity centrifugation on a sucrose gradient or by preparative gel electrophoresis, so as to give a random population of fragments of about 20 kb, which are suitable for insertion into the phage λ vector. Packaging *in vitro* ensures that an appropriately large number of independent recombinants can be recovered, which will give an almost completely representative library. In Maniatis' strategy, the use of two different restriction endonucleases with completely unrelated recognition sites, *Hae*III and *Alu*I, assists in obtaining fragmentation that is nearly random. These enzymes both produce blunt ends, and the cloning strategy requires linkers (see Fig. 6.2). A convenient simplification can be achieved by using a *single* restriction endonuclease which cuts frequently, such as *Sau*3AI. This will create a partial digest that is slightly less close to random than that achieved with a pair of enzymes. However, it has the great advantage that the *Sau*3AI fragments can be readily inserted into high-capacity phage λ vectors, such as λEMBL3 (see Chapter 5), which have been digested with *Bam*HI (Fig. 6.3). This is because *Sau*3AI and *Bam*HI create the same cohesive ends (see p. 31). These partial digestion methods, coupled with packaging the phage λ recombinants, have been widely employed strategies for creating genomic DNA libraries.

In place of the phage λ vectors, cosmid vectors may be chosen. These also have the high efficiency afforded by packaging *in vitro* and have an even higher capacity than any phage λ vector. However, there are two drawbacks in practice. First, most workers find that screening libraries of phage λ recombinants by plaque hybridization gives cleaner results than screening libraries of bacteria containing cosmid recombinants by colony hybridization (see Chapter 7). Plaques usually give less of a background hybridization than do colonies. Second, it may be desired to retain and store an amplified genomic library. With phage, the initial recombinant DNA population is packaged and plated out. It can be screened at this stage. Alternatively, the plates containing the recombinant plaques can be washed to give an *amplified* library of recombinant phage. The amplified library can then be stored almost indefinitely; phage have a long shelf-life. The amplification is so great that samples of this amplified library could be plated out and screened with different probes on hundreds of occasions. With bacterial colonies containing cosmids it is also possible to store an amplified library (Hanahan & Meselson 1980), but bacterial populations cannot be stored as readily as phage populations. There is often an unacceptable loss of viability when the bacteria are stored.

A word of caution is necessary when considering the use of any amplified library. This is the possibility of *distortion*. Not all recombinants in a population will propagate equally well, e.g. variations in target DNA size or sequence may affect replication of a recombinant phage, plasmid or cosmid. Therefore, when a library is put through an amplification step particular recombinants may be increased in frequency, decreased in frequency, or lost altogether. Development of modern vectors and cloning strategies has simplified library construction to the point where many workers now prefer to create a new library for each screening, rather than risk using a previously amplified one.

The ease with which random libraries can be created and screened, and the possibility of chromosome walking, means that the shotgun approach has now become very widely adopted. However, as an alternative approach we could obtain a partially purified DNA fraction which is enriched in the desired sequence. The task of screening would then be diminished correspondingly. This approach is now rather outdated but it was necessary at a time before packaging *in vitro* had been developed, because then it was not possible readily to produce enough independent clones for a complete library. One method that was employed very successfully for such

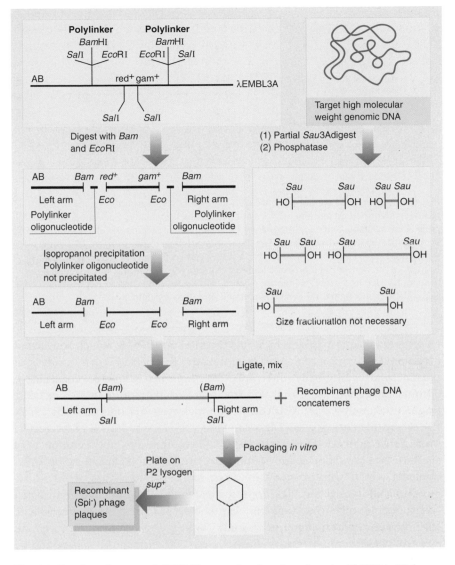

Fig. 6.3 Creation of a genomic DNA library using the phage λ vector EMBL3A. High-molecular-weight genomic DNA is partially digested with *Sau*3A. The fragments are treated with phosphatase to remove their 5'-phosphate groups. The vector is digested with *Bam*HI and *Eco*RI which cut within the polylinker sites. The tiny *Bam*HI/*Eco*RI polylinker fragments are discarded in the isopropanol precipitation, or alternatively the vector arms may be purified by preparative agarose gel electrophoresis. The vector arms are then ligated with the partially digested genomic DNA. The phosphatase treatment prevents the genomic DNA fragments from ligating together. Non-recombinant vector cannot reform because the small polylinker fragments have been discarded. The only packageable molecules are recombinant phages. These are obtained as plaques on a P2 lysogen of *sup*⁺ *E. coli*. The Spi⁻ selection ensures recovery of phage lacking *red* and *gam* genes. A *sup*⁺ host is necessary because, in this example, the vector carries amber mutations in genes *A* and *B*. These mutations increase biological containment, and can be applied to selection procedures such as recombinational selection (see Chapter 7), or tagging DNA with a *sup*⁺ gene (see Chapter 16). Ultimately, the foreign DNA can be excised from the vector by virtue of the *Sal*I sites in the polylinker. (*Note*: Rogers *et al.* 1988 have shown that the EMBL3 polylinker sequence is not exactly as originally described. It contains an extra sequence with a previously unreported *Pst*I site. This does not affect most applications as a vector.)

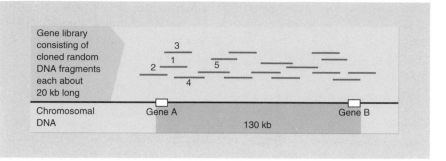

Fig. 6.4 Chromosome walking. It is desired to clone DNA sequences of gene *B*, which has been identified genetically but for which no probe is available. Sequences of a nearby gene *A* are available in cloned fragment 1. Alternatively, a sequence close to gene *B* could be identified by *in situ* hybridization to *Drosophila* polytene chromosomes. In a large, random genomic DNA library many overlapping cloned fragments are present. Clone 1 sequences can be used as a probe to identify overlapping clones 2, 3 and 4. Clone 4 can, in turn, be used as a probe to identify clone 5, and so on. It is, therefore, possible to walk along the chromosome until gene *B* is reached. (See text for details.)

enrichment is chromatography upon the medium RPC-5, which consists of a quaternary ammonium salt, the extractant, supported upon a matrix of plastic beads. Originally developed for high-resolution reversed-phase chromatography of tRNAs, it will fractionate milligram quantities of DNA fragments generated by restriction endonucleases (Hardies & Wells 1976). Leder and his co-workers (Tilghman *et al.* 1977) loaded an *Eco*RI digest of total mouse genomic DNA onto an RPC-5 column, and upon elution with a concentration gradient of sodium acetate, obtained a series of DNA fractions which were assayed for their ability to hybridize with mouse globin cDNA. A fraction was identified as being substantially enriched in a DNA fragment bearing β-globin sequences. The discrimination on the RPC-5 column chromatography is only slightly dependent upon fragment size, so that additional enrichment (to about 500-fold) could be obtained by combining it with preparative gel electrophoresis. The final enriched fraction was ligated into the λWES.λB vector, and out of about 4300 plaques obtained by transfection, three positive clones were detected.

E. coli hosts for library construction in phage vectors

Genomic DNA libraries in phage λ vectors are expected to contain most of the sequences of the genome from which they have been derived. However, anecdotal reports of sequences that cannot be found in libraries of eukaryotic genomes are not uncommon. There is also the related observation that deletions can occur during the cloning of mammalian DNA. Only some of these deletions are preventable by growth on an *E. coli recA* recombination-deficient host.

Leach and Stahl (1983) have shown that inverted repeat sequences (large palindromes) cannot be cloned in phage λ vectors unless special host strains are used. Wyman *et al.* (1985) have extended these observations by showing that a host with mutant *rec*B, *rec*C and *sbc*B recombination-function genes will propagate many phage λ recombinants containing human DNA fragments that would not be propagated on other host strains.

Chromosome walking (Fig. 6.4)

Walking along the chromosome is a term used to describe an approach which allows the isolation of gene sequences whose function is quite unknown but whose genetic location is known. The principle is as follows.

• A cloned genomic fragment must be found as a starting point for the walk. This should be as close as possible to the suspected destination point. In the human genome the starting point may be a restriction fragment-length protein (RFLP) sequence that is closely linked to a disease locus. In *Drosophila* the starting point may be found cytologically by *in situ* hybridization to polytene chromosomes.

• A random set of cloned genomic DNA is localized in this way, and one is chosen whose location on the chromosome in question is closest to the map position of the mutation under investigation.

• The genomic library is then screened with this chosen clone as probe to identify other clones containing DNA with which it reacts and which represent clones overlapping with it. The overlap can be to the left or to the right.

• Repetition of this single walking step along the chromosome.

The walk can potentially occur in both directions along the chromosome. A large scale map of the walk can be made by assembling contiguous clones: this will be done in combination with methods for analysing very large DNA regions, discussed below (p. 109).

A possibility that has to be recognized arises from the existence of repeated DNA sequences. These may occur dispersed at several places in the genome and could disrupt the orderly progress of the walk. For this reason the probe used for stepping from one genomic clone to the next must be a unique sequence clone, or a subclone which has been shown to contain only a unique sequence.

Once the destination of the walk is reached it is important to be able to recognize that fact. In a most spectacular application of walking to the Human Genome Project – the identification of the cystic fibrosis gene on chromosome 7 – a combination of the following criteria was considered in trying to define whether a cloned DNA segment might contain the cystic fibrosis gene (Rommens *et al.* 1989, Riordan *et al.* 1989, Kerem *et al.* 1989).

• Detection of cross-hybridizing sequences in other species. This relies on the tendency for gene sequences to be conserved in evolution, whereas 'spacer' is not well-conserved.

• Identification of 'CpG islands', sequences rich in the CG dinucleotide, which often mark the 5′ upstream region of vertebrate genes (Bird 1986).

• Occurrence of transcripts of the segment in normal tissues which are affected in the diseased state.

- Isolation of cDNAs corresponding to the segment.
- Identification of open reading frames in the nucleotide sequence of the segment.

Chromosome walking is simple in principle, but walking on a large scale is technically demanding. For large distances, walking is usually combined with chromosome jumping (see below). Jumping overcomes difficulties posed by highly repeated, unclonable, or unstable genomic segments (Rommens *et al.* 1989). Chromosome walking has become a key technique for the human genome project. It has also been very important in studies of the *Drosophila* genome. Here the advanced genetics of *Drosophila* comes to the rescue, for combined with the inherent usefulness of polytene chromosome hybridization are the numerous inversions and translocations that may be exploited to reduce the distance to be walked. In one of the first applications of this technology, Hogness and his co-workers (Bender *et al.* 1983) cloned DNA from the *Ace* and *rosy* loci and the homeotic *Bithorax* gene complex in *D. melanogaster*.

The existence of the polytene chromosomes in *Drosophila* permits a different, more direct, approach. It has, by a tremendous technical *tour de force*, been found possible to physically excise a region of such a salivary gland chromosome by micromanipulation, and thence to extract its DNA, restrict it and ligate it to a phage λ vector, all within a microdrop under oil, and thereafter obtain clones with reasonable efficiency (Scalenghe *et al.* 1981). With this technique it should be possible to isolate clones from any desired region of the genome of the fruit fly. The main interest in these micromanipulation experiments lies not in the ability to clone sequences from particular regions of the *Drosophila* genome – this can be achieved by other means such as chromosome walking – but rather in demonstrating that minute quantities of DNA can be cloned successfully.

Following the early demonstrations of the power of chromosome walking, improved vectors and strategies have been developed in response to the need for easier chromosome walking. For example, two phage λ derivatives, λDASH and λFIX, have been designed by Stratagene Cloning Systems. The map of the λDASH genome is shown in Fig. 6.5. This replacement vector can accommodate inserts of 9–22 kb and includes multiple cloning sites. These sites define the stuffer fragment which bears *red*[+] *gam*[+] genes and therefore permits Spi selection. In these regards, λDASH resembles EMBL3 (p. 104). Immediately adjacent to the cloning sites are T3 and T7 promoters for the T3 and T7 RNA polymerases. These promoters allow RNA probes to be made from insert sequences (see p. 96). Importantly, from the point of view of chromosome walking, *end-specific* RNA probes will be produced if the recombinant DNA to be transcribed into RNA is first digested with a restriction endonuclease that cuts relatively frequently (e.g. a tetranucleotide target site), and which thus leaves only a short insert DNA fragment attached to the T3 and T7 promoters. These probes are ideal for probing a library for the next step in a chromosome walk and have the great advantage that they can be made conveniently without recourse to subcloning.

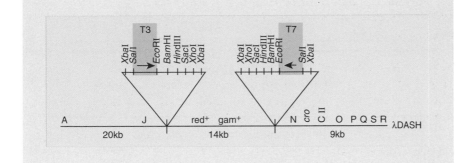

Fig. 6.5 The replacement vector λDASH can accommodate inserts of 9–22 kb. In the arms, immediately adjacent to the cloning sites, are promoters specific for T3 and T7 RNA polymerases. Thus RNA probes can be made from inserted sequences without subcloning. End-specific RNA probes are generated from recombinant DNA pre-digested with a restriction enzyme which has a tetranucleotide target site and which leaves only a short insert sequence attached to the promoter.

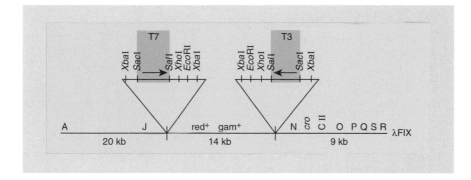

Fig. 6.6 The replacement vector λFIX is similar to λDASH.

λFIX (Fig. 6.6) is similar to λDASH except that it incorporates *Xho*I sites situated so as to take advantage of a cloning strategy that prevents the ligation of vector arms without included foreign DNA, and that eliminates multiple inserts. The principle of this strategy is to digest the vector with *Xho*I and then partially fill in the sticky ends of the *Xho*I site to leave dinucleotide 5′ overhangs (Fig. 6.7). Such filled-in ends cannot re-associate. Similarly, partially filled-in *Sau*3A ends cannot re-associate, but *can* combine with the vector bearing the partially filled-in *Xho*I ends, and hence can be joined with the vector arms.

Analysing very large DNA sequences and long-distance chromosome walking

Conventional cloning techniques cannot readily accept chromosome walks of thousands of kilobases, yet distances of this magnitude commonly separate linked marker mutations in mammals. As a rule of thumb, about

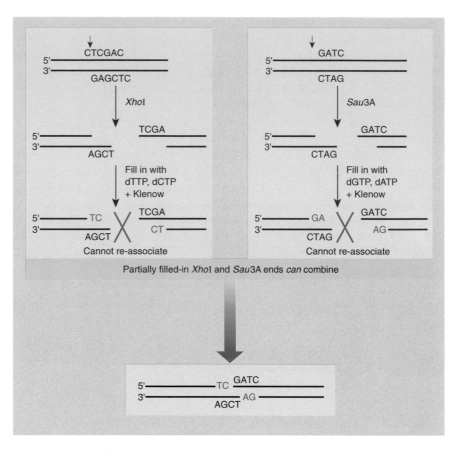

Fig. 6.7 Partially filled-in *Xho*I sites cannot self-ligate but can accommodate partially filled-in, non-self-ligatable *Sau*3A ends.

1000 kb corresponds to a recombination distance of 1 cm (1% recombination) in humans. The problems of analysing (mapping) and walking such large distances have been largely overcome by progress in three areas.

• The discovery of restriction endonucleases, or combinations of methylases and endonucleases, which cut DNA at very infrequent sites.

• The development of *pulsed-field gel electrophoresis* and *field-inversion gel electrophoresis* to resolve very large DNA fragments.

• The deployment of specialized *chromosome-jumping* strategies.

These developments are described in the following sections. The reader is also referred to the use of yeast artificial chromosomes as a means of cloning very large DNA fragments (Chapter 13).

Cutting DNA at very rare target sites

Type II restriction endonucleases that have octanucleotide-recognition sequences are expected to cut DNA very infrequently. Examples of such enzymes are *Not*I (GCGGCCGC), *Sfi*I (GGCCNNNNNGGCC), and *Pac*I (TTAATTAA). Other enzymes with similarly large recognition sequences will probably be discovered and extend this list. In addition, there are restriction endonucleases with shorter recognition sequences that have the property of being uncommon sequences in mammalian DNA. Examples of this category are *Nru*I (TCGCGA) and *Bss*HII (GCGCGC). Their target sequences include two CG dinucleotides. The CG dinucleotide is rare in mammalian DNA (except at CpG islands) and the target sites for these enzymes are consequently very infrequent.

Additional specificities in target sites may be created by combining the specificity of certain methylases with that of restriction endonucleases. An example of this approach involves the restriction endonuclease, *Dpn*I, which cuts DNA at the sequence G–^{m}A–T–C to produce flush ends. This enzyme is unusual in that it requires the methylation of the adenines in *both* strands of the DNA in order for cleavage to occur. Thus DNA from *dam*$^+$ *E. coli* is cleaved by *Dpn*I, whereas DNA from *dam*$^-$ mutants is not (see Chapter 3). McClelland *et al.* (1984) exploited this property by combining it with the specificity of the modification methylases M.Taq I or M.Cla I. The enzyme M.Taq I methylates both strands of the sequence TCGA to create T–C–G–^{m}A. Thus in DNA that is not previously methylated at these sites, cleavage by *Dpn*I will only occur at the following octanucleotide sequence: T–C–G–^{m}A–T–C–G–A. Similarly, M.Cla I methylates both strands of the sequence A–T–C–G–A–T to produce A–T–C–G–^{m}A–T, and so in combination with *Dpn*I cleavage will only occur at the decanucleotide sequence ATCGATCGAT.

Pulsed-field gel electrophoresis and field-inversion gel electrophoresis

A new type of gel electrophoresis was developed by Schwartz and Cantor (1984) which resolves DNA molecules up to 2000 kb in length. The technique uses conventional agarose and conventional buffers. The innovation is the use of alternately pulsed, perpendicularly (i.e. orthogonally) oriented electrical fields. In their demonstration of the power of this technique, Schwartz and Cantor separated intact yeast, *Saccharomyces cerevisiae*, chromosomal DNA. Each chromosome was evidently a single piece of DNA and could be analysed further by Southern blotting the gel. The separation appears to depend upon the electrical perturbation of the orientation of the DNA, and on the degree of extension of long DNA molecules. The relaxation time of such molecules in free solution is a very sensitive function of the molecular weight. This orthogonal pulsed-field gel electrophoresis (PFGE) technique has been further developed (Carle & Olson 1984), but has the disadvantages that the DNA samples do not run in straight-line tracks.

These difficulties have been overcome in field-inversion gel electrophoresis (FIGE) (Carle *et al.* 1986). This produces good resolution up to 2000 kb without the need for a complicated perpendicular-field gel apparatus. FIGE uses a conventional electrophoresis apparatus with an electrical field that pulses forward–reverse combined with a pause between each phase. The pulses are in the form of a time-ramp. The DNA tracks run in straight lines and are comparable across the gel, thus making

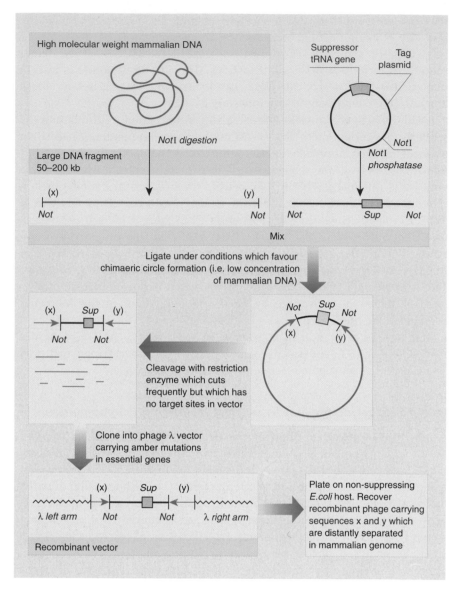

Fig. 6.8 Jumping library construction. This example illustrates the use of *Not*I-digested mammalian DNA. A variant of this procedure could use mammalian DNA which has been *partially* digested with an enzyme that cuts more frequently than *Not*I. Such a procedure would produce a jumping library containing overlapping DNA fragments.

calibration with markers and the interpretation of Southern blots much simpler than in orthogonal PFGE.

Chromosome jumping strategies

Chromosome jumping, as described by Collins and Weissman (1984) and Poustka and Lehrach (1986), depends upon the circularization of very large genomic DNA fragments, followed by cloning DNA from the region covering the closure site of these circles, thus bringing together DNA sequences that were originally located a considerable distance apart in the genome. These cloned DNAs from the closure sites make up a 'jumping library'. Such a jumping library greatly speeds up the process of long-range chromosome walking.

Figure 6.8 (Poustka *et al.* 1987) shows a strategy for creating a jumping library using *Not*I-digested human genomic DNA. This enzyme cuts only rarely in mammalian DNA, and so drastically reduces the size of the library required to cover the mammalian genome. This strategy has the additional advantage that analysis of jumps is easy by FIGE analysis and Southern blotting of *Not*I-digested genomic DNA. However, with a complete digest such as this, overlaps between *Not*I fragments have to be obtained indirectly: one solution is to use 'linking clones' containing conventionally cloned fragments with an internal *Not*I site (Poustka *et al.* 1987). A variant of the strategy shown in Fig. 6.8 employs *partial* digestion with a restriction endonuclease to generate large genomic DNA fragments (Collins & Weissman 1984). This technique has been applied to a jump of 100 kb in the cystic fibrosis locus (Collins *et al.* 1987) and to a jump of 200 kb in the region of the Huntington's disease gene (Richards *et al.* 1988). See Box 6.1.

Box 6.1 A landmark publication. Identification of the cystic fibrosis gene: chromosome walking and jumping

Cystic fibrosis (CF) is a relatively common severe autosomal recessive disorder. Until the CF gene was cloned, there was little definite information about the primary genetic defect. The cloning of the CF gene was a breakthrough for studying the biochemistry of the disorder (abnormal chloride channel function), for providing probes for pre-natal diagnosis, and for potential treatment by somatic gene therapy or other means.

The publication is especially notable for the generality of the cloning strategy. In the absence of any direct functional information about the CF gene, the chromosomal location of the gene was used as the basis for its cloning. Starting from markers identified by linkage analysis as being close to the CF locus on chromosome 7, a total of about 500 kb were encompassed by a combination of chromosome walking and jumping. Jumping was found to be very important to overcome problems caused by 'unclonable' regions which halted the sequential walks.

In this work, large numbers of clones were involved, obtained from several different phage and cosmid genomic libraries. Among these libraries were one prepared by the Maniatis strategy in the λCharon 4A vector (Fig. 6.2), and several prepared using the λDASH and λFIX vectors (Figs 6.5 and 6.6) after partial digestion of human genomic DNA with *Sau*3AI. Also several cosmid libraries were constructed.

Cloned regions were aligned with a map of the genome in the CF region, obtained by long-range restriction mapping (using rare cutters) and pulsed-field gel electrophoresis. The actual CF gene was detected in this cloned region by a number of criteria that are discussed in the main text.

From: Rommens *et al.* (1989) *Science* 245: 1059–1065.

cDNA cloning

Cloned eukaryotic cDNAs have their own special uses, which derive from the fact that they lack the intron sequences that are usually present in the corresponding genomic DNA. Introns are non-coding sequences that often occur within eukaryotic gene sequences. They can be situated within the coding sequence itself, where they then interrupt the co-linear relationship of the gene with the encoded polypeptide. They may also occur in the 5′ or 3′ untranslated regions of the gene, but in any event they are copied into RNA by RNA polymerase when it transcribes the gene. The initial, primary transcript is a precursor to mRNA. It goes through a series of processing events in the nucleus before appearing in the cytoplasm as mature mRNA. These events include the removal of intron sequences by a process called *splicing*. When cDNA is derived from mRNA it therefore lacks intron sequences. Since removal of eukaryotic intron transcripts by splicing does not occur in bacteria, eukaryotic cDNA clones find application where bacterial expression of the foreign DNA is necessary, either as a prerequisite for detecting the clone (see Chapter 7), or because the polypeptide product is the primary objective. Also, where the sequence of the genomic DNA is known, the position of intron/exon boundaries can be assigned by comparison with the cDNA sequence.

A second situation where cDNA cloning is carried out involves the analysis of temporally regulated gene expression in development, or tissue-specific gene expression. By using the differential screening procedure (see Chapter 7) it is possible to screen a cDNA clone library to identify cDNA clones derived from mRNA molecules present in one cell type but absent in another cell type.

As with genomic DNA, it may rarely be appropriate to isolate cDNA from a purified mRNA species. Much more commonly a cDNA clone library may be prepared and screened for particular sequences. Before proceeding further, it is necessary to consider the nature of mRNA populations in tissues. In many tissues and cultured cells mRNAs are present at widely different *abundances*, i.e. some mRNA types are present in large numbers per cell, others may be present at just a few copies per cell. Table

Table 6.2 Abundance classes of typical mRNA populations

Source	Number of different mRNAs	Abundance (molecules/cell)
Mouse liver cytoplasmic poly(A)$^+$	9	12 000
	700	300
	11 500	15
Chick oviduct polysomal poly(A)$^+$	1	100 000
	7	4 000
	12 500	5

References: mouse (Young *et al.* 1976); chick oviduct (Axel *et al.* 1976).

6.2 gives some representative examples. Notice that in the chick oviduct one mRNA type is superabundant. This is the mRNA encoding ovalbumin, the major eggwhite protein. Therefore, this mRNA population is naturally so enriched in ovalbumin mRNA that cloning the ovalbumin cDNA presents no problem in screening. The clones could be identified by screening a small number of recombinants; the hybrid released translation procedure would be appropriate (see Chapter 7).

Another appropriate strategy for obtaining abundant cDNAs is to clone the cDNA directly in an M13 vector such as M13 mp8. A set of clones can then be sequenced immediately and identified on the basis of the polypeptide that each encodes. A successful demonstration of this *shotgun sequencing* strategy is given by Putney *et al.* (1983) who determined DNA sequences of 178 randomly chosen muscle cDNA recombinants. Complete amino acid sequences were available for 19 abundant muscle-related

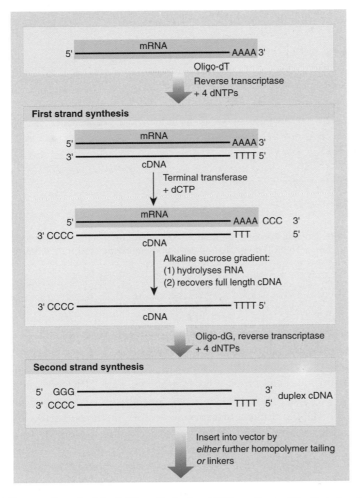

Fig. 6.9 Improved method for full-length duplex cDNA synthesis. The first strand is tailed with oligo(dC) so as to allow priming of the second-strand synthesis by oligo(dG).

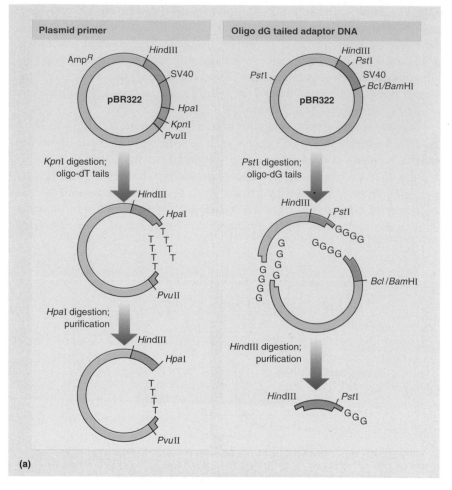

Fig. 6.10 The Okayama and Berg method of cDNA cloning. (a) Preparation of plasmid primer and adaptor DNA. The unshaded portion of each ring is pBR322 DNA, and the shaded or stippled segments are from SV40 DNA.

proteins. Altogether, they were able to identify clones corresponding to 13 of these 19 proteins, including interesting protein variants.

For cDNA clones in the low abundance class it is usual to construct a cDNA library. Typically, 10^5 clones will be sufficient for low-abundance mRNAs from most cell types. Once again the high efficiency obtained by packaging *in vitro* makes phage λ vectors attractive for obtaining large numbers of cDNA clones. Insertional vectors such as λgt10, λNM1149, λZAP or λgt11 (see Chapter 7) are particularly well suited for such cDNA cloning.

Is it worth enriching for a particular mRNA before cloning? Only in special circumstances is a ready purification possible and attractive. Where enrichment has been carried out, the most widely-used procedure has been size fractionation of mRNA on agarose gel electrophoresis systems

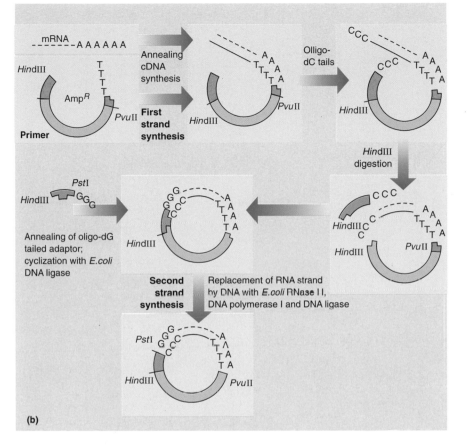

Fig. 6.10 *continued.* (b) Steps in the construction of plasmid–cDNA recombinants. The designations for the DNA segments are as mentioned in (a). In up-to-date form, the adaptor would not be isolated as shown, but synthesized chemically.

(see for example Pennica *et al.* 1983). One strategy for cDNA cloning that is selective, and which has been used frequently, employs specific priming of cDNA synthesis with oligonucleotide primers (either unique or degenerate in sequence – see Chapter 7, Box 7.1). Selective priming is also embodied in various PCR strategies that are discussed in Chapter 10. In general, the most commonly-used primary screening technique involves colony or plaque hybridization with radioactive or immunochemical probes. This is applicable to very large libraries and the effort involved is largely independent of the number of recombinants to be screened. There is, therefore, usually little to be gained by attempting to enrich the starting mRNA in order to reduce the number of recombinants to be screened. The isolation of clones by such screening procedures has become a commonplace technical feat that effectively performs a purification impossible by any other means.

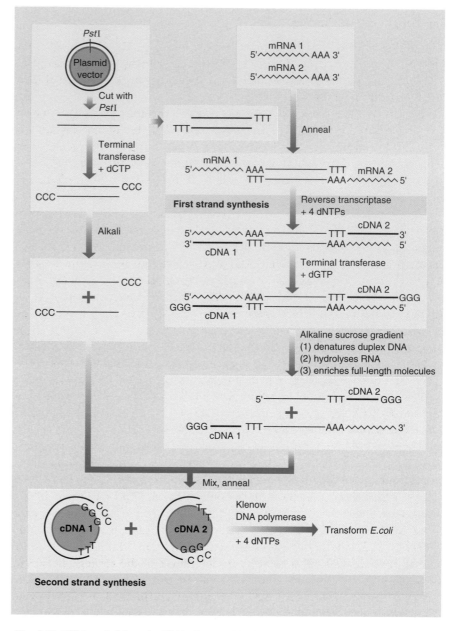

Fig. 6.11 Efficient full-length cDNA cloning (Heidecker & Messing 1983). The mRNA is annealed to linearized and oligo(dT) tailed plasmid DNA, which then primes synthesis of the first cDNA strand. Oligo(dG) tails are added to the cDNA-plasmid molecules, which are then centrifuged through an alkaline sucrose gradient. This step removes small molecules, hydrolyses the mRNA and separates the two cDNAs which were formerly attached to the same duplex plasmid. Denatured, oligo(dC) tailed plasmid DNA is added (in excess) and conditions adjusted to favour circularization by the complementary homopolymer tails. The excess oligo(dC) tailed plasmid may simply renature, but cannot circularize. The circular molecules have a free 3-hydroxyl on the oligo(dC) tail which primes second-strand synthesis of the cDNA to create duplex recombinant plasmids which transform *E. coli*. Clones can be obtained with the cDNA inserted in both orientations.

Full-length cDNA cloning

In Chapter 3 some early strategies for synthesizing cDNA molecules for cloning were described. These involved self-priming of second-strand synthesis and therefore led to the loss of some sequence corresponding to the 5' end of the mRNA.

Several strategies were developed to overcome this difficulty. One of the simplest of these is shown in Fig. 6.9. The dC tailing of single-stranded cDNA followed by oligo(dG) priming of second-strand synthesis does not lead to hairpin formation, nuclease S1 treatment is not required, and consequently this is an effective method for generating full-length cDNA clones (Land *et al.* 1981).

Two further methods, shown schematically in Figs 6.10 and 6.11, have been devised to eliminate the use of nuclease S1. Additionally, in both methods the oligo(dT) sequence for priming the first-strand cDNA synthesis is linked to the vector DNA in a prior reaction. Both methods have been reported to promote full-length cDNA cloning with a very high efficiency (Okayama & Berg 1982, Heidecker & Messing 1983). It is thought that full-length reverse transcripts are obtained *preferentially* because in each case an RNA–DNA hybrid molecule, which is the result of first-strand synthesis, is the substrate for a terminal transferase reaction. A cDNA that does not extend to the end of the mRNA will present a shielded 3 hydroxyl group, which is a poor substrate for tailing.

It will also be noticed that the Okayama and Berg strategy employs a second-strand synthesis step in which the RNA strand is replaced in a DNA polymerase reaction that is primed by nicking the RNA with RNase H. The second-strand synthesis occurs at these nicks by a nick-translation type of reaction. This type of second-strand reaction has been exploited in efficient cDNA library cloning schemes that are simpler than that of Okayama and Berg, e.g. those developed by Gubler and Hoffman (1983), and Lapeyre and Amalric (1985). The Gubler and Hoffman protocol and modifications of it have been popular. A particular advantage of the Okayama and Berg strategy is that the cDNA is inserted into the vector in a defined orientation. This is useful in derivatives of the strategy in which the cDNA is inserted into a vector adjacent to a phage T3, T7 or SP6 promoter, because then an RNA copy can be synthesized *in vitro* from a defined cDNA strand with purified phage RNA polymerase (see Chapter 5).

The use of random primers in cDNA cloning

In the examples discussed so far, cDNA synthesis has been initiated on poly(A)$^+$ mRNA by priming with oligo(dT) sequences. There are three limitations to this approach.
• Not all RNAs bear a 3'-terminal poly(A) sequence. It is possible to add a poly(A) sequence *in vitro* with purified poly(A) polymerase, but this can be problematical.
• Large mRNAs can be difficult to deal with because it is not reasonable

to expect to synthesize and clone sequences which lie further than a few kilobases from the oligo(dT) primer.

- 3'-end bias. Because priming occurs at the 3'-end of poly(A)$^+$ mRNAs and because cDNA synthesis is often incomplete, cDNA libraries are enriched in 3'-terminal sequences.

The first two limitations are common in cloning genomic RNAs from RNA viruses. The third limitation may be important where cDNA libraries are to be made in a vector such as λgt11 or λZAP and screened for expression of fusion polypeptides (Chapter 7). These limitations can often be overcome by priming the first-strand cDNA synthesis not with oligo(dT), but rather with 'random' oligonucleotide primers. Usually the primer consists of a mixture of all possible chemically synthesized hexadeoxynucleotides. These hybridize at random sites along the RNA, prime cDNA synthesis, and generate cDNA sequences of a sufficient size to be useful when cloned. Commonly a Gubler and Hoffman-type second-strand step is employed.

Genomic and cDNA libraries versus PCR

In this chapter we have described methods for constructing genomic and cDNA libraries. Efficient methods exist for ensuring that large, representative, libraries can be constructed. From such libraries, specific genomic clones, or cDNA clones, can be isolated, using strategies described in the next chapter. However there are alternative routes to obtaining specific genomic or cDNA clones, which are based upon the polymerase chain reaction. For many applications a PCR-based approach is quicker and simpler than library construction and screening. Even in circumstances where a suitable library already exists, a PCR-based approach may be attractive. In view of this, does PCR make library construction obsolete, and most especially, does cDNA cloning by PCR supersede cDNA cloning from libraries? These issues are addressed in Chapter 10.

7 Recombinant selection and screening

Introduction

The task of isolating a desired recombinant from a population of bacteria or phage depends very much upon the cloning strategy that has been adopted; for instance, when a cDNA derived from an abundant mRNA is to be cloned, the task is relatively simple – only a small number of clones need to be screened. Isolating a particular single-copy gene sequence from a complete mammalian genomic library requires techniques in which hundreds of thousands of recombinants can be screened.

In this chapter we give an overview of the general principles employed in recombinant selection and screening procedures, under the headings of genetic, immunochemical, South-western, nucleic acid hybridization and recombinational methods. As we shall see, nucleic acid hybridization with labelled probes is the most generally applicable method.

Genetic methods

Selection for presence of vector

When combined with microbiological techniques, genetic selection is a very powerful tool since it can be applied to large populations. All useful vector molecules carry a selectable genetic marker or property. Plasmid and cosmid vectors carry drug-resistance or nutritional markers, and in the case of phage vectors plaque formation is itself the selected property. Genetic selection for presence of the vector is a prerequisite stage in obtaining the recombinant population. As we have seen, this can be re-fined to distinguish recombinant molecules and non-recombinant, parental vector. Insertional inactivation of a drug-resistance marker, or of a gene such as β-galactosidase for which there is a colour test, are examples of this (see p. 67). With certain replacement-type lambda vectors, and with

[121]

cosmid vectors, size selection by the phage particle selects recombinant formation.

Selection of inserted sequences

If an inserted foreign gene in the desired recombinant is expressed, then genetic selection may provide the simplest method for isolating clones containing the gene. Cloned *E. coli* DNA fragments carrying biosynthetic genes can be identified by complementation of non-revertible auxotrophic mutations in the host *E. coli* strain. A related early example comes from the work of Cameron *et al.* (1975) who cloned the *E. coli* DNA ligase gene in a phage λgt.λB vector. They exploited the inability of λ*red*⁻ phage (the vector is *red*⁻ by deletion of the C fragment) to form plaques on *E. coli lig* ts at the permissive temperature, whereas λ*red*⁻ phage will form plaques on *E. coli* Lig⁺. Recombinant phage carrying the wild-type ligase function could therefore be selected simply by their ability to form plaques through complementation of the host deficiency when plated on *E. coli lig* ts.

It has been found that certain eukaryotic genes are expressed in *E. coli* and can complement auxotrophic mutations in the host bacterium. Ratzkin and Carbon (1977) inserted fragments of yeast DNA, obtained by mechanical shearing, into the plasmid Col E1 using a homopolymer tailing procedure. They transformed *E. coli his* B mutants with recombinant plasmid and, by selecting for complementation, isolated clones carrying an expressed yeast *his* gene.

A similar approach has even been applied successfully to cloned mouse sequences. Chang *et al.* (1978) constructed a population of recombinant plasmids containing cDNA that was derived from an unfractionated mouse cell mRNA preparation in which dihydrofolate reductase (DHFR) mRNA was present. Mouse DHFR is much less sensitive to inhibition by the drug trimethoprim than is *E. coli* DHFR, so that by selecting transformants in medium containing the drug, clones were isolated in which resistance was conferred by synthesis of the mouse enzyme. This was an early example of expression of a mammalian structural gene in *E. coli*. The factors affecting expression of heterologous genes are complex, and an efficient selection procedure was required in order to identify clones actually synthesizing mouse DHFR amongst those containing non-expressed DHFR cDNA.

Immunochemical methods

Immunochemical detection of clones synthesizing a foreign protein has also been successful in cases where the inserted gene sequence is expressed. A particular advantage of the method is that genes that do not confer any selectable property on the host can be detected, but it does of course require that specific antibody is available.

During the early development of recombinant DNA technology, a number of laboratories published similar immunochemical detection methods (Skalka & Shapiro 1976, Ehrlich *et al.* 1978a). The method of Broome and

Gilbert is one of these and is discussed here because it is illustrative, even though it is now rather out of date because the plaque-based methods (using λgt11 or λZAP) have superseded it, and the radiolabelled IgG has largely been replaced by non-radioactive detection methods. It depends upon three points:

• an immune serum contains several IgG types that bind to different determinants on the antigen molecule;

• antibody molecules absorb very strongly to plastics such as polyvinyl, from which they are not removed by washing;

• IgG antibody can be readily radiolabelled with ^{125}I by iodination *in vitro*.

These properties are exploited in the following way. First, transformed cells are plated on agar in a conventional Petri dish. A replica plate must also be prepared because subsequent procedures kill these colonies. The bacterial colonies are then lysed in one of a number of ways — by exposure to chloroform vapour, by spraying with an aerosol of virulent phage, or by using a host bacterium that carries a thermo-inducible prophage. This releases the antigen from positive colonies. A sheet of polyvinyl that has been coated with the appropriate antibody (unlabelled) is applied to the surface of the plate, whereupon the antigen complexes with the bound IgG. The sheet is removed and exposed to ^{125}I-labelled IgG. The ^{125}I-IgG can react with the bound antigen via antigenic determinants at sites other than those involved in the initial binding of antigen to the IgG-coated sheet, as shown in Fig. 7.1. Positively reacting colonies are detected by washing the sheet and making an autoradiographic image. The required clones can then be recovered from the replica plate.

Two further aspects of the immunochemical method deserve mention. First, detection of altered protein molecules is possible providing that the alteration does not prevent cross-reaction with antibody. Thus Villa-

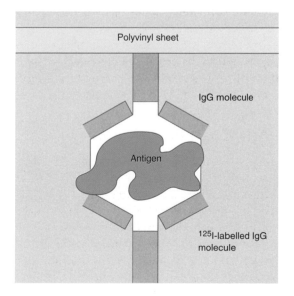

Fig. 7.1 Antigen–antibody complex formation in the immunochemical detection method of Broome and Gilbert. (See text for details.)

Polyvinyl sheet

IgG molecule

Antigen

^{125}I-labelled IgG molecule

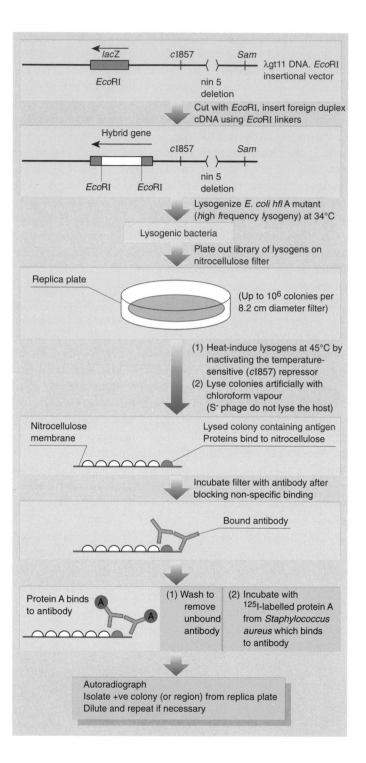

Komaroff *et al.* (1978) isolated *E. coli* clones containing cDNA sequences from rat preproinsulin mRNA. The cDNA was inserted by dG–dC homopolymer tailing at the *Pst*I site of pBR322. Using anti-insulin antibody, they isolated a clone which expressed a fused protein composed of the *N*-terminal region of β-lactamase (from pBR322) and a region of the proinsulin protein linked through a stretch of six glycine residues encoded by $d(G)_{18}$ of the joint (see p. 143). Second, the two-site detection method is particularly suited to the detection of novel genetic constructions; for instance, by coating polyvinyl discs with IgG prepared from an immune serum directed against one protein, and detecting the immobilized antigen with ^{125}I-antibodies directed against another protein, only hybrid polypeptide molecules, synthesized as a result of DNA recombination, would produce an autoradiographic response.

A very efficient exploitation of the immunochemical detection method involves the phage λ expression vector λgt11 (Young & Davis 1983). This vector carries the *E. coli lacZ* gene. A unique *Eco*RI site is located within the β-galactosidase coding region. Recombinant libraries can be constructed in which eukaryotic cDNA has been inserted, by means of linkers, into the *Eco*RI sites. In such recombinants the β-galactosidase is insertionally inactivated and, depending upon the translational phase at the fusion junction, hybrid proteins are expressed. In a population of cDNA recombinants in which duplex cDNA has been synthesized by any of the common methods, we can expect a proportion of recombinants containing any particular cDNA to be in phase. The vector can accept up to 8.3 kb of insert DNA and complete cDNA libraries containing large numbers of independent recombinants can be constructed readily because of the efficiency endowed by packaging *in vitro*.

Immunochemical screening of the library can be carried out upon colonies of induced lysogenic bacteria or, as is now more common, the screening is carried out on plaques of the recombinant phage. The original approach with induced lysogens is shown in Fig. 7.2. In this approach the library of recombinant λgt11 is first used to lysogenize *E. coli*. This is efficiently carried out with a *hfl*A (high frequency of lysogeny) mutant of *E. coli*. Lysogens produce detectable amounts of hybrid proteins upon

Fig. 7.2 (*facing page*) Immunochemical screening applied to induced lysogens with the expression vector λgt11 and its derivatives. The duplex cDNA is inserted within the *lacZ* gene. In a proportion of recombinants the insertion will be in the correct translational reading phase so as to direct the synthesis of a hybrid protein that will be detected by reaction with antibody raised against the required protein. The functions of the *cI* and *S* genes are discussed in Chapter 5. The *hfl*A mutation of *E. coli* results in a very high frequency of lysogenization by phage λ. These lysogens express detectable amounts of hybrid protein when they are induced by raising the temperature so as to inactivate the temperature-sensitive *cI* repressor carrying the c1857 mutation. The frequency of lysogenization is high on the *hfl*A strain, but some non-lysogens will be present on the filter. The procedure can be improved by incorporating a drug-resistance marker, e.g. ampicillin resistance (Kemp *et al.* 1983) or kanamycin resistance, borne on transposon Tn5, into the vector. Lysogenic bacteria can then be selected in the presence of the drug.

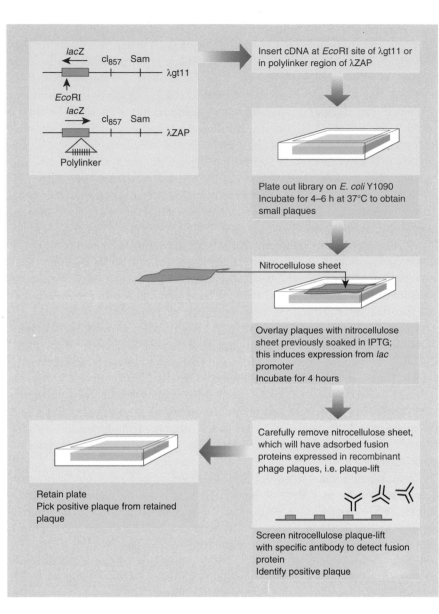

Fig. 7.3 Immunochemical screening of λgt11 or λZAP recombinant plaques.

induction. It has been claimed that up to 10^6 colonies can be screened on a single 8.2-cm diameter filter. This immunochemical approach has largely been replaced by a very similar but much more convenient method that is carried out on phage plaques. In this procedure the library is plated out at moderately high density (up to 5×10^4 plaques per 9 cm square plate), with *E. coli* strain Y1090 as host (Fig. 7.3). This *E. coli* strain over-produces the *lac* repressor and ensures that no expression of cloned sequences (which may be deleterious to the host) takes place until the inducer IPTG is presented to the infected cells. Y1090 is also deficient in the *lon* protease, hence increasing the stability of recombinant fusion proteins. Fusion proteins expressed in plaques are absorbed onto a nitrocellulose membrane overlay and this membrane is then processed for antibody screening. When a positive signal is identified on the membrane, the positive plaque can be picked from the agar plate (a replica is not necessary) and hence recombinant phage can be isolated.

South-western screening for DNA-binding proteins

In the previous section we have seen how fusion proteins may be detected immunochemically when expressed in plaques produced by recombinant λgt11 or λZAP. A closely-related approach has been used very successfully for screening and isolation of clones expressing fusion proteins where the foreign sequence encodes a DNA-binding protein that binds specifically to a particular DNA sequence. The method involves a nitrocellulose membrane 'plaque-lift' on which expressed fusion proteins are adsorbed. The procedure for performing the plaque-lift is the same as that just described. However, the screening is carried out by incubating the membrane with a radiolabelled *duplex* DNA oligonucleotide containing the sequence for which the DNA-binding protein is specific. Conditions are chosen in which non-specific binding is minimized. Positively reacting plaques are identified by autoradiography of the filter. This technique therefore uses a radiolabelled DNA to detect polypeptide on the nitrocellulose, and has been called a 'South-western' procedure. It has been spectacularly successful in the isolation of clones expressing cDNA sequences corresponding to certain mammalian transcription factors (Vinson *et al.* 1987, Staudt *et al.* 1988). Clearly the procedure can only be successful (a) where the binding activity is due to a single polypeptide chain, and (b) where it can be expressed in functional form when in a fusion polypeptide. (Note that several *different* fusions of any particular polypeptide with the β-galactosidase polypeptide are likely to be created in any random cloning of a particular cDNA in a single cDNA library.) It may also be significant that success has been obtained on some occasions where random priming has been used to generate the cDNA first strand. It is also clear that the affinity of the polypeptide for the specific DNA sequence must be high. The procedure has been found most efficient when the oligonucleotide containing the binding sequence has been ligated into multimeric form. This may mean that a single DNA multimer may be bound by more than one fusion polypeptide molecule on the filter, hence greatly increasing the average dissociation time.

Nucleic acid hybridization methods

Grunstein and Hogness (1975) developed a screening procedure to detect DNA sequences in transformed colonies by hybridization *in situ* with radioactive 'probe' RNA. Their procedure can rapidly determine which colony amongst thousands contains the required sequence. A modification of the method allows screening of colonies plated at a very high density (Hanahan & Meselson 1980).

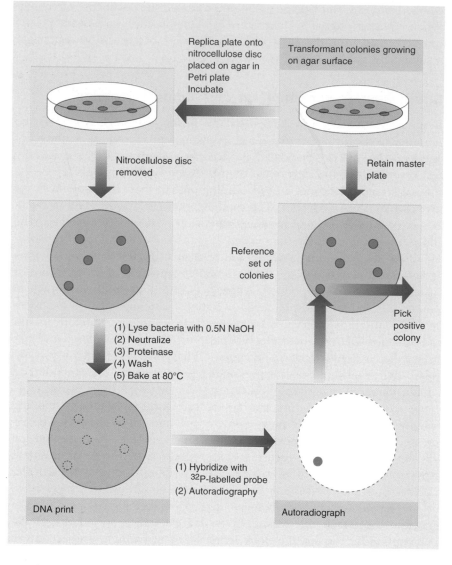

Fig. 7.4 Grunstein–Hogness method for detection of recombinant clones by colony hybridization.

The colonies to be screened are first replica plated onto a nitrocellulose filter disc that has been placed on the surface of an agar plate prior to inoculation (Fig. 7.4). A reference set of these colonies on the master plate is retained. The filter bearing the colonies is removed and treated with alkali so that the bacterial colonies are lysed and their DNAs are denatured. The filter is then treated with proteinase K to remove protein and leave denatured DNA bound to the nitrocellulose, for which it has a high affinity, in the form of a 'DNA-print' of the colonies. The DNA is fixed firmly by baking the filter at 80°C. The defining, labelled RNA is hybridized to this DNA and the result of this hybridization is monitored by autoradiography. A colony whose DNA print gives a positive autoradiographic result can then be picked from the reference plate.

Variations of this procedure can be applied to phage plaques (Jones & Murray 1975, Kramer *et al.* 1976). Benton & Davis (1977) devised a method in which the nitrocellulose filter is applied to the upper surface of agar plates, making direct contact between plaques and filter. The plaques contain phage particles as well as a considerable amount of unpackaged recombinant DNA. Both phage and unpackaged DNA bind to the filter and can be denatured, fixed, hybridized, etc. This method is the original 'plaque-lift' procedure and has the advantage that several identical DNA prints can easily be made from a single-phage plate: this allows the screening to be performed in duplicate, and hence with increased reliability, and also allows a single set of recombinants to be screened with two or more probes. The Benton and Davis plaque-lift procedure must be the most widely applied method of library screening, successfully applied in thousands of laboratories to the isolation of recombinant phage by nucleic acid hybridization. Figure 7.5 shows a scheme of the procedure. This is the prototype for a variety of plaque-lift screening procedures previously illustrated in this chapter.

The great advantage of the hybridization methods is generality. They do not require expression of the inserted sequences and can be applied to any sequence, provided a suitable probe is available.

The probe problem

Screening procedures which rely on nucleic acid hybridization are, as we have seen, general in application and powerful. Using these procedures it is now possible easily to isolate any gene sequence from virtually any organism, *if a probe is available*. The problem of gene isolation, then, is a problem of obtaining a suitable probe. There are several solutions to this.

• In certain specialized cell types or tissues, particular mRNAs are abundant or super-abundant. The corresponding cDNA clones can be isolated by screening small numbers of recombinants directly by sequencing. These cDNA probes can then be used to isolate genomic sequences.

• Nucleic acid sequences encoding certain proteins have been sufficiently conserved in evolution such that cross-species nucleic acid hybridization is possible. Examples of effective heterologous probes include histone sequences [sea urchin vs. *Xenopus* (Old *et al.* 1982b)], actin sequences [*Dictyostelium* vs. *Xenopus* (Cross *et al.* 1984)], and β-nerve growth factor [mouse vs. man (Ullrich *et al.* 1983)]. This means that the particular biological advantages of one experimental system can be exploited to isolate a gene sequence, which may then provide a probe for corresponding genes in other organisms.

• An oligonucleotide, which needs to be only about 14–20 nucleotides long, and which corresponds to a part of the sequence encoding the protein in question, can be synthesized chemically. This requires that short, oligopeptide, stretches of amino acid sequence must be known. The DNA sequence (or its complement) encoding the protein can then be deduced from the genetic code. However, a problem arises here because of the degeneracy of the code. Most amino acids are encoded by more than one codon. For this reason, oligopeptide sequences known to contain tryptophan or methionine residues are particularly valuable, because these

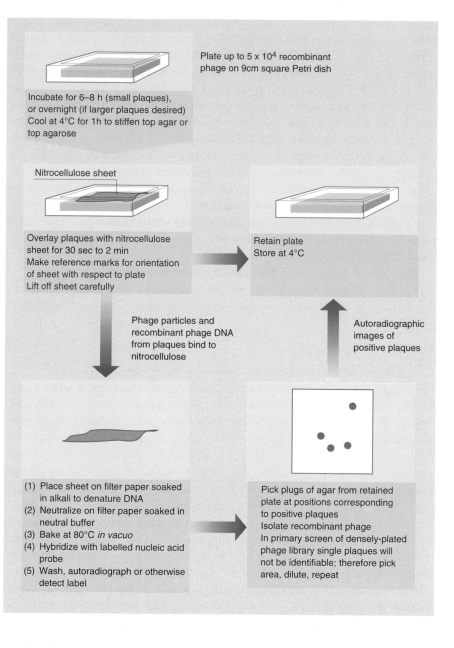

Fig. 7.5 Benton and Davis'
plaque-lift procedure.

two amino acids have single codons, and the number of possibilities is
thereby reduced. Thus, for example, the oligopeptide His-Phe-Pro-Phe-
Met may be identified and chosen to provide a probe sequence. This
oligopeptide could be encoded by the following sequences:

These 32 different sequences do not have to be synthesized individually. It is possible to perform a mixed addition reaction for each polymerization step where the sequence is degenerate. Therefore only one, mixed probe is prepared and radiolabelled. This mixed probe method was originally devised by Wallace and co-workers (Suggs *et al.* 1981). To cover all codon possibilities degeneracies of 64-fold (Orkin *et al.* 1983) or even 256-fold (Bell *et al.* 1984) have been employed successfully. What length of oligodeoxynucleotide is required for reliable hybridization? Even though 11-mers can be adequate for Southern blot hybridization (Singer-Sam *et al.* 1983) longer probes are necessary for good colony hybridization. Mixed probes of 14 nucleotides length have been successful (see Box 7.1); 16-mers are typical (Singer-Sam *et al.* 1983). (See Chapter 2.)

An alternative strategy for synthetic oligodeoxynucleotide probes employs a single longer probe, rather than a shorter mixed probe. Here the uncertainty at each codon is largely ignored and instead increased probe length confers specificity. This strategy is examined theoretically by Lathe (1985), and has been applied to sequences coding for human coagulation Factor VIII (Wood *et al.* 1984, Toole *et al.* 1984) and the human insulin receptor (Ullrich *et al.* 1985).

Differential screening

This is a variant of the nucleic acid hybridization method that is particularly suitable for isolating tissue-specific or developmentally-regulated cDNA sequences or clones derived from mRNAs that are induced by particular treatments.

Box 7.1 A landmark publication. Cloning and expression of human tissue-type plasminogen activator cDNA in *E. coli*

Human blood plasma contains an enzymatic system for dissolving blood clots. The enzyme, plasmin, degrades the fibrin network of the clot, forming soluble products. Plasmin is produced from its precursor, plasminogen, by limited proteolysis brought about by plasminogen activators. Because of the clinical potential of plasminogen activators in thrombolytic therapy, the cloning and expression of tissue plasminogen activator was highly significant. In addition, the *Nature* (1983) publication* has been cited extensively because it exemplifies a number of cDNA cloning techniques, and because it featured in a celebrated patent dispute. The strategy adopted for this cloning was as follows.

• A melanoma cell line was known to produce tPA. The tPA mRNA was expected to be non-abundant.
• Poly(A)$^+$ RNA was prepared from the cell line. This RNA was fractionated by electrophoresis through a urea-agarose gel, and fractions were assayed by translation in a rabbit reticulocyte lysate, supplemented

continued

Box 7.1 *continued*

with dog pancreas microsomes (to glycosylate the translation product). Fractions enriched in tPA mRNA were detected by immunoprecipitation of the translation products with a specific antibody.

• The enriched mRNA was used to prepare cDNA by oligo(dT) priming.

• Double-stranded cDNA was fractionated to eliminate short cDNAs (less than 350 base-pairs). The cDNA was tailed with dC, annealed with dG-tailed *Pst*I-cut plasmid pBR322, and used to transform *E. coli*. About 4600 transformants were obtained.

• Part of the tPA amino acid sequence was known to be Trp-Glu-Tyr-Cys-Asp. On the basis of this, an eightfold redundant 14-mer oligodeoxynucleotide (including all possible sequences complementary to mRNA in this region) was synthesized, radiolabelled, and used to screen the transformants by colony hybridization. A positive colony of 2304 nucleotides was isolated which encoded 508 amino acids of tPA. This clone did not contain a full-length coding region: it lacked amino terminal coding sequences.

• To obtain cDNA clones for the missing 5′ portion of the tPA mRNA, a unique 16-mer deoxyoligonucleotide, complementary to a region near the 5′ end of the original clone, was used to prime cDNA synthesis on mRNA, and hence to produce about 1500 transformants. In order to screen these clones two probes were used. One was the synthetic 16-mer, the other was a region of tPA genomic DNA that had been isolated by screening a genomic library (in a phage λ vector) using the original cDNA as probe. A partial cDNA was found that contained the missing sequence.

• In order to express tPA in *E. coli*, the two partial cDNAs were used to assemble a complete mature tPA coding region. The 'prepro' peptide-coding sequence was removed and replaced by a synthetic sequence that contained an initiator ATG codon preceding the mature sequence. The construct was ligated into an expression vector.

• Active tPA was synthesized in *E. coli*. For clinical use, tPA is expressed in a mammalian expression system which glycosylates the protein.

• The strategy adopted by Pennica *et al*.* has features that, after the passing of a decade, now appear out of date. Today, a phage vector such as λZAP would probably replace the plasmid vector. Many researchers would exploit the ability to create and screen a very large number of recombinants in such a vector, and so would probably dispense with the mRNA enrichment, producing a complete melanoma cell cDNA library. By adopting modern cDNA synthesis techniques (such as random priming and a Gubler–Hoffman type protocol for second strand synthesis) and by screening a large library, a full-length coding region should be obtainable in one clone. Finally, an attractive alternative to cDNA library construction is a strategy based on the polymerase chain reaction. For example, the RACE protocols (see Chapter 10), based on the small region of peptide sequence given above to provide specific primers, could be used to obtain a full-length cDNA.

* From: Pennica *et al*. (1983) *Nature* 301: 214–221.

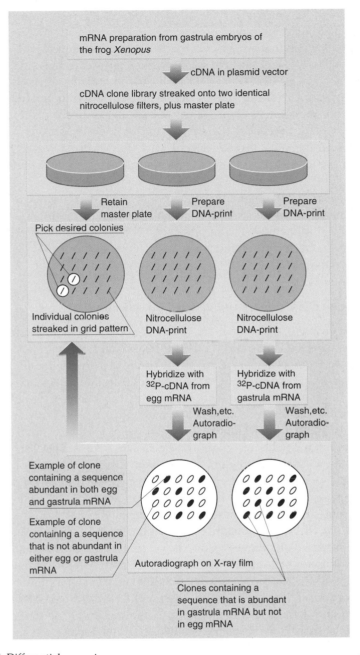

Fig. 7.6 Differential screening.

Let us consider, for example, the isolation of cDNAs derived from mRNAs which are abundant in the gastrula embryo of the frog *Xenopus* but which are absent, or present at low abundance, in the egg. A cDNA clone library is prepared from gastrula mRNA. Replica filters carrying identical sets of recombinant clones are then prepared (Fig. 7.6). One of these filters is then probed with ^{32}P-labelled mRNA (or cDNA) from gastrula embryos, and one with ^{32}P-labelled mRNA (or cDNA) from the egg. Some colonies will give a positive signal with both probes; these represent cDNAs derived from mRNA types that are abundant at both stages of development. Some colonies will not give a positive signal with either probe; these correspond to mRNA types present at undetectably low abundance in both tissues. This is a feature of using probes derived from mRNA populations: only abundant or moderately abundant sequences in the probe carry a significant proportion of the label and are effective in hybridization. Importantly, some colonies give a positive signal with the gastrula probe, but not with the egg probe. These should, therefore, correspond to the required sequences.

Such differential screening has been applied to the analysis of the development of *Xenopus* by Dworkin and Dawid (1980) and to the slime mould *Dictyostelium* by Williams and Lloyd (1979). In both instances it was estimated that a fivefold difference in mRNA abundance could be detected in this procedure.

Subtractive cDNA cloning and differential screening

As we have seen, differential screening is applicable in many biological situations where it is desirable to isolate cDNAs derived from mRNAs that are induced by a particular treatment. In order to increase the efficiency of the procedure, it is beneficial to be able to create a cDNA library that is enriched in the desired sequences. This can be achieved by subtractive hybridization, the essence of which is to remove those cDNA sequences that are ubiquitous or not induced.

An example of this technique is the isolation of rat cDNAs that are induced in a particular region of the brain (the hippocampus dentate gyrus, DG) that is involved in learning and memory (Nedivi *et al.* 1993). The inducing stimulus was kainate, a glutamate analogue that induces seizures and memory-related synaptic changes. Poly(A)$^+$ RNA was extracted from the DG of kainate-treated animals and used for first-strand cDNA synthesis. Ubiquitous sequences present in the activated DG cDNA preparation were hybridized with an excess of poly(A)$^+$ RNA from total rat brain. This RNA had previously been biotinylated (using a photobiotinylation procedure) and so hybrids, and excess RNA, could be removed using a streptavidin extraction method (Sive & StJohn 1988). The unhybridized cDNA was then converted into double-stranded form by conventional methods, and used to construct the subtracted cDNA library in λZAP. This subtracted library was differentially screened using radiolabelled cDNA from activated and non-activated DG as the differential probes. A large number of activated DG clones were isolated, of which 52 were partially sequenced. One-third of these clones corresponded to known genes; the remainder were new.

Recombinational probe

This ingenious and powerful method is based upon homologous recombination in the *E. coli* host (Seed 1983). The probe sequence here is inserted into a specially-constructed plasmid vector, πVX. This is a very small plasmid of 902 base-pairs, derived from the Col E1 replicon, which contains a convenient polylinker sequence and a suppressor tRNA gene, *sup* F (Fig. 7.7). Genomic phage λ libraries are propagated on recombination-proficient *E. coli* containing the probe-πVX recombinant plasmid. Phage-carrying sequences homologous to the probe acquire an integrated copy of the plasmid by homologous recombination. Phage-bearing integrated probe-πVX can then be recovered and isolated by growth under appropriate selective conditions. This is most easily achieved by using a phage λ vector carrying an amber mutation suppressible by *sup* F (e.g. EMBL3 derivatives, see Chapter 5). By finally plating on a non-suppressing *E. coli*, only those phage that have integrated a *sup* F gene, by virtue of homology with the probe, can form plaques.

This method can be applied readily to very large numbers ($>5 \times 10^6$) of phage in a genomic library, and has the advantage of being very quick, providing that the probe-πVX recombinant plasmid has been constructed at a prior stage. The shortest probe segment giving high recombination has been found to be about 60 base-pairs long. A certain amount of sequence divergence can be tolerated in the homologous recombination event, but it has been shown that the probe does discriminate between a perfectly homologous sequence and one with 8.2% sequence divergence (Seed 1983).

Hybrid released translation (HRT) and hybrid arrested translation (HART)

These methods enable a cloned DNA to be correlated with the protein(s) which it encodes. Hybrid *released* translation (HRT) is a direct method in which cloned DNA is bound to a nitrocellulose filter and hybridized with an unfractionated preparation of mRNA or even total cellular RNA. The filter is then washed, and hybridized messenger is eluted by heating in low-salt buffer or in a buffer containing formamide. The recovered mRNA is then translated in a cell-free translation system such as that derived from wheat germ or rabbit reticulocytes, and radiolabelled polypeptide products whose synthesis it directs are analysed on an appropriate gel electrophoresis system (Fig. 7.8).

Hybrid *arrested* translation (HART) is now less widely used than HRT. It is based upon the fact that an mRNA will not direct the synthesis of protein in a cell-free translation system when it is in a hybrid form with its DNA complement (Paterson *et al.* 1977). In a model experiment, rabbit globin mRNA, which is a mixture of mRNA species encoding α- and β-globin polypeptides, was mixed with denatured DNA of the recombinant plasmid pβG1 which contains a rabbit β-globin cDNA sequence. Conditions were chosen so that DNA–RNA hybridization was favoured whilst

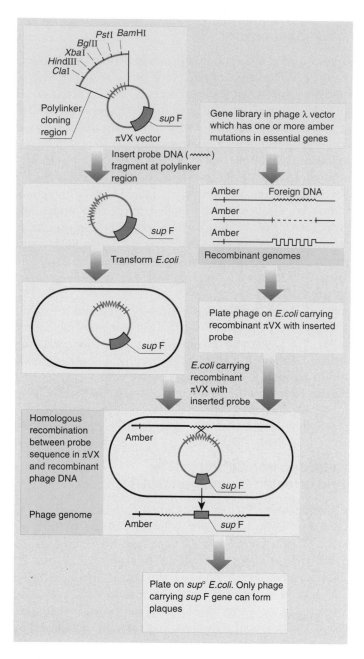

Fig. 7.7 Recombinational probing. (See text for details.)

suppressing re-association of the plasmid DNA (i.e. high formamide concentration). The nucleic acids were recovered from the hybridization mixture and added to a cell-free translation system prepared from wheat germ. ^{35}S-methionine was included so that synthesis of radioactive polypeptides could be analysed by polyacrylamide gel electrophoresis and

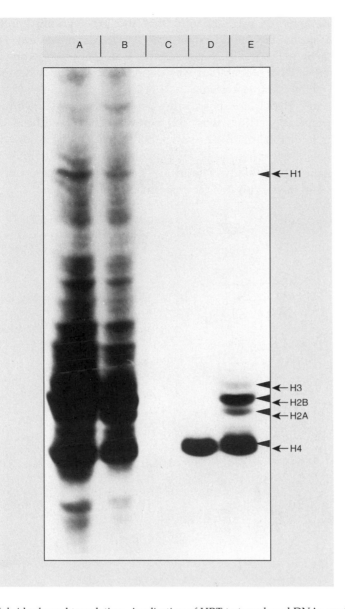

Fig. 7.8 Hybrid released translation. Application of HRT to two cloned DNAs containing histone sequences of *X. laevis*. The figure shows an autoradiograph of a polyacrylamide gel analysis of radiolabelled polypeptides synthesized in a wheat germ cell-free translation system with ^{3}H-lysine. The tracks correspond to the following RNA additions to the cell-free system: A and B, total RNA from *X. laevis* ovary; C, no added RNA; D, RNA released from a filter bearing histone H4 cDNA sequences cloned in pAT153; E, RNA released from a filter bearing H3, H2B, H2A and H4 sequences clustered in a genomic DNA fragment cloned in λWES. These filters had been hybridized with total ovary RNA. Positions of marker histones are indicated.

autoradiography. It was apparent that the pre-hybridized mRNA was rendered inactive by hybridization with its DNA complement. Full translational activity of the β-globin mRNA was recovered when the pre-hybridized mRNA preparation was dissociated by a brief heating before addition to the cell-free system.

HRT and HART are potentially powerful methods. Unfractionated mRNA preparations from most cell types direct the synthesis of several hundreds of distinct abundant polypeptides that can be resolved on two-dimensional electrophoresis systems. It should be quite feasible to isolate recombinant cDNAs derived from such a mixed population of mRNAs, and correlate them with their corresponding polypeptides.

8 Expression in *E. coli* of cloned DNA molecules

Introduction

Synthesis of a functional protein depends upon transcription of the appropriate gene, efficient translation of the mRNA and, in many cases, post-translational processing and compartmentalization of the nascent polypeptide. A failure to perform correctly any one of these processes can result in the failure of a given gene to be expressed. Transcription of a cloned insert requires the presence of a promoter recognized by the host RNA polymerase and, ideally, a transcription terminator at the 3' end of the gene. The key structural features of both are known and are discussed in more detail later (p. 152). Efficient translation requires that the mRNA bears a ribosome binding site (rbs). In the case of an *E. coli* mRNA a rbs includes the translational start codon (AUG or GUG) and a sequence that is complementary to bases on the 3' end of 16S ribosomal RNA. Shine and Dalgarno (1975) first postulated the requirement for this homology and various S–D sequences, as they are often called, have been found in almost all *E. coli* mRNAs examined.

Many proteins when made in their natural host undergo post-translational modification, e.g. glycosylation, amidation, phosphorylation or myristylation. Such modifications to proteins do not occur in *E. coli*. Although this may be a significant problem in particular cases, such modifications usually are not essential for biological activity. Of course, gene manipulation can be used to introduce such activities into *E. coli*. Thus myristylation of a protein kinase was obtained in *E. coli* by introducing a yeast gene encoding *N*-myristyl transferase (Duronio *et al.* 1990).

Another important aspect of gene expression is the stability of the

[139]

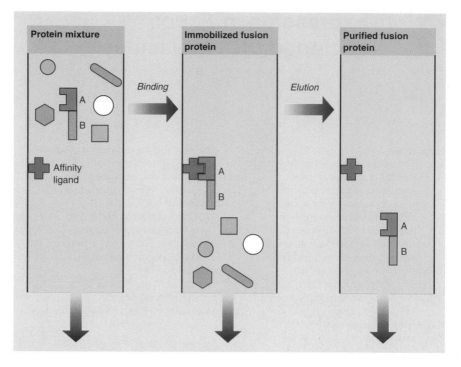

Fig. 8.1 Purification of a fusion protein by affinity chromatography. A crude extract from a culture expressing the fusion protein AB is passed down a column containing a ligand for the affinity handle A. The fusion protein AB binds to the ligand and all other cellular proteins are eluted in the unbound fraction. The fusion protein is then eluted in almost pure form by addition of an appropriate buffer.

protein, particularly if *E. coli* is not its usual host. It is known that the short polypeptides encoded by genes which have undergone nonsense mutations are rapidly degraded in *E. coli* while the wild-type proteins are stable. It can be envisaged that the foreign proteins would be rapidly degraded in the new host if their configuration or amino acid sequence did not protect them from intracellular proteases. The rate at which the foreign protein forms its natural three-dimensional structure also may be important. For a long time it was thought that the final tertiary structure of a protein forms spontaneously from an amino acid sequence. Recently, however, a set of proteins called *chaperones* have been identified that are required for the formation of the correct tertiary structure of many proteins. There is no guarantee that *E. coli* has the correct chaperones for all foreign proteins.

From the above discussion it is clear that the first requirement for expression in *E. coli* of a structural gene, inserted *in vitro* in a DNA molecule, is that the gene be placed under the control of a promoter. Ideally, that promoter should be one that functions effectively in *E. coli*. When this is done the protein product may be synthesized from its own *N*-terminus, but more often than not a fusion protein is produced. There can be advantages in synthesizing a fusion protein rather than the native molecule. For example, the *N*-terminal fusion may stabilize the protein

Table 8.1 Chemical and enzymatic agents that have been used to cleave fusion proteins site specifically

Cleavage method	Recognition sequence	Reference
Cyanogen bromide	↓ -Met-X-	Itakura *et al.* 1977
Formic acid	↓ -Asp-Pro-	Nilsson *et al.* 1985
Hydroxylamine	↓ -Asn-Gly-	Moks *et al.* 1987
Collagenase	↓ -Pro-Val-Gly-Pro-	Germino & Bastia 1984
Factor Xa	↓ -Ile-Glu-Gly-Arg-X	Smith & Johnson 1988
Thrombin	↓ -Gly-Pro-Arg-X	Smith & Johnson 1988
Trypsin	↓ -Arg-X	Varadarajan *et al.* 1985
Subtilisin	↓ -Gly-Ala-His-Arg-X	Carter & Wells 1987

The cleavage site is indicated by the arrow and X can be any amino acid residue.

from proteolytic degradation, as shown for somatostatin (p. 62). Second, *E. coli* protein at the *N*-terminus can be used to facilitate purification of the foreign protein activity (see Fig. 8.1) or as a reporter gene to monitor expression (see p. 143). It should be noted that in some cases it may be possible to cleave the fusion protein *in vitro* to yield native protein. One example already discussed is somatostatin (p. 65) and other examples of how this can be done are given in Table 8.1.

Positioning cloned inserts in the correct translational reading frame: expression of fusion proteins

In plasmid-based systems, expression of a gene can be achieved in two ways: by cloning a DNA cartridge containing the necessary regions for efficient transcription and translation upstream of the gene in question (Backman *et al.* 1976), or by cloning such a gene in a specially-designed expression vector. Although both approaches have been utilized, it is the second which is used more often. In the early days of *in vitro* gene manipulation, quite sophisticated experimental protocols were necessary to achieve gene fusions dictated by available restriction sites in the target gene and in the gene to be cloned. Today, it is much simpler because of the availability of DNA fragments bearing multiple cloning sites (see p. 67) and the availability of a wide range of expression vectors (Slauch & Silhavy 1991) which should meet most needs. The most widely used promoters are *lac* UV5, *trp* (tryptophan biosynthesis), *lpp* (lipoprotein synthesis) and λP_L and λP_R which are both strong and regulatable. The UV5 mutation renders the *lac* system insensitive to catabolite repression.

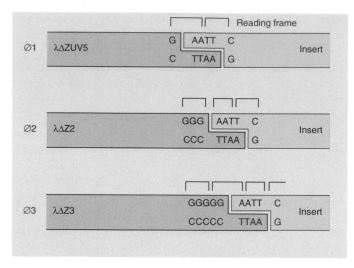

Fig. 8.2 The different reading frames with respect to the translation initiation site of the *lacZ* gene presented by three λ vectors.

There are two tests that can be used to show that expression of a cloned gene is under the control of the selected *E. coli* promoter. First, expression of the foreign gene should be obtained when it is placed in the correct orientation relative to the promoter but not when it is in the wrong orientation. Second, expression of the gene should be enhanced by those environmental conditions that normally lead to activation of the promoter, e.g. addition of an inducer such as IPTG with *lac*-based expression vectors or starvation for tryptophan with *trp*-based vectors.

Most of the available fusion vectors result in the expression of a protein where the amino-terminus is composed of the target protein and the carboxy-terminus is derived from the cloned gene. Some, however, are used for the creation of carboxy-terminal fusions with the foreign protein at the amino-terminus. These latter constructs are most often used for antibody production.

When expression of a foreign gene is dependent on its fusion to an *E. coli* gene, it is important that the correct translational reading frame is maintained. Again, the availability of multiple cloning sites now makes this relatively easy. So does the availability of specialist vectors. For example, two groups have developed sets of three vectors each having a single *Eco*RI restriction site in which the cloned gene can be placed in each of the three possible reading frames (Fig. 8.2) relative to the translation initiation site of the *lacZ* (Charnay *et al*. 1978) or *trp* genes (Tacon *et al.* 1980). Other vectors have been designed to facilitate the detection of inserts in the correct translational reading frame (Slauch & Silhavy 1991). For example, in plasmid pMR100, the *lac* promoter directs transcription of the NH$_2$-terminus of the *lacI* gene followed by a polycloning site and then an out-of-frame *lacZ* gene. Insertion of DNA into the polycloning site such that the frameshift is corrected allows production of a functional β-galactosidase (Weinstock *et al.* 1983). The *lacZ* gene thus acts as a *reporter* of biological activity. Other reporters which have been used include alkaline phosphatase (Brickman & Beckwith 1975), luciferase (Engegrecht *et al.* 1985), galactokinase (Russell & Bennett 1982) and chloramphenicol acetyl transferase.

An alternative strategy for obtaining fusion proteins: cloning in the *Pst*I site of pBR322

An alternative method of putting an insert under the control of an *E. coli* promoter without the concomitant problem of translational reading frame adjustment has been used by Villa-Komaroff *et al.* (1978). They cloned cDNA transcripts of rat preproinsulin mRNA at the unique *Pst*I site of pBR322 (see Fig. 4.8). This site lies in the Ap^R gene at a position corresponding to amino acid residues 183 and 184. Consequently, an insert at the *Pst*I site should result in the production of a fused gene product. To effect this insertion, use was made of the homopolymer tailing procedure (Fig. 8.3). The advantage of this method is that in each recombinant molecule the lengths of the repeating G:C base-paired joints may be different and at least some of them will be in the correct reading frame.

Manipulation of cloned genes to achieve expression of native proteins

So far we have described methods for placing a cloned DNA fragment under the control of an *E. coli* promoter. In many instances these methods result in the production of a hybrid protein carrying amino-terminal amino acids from β-lactamase, β-galactosidase, or the *trp* E gene product. It would be more satisfactory if the gene–promoter fusion always produced a native protein.

Tacon *et al.* (1983) have described the construction of two plasmid vectors which facilitate the expression of native proteins. Both make use of the *trp* promoter and associated S–D sequence. In one of them, pWT551, the gene to be cloned is inserted at a *Hind*III site but expression requires the presence of an initiator ATG or GTG codon at the start of the coding region. The second vector, pWT571, has an *Eco*RI site overlapping the initiation codon (ATG AATTC). Cloning a gene in the correct reading frame in this site will result in expression of a protein having the sequence fMet-Asn at the *N*-terminus. Alternatively, digestion with endonuclease *Eco*RI followed by S1 nuclease to remove the 5′-AATT extensions permits blunt-end ligation to the initiation codon of a fragment encoding a protein lacking *N*-formylmethionine.

As with vectors for the production of fusion proteins, a whole series of similar vectors with specialist properties has been developed. Representative examples have been described by Stark (1987) and Zaballos *et al.* (1987).

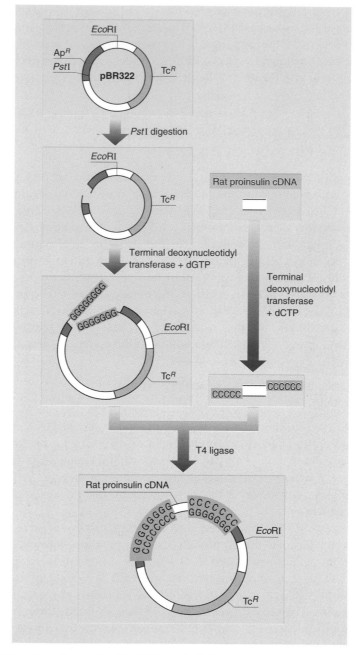

Fig. 8.3 Insertion of rat proinsulin cDNA at the *Pst*I site of pBR322.

Secretion of proteins

Gram-negative bacteria such as *E. coli* have a complex wall–membrane structure comprising an inner, cytoplasmic membrane separated from an outer membrane by a cell wall and periplasmic space. Secreted proteins may be released into the periplasm or integrated into or transported across the outer membrane. In *E. coli* it has been established that protein export through the inner membrane to the periplasm or to the outer membrane is achieved by a universal mechanism known as the general export pathway (GEP). This involves the *sec* gene products (for review see Pugsley 1993). Proteins which enter the GEP are synthesized in the cytoplasm with a signal sequence at the *N*-terminus. This sequence is cleaved by a signal or leader peptidase during transport. A signal sequence has three domains: a positively-charged amino-terminal region, a hydrophobic core consisting of 5–15 hydrophobic amino acids, and a leader peptidase cleavage site. A signal sequence attached to a normally cytoplasmic protein will direct it to the export pathway but the three-dimensional structure of the hybrid protein may not permit secretion to be completed. What exactly constitutes an export-competent state is not fully understood. Proteins can leave the export pathway by being degraded, by folding into their mature three-dimensional structure or by precipitating. Also, certain exported proteins require cytoplasmic chaperones, e.g. the *sec*B protein, for their export (Johnson *et al.* 1992).

The earliest experiment on engineering the secretion of foreign proteins by *E. coli* was undertaken by Villa-Komaroff *et al.* (1978). They inserted the rat proinsulin gene at the *Pst*I site of pBR322 (see p. 60 and Fig. 8.3) and, as expected, the β-lactamase–proinsulin fusion protein was found in the periplasmic space. Talmadge and Gilbert (1980) constructed a set of vectors in which the unique *Pst*I site was moved such that it lies within or near the pre-β-lactamase signal sequence. The genes for rat proinsulin and rat preproinsulin were cloned into the *Pst*I sites of these different plasmids resulting in the formation of hybrid β-lactamase (prokaryotic) and insulin (eukaryotic) signal sequences. By comparing the levels of insulin antigen in the *E. coli* periplasmic space, Talmadge *et al.* (1980) were able to deduce the structural requirements of a signal sequence and to show that a eukaryotic signal sequence can function in a prokaryote. Since then, many attempts have been made to secrete from *E. coli* either normal *E. coli* cytoplasmic proteins or foreign proteins. Success has been mixed, e.g. Itoh *et al.* (1981) were unable to get secretion of β-galactosidase. Secretion from *E. coli* is still a hit-or-miss proposition and is most likely to work with a naturally-secreted protein (Goeddel 1990).

A protein exported to the periplasm may be degraded by periplasmic proteases or precipitate if it is not folded properly and efficiently. Not surprisingly there are a number of periplasmic chaperones of which the *Dsb*A family of disulphide oxidases is the best studied (Bardwell *et al.* 1991). *Dsb*A facilitates disulphide-bond formation in the periplasm of *E. coli* and its activity is crucial for the stability of alkaline phosphatase. In a *dsb*A mutant, disulphide bonds form slowly and alkaline phosphatase is degraded.

There are three mechanisms whereby proteins cross both membranes (Fig. 8.4). By far the commonest mechanism is the general secretory pathway (GSP) which is a two-step process. The secreted proteins enter the periplasm via the GEP then cross the outer membrane by a separate

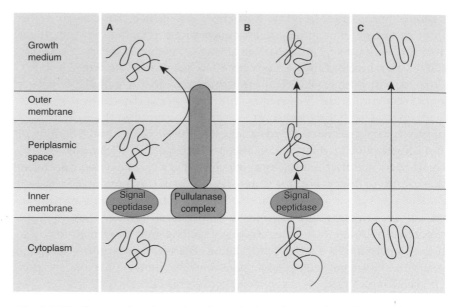

Fig. 8.4 The three modes of secretion of proteins into the growth medium. A, the general secretory pathway. B, secretion of IgA protease. C, the signal peptide-independent pathway. In both A and B, the protein to be secreted is synthesized as a preprotein with an *N*-terminal signal peptide which is cleaved during export to the periplasm.

pathway (Pugsley 1993). For any particular protein to cross the outer membrane, a large number of other protein-specific gene products are required. Thus 14 *pul* genes are required for pullulanase secretion and at least 4 *out* genes are known to be involved in pectinase secretion. Although the *out* genes have sequence homology with the *pul* genes they cannot complement *pul* mutations. Thus the two secretory systems are similar but specific. Proteins with homology to the *out* and *pul* genes have been detected in other bacteria (Wandersman 1992). When β-lactamase was fused to a large (>800 amino acids) *N*-terminal fragment of pullulanase, the β-lactamase activity was found attached to the surface of *E. coli* cells provided the pullulanase secretion genes were functional (Kormacher & Pugsley 1990). On prolonged incubation the β-lactamase was released into the medium.

A second mode of secretion is for proteins to cross the inner membrane and enter the outer membrane by the GEP and then be released into the medium without the help of other proteins. The best example of this is the IgA protease of *Neisseria gonorrhoeae*. When the *iga* gene is cloned and expressed in *E. coli* it directs the cytoplasmic synthesis of a large protease precursor and the secretion of the authentic mature protein. It is believed that sequences at the *C*-terminus of the protein are imported for insertion into and movement through the outer membrane. Klauser *et al.* (1992) have shown that fusion of the *Vibrio cholera* toxin B (CtxB) subunit and other proteins to the *C*-terminal domain of the IgA protease permits their transfer to the outer surface of *E. coli*. Transport of the CtxB domain across the outer membrane was maximized when reducing agents were present in the growth medium. This suggests that the formation of disulphide bonds and protein folding in the periplasmic space can prevent export.

A number of proteins are embedded in the outer membrane but have surface-exposed loops. In general, insertions of foreign sequences within these loops are well tolerated. However, only short sequences (up to 50 amino acids) can be displayed on the surface: larger sequences perturb localization (Charbit *et al.* 1988, Newton *et al.* 1989). An alternative surface expression system was developed by Francisco *et al.* (1992). This makes use of a lipoprotein *N*-terminal targeting sequence and the *C*-terminal fragment of the outer-membrane protein OmpA. When these were fused with a mature β-lactamase sequence, β-lactamase activity was found in the outer membrane. The same system has been used to target other proteins such as a single-chain Fv antibody and a cellulose binding protein (Georgiou *et al.* 1993).

The third method of secretion is via the so-called *signal peptide-independent pathway* which is widespread among Gram-negative bacteria. Proteins which use this pathway are synthesized without a signal sequence and export is mediated by ATP-dependent proteins known as ABC-transporters (for review see Kuchler 1993). In *E. coli* the α-haemolysin and colicin V are known to use this pathway as do a wide variety of proteins in other bacteria. Kenny *et al.* (1991) were able to obtain secretion of prochymosin, chloramphenicol acetyl transferase and, interestingly, β-galactosidase by fusing them to the *C*-terminal domain of haemolysin.

Stability of foreign proteins in *E. coli*

There are a number of examples of the instability of cloned gene products. Details of the degradation of somatostatin were given earlier (see p. 62). Human fibroblast pre-interferon and interferon also are known to be unstable (Taniguchi *et al.* 1980). Talmadge and Gilbert (1982) have shown that cellular location can affect stability. They measured the half-life of rat proinsulin in *E. coli* and found it to be 2 minutes when located in the cytoplasm but over tenfold greater in the periplasm.

Various strategies have been developed to cope with the instability of foreign proteins in *E. coli*. In the case of somatostatin, degradation was prevented by producing a fused protein consisting of somatostatin and β-galactosidase and subsequently cleaving off the somatostatin with cyanogen bromide. Although there are a number of examples in the literature of the stabilization of proteins in this way, the disadvantage is that the subsequent production of a native protein *in vitro* is not always possible (see p. 65). An alternative approach is to use certain mutants of *E. coli* which have a reduced complement of intracellular proteases. One particularly useful mutant lacks the *lon* protease, a DNA-binding protein with ATP-dependent proteolytic activity. Two disadvantages of *lon* mutants are that they are very mucoid and can form long, non-septate filaments which are

Box 8.1 Heat shock proteins and chaperones

In almost all kinds of cell subjected to a sudden rise in temperature (heat shock) certain proteins begin to be synthesized much faster than usual. In the case of *E. coli* 17 new proteins are made and some of these heat shock proteins (Hsps) are closely related to Hsps from eukaryotes. For example, the 66 kDa *E. coli* heat shock protein encoded by the *DnaK* gene has over 50% homology at the nucleotide level with the corresponding *Hsp70* gene from *Drosophila*. Although we know that some heat shock proteins of *E. coli* are required for phage λ-growth, e.g. GroEL, GroES and DnaK, their normal function appears to be as molecular chaperones.

Protein folding and the assembly of multimeric structures do not occur spontaneously but are facilitated by chaperones. These chaperones bind transiently and non-covalently to nascent polypeptides and unfolded or non-assembled proteins aiding in protein biogenesis in two ways. First, they block non-productive protein–protein interactions and, second, they mediate the folding of proteins to their native state by sequestering folding intermediates allowing the concerted folding by domains and assembly of oligomers.

The two major families of chaperones are the 70 kDa family of heat shock proteins (Hsp70s) and the 60 kDa family (chaperonins or Cpn60s). The Hsp70s are found in all bacteria and all cellular compartments of eukaryotes and bind partially folded proteins. They appear to bind nascent chains in the process of protein synthesis and completed polypeptides upon release from ribosomes. In eukaryotes, Hsp70s may facilitate translocation of proteins from the cytosol into organelles. The Cpn60s have been shown to bind unfolded proteins and they prevent their aggregation while correct folding takes place. In addition to the Hsp70s and Cpn60s, other chaperones have been identified, e.g. disulphide isomerases, proline isomerases, etc.

fragile and can be non-viable. Mucoidy can be eliminated by inactivating the *galE* gene which is essential for capsule biosynthesis. Filamentation can be minimized by keeping yeast extract out of the growth medium or by introducing the *sul*A mutation (Gottesman 1990).

Cells defective in the sigma factor which is encoded by the *htpR* gene (σ32) and which is necessary for the heat shock response have been used to increase the accumulation of cloned proteins (Buell *et al.* 1985). The *clp* gene encodes a second ATP-dependent protease and mutations in this gene may further enhance the stability of foreign proteins seen in *lon* mutants (Gottesman 1990). Mutations in other proteases have not proved useful for the stabilization of cytoplasmic proteins.

A number of proteases may be important in the turnover of secreted proteins. For example, *deg*P mutations (Strauch & Beckwith 1988) inactivate a periplasmic protease which can cleave alkaline phosphatase-fusion proteins. The degP protease may be able to degrade cytoplasmic proteins released during cell breakage. This certainly can happen with the ompT protease (Sedgwick 1989) which is located in the outer membrane.

The extent of proteolysis is greatly affected by the culture conditions used with the recombinant (Enfors 1992). Stress factors known to increase the rate of proteolysis include nutrient starvation, particularly nitrogen starvation, and growth conditions which favour the heat shock response. It is worth noting that higher and reproducible yields are usually possible in bioreactors where the dissolved oxygen content and pH value of the medium are tightly controlled. Batch-feeding of the carbon substrate also can be practised to ensure a high nitrogen to carbon ratio.

For any given protein a variety of factors singly or in combination may modulate half-life *in vivo*. Among such factors are the flexibility, accessibility and sequence of the *N*- and *C*-termini, the presence of chemically blocking amino-terminal groups such as the acetyl group, and the exposure on the surface of the folded protein of protease cleavage sites. Bachmair *et al.* (1986) have shown that in yeast the half-life of proteins derived from β-galactosidase varies from 3 minutes to 20 hours, depending on the amino acid at the *N*-terminus. It is not known if a similar phenomenon occurs in *E. coli*, but if it does it could explain the great variation in stability of different heterologous proteins made from similar gene constructs.

Detecting expression of cloned genes

When it is known that a cloned DNA fragment codes for a particular protein or RNA species, then suitable tests can be devised to determine if that gene is expressed in its new host. Thus somatostatin or insulin can be detected by sensitive radioimmunoassays, 5S RNA by hybridization to suitable probes, and enzymes either by appropriate assay procedures or by complementation of auxotrophic mutations in a suitable host. But how do we detect expression of cloned genes when the function of these genes is not known? The answer is to look for the synthesis of novel proteins or RNA species in cells carrying the recombinant molecules. However, with wild-type cells the detection of new proteins or RNA species is almost impossible because of the concurrent expression of the host genome. Three approaches have been devised to circumvent this problem. Two of them are *in vivo* methods: expression of plasmid-borne genes can be detected by the use of *mini-cells* or by UV irradiation of the host cell. The latter method is known as the *maxi-cell* method and is also applicable to genes cloned in λ vectors. The third method is a coupled *in vitro* transcription and translation system.

Mini-cells are small, spherical, anucleate cells produced continuously during the growth of certain mutant strains of bacteria. Because of their size difference, mini-cells and normal-sized cells can be separated easily on sucrose gradients. Mini-cells purified in this way from plasmid-free parents can be shown to contain normal amounts of protein and RNA but to lack DNA. *In vivo*, these mini-cells do not incorporate radioactive precursors into RNA or protein. By contrast, mini-cells produced from plasmid-carrying parents contain significant amounts of plasmid DNA. Plasmid-containing mini-cells are capable of RNA and protein synthesis and would appear to be ideal for detecting the expression of genes carried by recombinant plasmids.

In general, there is a correlation between the genotype of a plasmid and the polypeptides synthesized by mini-cells containing that plasmid. However, there are complications, since the introduction of deletions or DNA insertions into plasmids can have unpredictable results. For example, a 1 MDa deletion in a Col E1-derived plasmid prevented the synthesis in mini-cells of polypeptides of 56 000, 42 000, 30 000 and 28 000 daltons. The explanation of this result was apparent after the demonstration that the latter three polypeptides were degradation products of Col E1 (Meagher *et al.* 1977). The insertion into a kanamycin-resistance gene of a DNA fragment containing the *Eco*RI methylase gene caused the synthesis in

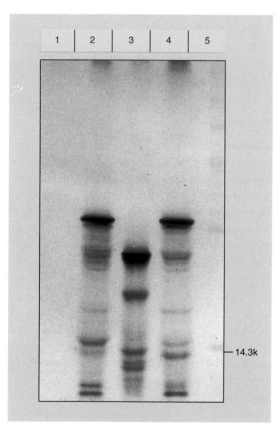

Fig. 8.5 Autoradioagraph produced after transcription and translation of plasmid DNA *in vitro*. Plasmid DNA was transcribed and translated *in vitro* in the presence of ^{35}S-methionine, the protein products separated by SDS-polyacrylamide electrophoresis and the individual proteins detected by autoradiography. Track 1, DNA-free control; track 2, plasmid pAT153; track 3, plasmid pACYC184; track 4, plasmid pWT111; track 5, molecular-weight marker protein. Tracks 2–4 were loaded with equivalent amounts of radioactivity (500 000 c.p.m.) and the gel was exposed to the photographic film for 3.5 h. The major band in tracks 2 and 4 is β-lactamase and in track 3 chloramphenicol acetyltransferase. Note that many truncated polypeptides are produced but that these are minor constituents.

mini-cells of *Eco*RI methylase and two other polypeptides. Again, this result could be explained since it was known that the kanamycin-resistance protein was inactivated by the insertion and that there was a portion of another gene on the inserted DNA fragment. These facts permit the prediction that two new polypeptides would be produced. When the DNA insertion is of unknown function, e.g. randomly-cleaved eukaryotic DNA sequences, interpretation of the expression of the DNA fragments in mini-cells could still be complicated.

The maxi-cell method (Sancar *et al.* 1979) is based on two observations. First, when irradiated with UV light, *E. coli rec*A *uvr*A cells stop DNA synthesis and chromosomal DNA is extensively degraded so that only a small amount remains several hours later. Second, if these cells contain a ColE1-like multicopy plasmid (e.g. pBR322), the plasmid molecules that do not receive a UV hit continue to replicate with plasmid DNA levels increasing about tenfold by 6 hours post-irradiation. If a radioactive amino acid is added a few hours after irradiation, the vast majority of the proteins which are labelled are plasmid-encoded.

The expression of genes cloned in λ vectors can be studied by infection of bacteria previously irradiated so severely with UV light that their own gene expression is effectively eliminated by DNA damage. Under these conditions the products of the cloned genes are seen against a background of λ-specified proteins. If the expression of the DNA insert is independent of λ promoters then even this background can be eliminated by infecting UV-irradiated cells lysogenized with a non-inducible mutant of λ (Newman *et al.* 1979). Under these conditions, sufficient λ repressor is present in the cells to prevent transcription of phage genes.

Bacterial cell-free systems for coupled transcription and translation of DNA templates were first described by Zubay and colleagues as early as 1967. These have been modified (Pratt *et al.* 1981) to permit expression *in vitro* of genes contained on bacterial plasmids or bacteriophage genomes. Since DNA fragments generated by restriction endonuclease cleavage can be used to programme polypeptide synthesis, it is possible to readily assign polypeptides to small regions of the coding template. There are two other advantages to the system. First, incorporation of radioactive label into protein is far more efficient than is possible using *in vivo* methods. This makes the system very sensitive and ^{35}S-labelled polypeptides can be identified by polyacrylamide gel electrophoresis after a few hours autoradiography (Fig. 8.5). Second, DNA from other prokaryotes can be expressed efficiently. This is particularly useful when cloning in organisms other than *E. coli* in which the *mini-cell* and *maxi-cell* techniques have not been developed. The Zubay method is not ideal for use with linear templates, e.g. those produced by PCR synthesis, unless the extract is prepared from nuclease-deficient cells (Lesley *et al.* 1991). Better still, the source of extract also should lack proteinases such as the *ompT* and *lon* gene products (see p. 147).

Table 8.2 Factors affecting the expression of cloned genes

Promoter strength
Translational initiation sequences
Codon choice
Secondary structure of mRNA
Transcriptional termination
Plasmid copy number
Plasmid stability
Host cell physiology

Maximizing the expression of cloned genes

In the previous section we discussed those features of gene expression in *E. coli* which need to be considered if detectable synthesis of a cloned gene product is to be obtained. However, much of the interest surrounding *in vitro* gene manipulation concerns the commercial applications, e.g. production of vaccines or human hormones in *E. coli*. Here detectable expression is not sufficient; rather it must be maximized. The key factors affecting the level of expression of a cloned gene are shown in Table 8.2 and, as will be seen in the following section, most of them exert their effect at the level of translation.

Constructing the optimal promoter

A large number of promoters for *E. coli* have been analysed (Hawley & McLure 1983, Harley & Reynolds 1987) and the most recent compilation gives the sequence of 263 of them. Clearly we would like to use this information to develop the most efficient promoter possible. Comparison of many promoters has led to the formulation of a consensus sequence which consists of the −35 region (5′-TTGACA-) and the −10 region or Pribnow box (5′-TATAAT), the transcription start point being assigned position +1. Of the four promoters used most widely in expression vectors none shows absolute identity with the consensus sequence (Fig. 8.6). In trying to identify the strongest promoter it is essential to have a measure of the relative efficiencies of the different candidates and a suitable system has been devised (Russell & Bennett 1982, de Boer *et al.* 1983a). The promoter to be tested is placed in front of a promoter-less galactokinase (*galK*) gene carried on a plasmid and the level of galactokinase synthesized in a GalK⁻ host used as a measure of promoter strength. Using the galactokinase system it was shown that promoter strength is directly proportional to the degree of similarity with the consensus sequence. Thus it is not surprising that compared with the *lac* promoter the two synthetic hybrid promoters *tac*I and *tac*II (Fig. 8.6) were 11 and 8 times stronger respectively. While the −35 and −10 regions show the greatest conservation across promoters and also are the sites of nearly all mutations which affect transcriptional strength, other bases flanking the −35 and −10 regions occur at greater than random frequency and can affect promoter activity (Hawley & McLure 1983, Dueschle *et al.* 1986, Keilty & Rosenberg 1987).

	-35 Region		-10 Region
	1 2 3 4 5 6 7 8 9 10 11 12 13 14 15 16 17		
CONSENSUS	· · · T T G A C A · · · · · · · · ·	· · · · · · · · T A T A A T · ·	
lac	G G C T T T A C A C T T T A T G C T T	C C G G C T C G T A T A T T G T	
trp	C T G T T G A C A A T T A A T C A T	C G A A C T A G T T A A C T A G	
λP_L	G T G T T G A C A T A A A T A C C A	C T G G C G G T G A T A C T G A	
rec A	C A C T T G A T A C T G T A T G A A	G C A T A C A G T A T A A T T G	
tacI	C T G T T G A C A A T T A A T C A T	C G G C T C G T A T A A T G T	
tacII	C T G T T G A C A A T T A A T C A T	C G A A C T A G T T T A A T G T	

Fig. 8.6 The base sequence of the −10 and −35 regions of four natural promoters and two hybrid promoters.

Furthermore, the distance between the −35 and −10 regions is important. Hawley and McClure analysed a number of mutations which affect this spacing. In all cases the promoter was stronger if the spacing was moved closer to 17 base-pairs and weaker if moved further away from 17 base-pairs.

Because of the increased strength of the *tac* promoters relative to the *lac* and *trp* promoters, Amann *et al.* (1983) have cloned a *tac* promoter on a series of plasmid vectors that facilitates the expression of cloned genes. These vectors contain various cloning sites followed by transcription signals. In addition, Amann *et al.* (1983) have described plasmids that facilitate the conversion of the *lac* promoter to the *tac* promoter. Maximal transcription of a cloned gene may require more than just −35 and −10 regions that are close to the consensus sequence. With some promoters, DNA sequences 10–100 base-pairs upstream from the −35 site can act as 'upstream activators'. The deletion and/or insertion of DNA sequences can distort these activators and decrease transcription (Lamond & Travers 1983, Bossi & Smith 1984, Gourse *et al.* 1986). Upstream activation is associated with sequences rich in A and T, which result in DNA curvature (Plaskon & Wartell 1987).

All other things being equal, expression from a strong promoter can result in 20–40% of total cell protein being the cloned gene product. During the exponential phase of growth this level of expression puts such an energetic strain on the cell that it leads to the rapid selection of mutants which produce lower amounts of protein or no protein at all. Thus it is important that only controllable promoters are used. For example, expression from the *trp* promoter can be induced by tryptophan starvation or by growth in the presence of indoleacrylic acid. Similarly, expression of the *lac* and *tac* promoters can be induced with IPTG. One drawback of this latter system is the leakiness of the transcriptional regulation, particularly where there are multiple copies of the (plasmid-borne) promoter and a single (chromosomal) copy of the *lacI* gene. To overcome this problem vectors have been constructed which carry a copy of the *lacI* gene as well as the *lac* or *tac* promoters. With them there is improved regulation because sufficient repressor is produced to bind to all the promoters (Stark 1987, Amann *et al.* 1988).

A series of vectors has been described (Rosenberg *et al.* 1987, Schoepfer 1993) in which gene-specific transcription is mediated by the phage T7 polymerase. Gene 1 of the phage T7 encodes an RNA polymerase that is responsible for the expression of most of the T7 genome. Unlike the *E. coli* RNA polymerase, the one from T7 is a single polypeptide that recognizes a unique 23-nucleotide DNA sequence within T7 promoters. Thus the expression vectors comprise the foreign gene insert under the control of a T7 promoter and the T7 polymerase gene under the control of the *lac* promoter. Control is further enhanced by inserting the *lac* operator sequence into the T7 promoter (Studier *et al.* 1990). Using this system a 10^5-fold increase in expression is seen following IPTG induction.

Optimizing translation initiation

As well as giving clues to the essential structural features of promoters, DNA sequence analysis has yielded considerable information about those factors which affect translational initiation. Details of these sequences can be found in the compilation of Stormo *et al.* (1982). Central to all of them is the Shine–Dalgarno (S–D) sequence or ribosome binding site (rbs). The degree of complementarity of this sequence with the 16S rRNA can affect the rate of translation (de Boer & Hui 1990). Perfect complementarity occurs with the sequence 5'-GGAGG-3' and a single base change within this region can reduce translation tenfold.

Using a gene expression system which includes a portable S–D region, de Boer *et al.* (1983b) have examined the effect on translation of varying the sequence of the four bases that follow the S–D region. The presence of four A residues or four T residues in this position gave the highest translational efficiency. Translational efficiency was 50% or 25% of maximum when the region contained, respectively, four C residues or four G residues.

The composition of the triplet immediately preceding the AUG start codon also affects the effciency of translation. For translation of β-galactosidase mRNA the most favourable combinations of bases in this triplet are UAU and CUU. If UUC, UCA or AGG replaced UAU or CUU the level of expression was 20-fold less (Hui *et al.* 1984).

The codon composition following the AUG start codon also can affect the rate of translation. For example, a change in the third position of the fourth codon of a human γ-interferon gene resulted in a 30-fold change in the level of expression (de Boer & Hui 1990). Also, there is a strong bias in the second codon of many natural mRNAs which is quite different from the general bias in codon usage. Highly expressed genes have AAA (Lys) or GCU (Ala) as the second codon. Devlin *et al.* (1988) changed all the G and C nucleotides for the first four codons of a granulocyte colony-stimulating factor gene and expression varied from undetectable to 17% of total cell protein.

Based on this information the reader can be forgiven for thinking that it would be easy to construct a consensus ribosome binding site using oligonucleotide synthesis. The first problem is that for maximal expression of a cloned gene, the ribosome binding site for that gene must be close to the promoter. The second problem is highlighted by experiments where the distance between the gene and the promoter was varied.

Gene–promoter separation: the effect of mRNA secondary structure

Cleavage of λ*plac* 5.1 DNA with a combination of *Eco*RI and *Alu*I restriction endonucleases produces a mixture of fragments one of which is 95 base-pairs long and contains the *lac* promoter and β-galactosidase gene S–D sequence. Roberts *et al.* (1979b) constructed a series of recombinant plasmids in which this promoter-bearing fragment was located at varying distances in front of the λ*cro* gene. Nine different recombinant plasmids were selected, transformed into *E. coli*, and the level of *cro* protein in each clone measured. The DNA sequence across each of the *lac–cro* fusions was also determined. The results obtained are summarized in Fig. 8.7. The most striking feature of these results was the enormous difference (>2000-fold) in the levels of *cro* protein produced by the different recombinants. Since the same promoter is being used in each plasmid, it is uniform in each case. Iserentant and Fiers (1980) have constructed secondary structure models for the RNA transcripts produced by the plasmids of Roberts *et al.* (1979b). From an analysis of these structures they concluded that for good expression the initiation and, to a lesser extent, the S–D site must be accessible (Fig. 8.8). Thus the initiation of translation involves interaction between an activated 30S ribosomal subunit and the 5′ terminal region of a mRNA *which is already folded* in a specific secondary structure. Confirmation that the sequence between the promoter and the ATG codon affects the level of gene expression has been provided by Chang *et al.* (1980), Shepard *et al.* (1982) and Gheysen *et al.* (1982). These latter workers were able to confirm the conclusions of Iserentant and Fiers (1980) that the varying levels of expression are a reflection of secondary structure of the mRNA.

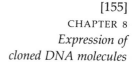

Fig. 8.7 Effect on *cro* protein production of gene promoter separation. Shown is a portion of the sequence of pTR161 extending from the *lac* promoter to the startpoint of translation of the *cro* gene. Also shown is the extent of the deletion in eight derivatives of pTR161. The figures on the brackets indicate the amount of *cro* protein synthesized relative to pTR161.

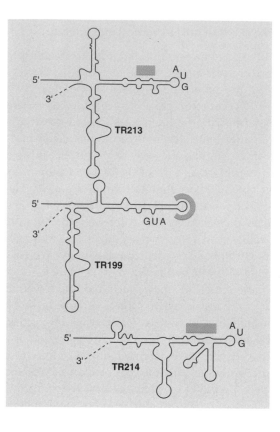

Fig. 8.8 Postulated secondary structure on the *cro* mRNA synthesized by three of the plasmids shown in Fig. 8.7. Note that for clone TR199, in which *cro* is poorly expressed, the initiation codon is base-paired in a stem structure. The solid pink bars indicate the location of the S–D sequence.

From an analysis of 123 naturally-occurring prokaryotic mRNAs, Ganoza *et al.* (1987) have concluded that sequences 5′ to the initiation codon have little self-pairing and do not pair extensively with the proximal coding region.

From the foregoing it might be thought that mRNA secondary structure alone accounts for the observation of Roberts *et al.* (1979b) that the distance between the gene and its promoter affects the efficiency of translation. However, length *per se* also may be important. Bingham and Busby (1987) measured the effect on translation of varying the distance between the 5′ end of the message and the S–D sequence. When this distance fell below 15 bases there was a marked decrease in translational efficiency.

Based on the above work a strategy for maximizing translation has been devised by Gold and Stormo (1990). A vector is selected with a polylinker downstream from a strong, controllable promoter. The gene of interest is cloned into the polylinker using the closest restriction site 3′ to the AUG and another restriction site 3′ to the chain-terminating codon. That construct will have no rbs or initiation codon and will also be missing a small number of codons. Synthetic DNA with the sequence shown in Fig. 8.9 then is used to expand this construct to provide the appropriate rbs and flanking nucleotides.

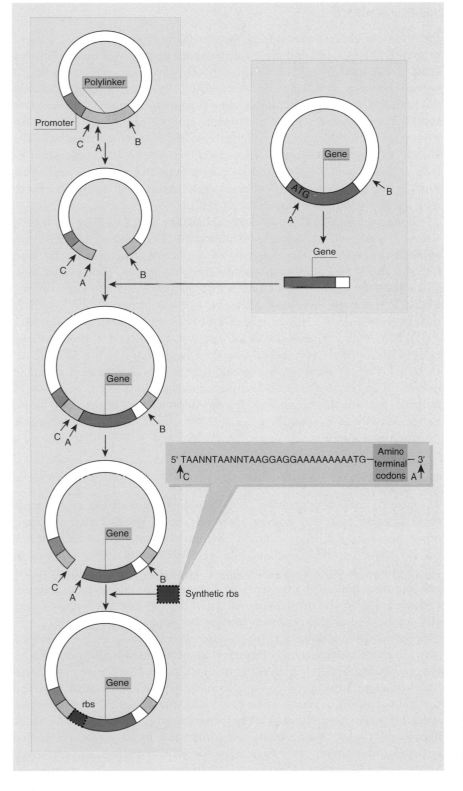

5' TAANNTAANNTAAGGAGGAAAAAAAAATG — Amino terminal codons — 3'

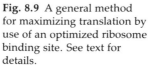

Fig. 8.9 A general method
for maximizing translation by
use of an optimized ribosome
binding site. See text for
details.

Stability of mRNA

Provided the various elements of the translational machinery are present in excess, the rate of synthesis of a particular protein will depend on the steady-state level of mRNA in the cell. This level is a reflection of the balance between synthesis and degradation. By using strong promoters the rate of synthesis is maximized, but a similar effect could be achieved by reducing the degradation of messenger. Degradation of mRNA usually proceeds by a combination of endonuclease and 3'-exonuclease attack. The stability of a given message *in vivo* reflects its susceptibility to digestion by these enzymes which, in turn, depends upon mRNA sequence and structure, as well as the association of macromolecules (e.g. ribosomes) with the message (for review see Higgins *et al.* 1992).

Exonucleolytic attack of mRNA is mediated by two 3' → 5' exonucleases, RNase II and polynucleotide phosphorylase. Degradation by these enzymes is impeded by the presence of secondary structures of the kind found at the 3' end of most messenger RNA molecules. Thus exonuclease attack must be preceded by endonuclease cleavage and the key enzyme, RNase E, has been identified. Mutants with a defective RNase E activity turn over mRNA at a slower rate (Mudd *et al.* 1990, Babitzke & Kushner 1991). Bouvet and Belasco (1992) have shown that RNase E preferentially cleaves RNAs that have several unpaired nucleotides at the 5' end. Although RNase E plays an important role in the degradation of many mRNAs it is certainly not the only endonuclease involved: in RNase E mutants, mRNA still turns over. Other endonucleases which may be involved are RNase K (Lundberg *et al.* 1990) and RNase M and RNase R (Strivastava *et al.* 1992).

The effect of codon choice

The genetic code is degenerate, and hence for most of the amino acids there is more than one codon. However, in all genes, whatever their origin, the selection of synonymous codons is distinctly non-random (for reviews see Ernst 1988, Kurland 1987 and McPherson 1988). The bias in codon usage has two components: correlation with tRNA availability in the cell, and non-random choices between pyrimidine-ending codons. Ikemura (1981a) measured the relative abundance of the 26 known tRNAs of *E. coli* and found a strong positive correlation between tRNA abundance and codon choice. Later, Ikemura (1981b) noted that the most highly-expressed genes in *E. coli* contain mostly those codons corresponding to major tRNAs but few codons of minor tRNAs. By contrast, genes that are expressed less well use more suboptimal codons. Kurland (1987) has detailed how the preferential use of a codon subset matching the major tRNA isoacceptor species is a sensible cellular strategy for optimizing translational kinetics at high growth rates. This explains why there is a major codon preference strategy and requires no explanation for which codons are preferred. McPherson (1988) believes that the codons are utilized according to mistranslational constraints. This is done in a fashion that favours replacement of one amino acid with a functionally conservative substitute in the event of codon – anticodon mispairing. It should be noted that the bias in codon usage even extends to the stop codons (Sharp & Bulmer 1988). UAA is favoured in genes expressed at high levels whereas UAG and UGA are used more frequently in genes expressed at a lower level.

Where maximal expression of a chemically synthesized gene is desired it would be an easy matter to optimize codon choice in accordance with the observations of Ikemura (1981a,b). However, the correlation of cellular tRNA level with codon frequency may be more complex. A shortage of a charged tRNA during protein synthesis can lead to misincorporation of amino acids and frame-shifting. Thus the arrangement of the successive codons in the coding region of a natural mRNA as well as its corresponding overlapping triplets may have evolved in order to minimize such frameshifts *in vivo*. Not surprisingly, a study by Grosjean and Fiers (1982) of the codon usage of several highly expressed mRNAs showed that most codons that are absent in the normal reading frame frequently are present in the two non-reading frames.

Analysis of the genetic code shows that triplets X_1X_2C and X_1X_2U always code for the same amino acid. Generally speaking, these two triplets are decoded by the same tRNA. Gouy and Gautier (1982) and Grosjean and Fiers (1982) have pointed out that there can be bias in the choice of pyrimidine at the third position of the codon and that the bias is correlated with gene expressivity. Thus, in highly expressed genes, if the first two bases of the codon are both A or U, then C is the favoured third base, whereas if the first two bases are G or C, the preferred third base is U. The explanation for this preference for a particular pyrimidine is equalization of the codon–anticodon interaction energy. Selection of codons which have a very strong or very weak energy of interaction with their corresponding anticodon leads to a decrease in the efficiency of translation.

Given the constraints on codon usage and distribution noted above it might be anticipated that the rate of translation, and hence the accumulation of product, might be limited when attempts are made to get high level expression from foreign or synthetic genes. Robinson *et al.* (1984) found that replacing rare codons with more common ones could improve expression. By contrast, Hoekma *et al.* (1987) observed only a threefold decrease in the amount of protein made from each mRNA molecule when they substituted rare codons for up to 39% of the optimal codons in a highly expressed gene.

The accuracy of translation, as opposed to its rate, also may be affected by codon choice. Significant substitution of lysine for arginine occurred at the rare AGA codon during high-level expression of heterologous genes in *E. coli* (McPherson 1988). Rosenberg *et al.* (1993) have shown that ribosomes stalled at rare codons can frameshift, hop or terminate, affecting both the level of expression and quality of heterologous protein.

Gene synthesis revisited

Improvements in the technology for the chemical synthesis of oligonucleo-tides means that today it is a very simple and quick matter to construct a gene with any desired sequence. However, from the foregoing discussion it should be apparent that generating the optimal sequence is much more difficult. With promoters, it is not just the -35 and -10 regions that are important, so too are the base sequences surrounding these sites. If the promoter is a controllable one then the base sequence can influence the fine-control mechanism known as attenuation (Yanofsky & Kolter 1982). The base sequence in the 5′ untranslated region can affect the rate of translation by its effect on mRNA stability and ribosome binding. Finally, the rate and fidelity of translation can be influenced by both codon choice and codon context. Clearly, gene synthesis may be easy, optimization is much more difficult, particularly since we do not know all the rules!

The effect of plasmid copy number

A major rate-limiting step in protein synthesis is the binding of ribosomes to mRNA molecules. Since the number of ribosomes in a cell far exceeds any one class of messenger, one way of increasing the expression of a cloned gene is to increase the number of the corresponding transcripts. Two factors affect the rate of transcription: first, the strength of the pro-moter as described earlier, and second, the number of gene copies. The easiest way of increasing the gene dosage is to clone the gene of interest on a high-copy-number plasmid.

Most of the high-copy-number vectors in common use, e.g. pBR322, are derivatives of ColE1. Two negatively-acting components are known to be involved in the control of ColE1 replication (Fig. 8.10). One is a 108 nucleotide untranslated RNA molecule called RNA I (Tomizawa & Itoh 1981), and the other is a protein repressor called ROP (Cesarini *et al.* 1982). RNA II is a plasmid-encoded RNA molecule which is processed by RNase H to give a 555 nucleotide primer for the initiation of ColE1 replication (Tomizawa & Itoh 1982). RNA II transcripts can adopt two alternative

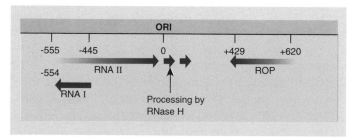

Fig. 8.10 The replication control region of plasmid ColE1. *ROP* controls the transcription of RNA II and RNA I inhibits the processing of RNA II by RNase H. The figures represent the base-pair coordinates measured from the origin of replication.

conformations and the choice between these two is made by RNA I. When RNA I is absent the RNA II has a conformation which permits binding to the origin of replication and priming of DNA synthesis. In the presence of RNA I, the RNA II takes a different shape which is incompatible with initiation of DNA synthesis (for review, see Cesareni *et al.* 1991). ROP is a 63 amino acid polypeptide which enhances the inhibitory activity of RNA I. Deletion of the *ROP* gene, as in pAT153 (Twigg & Sheratt 1980), or mutations in RNA I (Muesing *et al.* 1981) result in increased copy numbers. However, it should be noted that the host cell genetic background can also affect copy number (Nugent *et al.* 1983, Seelke *et al.* 1987).

Plasmids have been constructed in which copy number is controlled by a regulatable promoter (Yarranton *et al.* 1984). These plasmids consist of two origins of replication, one of which is active at 30°C and associated with a partition sequence (see p. 162) to ensure maintenance at low copy number. The other origin has the promoter for RNA II synthesis replaced by the λ P_L promoter. In strains carrying the thermolabile λ repressor gene mutation *cl857*, the second origin of replication is functional only when the repressor is inactivated by temperature induction. At temperatures above 38°C, uncontrolled plasmid DNA replication occurs, resulting in the accumulation of a high plasmid copy number per cell.

Mutant strains may also arise which result in a drop in copy number of plasmids that they harbour. These 'low-cop' mutations are chromosomal in origin and may be counter-selected using increased antibiotic concentrations. The use of high antibiotic concentrations, whilst ensuring a population of 'high-cop' plasmids, will not ensure that a high-cop plasmid which no longer produces the recombinant protein does not arise.

Plasmid low-cop mutants may arise which have a slower replication rate than normal. If such a mutation arose *in vivo* in one plasmid molecule out of a total cellular population of at least 50 (in the case of pBR322) then it is very unlikely that this mutant would ever take over in the population, since the normal plasmids would replicate to compensate for the mutant and maintain the normal cellular copy number of pBR322. Under some circumstances, however, low-cop plasmid mutants may be constructed *in vitro*.

Transcription termination

The presence of transcription terminators at the ends of cloned genes is important for a number of reasons. First, the synthesis of unnecessarily long transcripts will increase the energy drain on a cell which is expected to produce large amounts of non-essential protein. Second, undesirable secondary structures may form in the transcript, which could reduce the efficiency of translation. Finally, promoter occlusion may occur, i.e. transcription from the promoter of the cloned gene may interfere with transcription of another essential or regulatory gene. Thus Stueber and Bujard (1982) found that certain strong promoters led to a decrease in plasmid copy number because of read-through into the *ROP* gene and interference with plasmid replication. Copy number control, and hence gene expression, was restored to normal by the inclusion of transcriptional terminator at an appropriate point. In some instances strong terminators are required for shotgun cloning foreign DNA in *E. coli*. In their absence, strong promoters on the cloned fragments read through into vector sequences and cause plasmid instability (see also below) by interfering with replication (Stassi & Lacks 1982, Chen & Morrisson 1987).

Plasmid stability

Having maximized the expression of a particular gene it is important to consider what effects this will have on the bacterium harbouring the recombinant plasmid. Increases in the levels of expression of recombinant genes lead to reductions in cell growth rates and may result in morphological changes such as filamentation and increased cell fragility. If a mutant arises which has either lost the recombinant plasmid, or has undergone structural rearrangement so that the recombinant gene is no longer expressed, or has a reduced plasmid copy number, then this will have a faster growth rate and may quickly take over and become predominant in the culture (Fig. 8.11).

Segregative instability

The loss of plasmids due to defective partitioning is called segregative instability. Naturally-occurring plasmids are stably maintained because they contain a partitioning function, *par*, which ensures that they are accurately segregated at each cell division. Such *par* regions are essential for the stability of low-copy-number plasmids. The higher copy-number plasmid ColE1 also contains a *par* region, but this region is deleted in pBR322 which is segregated randomly at cell division. Although the copy number of pBR322 is high and the probability of plasmid-free cells arising is very low, under certain conditions such as nutrient limitation or during rapid host cell growth, plasmid-free cells may arise (Jones *et al.* 1980, Nugent *et al.* 1983). This problem can be obviated by maintaining antibiotic selection. However, this may not be a desirable solution for large-scale culture because of cost and waste disposal considerations. The *par* region from a plasmid such as pSC101 may be cloned into pBR322-type vectors, thus stabilizing the plasmid (Primrose *et al.* 1983, Skogman *et al.* 1983).

Recent work indicates that DNA superhelicity is involved in the partitioning mechanism (Miller *et al.* 1990). pSC101 derivatives lacking the *par* locus show decreased overall superhelical density as compared with DNA of wild-type pSC101. Partition-defective pSC101 derivatives, and similar mutants of unrelated plasmids, are stabilized in *E. coli* by *topA* mutations which increase negative DNA supercoiling. Conversely, DNA gyrase inhibitors and mutations in DNA gyrase increase the rate of loss of *par*-defective pSC101 derivatives. Beaucage *et al.* (1991) have found that the stability of *par*-deleted pSC101 derivatives is restored by introducing certain

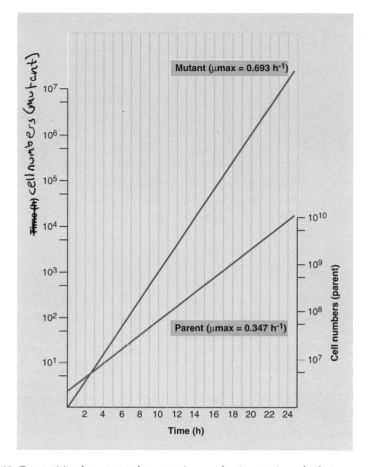

Fig. 8.11 Competition between a slow-growing production strain and a faster-growing mutant generated from it. At time zero there are 2×10^6 cells/ml of the parent strain and a single mutant cell/ml. Notice that when the parent cell density is 10^{10} cells/ml the mutant cell density is 10^7 cells/ml. On the next subculture the mutant cells will become predominant.

adventitious bacterial promoters onto the plasmid. Stabilization requires active transcription from the inserted promoter and is affected by the site and orientation of the insertion, the length of the nascent transcript and DNA gyrase activity. The observation that promoter-induced DNA supercoiling can mimic the effects of the *par* locus confirms that superhelicity generation is intrinsic to *par* function. As well as controlling plasmid stability, the *par* region also affects copy number (Manen *et al.* 1990).

An alternative strategy to counter segregative instability has been described by Rosteck and Hershberger (1983). Unlike the use of a *par* region which *prevents* plasmid loss, their method *counterselects* plasmid-free cells. In their method the plasmid carrying the gene of interest also carries the λ*cI* gene which encodes the λ repressor. Host cells carrying the chimaeric plasmids are then lysogenized with a repressor-defective mutant of bacteriophage λ. Loss of the plasmid from the lysogens causes concomitant loss of the λ repressor and hence cell death because the prophage is induced to enter the lytic growth cycle.

Plasmid instability may also arise due to the formation of multimeric forms of a plasmid. The mechanism that controls the copy number of a plasmid ensures a fixed number of plasmid origins per bacterium. Cells containing multimeric plasmids have the same number of plasmid origins but fewer plasmid molecules, which leads to greater instability if these plasmids lack a partitioning function. These multimeric forms are not seen with ColE1 which has a natural method of resolving the multimers back to monomers. It contains a highly recombinogenic resolution site that resolves multimers in a way analogous to the resolution of transposon-induced co-integrates (Summers & Sherrat 1984). The resolution site of ColE1 has been cloned into pBR322-type plasmids and has eliminated problems due to multimerization.

Structural instability

Structural instability of plasmids may arise by deletion, insertion or rearrangements of DNA. Some of the earliest reports of deletions were in chimaeric plasmids which can replicate in both *E. coli* and *B. subtilis* (for reviews see Ehrlich *et al.* 1981, Kreft & Hughes 1981). Spontaneous deletions have now been observed in a wide range of plasmid, virus and chromosomal DNAs. A common feature of these deletions is the involvement of homologous recombination between short direct repeats (Jones *et al.* 1982). Artificial plasmids with multiple tandem promoters are particularly prone to deletion formation (Nugent *et al.* 1983) as are palindromes (Hagan & Warren 1983). However, deletion formation can occur between two sites with no homology (Michel & Ehrlich 1986). Deletion 'hot spots' also can be identified and Bierne *et al.* (1991) have shown that arrest of DNA synthesis creates a very efficient deletion hot spot. This can happen when a repressor, e.g. LacI, binds to the cognate operator (Vilette *et al.* 1992). Another deletion hot spot can be created by cloned inducible promoters. Transcription dramatically affected deletions in an orientation-dependent way such that 95% of deletion endpoints were localized downstream from the inserted promoter when it faced the major plasmid transcripts. Vilette *et al.* (1992) have proposed that deletion events occur in those plasmid regions rendered positively supercoiled by convergent transcription.

As well as homologous recombination between sites on a plasmid, structural instability may be mediated by insertion (IS) elements or transposons resident in the host chromosome or on plasmids. Both of these elements can cause spontaneous mutations by their insertion, adjacent deletion or inversion of DNA. There are many reports of plasmid instability due to insertion of IS elements from the chromosome. For example, transformation of *tyr*R strains of *E. coli* with multicopy plasmids carrying the tyrosine operon gave rise to modified plasmids with either insertions or deletions (Rood *et al.* 1980). These effects were due to IS1 insertion.

Although we are beginning to have an understanding of the factors controlling deletion formation, we have insufficient knowledge to eliminate it totally. Nevertheless, the work of Ehrlich and colleagues (Bierne *et al.* 1991, Vilette *et al.* 1992) has provided very useful pointers as to factors to be considered in the design of high expression plasmids. Despite the fact that repressor binding might create a deletion hot spot it is still good practice to minimize expression of the cloned gene until it is absolutely necessary. This strategy minimizes the metabolic load on the cell and minimizes the selection of deletants. Maintaining a low vector copy number also helps provided it does not lead to segregational loss. Seehous *et al.* (1992) have devised an additional safeguard in the form of a vector which counterselects at least some of the deletions which can arise. They cloned DNA for their gene of interest (single-chain antibody) into the gene encoding β-lactamase at the 3'-terminus of its signal sequence. This construct generates a fusion protein with β-lactamase activity which protects the host bacterium from ampicillin. Spontaneous deletants which have the β-lactamase coding sequence out of frame are counterselected by the ampicillin.

Host-cell physiology can affect the level of expression

All of the factors affecting gene expression which we have discussed so far can be explained in terms of the sequence of bases in a nucleic acid and the way these sequences interact with a particular protein or protein complex. However, gene expression also is controlled by a less tangible factor – the physiology of the host cell. Factors which will be important include the choice of nutrients and the way in which they are supplied to the culture and environmental parameters such as temperature and dissolved oxygen. So far there has been no systematic study of the effect of different growth conditions on the synthesis of foreign proteins in *E. coli*. The few studies which have been done are outlined below and emphasize the importance of physiology.

One of the problems associated with the over-production of proteins in *E. coli* is the sequestration of the product into insoluble aggregates or 'inclusion bodies' (Fig. 8.12). They were first reported in strains of *E. coli* over-producing insulin chains A and B (Williams *et al.* 1982). At first their formation was thought to be restricted to the over-expression of heterologous proteins in *E. coli* but they can form in the presence of high levels of normal *E. coli* proteins, e.g. the product of the *envZ* gene (Masui *et al.* 1984) and subunits of RNA polymerase (Gribskov & Burgess 1983). Two parameters which can be manipulated to reduce inclusion body formation are temperature and growth rate. There are a number of reports which show that lowering the temperature of growth increases the yield of correctly folded, soluble protein (Schein & Noteborn 1988, Takagi *et al.* 1988, Schein 1991). Media compositions and pH values that reduce the growth rate also reduce inclusion-body formation (Schein 1991). Chaperones also can reduce the formation of inclusion bodies. Over-production of the Dnak protein

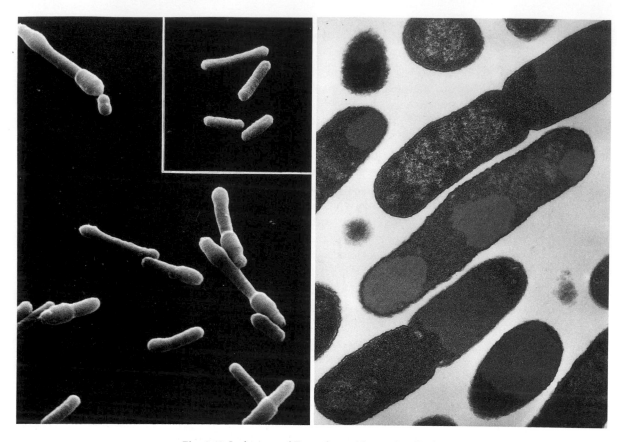

Fig. 8.12 Inclusions of Trp polypeptide–proinsulin fusion protein in *E. coli*. (*Left*) Scanning electron micrograph of cells fixed in the late logarithmic phase of growth; the inset shows normal *E. coli* cells. (*Right*) Thin section through *E. coli* cells producing Trp polypeptide–insulin A chain fusion protein. (Photographs reproduced from *Science* courtesy of Dr D.C. Williams (Eli Lilly & Co.) and the American Association for the Advancement of Science.)

simultaneously with human growth hormone decreased the size and number of inclusion bodies (Blum *et al.* 1992). The GroEL and GroES proteins are necessary for the correct assembly of other proteins over-expressed in *E. coli*, e.g. ribulose bisphosphate carboxylase (Van Dyk *et al.* 1989). Minor amino acid alterations also can affect solubility: cysteine-to-serine changes in fibroblast growth factor minimized inclusion-body formation (Rinas *et al.* 1992). La Vallie *et al.* (1993) have found that many heterologous proteins produced as insoluble aggregates are synthesized in soluble form when expressed as thioredoxin fusion proteins.

A different effect of physiology on gene expression was noted by Klotsky and Schwartz (1987). They found that plasmid copy number varied by a factor of three- to fourfold when cells were grown on a mineral salts medium supplemented with different carbon sources. The level of β-lactamase activity synthesized from a plasmid-borne gene mimicked the plasmid copy number.

9 Analysing DNA sequences

Introduction

DNA sequencing is a fundamental capability for modern gene manipulation. Knowledge of the sequence of a DNA region may be an end in its own right, perhaps in understanding an inherited human disorder. In any event, sequence information is a prerequisite for planning any substantial manipulation of the DNA; for example, a computer search of the sequence for all known restriction endonuclease target sites will provide a complete and precise restriction map.

Techniques for large-scale DNA sequencing became available in the late 1970s. The Maxam and Gilbert technique, which relies on base-specific chemistry, was popular for a time. But chain-terminator techniques soon gained popularity. There have been many modifications of the chain-terminator principle, and automated sequencing has been developed from it. In this chapter, these technological developments are described, and the basic principles of sequence data analysis are discussed.

DNA sequencing by the Maxam and Gilbert method

This method for DNA sequencing makes use of chemical reagents to bring about base-specific cleavage of the DNA. For large-scale sequencing it is now less favoured than the enzymatic, 'dideoxy', method. However, it still finds application and illustrates principles of polyacrylamide gel electrophoresis as applied to sequence determination.

In the Maxam and Gilbert method (Maxam & Gilbert 1977), the starting point is a defined DNA restriction fragment. The DNA strand to be sequenced must be radioactively labelled at one end with a ^{32}P-phosphate group. [A detailed practical account of the entire sequencing procedure, including end-labelling methods, is available (Maxam & Gilbert 1980).] The base-specific cleavages depend upon the following points.
• Chemical reagents have been characterized which alter one or two bases in DNA (Table 9.1). These are base-specific reactions; for example, dimethylsulphate methylates guanine (at the N7 position).

Table 9.1 Reagents for Maxam and Gilbert DNA sequencing

Base specificity	Base reaction	Altered base removal	Strand cleavage
G	Dimethylsulphate	Piperidine	Piperidine
G + A	Acid	Acid-catalysed depurination	Piperidine
T + C	Hydrazine	Piperidine	Piperidine
C	Hydrazine + NaCl	Piperidine	Piperidine
A > C	NaOH	Piperidine	Piperidine

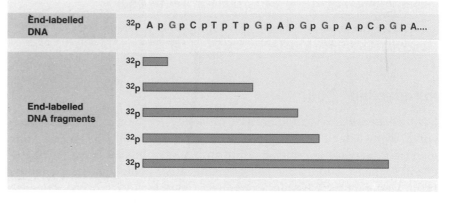

Fig. 9.1 Chemical cleavage of hypothetical DNA at G residues. A nested set of end-labelled DNA fragments is produced by limited reaction of an end-labelled DNA with G-specific reagents. Other fragments are produced, but only the terminal fragments bear the label.

- An altered base can then be removed from the sugar-phosphate backbone of DNA (Table 9.1).
- The strand is cleaved with piperidine at the sugar residue lacking the base. This cleavage is dependent upon the previous step.

When each of the base-specific reagents is used in a limited reaction with end-labelled DNA, a nested set of end-labelled fragments of different lengths is generated. It is important to emphasize that the base-specific reactions are deliberately limited to give about one, or a few, cleavages per molecule. This is illustrated in Fig. 9.1 where the nested set of fragments produced by the G-specific reaction is given as an example. Sets of fragments are produced by reacting the DNA with each of the reagents separately (Table 9.1). All five reactions (Table 9.1) may be performed; the fifth reaction gives redundant information but is confirmatory. These sequencing reactions are analysed by running the four or five samples side by side on a sequencing gel.

A sequencing gel is a high-resolution gel designed to fractionate single-stranded (denatured) DNA fragments on the basis of their length. Such gels routinely contain 6–20% polyacrylamide and 7M urea. The urea is a denaturant whose function is to minimize DNA secondary structure which affects electrophoretic mobility. The gel is run at sufficient power to heat up to about 70°C. This also minimizes DNA secondary structure. The

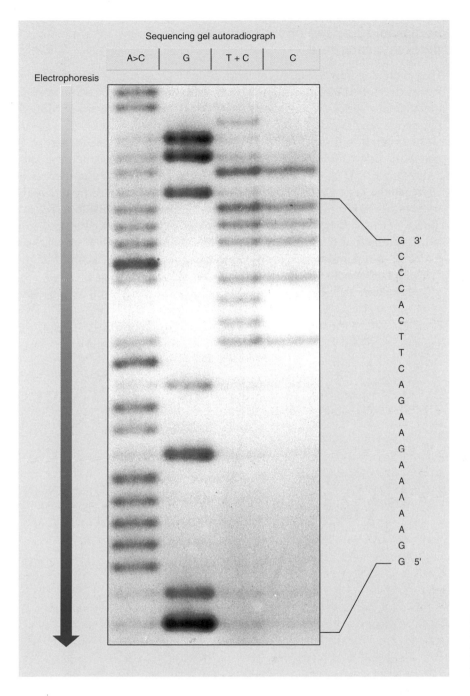

Fig. 9.2 Enlargement of a Maxam and Gilbert autoradiography. (Photograph courtesy of N. Warburton.)

labelled DNA bands obtained after such electrophoresis are revealed by autoradiography on large sheets of X-ray film. The sequence can then be read directly from the sequencing ladders in the adjacent base-specific tracks (Fig. 9.2).

Sequencing by the chain-terminator or dideoxy procedure

This method is now favoured for large-scale DNA sequence determination. It was developed in conjunction with cloning the DNA to be sequenced in the M13 mp series of single-stranded vectors (or alternatively pEMBL or Bluescript vectors may be used: see Chapter 5).

In order to appreciate the elegance of the combined technology it is necessary first to understand the chain-terminator DNA sequencing procedure (Sanger *et al.* 1977). This procedure capitalizes on two properties of DNA polymerase. First, its ability to synthesize faithfully a complementary copy of a single-stranded DNA template. Second, its ability to use 2',3'-dideoxynucleoside triphosphates as substrates (Fig. 9.3). Once the analogue is incorporated at the growing point of the DNA chain, the 3' end lacks a hydroxyl group and no longer is a substrate for chain elongation; thus the growing DNA chain is terminated.

In practice, the Klenow fragments of DNA polymerase I, which lacks

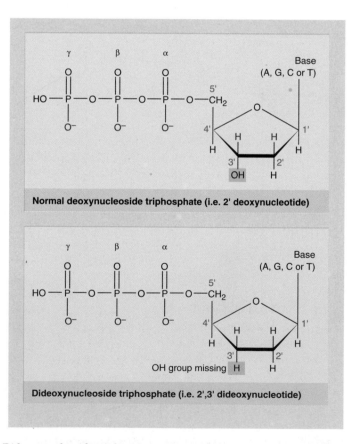

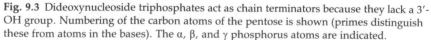

Fig. 9.3 Dideoxynucleoside triphosphates act as chain terminators because they lack a 3'-OH group. Numbering of the carbon atoms of the pentose is shown (primes distinguish these from atoms in the bases). The α, β, and γ phosphorus atoms are indicated.

the $5' \rightarrow 3'$ exonuclease activity of the intact enzyme, is used to synthesize a complementary copy of the single-stranded target sequence. Initiation of DNA synthesis requires a primer and usually this is a chemically synthesized oligonucleotide which is annealed close by.

DNA synthesis is carried out in the presence of the four deoxynucleoside triphosphates, one or more of which is labelled with ^{32}P, and in four separate incubation mixes containing a low concentration of one each of the four dideoxynucleoside triphosphate analogues. Therefore, in each reaction there is a population of partially synthesized radioactive DNA molecules, each having a common 5'-end, but each varying in length to a base-specific 3'-end. After a suitable incubation period, the DNA in each mixture is denatured, electrophoresed side by side, and the radioactive bands of single-stranded DNA detected by autoradiography. The sequence can then be read off directly from the autoradiograph as shown in Figs 9.4 and 9.5.

The sharpness of the autoradiographic images can be improved by replacing the ^{32}P-radiolabel with the much weaker β-emitter ^{35}S. This improvement is achieved by including an α-^{35}S-deoxynucleoside triphosphate (Fig. 9.6) as a supplement to the normal deoxynucleoside triphosphates in the sequencing reaction. The α-^{35}S-deoxynucleoside triphosphate is accepted and incorporated by the DNA polymerase. Other technical improvements to Sanger's original method have been made. Some workers prefer to use DNA polymerases other than the Klenow fragment of *E. coli* DNA polymerase I. Natural or modified forms of the phage T7 DNA polymerase ('sequenase') have found favour, as has the DNA polymerase of the thermophilic bacterium *Thermus aquaticus* (Taq DNA polymerase). The T7 DNA polymerase is more processive than Klenow polymerase, i.e. it is capable of polymerizing a longer run of nucleotides before releasing from the template. Its incorporation of chain terminator nucleotides is less affected by local nucleotide sequences, and so the sequencing ladders comprise a series of bands with more even intensities than those obtained with Klenow polymerase. The Taq DNA polymerase can be used in a chain termination reaction carried out at high temperature (65–70°C). This minimizes chain termination artefacts caused by secondary structure in the DNA.

The DNA to be sequenced is first cloned into one of the clustered cloning sites in the *lac* region of M13 mp series of vectors. Recombinants are detected by the formation of white plaques on media containing IPTG and Xgal (Fig. 9.7). Virus is isolated from these white plaques, stocks prepared and the single-stranded viral DNA extracted for sequencing. A feature of the M13 mp series is that cloning into the same, specific region of the genome obviates the need for many different primers, since a single primer can be used for all inserts. The original primer was a short restriction fragment which was cloned in pBR322 and which was complementary to a region of the *lacZ* gene immediately adjacent to the righthand *Eco*RI insertion site (Anderson *et al.* 1980). Messing *et al.* (1981) have developed a more suitable 15-base synthetic oligonucleotide which primes just to the

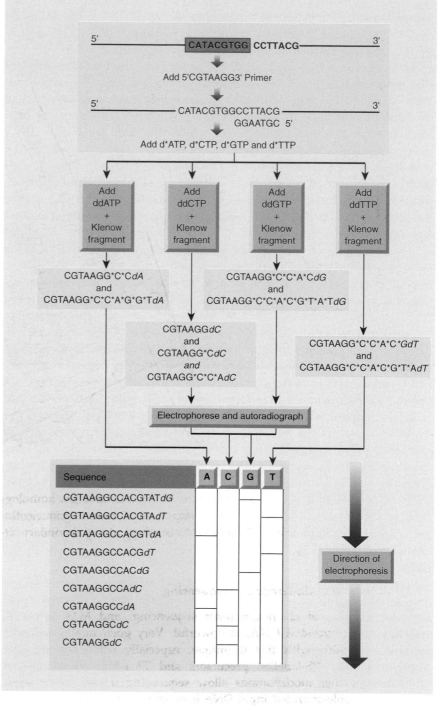

Fig. 9.4 DNA sequencing with dideoxynucleoside triphosphates as chain terminators. In this figure, asterisks indicate the presence of ^{32}P and the prefix 'd' indicates the presence of a dideoxynucleoside. At the top of the figure the DNA to be sequenced is enclosed within the box. Note that unless the primer is also labelled with a radioisotope the smallest band with the sequence CGTAAGGdC will not be detected by autoradiography as no labelled bases were incorporated.

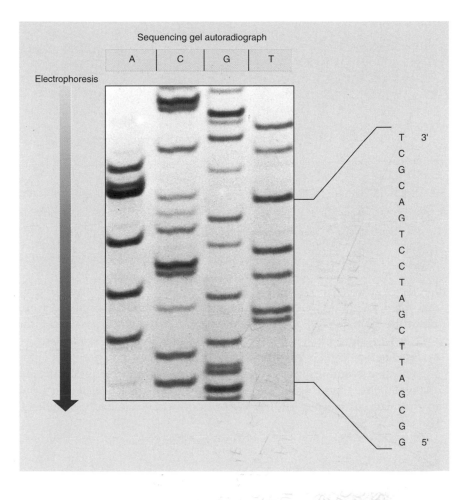

Fig. 9.5 Enlarged autoradiograph of a sequencing gel obtained with the chain terminator DNA sequencing method.

right of the polylinker. This was subsequently found to have low homology with a second site in M13, so that an improved 15-base oligonucleotide primer has been synthesized. This has no homology at any secondary site (Norrander *et al.* 1983).

Modifications of chain-terminator sequencing

The combination of chain-terminator sequencing, and M13 vectors to produce single-stranded DNA, is powerful. Very good quality sequence data is obtainable with this technique, especially when the improvements given by ^{35}S-labelled precursors and T7 DNA polymerase are exploited. Further modifications allow sequencing of 'double-stranded' DNA, i.e. double-stranded input DNA is denatured by alkali, neutralized, and one strand is then annealed with a specific primer for the actual chain-terminator sequencing reactions. This approach has gained in popularity as the convenience of having a universal primer has grown less important with the widespread availability of oligonucleotide synthesizers. With

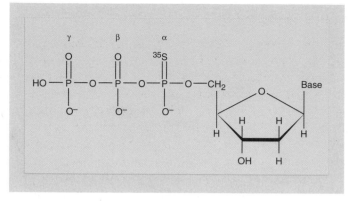

Fig. 9.6 Structure of α-^{35}S-deoxynucleoside triphosphate.

this development, the chain-terminator technique was liberated from its attachment to the M13 cloning system: for example, PCR-amplified DNA segments (Chapter 10) can be sequenced directly. One variant of the double-stranded approach, often employed in automated sequencing, is 'cycle-sequencing'. This involves a *linear* amplification (compare PCR) of the sequencing reaction using 25 cycles of denaturation, annealling of a specific primer to one strand only, and extension in the presence of Taq DNA polymerase plus labelled (radioactive or fluorescent) dideoxynucleotides.

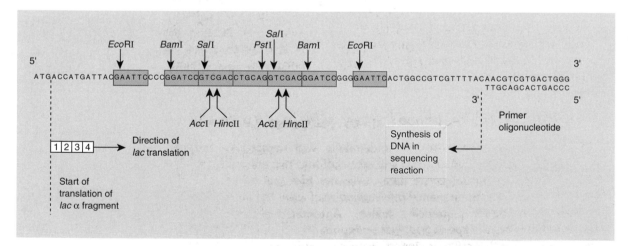

Fig. 9.7 Sequence of M13 mp7 DNA in the vicinity of the multipurpose cloning region. The upper sequence is that of M13 mp7 from the ATG start codon of the β-galactosidase α-fragment, through the multipurpose cloning region, and back into the β-galactosidase gene. The horizontal bars indicate the recognition sites for the enzymes shown. The short sequence at the right hand is that of the primer used to initiate DNA synthesis across the cloned insert. The numbered boxes correspond to the amino acids of the β-galactosidase fragment.

Automated sequencing

In manual sequencing, the DNA fragments are radiolabelled in four chain-termination reactions, separated on the sequencing gel in four lanes, and detected by autoradiography. This approach is not well suited to automation. To automate the process, it is desirable to acquire sequence data in real-time by detecting the DNA bands, within the gel, during the electrophoretic separation. Detecting the bands within the gel is not trivial, as there are only about 10^{-15} to 10^{-16} moles of DNA in each band. The solution to the detection problem is to use fluorescence methods.

In the first generation of fluorescence techniques, fluorescent tags were attached to the sequencing primer (Gocayne *et al.* 1987, Brumbaugh *et al.* 1988). By attaching four different fluorescent dyes to the primer, each of which can be discriminated by its fluorescent spectrum, it is possible to electrophorese the products of all four chain-terminator reactions together in one lane of a sequencing gel. The DNA bands are detected by their fluorescence as they electrophorese past a detector. If the detector is made to scan horizontally across the base of a slab gel, many separate sequences can be scanned, one sequence per lane (Fig. 9.8).

In the second generation of fluorescence techniques (Prober *et al.* 1987), the fluorescent tags are attached to the chain-terminating nucleotides. Each of the four chain-terminator nucleotides carries a spectrally distinct fluorophore. The tag is incorporated into the DNA molecule by the DNA polymerase, and accomplishes two operations in one step: it terminates synthesis, and it attaches the fluorophore to the end of the molecule. There are important advantages to this approach. First, conventional primers are used. Second, rather than having to do the enzymatic reactions in four separate tubes, they can be done in a single tube because the termination events also specifically attach one of the four fluorophores. As before, the DNA bands are detected as they electrophorese past a fluorescence detector.

Analysis of DNA sequence data by computer, i.e. *in silico*

Any substantial DNA sequence undertaking will necessitate computer analysis of the data. The analysis will often fall into two stages. In the first stage, compiling the sequence data, computer files will be more or less directly inputted with sequence information read from the autoradiograph of the gel or other sequencing system. Automated systems have the obvious advantage of speed and, just as importantly, they help to minimize the ever-present difficulty of clerical errors accumulating in sequence data. The computer will then be used to compile readings from several sequencing runs, to search for and identify overlaps and hence *contiguities* between runs, and to compile data from sequencing complementary strands of the DNA. The software will also identify inconsistencies, indicating sequencing errors, between sequences determined more than once on the same or complementary strands.

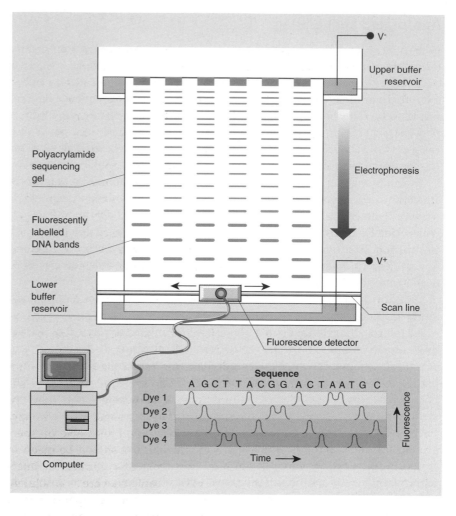

Fig. 9.8 Block diagram of an automated DNA sequencer; and idealized representation of the correspondence between fluorescence in a single electrophoresis lane and nucleotide sequence.

In the second stage, the analysis proper, the deduced sequence may be processed in a variety of ways.

● The sequence can be searched for all known restriction endonuclease target sites, thus producing a comprehensive and precise map.

● The sequence can be searched for features such as tandem repeats or inverted repeats. Inverted repeats lead to the possibility of hairpin-loop formation within one strand of the DNA or in RNA transcribed from it.

● The sequence can be conceptually translated into protein in all three possible reading frames on both strands – six frames in all. An 'open' reading frame, ORF, is a frame that does not include a termination codon. A long open reading frame may indicate that a previously unknown polypeptide coding region exists (Doolittle 1986).

● The DNA sequence itself, or the deduced polypeptide sequence, may be compared with a data bank of other sequences. Often, because of the degeneracy of the genetic code, similarities are found between two polypeptide sequences, which would not have been apparent had the comparison been carried out at the DNA level.

Similarities between two sequences can, in principle, arise by two routes: either by convergent evolution, or through their being related by descent from a common ancestral sequence. Convergent evolution implies that the two sequences have not arisen in evolutionary time through having had a common ancestral sequence, but that selection for a particular function in two lineages has converged on a particular structure or two related structures. By contrast, two sequences may remain similar in evolutionary time because of selection pressure limiting the scope for divergence.

This selection pressure need not necessarily act on whole proteins. It is possible for protein coding regions to be assembled *piecemeal*, with conserved domains fulfilling a particular function, e.g. DNA binding or nucleoside triphosphate binding, in polypeptides that are otherwise distinct in function. The fact that certain polypeptide sequence motifs or domains are often found as exon units is consistent with a piecemeal gene assembly model. So-called *exon-shuffling* is evident.

Whatever the evolutionary route to sequence similarity has been, any sequence similarity is taken to be an indication of similar function. It has been a feature of modern molecular biology that striking and unexpected sequence similarities have been discovered in a range of situations. The message seems to be that evolution is reluctant to discard a good idea for building up functional polypeptide domains. The sequence data banks are of sufficient size even at our present state of knowledge (and they are expanding very rapidly) to make a search of them for similarities with any newly discovered sequence an undertaking that has a reasonable prospect of turning up something of interest. Two final points should be made in this context. First, it is important that development continues in international arrangements for gaining access to data banks that are as absolutely up to date and as accurate as possible. The current rate of expansion is so great that a significant proportion of known sequences is, at any time, not accessible. Second, it has been the authors' experience that, in the cases where one has been in a position to check particular sequences in data banks or scientific journals, a disturbingly high proportion of entries contain errors. Errors are especially common in non-coding regions and appear to arise from a variety of causes ranging from outright sequencing errors to clerical errors.

10 The polymerase chain reaction

Introduction

The impact of the PCR upon molecular biology has been profound. The reaction is easily performed, and leads to the amplification of specific DNA sequences by an enormous factor. From a simple basic principle, many variations have been developed with applications throughout gene technology (Erlich 1989, Innis *et al.* 1990). Very importantly, the PCR has revolutionized pre-natal diagnosis by allowing tests to be performed using small samples of fetal tissue. In forensic science, the enormous sensitivity of PCR-based procedures is exploited in DNA profiling; following the publicity surrounding *Jurassic Park*, virtually everyone is aware of potential applications in palaeontology and archaeology.

In many applications of the PCR to gene manipulation, the enormous amplification is secondary to the aim of altering the amplified sequence. This often involves incorporating extra sequences at the ends of the amplified DNA.

Basic reaction

First we need to consider the basic polymerase chain reaction. The principle is illustrated in Fig. 10.1. The PCR involves two oligonucleotide primers, 17–30 nucleotides in length, which flank the DNA sequence that is to be amplified. The primers hybridize to opposite strands of the DNA after it has been denatured, and are oriented so that DNA synthesis by the polymerase proceeds through the region between the two primers. The extension reactions create two double-stranded target regions, each of which can again be denatured ready for a second cycle of hybridization and extension. The third cycle produces two double-stranded molecules that comprise precisely the target region in double-stranded form. By repeated cycles of heat denaturation, primer hybridization, and extension,

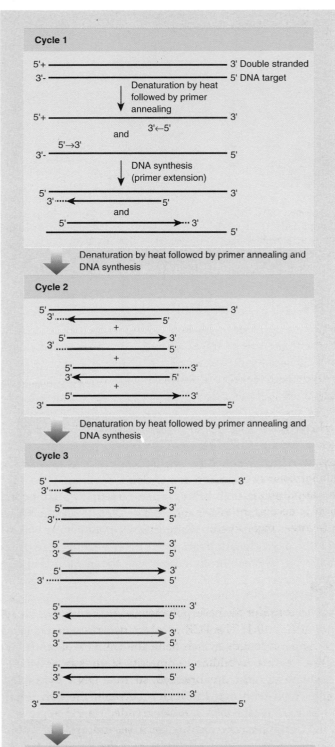

Fig. 10.1 The polymerase chain reaction. In cycle 1 two primers anneal to denatured DNA at opposite sides of the target region, and are extended by DNA polymerase to give new strands of variable length. In cycle 2, the original strands and the new strands from cycle 1 are separated, yielding a total of four primer sites with which primers anneal. The primers that are hybridized to the new strands from cycle 1 are extended by polymerase as far as the end of the template, leading to a precise copy of the target region. In cycle 3, double-stranded DNA molecules are produced (highlighted in colour) that are precisely identical to the target region. Further cycles lead to exponential doubling of the target region. The original DNA strands, and the variably-extended strands, become negligible after the exponential increase of target fragments.

Cycle number	Number of double-stranded target molecules
1	0
2	0
3	2
4	4
5	8
6	16
7	32
8	64
9	128
10	256
11	512
12	1024
13	2048
14	4096
15	8192
16	16,384
17	32,768
18	65,536
19	131,072
20	262,144
21	524,288
22	1,048,576
23	2,097,152
24	4,194,304
25	8,388,608
26	16,777,216
27	33,554,432
28	67,108,864
29	134,217,728
30	268,435,456

Fig. 10.2 Theoretical PCR amplification of a target fragment with increasing number of cycles.

there follows a rapid exponential accumulation of the specific target fragment of DNA. After 22 cycles, an amplification of about 10^6-fold is expected (Fig. 10.2), and amplifications of this order are actually attained in practice.

In the original description of the PCR method (Mullis & Faloona 1987, Sakai *et al.* 1988, Mullis 1990) Klenow DNA polymerase was used, and because of the heat denaturation step, fresh enzyme had to be added during each cycle. A breakthrough came with the introduction of Taq DNA polymerase (Lawyer *et al.* 1989) from the thermophilic bacterium *Thermus aquaticus*. The Taq DNA polymerase is resistant to high temperatures and so does not need to be replenished during the PCR (Erlich *et al.* 1988, Sakai *et al.* 1988). Furthermore, by enabling the extension reaction to be performed at higher temperatures, the specificity of the primer annealing is not compromised. As a consequence of employing the heat-resistant enzyme, the PCR could be automated very simply by placing the assembled reaction in a heating block with a suitable thermal cycling programme (see Box 10.1). The Taq DNA polymerase lacks a $3'-5'$ proofreading exonuclease activity. This lack appears to contribute to errors during PCR amplification due to misincorporation of nucleotides (Eckert & Kunkel 1990). Partly to overcome this problem, other thermostable DNA polymerases with improved fidelity have been sought and described (Cariello *et al.* 1991,

Box 10.1 The polymerase chain reaction achieves enormous amplifications, of specific target sequences, very simply

The reaction is assembled in a single tube, and then placed in a thermal cycler (a programmable heating/cooling block), as described below.

A typical PCR for amplifying a human genomic DNA sequence has the following composition. The reaction volume is 100 µl.

Input genomic DNA, 0.1–1 µg
Primer 1, 20 pmoles
Primer 2, 20 pmoles
20 mM Tris-HCl, pH 8.3 (at 20°C)
1.5 mM magnesium chloride
25 mM potassium chloride
50 µM each deoxynucleoside triphosphate (dATP, dCTP, dGTP, dTTP)
2 units Taq DNA polymerase

A layer of mineral oil is placed over the reaction mix to prevent evaporation.

The reaction is cycled 25–35 times, with the following temperature programme:

Denaturation	94°C, 0.5 min
Primer annealing	55°C, 1.5 min
Extension	72°C, 1 min

Typically, the reaction takes some 2—3 hours overall.

Notes:
• The optimal temperature for the annealing step will depend upon the primers used.
• The pH of the Tris-HCl buffer decreases markedly with increasing temperature. The actual pH varies between about 6.8 and 7.8 during the thermal cycle.
• The time taken for each cycle is considerably longer than 3 min (0.5 + 1.5 + 1 min), depending upon the rates of heating and cooling between steps.
• The PCR does not efficiently amplify sequences much longer than about 3 kb.

Mattila *et al.* 1991, Lundberg *et al.* 1991), although the Taq DNA polymerase remains the most widely used enzyme for PCR. In certain applications, especially where amplified DNA is cloned, it is important to check the nucleotide sequence of the cloned product to reveal any mutations that may have occurred during the PCR. The fidelity of the amplification reaction can be assessed by cloning, sequencing, and comparing several independently-amplified molecules.

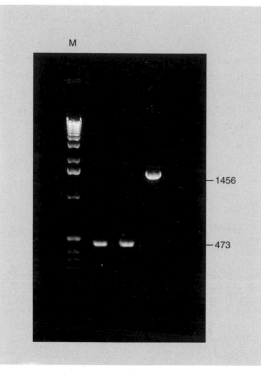

Fig. 10.3 PCR amplification of human genomic DNA sequences. Starting with a small amount of human genomic DNA (0.1 μg), obtained from a small blood sample, specific gene fragments of 1456 and 473 nucleotide pairs have been amplified by PCR, and characterized by agarose gel electrophoresis. M is a marker track displaying a range of DNA fragments of known size.

Amplification

There are two important fundamental aspects of the PCR. One aspect is the enormous amplification achieved. As an illustration of this, we can take a routine application of the PCR – the amplification of a human gene sequence. A typical input is 1 μg of human genomic DNA. (This can be obtained readily from blood since a 1 ml sample of blood provides about 50 μg of DNA.) The input DNA contains only 3×10^5 molecules of any single copy sequence, equivalent to 0.1 pg (or 10^{-13} g) of a target sequence 300 nucleotides long. PCR amplification of a sequence in this genomic DNA produces an amount of the target DNA (up to 1 μg) sufficient for direct application in any one of a wide range of molecular biological procedures, including direct DNA sequencing (Scharf *et al.* 1986, Wong *et al.* 1987, Innis *et al.* 1988). For example, a small sample (10 μl out of a typical 100 μl reaction) of the PCR-amplified product will produce a DNA band that is readily visualized by ethidium fluorescence after gel electrophoresis (Fig. 10.3). At the limit, even single target molecules can be amplified (Li *et al.* 1988). With such extreme sensitivity in PCR procedures, contamination of the sample or reagents must be carefully avoided.

Specificity

The second aspect is the specificity of the PCR. The target region is defined by the flanking primers, and the specificity derives from the specific hybridization of the primers under the annealing conditions set for the thermal cycle. The fact that the length of the target sequence is limited, in practice, to less than a few kilobases has a beneficial consequence for specificity. The limitation is beneficial because in some applications of PCR a primer may hybridize at points additional to the one intended: this is especially likely when degenerate primers are employed (see below). However, an unintended point of hybridization will often be irrelevant because it is not likely that a second primer molecule will hybridize near enough for amplification to occur.

Primers

The specificity of the PCR depends crucially upon the primers. The following factors are important in choosing effective primers.
- Primers should be 17 to 30 nucleotides in length.
- A GC content of about 50% is ideal. For primers with a low GC content, it is desirable to choose a long primer so as to avoid a low melting temperature.
- Sequences with long runs (i.e. more than three or four) of a single nucleotide should be avoided.
- Primers with significant secondary structure are undesirable.
- There should be no complementarity between the two primers.
The great majority of primers which conform with these guidelines can be made to work, although not all comparable primer sets are equally effective even under optimized conditions.

Degenerate primers

There are several applications where the use of *degenerate* primers is necessary. A degenerate primer is actually a mixture of primers, all of similar sequence but with variations at one or more positions. A common circumstance requiring the use of degenerate primers occurs when the primer sequences have to be deduced from amino acid sequences (Lee *et al.* 1988). Degenerate primers may also be employed to search for novel members of a known family of genes (Wilks 1989), or to search for homologous genes between species (Nunberg *et al.* 1989). When a degenerate primer is designed on the basis of an amino acid sequence, the degeneracy of the genetic code must be considered, as discussed on p. 130 for the design of degenerate oligonucleotide probes. Selection of amino acids with low codon degeneracy is desirable. A 128-fold degeneracy in each primer can be successful in amplifying a single copy target from the human genome (Girgis *et al.* 1988). Under such circumstances the concentration of any individual primer sequence is very low, so mismatching between primer and template must occur under the annealing conditions chosen. Since mismatching of the 3'-terminal nucleotide of the primer may prevent efficient extension, degeneracy at this position is to be avoided.

Empirical optimization of the PCR reaction conditions is especially important when degenerate primers are employed. Particular attention is paid to the annealing temperature. It is also desirable to employ a *hot-start protocol*, which entails adding the DNA polymerase after the heat denaturation step of the first cycle, the addition taking place at a temperature at or above the annealing temperature and just prior to the annealing step of the first cycle. The hot start overcomes the problem that would arise if the DNA polymerase were added to complete the assembly of the PCR reaction mixture at a relatively low temperature. At low temperature, below the desired hybridization temperature for the primer (typically in the region 45°–60°), mismatched primers will form and may be extended somewhat by the polymerase. Once extended, the mismatched primer is stabilized at the unintended position. Having been incorporated into the extended DNA during the first cycle, the primer will hybridize efficiently in subsequent cycles, and hence may cause the amplification of a spurious product.

Nested primers

In order to minimize further the amplification of spurious products, the strategy of *nested primers* may be deployed. Here the products of an initial PCR amplification are used to seed a second PCR amplification in which one or both primers are located internally with respect to the primers of the first PCR. Since there is little chance of the spurious products containing sequences capable of hybridizing with the second primer set, the PCR with these nested primers selectively amplifies the sought-after DNA.

Amplification by inverse PCR

A limitation of conventional PCR is the requirement for the two primers which define the region to be amplified. We may wish to amplify sequences that lie beyond a region for which a primer pair can be designed. Inverse PCR is useful here, because it allows the amplification of DNA flanking a region of known sequence (Ochman *et al*. 1988, Triglia *et al*. 1988). The method (Fig. 10.4) is based upon cutting DNA with a restriction endonuclease, and ligating the fragments intramolecularly to form circular molecules. Primers designed to extend outwardly from a known core sequence can then be used to amplify a linear fragment which comprises sequences from a specific circular molecule. The amplified sequences are those that flank the core sequence in the genome, their lengths depending upon the positions of the restriction sites on each side.

Repeated inverse PCR can be applied to chromosome walking (see p. 107). But the length of DNA that can be amplified by PCR is limited, constraining this strategy to relatively short steps along the chromosome.

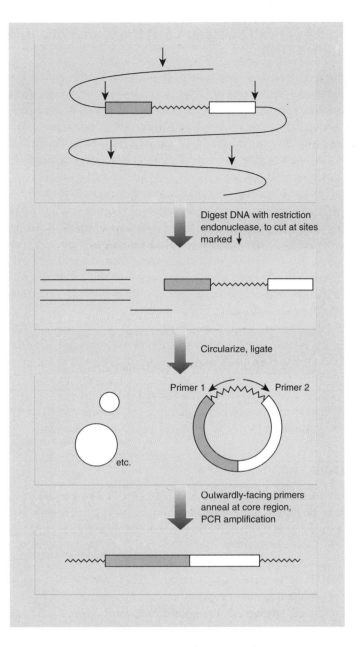

Fig. 10.4 Inverse PCR. The core region is indicated by the wavy line. Restriction sites are marked with arrows, and the left and right regions which flank the core region are represented by closed and open boxes. Primers are designed to hybridize with core sequences and are extended in the directions shown. PCR amplification generates a linear fragment containing left and right flanking sequences.

Incorporation of extra sequence at the 5'-end of a primer into amplified DNA

A PCR primer may be designed which, in addition to the sequence required for hybridization with the input DNA, includes an extra sequence at its 5'-end. The extra sequence does not participate in the first hybridization step, only the 3'-portion of the primer hybridizes, but it subsequently becomes incorporated into the amplified DNA fragment (Fig. 10.5). Because the extra sequence can be chosen at the will of the experimenter, great flexibility is available here.

A common application of this principle is the incorporation of restriction sites at each end of the amplified product. Figure 10.5 illustrates the addition of a *Hin*dIII site and an *Eco*RI site to the ends of an amplified DNA fragment. In order to ensure that the restriction sites are good substrates for the restriction endonucleases, four nucleotides are placed between the hexanucleotide restriction sites and the extreme ends of the DNA. The incorporation of these restriction sites provides one method for cloning amplified DNA fragments (see below).

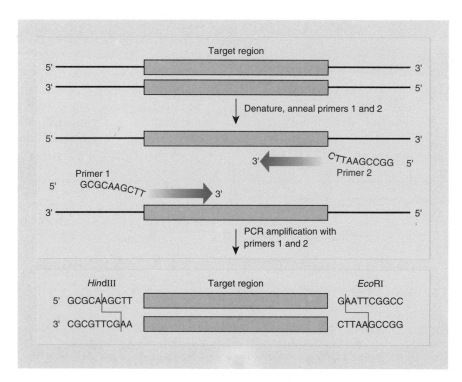

Fig. 10.5 Incorporation of extra sequence at the 5'-end of a primer. Two primers have sequences designed to hybridize at the ends of the target region. Primer 1 has an extra sequence near its 5'-end which forms a *Hin*dIII site (AAGCTT), and primer 2 has an extra sequence near its 5'-end which forms an *Eco*RI (GAATTC) site. Each primer has an additional 5'-terminal sequence of four nucleotides so that the hexanucleotide restriction sites are placed within the extreme ends of the amplified DNA, and so present good substrates for endonuclease cleavage.

A second application of the principle is the incorporation of a phage T7 (or T3, or SP6) promoter at the end of the amplified DNA. This leads to a DNA fragment that can be transcribed *in vitro* by phage RNA polymerase without the need for cloning steps. A third application is the construction of genes for chimaeric proteins (see p. 386).

Amplification of cDNA (RT-PCR), rapid amplification of cDNA ends (RACE) and cDNA libraries compared

Reverse transcription followed by the polymerase chain reaction (RT-PCR), leading to amplification of specific RNA sequences in cDNA form, is a sensitive means for detecting RNA molecules, a means for obtaining material for sequence determination, and a step in cloning a cDNA copy of the RNA. Various strategies can be adopted for first-strand cDNA synthesis: the reverse transcriptase reaction can be primed by the downstream PCR primer annealed to the RNA, or by random hexamers, or by an oligo(dT) primer at the poly(A) tail of mRNA (Kawasaki 1990). The second-strand of the cDNA is synthesized by the Taq DNA polymerase during the first cycle of the PCR.

Because of the sensitivity of RT-PCR, total cellular RNA is a suitable input in many circumstances, obviating the need for isolation of poly(A)$^+$ mRNA, and hence making the procedure very rapid. A potential problem can be caused by the presence of contaminating genomic DNA in the RNA preparation. This is because a target sequence in a trace of genomic DNA may be amplified, leading to a false result. In studying eukaryotic mRNAs, it is often desirable therefore to choose primers derived from different exons which are spaced several kilobases apart, so that genomic sequences cannot be amplified. Alternatively, the RNA can be treated with deoxyribonuclease to destroy any contaminating DNA.

Because of the speed with which RT-PCR can be carried out, it is an attractive approach for obtaining a specific cDNA sequence for cloning. By comparison, screening a cDNA library (which involves several rounds of plaque hybridization) is laborious—even presuming that a suitable cDNA library is already available and does not have to be constructed for the purpose. Quite apart from the labour involved, a cDNA library may not yield a cDNA clone with a full-length coding region because, as described on p. 119, generating a full-length cDNA clone may be technically challenging, particularly with respect to long mRNAs. Furthermore, the sought-after cDNA may be very rare even in specialized libraries. So, given that suitable primers can be devised, these are further reasons for adopting a cDNA cloning strategy involving RT-PCR.

Does this mean that cDNA libraries have been superseded? Despite the advantages of RT-PCR, there are reasons for constructing cDNA libraries. The first reason involves availability of starting material, and the permanence of the library. A sought-after mRNA may occur in a source that is not readily available, perhaps a small number of cells in a particular human tissue. A good-quality cDNA library has only to be constructed once from this tissue, to give a virtually infinite resource for future use. The specialized library is permanently available for screening. Indeed, the library may be used as a source of cDNA from which a specific cDNA can be obtained by PCR amplification. The second reason concerns screening strategies. The PCR-based approaches are dependent upon specific primers. But with cDNA libraries screening strategies are possible which are based upon expression, e.g. immunochemical screening, rather than nucleic acid hybridization.

Rapid amplification of cDNA ends (RACE)

It is possible to use RT-PCR to amplify a complete cDNA coding region, but in many applications there may be insufficient information for a straightforward strategy. Often only limited information will be available either from limited amino acid sequencing or from expected conservation of protein domains. In such circumstances a method called *rapid amplification of cDNA ends* (RACE) may be applicable (Frohman *et al.* 1988).

The RACE protocols generate cDNA fragments by using PCR to amplify sequences between a single region in the mRNA and either the 3'- or the 5'-end of the transcript. To use RACE it is necessary to know or to deduce a single stretch of sequence within the mRNA. From this sequence, specific primers are chosen which are oriented in the 3' and 5' directions, and which produce overlapping cDNA fragments. In the two RACE protocols, extension of the cDNAs from the ends of the transcript to the specific primers is accomplished by using primers that hybridize either at the natural 3' poly(A) tail of the mRNA, or at a synthetic poly(dA) tract added to the 5'-end of the first-strand cDNA (Fig. 10.6). Finally, after amplification, the overlapping RACE products can be combined if desired, to produce an intact full-length cDNA.

As might be anticipated, because only a single *specific* primer is used in each of the RACE protocols, the specificity of amplification may not be very great. This is especially problematical where the specific primer is degenerate. In order to overcome this problem, a modification of the RACE method has been devised which is based on using nested primers to increase specificity (Frohman & Martin 1989).

Cloning PCR products

As a means of simply isolating genomic DNA sequences or cDNA sequences for analysis, PCR generates sufficient material for analysis. But when extensive manipulation of the sequence is envisaged, or when large amounts of DNA are required, it is necessary to clone the PCR product. Amplified DNA fragments do not clone very efficiently. Factors discussed below may account for the difficulty.

The basic PCR scheme illustrated in Fig. 10.1 shows the production of a double-stranded amplified product. However the product of the reaction does not contain a large proportion of perfectly blunt-ended molecules, so

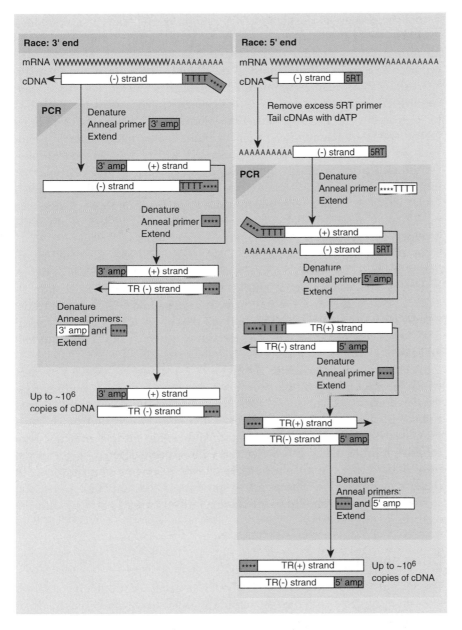

Fig. 10.6 Rapid amplification of cDNA ends (RACE) (Frohman *et al.* 1988). *3' Protocol.*
The mRNA is reverse transcribed using an oligo(dT$_{17}$) primer which has a 17 nucleotide
extension at its 5'-end. This extension, the anchor sequence, is designed to contain
restriction sites for subsequent cloning. Amplification is performed using the anchor 17-
mer (which has a T_m higher than oligo(dT$_{17}$)) and a primer specific for the sought-after
cDNA. *5' Protocol.* The mRNA is reverse transcribed from a specific primer. The resultant
cDNA is then extended by terminal transferase to create a poly(dA) tail at the 3'-end of
the cDNA. Amplification is performed with the oligo(dT$_{17}$)/anchor system as used for the
3' protocol, and the specific primer. Open boxes represent DNA strands being
synthesized; coloured boxes represent DNA from a previous step. The diagram is
simplified to show only how the *new* product from a previous step is used. Molecules
designated TR, truncated, are shorter than full-length (+) or (−) strands.

a 'polishing' step is desirable. This can be carried out in a combined exonuclease/repair reaction with Klenow polymerase in the presence of all four deoxynucleoside triphosphates. Taq DNA polymerase has been reported to remain tightly bound to the 3'-ends of the DNA so proteinase K treatment has been recommended (Crowe *et al.* 1990). For subsequent cloning it is necessary to remove excess primer, which is often accomplished by preparative gel electrophoresis as a means of purifying the amplified DNA. The DNA fragment can then be blunt-ligated into a suitable vector such as pBluescript (see p. 67) linearized with *Sma*I. The chemically-synthesized primers which form the 5'-end of each strand of the DNA fragment do not bear 5'-terminal phosphates, therefore if the linearized vector has been treated with phosphatase to prevent simple re-closure (see p. 39), it is essential that 5'-phosphate groups are added to the PCR product. This requires a polynucleotide kinase step.

A proportion of the amplified DNA fragments are not truly blunt-ended because Taq DNA polymerase has a tendency to add an extra A residue at the 3'-end of each strand, giving rise to a single-nucleotide 3'-extension at each end of the fragment. In order to take advantage of this, vectors have been devised which bear a T residue as a 3'-extension at each end, hence creating short cohesive termini with which PCR fragments can be ligated (Holton & Graham 1990, Marchuk *et al.* 1990).

An alternative approach to cloning a PCR product is to devise primers which incorporate restriction sites at the 5'-ends of the DNA fragment, as described above (Scharf *et al.* 1986). This approach has the advantage that cloning is directional if different restriction sites are created by the two primers. The DNA fragment is treated with restriction enzymes to create cohesive termini which can be ligated with suitably-cleaved vector in the conventional way. To ensure that the chosen restriction enzymes are not cleaving within any uncharacterized portion of the amplified DNA, the size of the fragment should be checked before and after treatment with the restriction endonucleases.

Other amplification systems

The PCR is based on the DNA polymerase-catalysed doubling of target DNA in each cycle. Alternative amplification systems can be based upon the ability RNA polymerase to generate multiple transcripts from a DNA template (Kwoh *et al.* 1989, Compton 1991). Because the amplification inherent in transcription is much more than twofold, fewer cycles of amplification are necessary. These systems involve phage T7 RNA polymerase for transcriptional amplification, and reverse transcriptase to regenerate DNA template for further amplification. While the transcription-based systems have yet to find wide application, there is scope for future development.

11 Changing genes: site-directed mutagenesis

Introduction

Mutants are an essential prerequisite for any genetic study and never more so than in the study of gene structure and function relationships. Classically, mutants are generated by treating the test organism with chemical or physical agents that modify DNA (mutagens). This method of mutagenesis has been extremely successful, as witnessed by the growth of molecular biology, but suffers from a number of disadvantages. First, any gene in the organism can be mutated and the frequency with which mutants occur in the gene of interest can be very low. This means that selection strategies have to be developed. Second, even when mutants with the desired phenotype are isolated there is no guarantee that the mutation has occurred in the gene of interest. Third, prior to the development of gene cloning and sequencing techniques there was no way of knowing where in the gene the mutation had occurred and whether it arose by a single base change, an insertion of DNA, or a deletion.

As techniques in molecular biology have developed, so that the isolation and study of a single gene is not just possible but routine, so mutagenesis has also been refined. Instead of crudely mutagenizing many cells or organisms and then analysing many thousands or millions of offspring to isolate a desired mutant, it is now possible to specifically change any given base in a cloned DNA sequence. This technique is known as *site-directed mutagenesis*. It has become a basic tool of gene manipulation for it simplifies DNA manipulations that in the past required a great deal of ingenuity and hard work, e.g. the creation or elimination of cleavage sites for restriction endonucleases. The importance of site-directed mutagenesis goes beyond gene structure–function relationships for the technique enables mutant proteins to be generated with very specific changes in particular amino acids (protein engineering). Such mutants facilitate the study of the mechanisms of catalysis, substrate specificity, stability, etc. Three different methods of site-directed mutagenesis have been devised: cassette

[191]

mutagenesis, primer extension and procedures based on the polymerase chain reaction (PCR). All three are described below.

In some cases the goal of protein engineering is to generate a molecule with an improvement in some operating parameter but it is not known what amino acid changes to make. In this situation a random mutagenesis strategy provides a route to the desired protein. However, methods based on gene manipulation differ from traditional mutagenesis in that the mutations are restricted to the gene of interest or a small portion of it. Finally, genetic engineering also provides a number of simple methods of generating chimaeric proteins where each domain is derived from a different protein.

It should not be forgotten that constructing the mutant DNA is only part of the task. The vector for expression, the expression system, strategies for purification and assay must also be considered before embarking on protein mutagenesis.

In this chapter we will thus concentrate on a wide range of methodological developments in mutagenesis. This range of techniques will continue to expand at a rapid rate, indicating the pervasiveness of mutagenesis in analysis of gene structure and function. Some specific examples of the way directed mutation has been used to 'improve' proteins are given in Chapter 17.

Cassette mutagenesis

In cassette mutagenesis, a synthetic DNA fragment containing the desired mutant sequence is used to replace the corresponding sequence in the wild-type gene. This method was used originally to generate improved variants of the enzyme subtilisin (Wells *et al.* 1985). It is a simple method for which the efficiency of mutagenesis is close to 100%. The disadvantages are the requirement for unique restriction sites flanking the region of interest and the limitation on the realistic number of different oligonucleotide replacements which can be synthesized. The latter problem can be minimized by the use of 'doped' oligonucleotides (see p. 201) or by the suppression of amber codons (see p. 203).

Primer extension: the single-primer method

The simplest method of site-directed mutagenesis is the single primer method (Gillam *et al.* 1980, Zoller & Smith 1983). The method involves priming *in vitro* DNA synthesis with a chemically-synthesized oligonucleotide (7–20 nucleotides long) that carries a base mismatch with the complementary sequence. As shown in Fig. 11.1 the method requires that the DNA to be mutated is available in single-stranded form, and cloning the gene in M13-based vectors makes this easy. However, DNA cloned in a plasmid and obtained in duplex form can also be converted to a partially single-stranded molecule that is suitable (Dalbadie-McFarland *et al.* 1982).

The synthetic oligonucleotide primes DNA synthesis and is itself incor-

[193]
CHAPTER 11
Changing genes

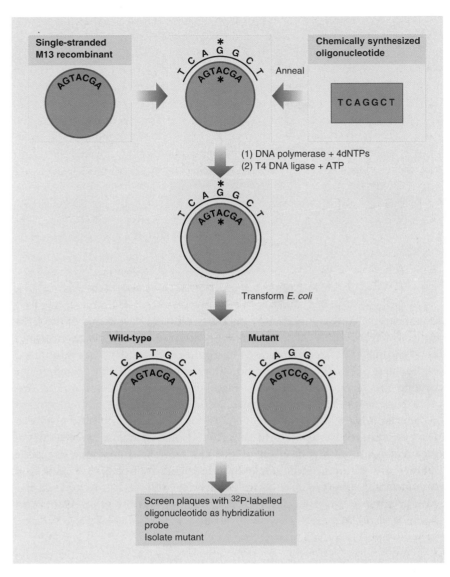

Fig. 11.1 Oligonucleotide-directed mutagenesis. Asterisks indicate mismatched bases. Originally the Klenow fragment of DNA polymerase was used but this has been largely replaced with T7 polymerase.

porated into the resulting heteroduplex molecule. After transformation of the host *E. coli*, this heteroduplex gives rise to homoduplexes whose sequences are either that of the original wild-type DNA or that containing the mutated base. The frequency with which mutated clones arise, compared with wild-type clones, may be low. In order to pick out mutants, the clones can be screened by nucleic acid hybridization (see Chapter 7) with ^{32}P-labelled oligonucleotide as probe. Under suitable conditions of stringency, i.e. temperature and cation concentration, a positive signal will be obtained only with mutant clones. This allows ready detection of the desired mutant (Wallace *et al.* 1981, Traboni *et al.* 1983). In order to check that the procedure has not introduced other adventitious changes, it is prudent to check the sequence of the mutant directly by DNA

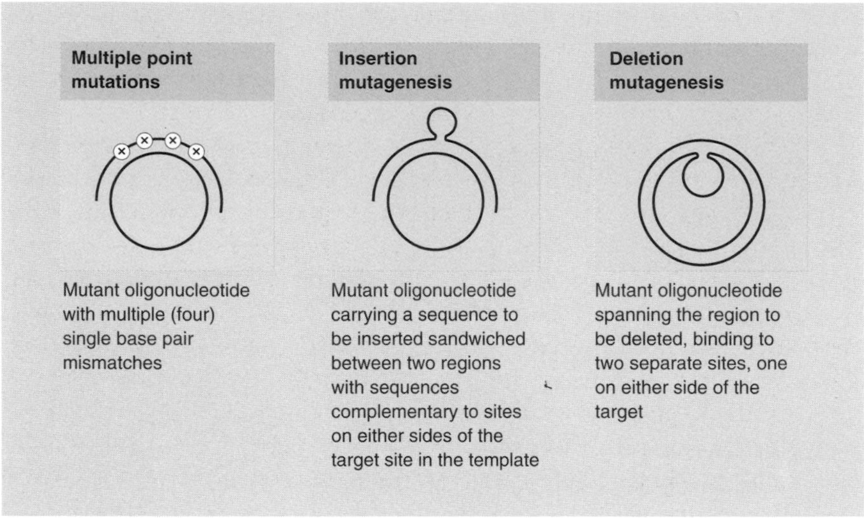

Fig. 11.2 Oligonucleotide-directed mutagenesis used for multiple point mutation, insertion mutagenesis, and deletion mutagenesis.

sequencing. This was a particular necessity with early versions of the technique which made use of *E. coli* DNA polymerase. The more recent use of the high-fidelity DNA polymerases from phages T4 and T7 has minimized the problem of extraneous mutations as well as shortening the time for copying the second strand. Also, these polymerases do not 'strand displace' the oligomer, a process which would eliminate the original mutant oligonucleotide.

A variation of the procedure (Fig. 11.2) outlined above involves oligonucleotides containing inserted or deleted sequences. As long as stable hybrids are formed with single-stranded wild-type DNA, priming of *in vitro* DNA synthesis can occur giving rise ultimately to clones corresponding to the inserted or deleted sequence (Wallace *et al.* 1980, Norrander *et al.* 1983).

Deficiencies of the single-primer method

The efficiency with which the single-primer method yields mutants is dependent upon several factors. The double-stranded heteroduplex molecules that are generated will be contaminated both by any single-stranded non-mutant template DNA that has remained uncopied, and by partially double-stranded molecules. The presence of these species considerably reduces the proportion of mutant progeny. They can be removed by sucrose gradient centrifugation, or by agarose gel electrophoresis, but this is time-consuming and inconvenient.

Following transformation and *in vivo* DNA synthesis, segregation of the two strands of the heteroduplex molecule can occur, yielding a mixed population of mutant and non-mutant progeny. Mutant progeny have to be purified away from parental molecules, and this process is complicated by the cell's mismatch repair system. In theory, the mismatch repair system should yield equal numbers of mutant and non-mutant progeny, but in practice mutants are counterselected. The major reason for this low yield of mutant progeny is that the methyl-directed mismatch repair system of *E. coli* favours the repair of non-methylated DNA. In the cell, newly synthesized DNA strands that have not yet been methylated are preferentially repaired at the position of the mismatch, thus preventing a mutation. In a similar way, the non-methylated *in vitro*-generated mutant strand is repaired by the cell so that the majority of progeny are wild-type (Kramer *et al*. 1984a). The problems associated with the mismatch repair system can be overcome by using host strains carrying the *mut*L, *mut*S or *mut*H mutations which prevent the methyl-directed repair of mismatches.

A heteroduplex molecule with one mutant and one non-mutant strand must inevitably give rise to both mutant and non-mutant progeny upon replication. It would be desirable to suppress the growth of non-mutants, and various strategies have been developed with this in mind. Two early methods were the gapped duplex method (Kramer 1984b) and the primer selection method (Carter *et al*. 1985). They have been superseded by two other methods which are both simpler and available commercially. The first of these is the method of Kunkel (1985). In this method phage are grown on a specialized host *before* mutagenesis. The host used is deficient in dUTPase (*dut*) and uracil glycosylase (*ung*). The *dut* mutation results in increased intracellular dUTP levels and the *ung* mutation permits the incorporation of deoxyuridine into the DNA in place of thymidine at some positions. Phage M13 grown in a *dut ung* host contain 20–30 uracil residues per genome and are unable to grow in an *ung*⁺ host. Thus a heteroduplex composed of a uracil containing parental strand and a mutant strand synthesized *in vitro* in the presence of dTTP will give rise to mostly mutant progeny when plated on a wild-type (*ung*⁺) host. Sequential mutations can be made using this system by growing the first mutant on a *dut ung* host before subjecting it to another round of mutagenesis.

All the methods referred to above generate heteroduplex molecules, each consisting of a mutant and a non-mutant strand, and then attempt to select for the mutant strand *in vivo*. This can result in considerable loss of efficiency and often requires mutant vectors and/or host strains. In addition, the heteroduplex molecules give rise to mixed plaques containing both mutant and wild-type progeny. Clones identified as mutants require further plaque purification and identification.

An alternative method makes use of the observation that certain restriction enzymes (e.g. *Ava*I, *Ava*II, *Ban*II, *Hin*dII, *Nci*I, *Pst*I and *Pvu*I) cannot cleave phosphorothioate DNA (Taylor *et al*. 1985). The mutant oligonucleotide is annealed to the single-stranded DNA template in the usual manner, but is then extended by DNA polymerase in the presence of a thionucleotide (Fig. 11.3). This generates a heteroduplex in which the mutant strand is phosphorothioated. After sealing the gap in the mutant strand with DNA ligase the heteroduplex is treated with *Nci*I, which cleaves *only* the parental strand. The parental strand is partially digested with exonuclease and then repolymerized (Fig. 11.4). This method permits strand

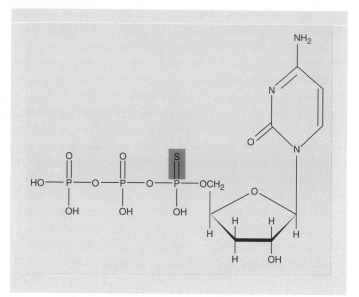

Fig. 11.3 Structure of a thionucleotide (dCTPαS).

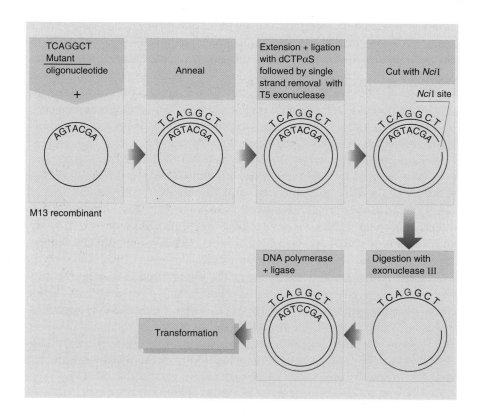

Fig. 11.4 *In vitro* strand
selection of mutants. The
DNA containing sulphur
nucleotides is shown in red.

selection *in vitro* and gives very high efficiencies of mutation, not just for point mutants, but for making deletions and insertions as well. Another advantage is that it does not require specialized hosts or phage vectors and thus can be used for repeated rounds of mutagenesis.

Originally nitrocellulose filters were used with the phosphorothioate method to remove single-stranded template DNA, but these have been replaced with phage T5 exonuclease. The T5 enzyme has single- and double-stranded exonuclease activities with a co-purifying single-stranded endonuclease (Sayers & Eckstein 1991). It can nick and digest the unwanted template DNA that would otherwise cause a non-mutant background. Nicked, double-stranded DNA is also a substrate. However, the closed circular mutant heteroduplex is resistant to digestion and passes through the reaction undamaged. T5 digestion is quicker and easier than removing the single-stranded DNA from the double-stranded DNA by filtration through nitrocellulose membranes, or by immobilizing one of them to a solid phase (see for example, Hultman *et al.* 1990).

A disadvantage of all of the primer extension methods is that they require a single-stranded template. Olsen and Eckstein (1990) have adapted the phosphorothioate method for use with a double-stranded template. Although highly efficient, the method is very laborious. Another plasmid based method also has been developed (Deng & Nickeloff 1992). This relies on denaturation of the double-stranded template to allow annealing of mutant oligonucleotides followed by primer extension. Amplification of the mutant is performed by two rounds of transformation.

PCR methods of site-directed mutagenesis

Early work on the development of the PCR method of DNA amplification showed its potential for mutagenesis (Scharf *et al.* 1986). Single bases mismatched between the amplification primer and the template become incorporated into the template sequence as a result of amplification (Fig. 11.5). Higuchi *et al.* (1988) have described a variation of the basic method which enables a mutation in a PCR-produced DNA fragment to be introduced anywhere along its length. Two primary PCR reactions produce two overlapping DNA fragments, both bearing the same mutation in the overlap region. The overlap in sequence allows the fragments to hybridize (Fig. 11.5). One of the two possible hybrids is extended by DNA polymerase to produce a duplex fragment. The other hybrid has recessed 5'-ends and since it is not a substrate for the polymerase, is effectively lost from the reaction mixture. As with conventional primer extension mutagenesis, deletions and insertions also can be created.

The method of Higuchi *et al.* (1988) requires four primers and three PCRs (a pair of PCRs to amplify the overlapping segments and a third PCR to fuse the two segments). Sarkar and Sommer (1990) have described a simpler method which utilizes three oligonucleotide primers to perform two rounds of PCR. In this method, the product of the first PCR is used as a *megaprimer* for the second PCR (Fig. 11.6).

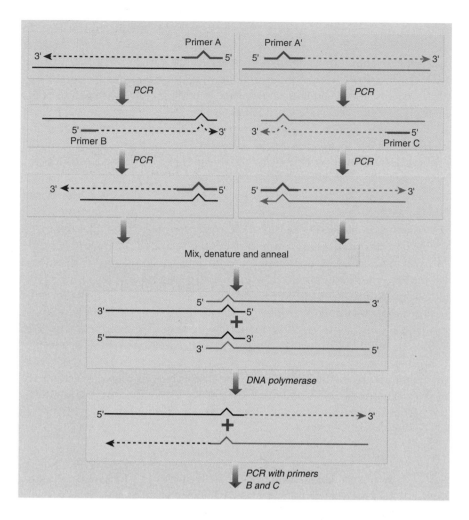

Fig. 11.5 Site-directed
mutagenesis by means of the
PCR. The steps shown in the
top-left corner of the diagram
show the basic PCR method
of mutagenesis. The bottom
half of the figure shows how
the mutation can be moved
to the middle of a DNA
molecule. Primers are shown
in red and primers A and A′
are complementary.

The advantage of a PCR-based mutagenic protocol is that the desired
mutation is obtained with 100% efficiency. There are two disadvantages.
First the PCR product usually needs to be ligated into a vector, although
Sarkar and Sommer (1990) have generated the mutant protein directly
using coupled *in vitro* transcription and translation. Second, *Taq* polymerase
copies DNA with low fidelity (see p. 203). Thus the sequence of the entire
amplified segment generated by PCR mutagenesis must be determined to
ensure that there are no extraneous mutations. Recently, two thermostable
polymerases with improved fidelity have been described (Cariello *et al.*
1991, Mattila *et al.* 1991, Lundberg *et al.* 1991).

Making unidirectional deletions

All the methods described above can be used to make both additions and
deletions. They are particularly suitable if we want to make only a few
addition or deletion mutants. However, where a whole family of mutants

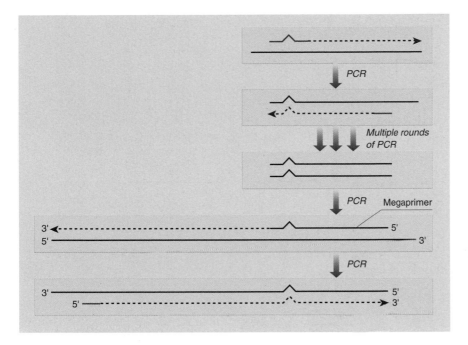

Fig. 11.6 The megaprimer method of mutagenesis. The mutant molecule produced in the early rounds of PCR acts as a primer ('megaprimer') for later rounds of PCR.

is desired then a corresponding family of oligonucleotide primers would be required. An alternative method is available for making a series of deletions of varying length (Yanisch-Perron *et al*. 1985, Barcack & Wolf 1986). The method is shown in Fig. 11.7 and makes use of the fact that α-thiophosphate-containing phosphodiester bonds are resistant to hydrolysis by the 3′ → 5′ exonucleolytic activity of phage T4 DNA polymerase. Linear duplex DNA molecules blocked at one 3′-terminus with a thiophosphate are then degraded from the other end with the exonuclease. Digestion for different lengths of time followed by treatment with nuclease S1 and ligation allows the preparation and recovery of a nested set of deletion mutants.

Construction of genes for chimaeric proteins

There are many reasons why we might want to synthesize chimaeric proteins. For example, to facilitate purification (see p. 140) or to create multifunction proteins which carry out a series of sequential synthetic steps. Originally, genes encoding chimaeric proteins were either synthesized *de novo* or by ligation of oligonucleotide cassettes encoding different functional domains. Both methods are time-consuming and tedious. An alternative method of building chimaeras, which involves domain-swapping, has been described by Clackson and Winter (1989). DNA sequences encoding the domain of interest are equipped with 'sticky feet' by PCR primer extension (Fig. 11.8). One strand of the PCR product then is used in a standard primer mutagenesis reaction using the other protein coding sequence as a

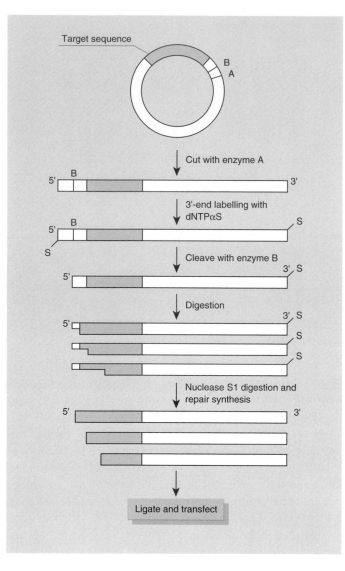

Fig. 11.7 Procedure for making unidirectional deletions in a DNA molecule. A and B represent recognition sequences for two different restriction endonucleases.

template. The efficiency of selection of the mutant can be improved by doing the primer extension in the presence of thionucleotides and enzymically destroying the parental strand (Wychowski *et al*. 1990). An alternative method of domain fusion has been used to produce a library of single-chain antibodies (Clackson *et al*. 1991). First the variable and light-chain DNAs were independently amplified by PCR. Subsequently, the two segments were joined by a third PCR fragment. When expressed, the resulting chimaera retained antigen-binding properties.

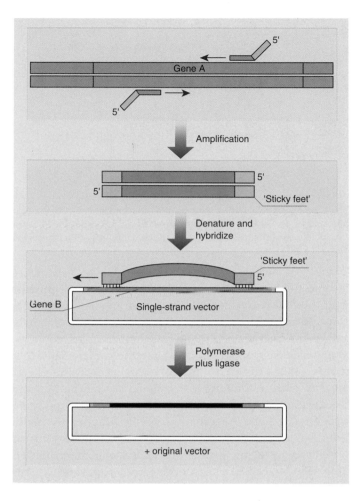

Fig. 11.8 The 'sticky feet' method of replacing a portion of one gene with a segment of DNA from another gene. Note that the primers used to amplify the segment of gene *A* have small lengths of DNA attached to them which are complementary to sequences in gene *B*.

Random mutagenesis

While rational design remains the goal of protein engineering, a great deal of effort continues to be put into modern extensions of the classical genetical approach of random mutagenesis and screening or selection. This approach is a powerful way of rapidly building a structure–function database in a protein system, and can be equally powerful in the isolation of molecules with improved properties. Random mutagenesis of a gene or gene segment is best carried out using *doped oligonucleotides*. These are synthesized using nucleotide precursors that are contaminated by a small amount of the other three nucleotides. Reidhaar-Olson and Sauer (1988) have used this method to probe functionally and structurally important amino acids in a bacteriophage repressor. Figure 11.9 shows the principle.

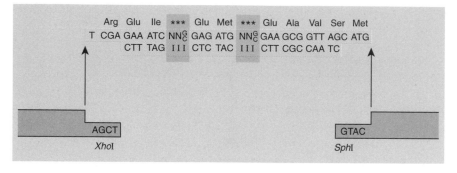

Fig. 11.9 Mutagenesis by means of doped oligonucleotides. During synthesis of the upper strand of the oligonucleotide, a mixture of all four nucleotides is used at the positions indicated by the letter N. When the lower strand is synthesized inosine (I) is inserted at the positions shown. The double-stranded oligonucleotide is inserted into the relevant position of the vector.

The first strand of the cassette is synthesized with equal mixtures of all four bases in the first two codon positions and an equal mixture of G and C in the third position. The resulting population of base combinations will include codons for each of the 20 naturally-occurring amino acids at each of the mutagenized residue positions. When the complementary strand is synthesized, inosine is inserted at each randomized position because it is able to pair with each of the four conventional bases. The two strands are annealed and the mutagenic cassette ligated into the gene of interest.

A constraint of the above method is that it is limited by the length of oligonucleotide that can be synthesized. One solution is to prepare a library of mutants by using a series of doped oligomers that cover the entire coding sequence when placed end to end. Hermes *et al.* (1990) used this method to isolate intragenic suppressors of a 'sluggish' mutant of triose phosphate isomerase. Five suppressor sites were identified. Although they were scattered throughout the sequence (at positions 10, 96, 97, 167 and 233) they were all close to the active site.

Another limitation of random mutagenesis is that even for a short oligonucleotide sequence, the resulting mutant library quickly becomes too large to be screened. For example, mutagenesis of a 12 amino acid peptide using NNN triplets results in a DNA complexity of 4^{36} and a protein complexity of 20^{12}. The use of NN (G, C) by Reidhaar-Olson and Sauer (1988), as described above, represents a slightly improved 'dope'. It also should be remembered that since different codons may translate to the same amino acid, each protein sequence does not appear equiprobably in the mutant library. Thus for a 12 amino acid sequence using NNN, polyleucine would appear 6^{12} times more frequently than polymethionine because there are six codons for leucine but only one for methionine. Arkin and Youvan (1992) have described a set of nucleotide mixtures that encode a subset of amino acids. For example, NTG/C encodes the hydrophobic amino acids phenylalanine, isoleucine, leucine, methionine and valine. Other mixtures can be designed that encode primarily hydrophilic, small amphipathic or aromatic amino acids. The advantage of this approach is that it reduces the degeneracy in a way that is most likely to be functionally acceptable.

An alternative method of generating mutants makes use of the fact that *Taq* DNA polymerase lacks a 3'–5' exonuclease activity and so is unable to proofread copied sequences. This results in a high error frequency in PCR amplified DNA (Keohavong & Thilly 1989, Eckert & Kunkel 1991), the average frequency being one mistake per 10 000 nucleotides per cycle. Lerner *et al.* (1990) made use of this spontaneous misincorporation to isolate random amino acid substitutions in the propeptide region of preprosubtilisin.

Selection of mutant peptides by phage display

Filamentous bacteriophages such as M13 (see p. 87) and fd have three to five copies of the gene III protein located at one end of the virus particle. This protein is essential for proper phage assembly and for adsorption to the F pilus of male strains of *E. coli*. When small DNA fragments are inserted into the middle of gene III the progeny phage carry the corresponding protein sequence (Parmley & Smith 1988). In the technique known as *phage display*, a library of variant peptides or proteins is displayed on the surface of the phage particles. This is achieved by inserting a random peptide DNA cassette (see above) into the gene III coding sequence. Particular phage displaying peptide motifs with, for example, antibody-binding properties can be isolated by affinity chromatography (Fig. 11.10). Several rounds of affinity chromatography and phage propagation can be used to further enrich for phage with the desired binding characteristics. In this way millions of random peptides have been screened for their ability to bind to an anti-peptide antibody or to streptavidin (Scott & Smith 1990, Devlin *et al.* 1990, Cwirla *et al.* 1990) and variants of human growth hormone with improved affinity and receptor specificity have been isolated (Lowman *et al.* 1991).

With the method described above, three to five copies of the peptide of interest are displayed on each virion, one per molecule of gene III protein. A variation of this method which results in only a single copy of the novel peptide being displayed has been developed (Bass *et al.* 1990). Yet another variation is to display the random peptide sequences as fusions to the major coat protein (gene VIII). The use of this polyvalent approach leads to the display of several hundred copies of the fusion protein per virion and is useful for selecting variants of low affinity.

For a detailed review of phage display the reader should consult Wells and Lowman (1992).

Suppression of amber mutations

Amber mutations are ones which result in premature peptide chain termination during translation. This happens because of the creation of a stop codon (AUG) in the gene sequence. Bacterial strains capable of suppressing

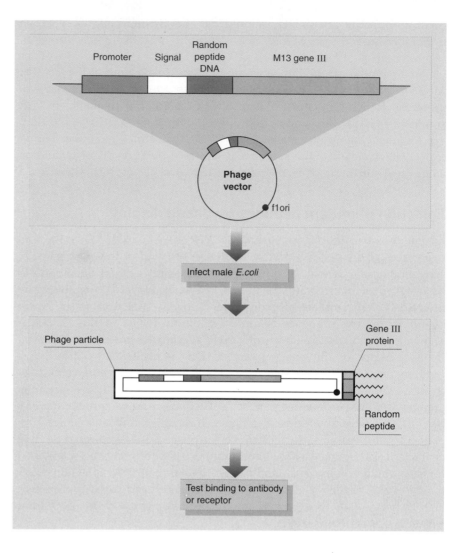

Fig. 11.10 The principle of phage display of random peptides.

amber mutations produce tRNA molecules which recognize the amber codon and insert an amino acid during protein synthesis. Normanly *et al.* (1990) have described a set of *E. coli* strains containing different amber suppressor tRNAs: each strain inserts a different amino acid at the amber codon. These strains can be used as follows. By site-directed mutagenesis an amber codon is introduced into the desired position of the gene which then is cloned in an expression vector. This vector is introduced into each of the different suppressor strains and the expressed protein isolated from each one. In a heroic example of this approach, Rennell *et al.* (1991) isolated 163 different amber mutants of phage T4 lysozyme. These mutant genes then were introduced into 13 suppressor strains to generate over 2000 variants of the enzyme. From the analysis of plaque formation, which is dependent on the activity of the T4 lysozyme, the tolerance to substitution could be correlated with structure and activity. Analysis of the data

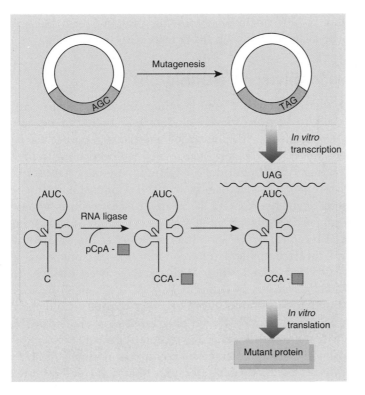

Fig. 11.11 Insertion of unusual amino acids into a protein by suppression of amber mutations *in vitro* using a chemically modified tRNA. The red box represents an unusual amino acid.

showed that lysine is the most frequently unacceptable substitution and alanine the least deleterious.

The advantage of the above method is that it permits isolation of a large number of protein variants for a single *in vitro* mutagenesis step, the creation of an amber codon. Nor is there any biological selection of the mutant proteins. The disadvantage of the method is that not all 20 amino acids can be inserted at the site of the amber codon.

Suppression of amber mutations also can occur in cell free transcription–translation systems. This feature can be used to incorporate non-natural amino acids into proteins. Plasmids carrying the amber mutation are added to a cell-free transcription–translation mix supplemented with a suppressor tRNA to which the desired non-natural amino acid has been chemically coupled (Fig. 11.11). This system has been used to prepare novel variants of β-lactamase (Noren *et al.* 1989) and lysozyme (Mendel *et al.* 1992) including a version of the latter which is light activated (Mendel *et al.* 1991). The major disadvantage of this method is that it generates only low amounts of modified protein. An alternative method of introducing non-natural amino acids has been developed by Smith *et al.* (1990). A cysteine substitution is first engineered into the protein which subsequently is treated with reagents that selectively react with the side chain of cysteine.

Systematic mutational analysis

Although the techniques of random mutagenesis and amber suppression can be used to identify functional domains, both can be laborious in practice. Consequently, a number of alternative approaches have been developed. *Homologue-scanning mutagenesis* has been used to identify segments of human growth hormone (hGH) that are important for binding to the hGH receptor (Cunningham *et al.* 1989). Chimaeras were constructed by replacing segments of hGH, 7–30 amino acid residues in length, with the corresponding regions from homologous hormones known not to bind to the hGH receptor. By determining the ability of the different chimaeras to bind to the receptor and different monoclonal antibodies, three key domains were recognized. These domains have been analysed further by *alanine-scanning mutagenesis* (Cunningham & Wells 1989) in which single alanine mutations were introduced at every residue within the domains. The alanine scan revealed a cluster of a dozen large side chains that when mutated to alanine each reduced the receptor-binding affinity by a factor of four. It also identified at least one side chain that normally hinders binding: when it was mutated to alanine, receptor affinity increased fourfold.

Homologue-scanning mutagenesis can only be used when there is a family of structurally-related proteins with different functions. Without the information which it gives, alanine-scanning mutagenesis would be tedious because alanine substitutions would be required at every residue in the protein. A simpler approach is to use *charged-to-alanine scanning mutagenesis* (Bass *et al.* 1991). In this method, charged residues that cluster within a window of four to eight amino acids are changed to alanine. Gibbs and Zoller (1991) constructed 60 such mutants of yeast cAMP-dependent protein kinase. The mutants were analysed *in vivo* by introduction into a yeast strain that required a functional protein kinase for growth, and *in vitro* by kinetic properties. Specific residues and regions of the enzyme were identified that are likely to be important in catalysis and in binding of ATP, functions that are common to all protein kinases. Additional regions were identified which are likely to be important in binding a protein substrate.

12 Cloning in bacteria other than *E. coli*

Introduction

For many experiments it is convenient to use *E. coli* as a recipient for genes cloned from eukaryotes or other prokaryotes. Transformation is easy and there is available a wide range of easy-to-use vectors with specialist properties, e.g. regulatable high-level gene expression. However, use of *E. coli* is not always practicable because it lacks some auxiliary biochemical pathways that are essential for the phenotypic expression of certain functions, e.g. degradation of aromatic compounds or plant pathogenicity. In such circumstances the genes have to be cloned back into species similar to those whence they were derived. Yet another reason for cloning in other organisms is that *E. coli* is not used in any of the traditional industrial fermentations. Rather, the bacteria used tend to be Gram-positive species such as *Bacillus* (amylases, proteases), actinomycetes (antibiotics) or coryneforms (amino acids, steroid transformations). However, cloning in these organisms presents a different set of problems compared to cloning in *E. coli*.

An essential prerequisite for cloning in any new organism is a method, preferably transformation, for introducing recombinant plasmids into the organisms of choice. Although efficient methods for the transformation of *E. coli* with plasmid DNA have been developed this is not the case with most other bacteria. There are four reasons why transformants may not be detected. First, failure to detect transformants could be due to a failure of the plasmids to enter the cell. Outside of *Bacillus* sp. (see p. 217) so little is known about the transformation process that there is no rational basis for protocol development in a new organism. Electroporation (p. 21) provides a possible alternative and with the commercial availability of the necessary equipment is being used increasingly with considerable success (Wirth *et al.* 1989).

A second source of failure could be restriction of the transformed plasmid DNA. In *E. coli* efficient transformation is dependent on the availability of restriction-defective recipients. Such mutants of non-enteric bacteria generally are not available and in their absence transformation is extremely difficult, if not impossible. For example, the frequency of transformation of *Pseudomonas putida* is 10^5 transformants/μg of plasmid RSF1010 DNA, a frequency tenfold lower than in *E. coli*, but only if the plasmid is prepared from another *P. putida* strain. Otherwise no transformants are obtained (Bagadasarian *et al.* 1979).

There are two other reasons why transformants might not be detected. The plasmid DNA might be taken up by the cell but fail to replicate. Alternatively, the plasmid-borne markers might not be expressed in their new environment. The solution here is to use a plasmid carrying an easily selectable marker and which is indigenous to the chosen organism. However, in many instances the indigenous plasmids are cryptic, as with most *Bacillus* sp. and/or too large to handle easily. Furthermore, considerable time and effort are required to develop a useful vector from a newly-isolated plasmid. Despite these problems, numerous groups of necessity have developed vectors. In recent years this task has become simpler with the ready availability of multiple cloning sites (polylinkers, see p. 67) and a wide variety of cloned genes.

Cloning in non-enteric Gram-negative bacteria has been facilitated by the availability of plasmids which have a broad host range and which can be subjugated as vectors. This eliminates the need to construct a vector system specific to every species of interest. In the case of Gram-positive *Bacillus* sp., rapid progress in the development of vectors was facilitated by the discovery that antibiotic-resistance plasmids from *Staphylococcus aureus* can replicate in them too and express their antibiotic resistance. The recent discovery that a number of plasmids can be *conjugally* transmitted to a wide range of unrelated bacteria and to eukaryotes (see Box 12.1) greatly expands the cloning options available.

For details of methods, vectors, etc. for cloning in specific Gram-negative and Gram-positive organisms not covered in this chapter, the reader is advised to consult the genus-specific sections in volume 204 of *Methods in Enzymology* which was published in 1991.

Broad host-range vectors for cloning in Gram-negative bacteria

Plasmids belonging to the *E. coli* incompatibility groups C, J, N, P, Q and W are capable of replication and remain more or less stable in diverse, unrelated Gram-negative bacteria. In particular, IncP and IncQ group plasmids display a very extensive host range which in the case of RSF1010, at least, extends to Gram-positive bacteria and eukaryotes (see Box 12.1). For this reason, P and Q group plasmids have been the most used as

Box 12.1 Barriers to interspecific gene transfer: a wall falls

Over the years there have been many attempts to introduce foreign DNA into bacterial cells. Early experiments were unsuccessful, most likely because DNA fragments were used rather than intact replicons (see p. 6). The advent of gene manipulation facilitated attempts at interspecific gene transfer because plasmids (replicons) containing easily selectable markers became available. Even then there was little success, leading to the view that plasmid maintenance and associated marker gene expression were restricted to closely related bacterial species or genera. Promiscuous plasmids were identified (see p. 208) which were transferable between a wide range of Gram-negative bacteria but not to Gram-positive bacteria. Similarly plasmids from the Gram-positive *Staphylococcus aureus* could be established in *Bacillus subtilis* (p. 221) but not in Gram-negative bacteria. Existing knowledge of the mechanisms of plasmid replication and gene expression provided satisfactory explanations for the failure of plasmids to breach the barrier between Gram-negative and Gram-positive bacteria.

The first crack in the wall came with the observation (Goze & Ehrlich 1980) that composite replicons of pBR322 and the staphylococcal plasmid pC194 can be established in *E. coli* under conditions where pBR322 is not maintained. More direct evidence comes from the later demonstration that small plasmids from Gram-positive bacteria can be established as autonomous replicons in Gram-negative hosts (Lacks *et al.* 1986, Leenhouts *et al.* 1991). More surprising was the demonstration by Trieu-Cuot *et al.* (1987) of *conjugative* transfer of a shuttle plasmid between *E. coli* and a variety of Gram-positive bacteria. To achieve this transfer, the plasmid was designed such that it contained an origin of replication for *E. coli* and a broad host-range origin of replication for Gram-positive bacteria. This conjugal transfer of recombinant plasmids from *E. coli* to Gram-positive bacteria has been confirmed by others (Mazodier *et al.* 1989, Schafer *et al.* 1990) and also has been shown to occur with the natural plasmid RSF1010 (Gormley & Davies, 1991). Conjugal transfer of a recombinant plasmid from the Gram-positive *Enterococcus faecalis* to *E. coli* also has been demonstrated (Trieu-Cuot *et al.* 1988).

It has long been known that a small number of bacteria have a very close association with plant tissue, e.g. the plant pathogen *Agrobacterium tumefaciens* and the nitrogen-fixing symbiont *Rhizobium*. In the case of *A. tumefaciens* it has been shown unequivocally that transfer of DNA occurs naturally between the bacterium and the infected plant cell (see Chapter 14). Such transfer of genetic material between a prokaryote and a eukaryote has been considered exceptional. Recently, however, Heinemann and Sprague (1989) and Sikorski *et al.* (1990) have shown that conjugative plasmids of bacteria can transfer DNA conjugally to the budding yeast *Saccharomyces cerevisiae* and the fission yeast *Schizosaccharomyces pombe*. While the role of sex pili is well established for bacterial conjugation, if and how they function in gene transfer to eukaryotes is not known.

Table 12.1 Properties of representative broad host-range plasmids

Plasmid	Incompatibility group	Size (kb)	Copy no.	Self-transmissible	Markers
RP4 (RK2, RP1)	P	60	1–3	Yes	$Ap^R Km^R Tc^R$
RSF 1010	Q	8.68	15–40	No	$Sm^R Su^R$
Sa	W	29.6	3–5	Yes	$Km^R Sm^R Cm^R SP^R$

vectors. A group W plasmid has been used extensively with one particular host, *Agrobacterium* (see p. 274). Representative plasmids from these three groups are shown in Table 12.1.

For P and Q group plasmids, host range is dependent upon the possession of a complete replication system in addition to incompatibility determinants. The structure of these replication functions has been elucidated and their control shown to be far more complex than those in narrow host range plasmids such as ColE1 (for review see Kues & Stahl 1989). They also differ in being distributed in several regions of the plasmid genome and this makes subjugation of the plasmid as a vector more difficult. For ColE1-type plasmids, and vectors such as pBR322 derived from it, there are indications that host range restriction depends on initiation of replication. For example, ColE1 cannot replicate in cell-free *Pseudomonas* extracts but it shows a partial replication ability when purified *E. coli* gyrase and DNA polymerase I are added. Therefore, in Gram-negative bacteria the absence of these enzymes may be responsible for the restriction of replication of ColE1-based plasmids to enteric bacteria and, surprisingly, *Legionella* sp. (Diaz & Staudenbauer 1982, Engleberg *et al.* 1988).

Members of groups P and W are self-transmissible, i.e. they carry *tra* genes encoding the conjugative apparatus. Considerable DNA coding capacity is required to specify conjugal transfer, which to a great extent explains the much larger size of RP4 and Sa relative to RSF1010. Group Q plasmids are not self-transmissible and usually are introduced into the recipient by transformation. However, if the donor cell also carries a narrow host range conjugative plasmid, it can be transferred to the recipient by conjugation. Thus group Q plasmids have a mobilization (*mob*) function and a compatible origin of DNA transfer (*ori*T or *nic*) which is the site of action of the *mob* nuclease. Vectors derived from different incompatibility groups can be stably maintained in the same cell. This permits a variety of genetic analyses, e.g. complementation studies between cloned sequences in a wide range of Gram-negative hosts.

The problem of poor transformation efficiency makes it unrealistic to attempt direct transformation of non-enteric bacteria with DNA from a ligation. The most satisfactory approach is to make use of *E. coli* as an intermediate host for transformation of the ligation mix and screening for recombinant plasmids. Once identified, the recombinant DNA of interest can be purified and transformed or conjugated into the desired host. The isolation of the recombinant DNA in this way acts as an amplification step, although it suffers from the disadvantage that screening based on gene function may not be possible.

Vectors derived from Q group plasmid RSF1010

Plasmid RSF1010 is a multicopy replicon which specifies resistance to two antimicrobial agents, sulphonamide and streptomycin. The plasmid DNA which is 8684 base-pairs long has been completely sequenced (Scholz *et al.* 1989). A detailed physical and functional map has been constructed (Bagdasarian *et al.* 1981, Scherzinger *et al.* 1984). The features mapped are the restriction endonuclease recognition sites, RNA polymerase binding sites, resistance determinants, genes for plasmid mobilization (*mob*), three replication proteins (Rep A, B and C) and the origins of vegetative (*ori*) and transfer (*nic*) replication.

Plasmid RSF1010 has unique cleavage sites for *Eco*RI, *Bst*EII, *Hpa*I, *Dra*II, *Nsi*I and *Sac*I and from the nucleotide sequence data is predicted to have unique sites for *Afl*III, *Ban*II, *Not*I, *Suc*II, *Sfi*I and *Spl*I. There are two *Pst*I sites, about 750 base-pairs apart, which flank the sulphonamide-resistance determinant (Fig. 12.1). None of the unique cleavage sites is located within the antibiotic-resistance determinants and none is particularly useful for cloning. Before the *Bst*, *Eco* and *Pst* sites can be used, another selective marker must be introduced into the RSF1010 genome. This need arises because the Sm^R and Su^R genes are transcribed from the same promoter (Bagdasarian *et al.* 1981). Insertion of a DNA fragment between the *Pst* sites inactivates both resistance determinants. Although the *Eco* and *Bst* sites lie outside the coding regions of the Sm^R gene, streptomycin resistance is lost if a DNA fragment is inserted at these sites unless the fragment provides a new promoter. Furthermore, the Su^R determinant which remains is a poor selective marker.

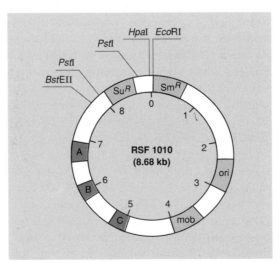

Fig. 12.1 The structure of plasmid RSF1010. The pale red tinted areas show the positions of the Sm^R and Su^R genes. The region marked *ori* indicates the location of the origin of replication. The *mob* function is required for conjugal mobilization by a compatible self-transmissible plasmid. A, B and C are the regions encoding the three replication proteins.

A whole series of vectors has been derived from plasmid RSF1010. The earliest vectors contained additional unique cleavage sites and more useful antibiotic resistance determinants. For example, plasmids KT230 and KT231 encode Km^R and Sm^R and have unique sites for *Hin*dIII, *Xma*I, *Xho*RI and *Sst*I which can be used for insertional inactivation. These two vectors have been used to clone in *P. putida* genes involved in the catabolism of aromatic compounds (Franklin *et al.* 1981). Vectors for the regulated expression of cloned genes also have been described. Some of these make use of the *tac* promoter (Bagdasarian *et al.* 1983, Deretic *et al.* 1987) or the phage T7 promoter (Davison *et al.* 1989) which will function in *P. putida* as well as *E. coli*. Another makes use of positively-activated twin promoters from a plasmid specifying catabolism of toluene and xylenes (Mermod *et al.* 1986). Expression of cloned genes can be obtained in a wide range of Gram-negative bacteria following induction with micromolar quantities of benzoate, and the product of the cloned gene can account for 5% of total cell protein.

Brunschwig and Darzins (1992) have constructed a two-component T7 expression system for use in *Pseudomonas* sp. This system was configured with the gene for phage T7 RNA polymerase in the chromosome, under the control of the inducible *lac* UV5 promoter, and the target gene on an RSF1010-based vector under control of a T7 RNA polymerase-responsive promoter. Upon induction, this system gave a 60-fold increase in gene expression with the gene product accounting for 20% of total soluble protein. The ability to specifically label proteins *in vivo* with ^{35}S-methionine following induction distinguishes this vector system from others described earlier. Other expression vectors permitting transcriptional and translational fusions of cloned genes to the *E. coli lacZ* coding region have been developed by Labes *et al.* (1990).

Plasmid vectors that contain the *cos* site (cosmids, see p. 82) and which can be packaged *in vitro* into bacteriophage λ particles are of considerable utility for cloning large segments of DNA and for constructing gene banks. Bagdasarian *et al.* (1981) have produced a cosmid derivative of RSF1010 called pKT247. Whereas normal RSF1010-derived vectors can be used to transform the recipient directly, *E. coli* has to serve as an intermediate host if cosmid cloning is employed. Cosmid pKT249 encodes Ap^R, Su^R and Sm^R and has unique sites for endonucleases *Eco*RI and *Sst*I. Frey *et al.* (1983) have described two improved cosmids. These new cosmids permit the selective cloning into their unique *Bam*HI site of 36-kb DNA fragments by a strategy that avoids the formation of polycosmids but does not require the cleaved vector to be dephosphorylated.

Sometimes it is desirable to introduce only a single copy of a gene into a new host, preferably into the chromosome. Methods are available for doing this in hosts which are genetically well-mapped (see p. 229 for an example in *Bacillus subtilis*). Targeted integration is more difficult in most other organisms. Hermesz *et al.* (1992) have developed a suitable method for the nitrogen-fixing symbiont *Rhizobium meliloti* using an integrative vector incorporating site-specific recombination elements of a temperate phage. Vectors carrying the phage-borne attachment site were constructed along with helper phages to provide the site-specific recombination functions *in trans*.

As indicated earlier, RSF1010-derived cloning vectors can be mobilized to other bacteria by certain conjugative plasmids. This characteristic is advantageous when the ultimate recipient is not transformable. However, for certain experiments regulatory authorities prefer vectors which have a low or non-existent frequency of conjugal transfer. For this reason Bagdasarian *et al.* (1981) constructed mobilization-defective derivatives of pKT230, pKT231 and pKT247, called pKT262, pKT263 and pKT264 respectively, by deleting the *mob* function (see Fig. 12.1). Three regions have been identified where deletion mutations affect mobilization but have the beneficial effect of increasing copy number up to fourfold (Frey *et al.* 1992).

Vectors derived from P group plasmids

The best-studied P group plasmid is RP4 (also known as RP1 and RK2). It specifies resistance to Ap, Km and Tc and has single cleavage sites for *Eco*RI, *Bam*HI (in the Ap^R gene), *Bgl*II, *Hpa*I and *Hind*III (in the Km^R gene). A study of transcriptional initiation in five species of Gram-negative bacteria has been made by Greener *et al.* (1992). Relatively little use has been made of RP4 as a cloning vehicle but Jacob *et al.* (1976) have used it for cloning DNA from *Rhizobium leguminosarum* and *Proteus mirabilis*, and Windass *et al.* (1980) for cloning the *E. coli* glutamate dehydrogenase gene in *M. methylotrophus*.

The basic problem with P group plasmids such as RP4 (RP1, RK2) is their size (60 kb). Attempts to produce smaller derivatives have been hindered by the fact that non-contiguous regions are required for replication (Fig. 12.2). Originally it was thought that three dispersed functions

Fig. 12.2 The structure of P group plasmid RK2. The red tinted areas indicate those regions of RK2 originally selieved to be essential for replication. The black areas indicate the location of genes encoding self-transmission of RK2. The site designated *rlx* must be present if conjugation-deficient derivatives of RK2 are to be mobilized by a compatible self-transmissible plasmid.

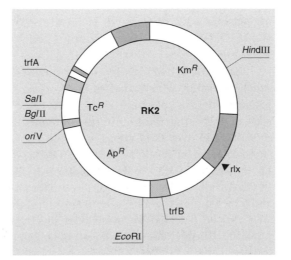

*ori*V, *trf*A and *trf*B were required. Later it was shown that a 0.7-kb segment containing the replication origin and a 1.8-kb fragment designated *trf*A* are the minimal requirements for replication (Schmidhauser *et al.* 1983). Smaller derivatives of RK2 have been obtained by partial digestion with *Hae*II. One derivative, pRK248, is only 10.5 kb in size and specifies Tc^R (Thomas *et al.* 1979). A more useful derivative, pRK2501, has been constructed by inserting a *Hae*II fragment specifying Km^R into pRK248. Plasmid pRK2501 has unique cleavage sites for *Sal*I, *Hind*III, *Xho*I, *Bgl*II and *Eco*RI, the first three being in either the Tc^R or Km^R genes. There are two problems associated with pRK2501. First, it is not stably maintained (Thomas *et al.* 1981). Second, it can only be transferred into recipients by transformation the conjugative functions of RK2 have been deleted. It cannot even be mobilized from *E. coli* to other recipients by other conjugative plasmids because a region of RK2 called *rlx*, a *cis*-acting function necessary for conjugal transfer, also has been removed.

An improved RK2-derived vector system has been constructed by Ditta *et al.* (1980). In this system the transfer and replication functions are located on different plasmids. Plasmid pRK290 contains a functional RK2 replicon and can be mobilized at high frequency by a helper plasmid because it retains the *rlx* locus. It encodes Tc^R and has single *Eco*RI and *Bgl*II sites for insertion of foreign DNA. The Km^R plasmid pRK203 consists of the RK2 transfer genes cloned onto a ColE1 replicon and its function is to mobilize RK290 into other hosts. Where the intended recipient of cloned genes is transformable, RK290 alone is used. If the recipient is not transformable, the vector RK290 containing the foreign DNA insert first is transformed into an *E. coli* strain carrying pRK2013 and then conjugated into the desired recipient. Many derivatives of RK290 have been constructed which have additional selectable markers and additional sites for gene cloning (for review see Schmidhauser *et al.* 1988). Cosmid vectors also have been constructed.

Most RK290-based vectors suffer from the same instability problem as pRK2501: they are not stably maintained. A *par* locus has been identified on plasmid RK2 (Roberts & Helinski 1992) and used to generate more stable RK2-derived cloning vectors (Crouzet *et al.* 1992, Weinstein *et al.* 1992).

Olsen *et al.* (1982) discovered a small (3.1 kb) multicopy, broad host-range plasmid which had arisen spontaneously from plasmid RP1. Presumably this plasmid retains the *ori*V and *trf*A* functions which appear to be the minimum requirements for replication of RK2. From this plasmid two derivatives were constructed. The first has two *Pst*I sites and can be used for cloning DNA where there is direct selection for the acquired trait, e.g. acquisition of antibiotic resistance or reversal of auxotrophy. The second plasmid contains an entire pBR322 molecule and consequently genes are inserted at the unique *Hind*III or *Bam*HI sites can be detected by insertional inactivation of the Tc^R marker. Like pRK2501, these two vectors can be used only if the intended host is transformable.

Vectors derived from group W plasmid Sa

Although a group W plasmid such as plasmid pSa (Fig. 12.3) can infect a wide range of Gram-negative bacteria, it has been developed mainly as a vector for use with the oncogenic bacterium *Agrobacterium tumefaciens* (see p. 274). Two regions of the plasmid have been identified as involved in conjugal transfer of the plasmid and one of them has the unexpected property of suppressing oncogenicity by *A. tumefaciens* (Tait *et al.* 1982). Information encoding the replication of the plasmid in *E. coli* and *A. tumefaciens* is contained within a 4-kb DNA fragment. Leemans *et al.* (1982) have described four small (5.6–7.2 MDa), multiply marked derivatives of pSa. The derivatives contain single target sites for a number of the common restriction endonucleases and at least one marker in each is subject to insertional inactivation. Although these Sa derivatives are non-conjugative they can be mobilized by other conjugative plasmids. Tait *et al.* (1983) also have constructed a set of broad host range vectors from pSa. The properties of their derivatives are similar to those of Leemans *et al.* (1982b) but one of them also contains the bacteriophage λ *cos* sequence and hence functions as a cosmid. Specialist vectors for use in *Agrobacterium* and which are derived from a natural *Agrobacterium* plasmid have been described by Gallie *et al.* (1988).

Factors affecting vector utilization

Quantitative data on the stability of maintenance of broad host-range vectors are limited given the frequent use of these plasmids in a range of Gram-negative bacteria. Quantitatively, the mini-RK2 vectors are known to

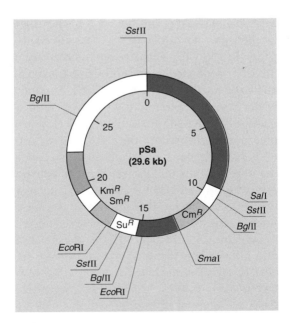

Fig. 12.3 The structure of plasmid Sa. The grey area encodes the functions essential for plasmid replication. The dark red areas represent the regions containing functions essential for self-transmission, the one between the *Sst* and *Sal* sites being responsible for suppression of tumour induction by *Agrobacterium tumefaciens*.

be segregationally unstable, as indicated above (p. 214), in the absence of a *par* function. In some cases the wild-type plasmid cannot be maintained in particular bacteria, e.g. RP4 in *Myxococcus xanthus* (Jaoua *et al.* 1987), Q group plasmids in *Bradyrhizobium japonicum* (Donnelly *et al.* 1987) and both P and Q group plasmids in *Bacteroides* (Shoemaker *et al.* 1986).

Analysis of the utility of different vectors is made difficult by the differential expression of plasmid-borne markers in different bacteria (Mermod *et al.* 1986, Labes *et al.* 1990). Comparative studies have shown that the optimal promoter in *E. coli* is not necessarily the best in unrelated bacteria. Furthermore, where regulatable promoters are used, different levels of inducer are needed in different bacteria. Also, there are indications that plasmid copy number can vary significantly from host to host, thus affecting expression levels, but few copy-number determinations have been done in bacteria other than *E. coli*.

Transposons as broad host-range vectors

Transposons are mobile genetic elements which can insert at random into plasmids or the bacterial chromosome independently of the host cell recombination system. In addition to genes involved in transposition, transposons carry genes conferring new phenotypes on the host cell, e.g. antibiotic resistance. Grinter (1983) devised a broad host-range cloning vector based on transposon Tn7 which encodes Tp^R and Sm^R. Unlike the vectors just described, the transposon vector permits the cloned genes to be stably inserted into the chromosome. Such a vector could have useful industrial applications for it does not put an extragenomic genetic load on the recipient cell.

Two compatible broad host-range plasmids form the basis of the original vector system. One of the plasmids is derived from RP1 but only encodes Tc^R. This plasmid is unstable and in the absence of selection for Tc^R is lost rapidly from host cells. A Tn7 derivative was inserted into this plasmid to generate pNJ5073. The Tn7 retains the original Tp^R and Sm^R markers but the DNA sequence which encodes transposition functions was replaced by a *Hin*dIII fragment containing the *E. coli trpE* gene. The second part of the vector system is plasmid pNJ9279 which encodes Km^R and is derived from plasmid R300b ($\equiv$ RSF 1010, see p. 211). It is used to provide the transposition functions missing from the Tn7 derivative.

If pNJ5073 alone is introduced into a bacterium then, as noted above, in the absence of antibiotic selection it will not be maintained stably. Not even the Tp^R and Sm^R markers will be retained, for the variant Tn7 is unable to transpose. If pNJ9297 also is present then it can provide in *trans* the necessary transposition functions: selection of Tp^R and Sm^R followed by screening for Tc^S will identify those cells in which the defective transposon has hopped into the chromosome. In this way Grinter (1983) was able to select strains of *M. methylotrophus* and *P. aeruginosa* which had incorporated the *E. coli trpE* gene into their chromosome. In theory, any gene can be inserted into the chromosome by replacing the *trpE* gene with an appropriate *Hin*dIII-generated DNA fragment.

The original vectors of Grinter (1983) have been much improved (Barry

1988). The Tn7 element has been marked with the *lac*YZ genes and transferred to the pUC-based high-copy-number vectors (see p. 67). Additional modifications include removal of restriction sites in the vector portion, the deletion of non-essential Tn7 and other sequences, the addition of new cloning sites within the Tn7-*lac* element and the incorporation of different promoters upstream from the *lac* genes to facilitate selection for transposition in diverse bacteria. The range of options to deliver DNA to the chromosome has been expanded with new carrier and delivery plasmids. The biocomponent system has been improved by employing IncQ plasmids as both helper and carrier and using a combination of incompatibility and instability to facilitate the isolation of chromosomal transposition derivatives.

More recently, Mahillon and Kleckner (1992) have used as a suicide vector for transposon delivery, a plasmid derived from *Bacillus* which is incapable of replicating in *E. coli* and other Gram-negative bacteria. The transposon cassette used by them also includes a polylinker carrying recognition sites for rare-cutting restriction endonucleases, e.g. *Not*I, *Sfi*I and *Pac*I, to facilitate physical mapping of chromosomal insertions.

Haas *et al.* (1993) have employed a different suicide vector strategy. The plasmid carrying the transposon, in this case a derivation of Tn1721, replicates in *E. coli* via an origin of replication derived from filamentous phage fd. Replication from this origin is strictly dependent on the bacteriophage fd gene 2 protein. Plasmid maintenance only occurs in strains in which the gene 2 protein is supplied from a recombinant λ prophage.

Cloning in *B. subtilis*

There are a number of reasons for cloning in *B. subtilis* (Harwood 1992). First, *Bacillus* spp. are Gram-positive and generally obligate aerobes compared with *E. coli* which is a Gram-negative facultative anaerobe. Thus the two groups of organisms have substantial differences in their biochemistry and metabolism. Second, *Bacillus* spp. are able to sporulate and consequently are used as models to study prokaryotic differentiation. The use of gene manipulation is facilitating these studies. Third, *Bacillus* spp. are widely used in the fermentation industry particularly for the production of exoenzymes. They can be tailored to secrete the products of cloned eukaryotic genes. Finally, from a biohazard point of view, *B. subtilis* is an extremely safe organism for it is not known to be pathogenic except in immunocompromised individuals. Indeed it is consumed in large quantities in the Far East.

Plasmid transformation

An essential feature of any cloning experiment involving plasmids is transformation of a recipient cell with recombinant DNA. Unlike *E. coli*, the genetically developed strains of *B. subtilis* are naturally transformable. Genetic competence develops as one of a global set of responses to nutrient limitation. The onset of competence is accompanied by the expression of several late gene products. Although all the regulatory circuits are not fully understood (Dubnau 1991, Errington 1993) the conditions for maximizing competence are well established. A highly transformable and restriction-deficient strain has been developed by Haima *et al.* (1987). The mechanism of DNA uptake and incorporation into homologous DNA is also understood in some detail (Venema 1979). Double-stranded DNA molecules are adsorbed to the cell surface of competent cells where they are subjected to exonuclease and endonuclease processing. One strand of the DNA molecule, about 6000–13 000 bases long, is internalized. Subsequent establishment of the DNA within the cell can occur by recombination with a homologous replicon in recombination-proficient (Rec$^+$) strains. Alternatively a Rec-independent pathway can be utilized which probably involves annealing of overlapping complementary strands derived from the same parental molecule, followed by repair synthesis (Michel *et al.* 1982).

Although it is very easy to transform *B. subtilis* with fragments of chromosomal DNA, there are problems associated with transformation by plasmid molecules. Ehrlich (1977) first reported that competent cultures of *B. subtilis* can be transformed with CCC plasmid DNA from *S. aureus* and that this plasmid DNA is capable of autonomous replication and expression in its new host. The development of competence for transformation by plasmid and chromosomal DNA follows a similar time course and in both cases transformation is first order with respect to DNA concentration suggesting that a single DNA molecule is sufficient for successful transformation (Contente & Dubnau 1979). However transformation of *B. subtilis* with plasmid DNA is very inefficient by comparison with chromosomal transformation for only one transformant is obtained per 10^3–10^4 plasmid molecules.

As will be seen later, much cloning in *B. subtilis* is done with bifunctional vectors that replicate in both *E. coli* and *B. subtilis*. Van Randen and Venema (1984) have shown that such bifunctional vectors can be transferred by replica-plating *E. coli* colonies containing them onto a lawn of competent *B. subtilis* cells. However, plasmid transformation by replica plating differed in one respect from plasmid transformation in liquid. Whereas chromosomal integration of plasmid-borne chromosomal alleles with concomitant loss of plasmids occurred frequently during regular plasmid transformation of Rec$^+$ *B. subtilis*, this was a rare event during plasmid transfer by replica plating.

An explanation for the poor transformability of plasmid DNA molecules was provided by Canosi *et al.* (1978). They found that the specific activity of plasmid DNA in transformation of *B. subtilis* was dependent on the degree of oligomerization of the plasmid genome. Purified monomeric CCC forms of plasmids transform *B. subtilis* several orders of magnitude less efficiently than do unfractionated plasmid preparations or multimers. Furthermore, the low residual transforming activity of monomeric CCC DNA molecules can be attributed to low level contamination with multimers (Mottes *et al.* 1979). Using a recombinant plasmid capable of replication in both *E. coli* and *B. subtilis* (pHV14, see p. 223). Mottes *et al.* (1979)

were able to show that plasmid transformation of *E. coli* occurs regardless of the degree of oligomerization, in contrast to the situation with *B. subtilis*. Oligomerization of linearized plasmid DNA by DNA ligase resulted in a substantial increase of specific transforming activity when assayed with *B. subtilis* and caused a decrease when used to transform *E. coli*. An explanation of the molecular events in transformation which generate the requirement for oligomers has been presented by de Vos *et al.* (1981). Basically, the plasmids are cleaved into linear molecules upon contact with competent cells just as chromosomal DNA is cleaved during transformation of *Bacillus*. Once the linear single-stranded form of the plasmid enters the cell it is not reproduced unless it can circularize, hence the need for multimers to provide regions of homology which can recombine. Michel *et al.* (1982) have shown that multimers, or even dimers, are not required provided part of the plasmid genome is duplicated. They constructed plasmids carrying direct internal repeats 260–2000 base-pairs long and found that circular or linear monomers of such plasmids were active in transformation.

Canosi *et al.* (1981) have shown that plasmid monomers will transform recombination proficient *B. subtilis* if they contain an insert of *B. subtilis* DNA. However the transformation efficiency of such monomers is still considerably less than that of oligomers. One consequence of the requirement for plasmid oligomers for efficient transformation of *B. subtilis* is that there have been very few successes in obtaining large numbers of clones in *B. subtilis* recipients (Keggins *et al.* 1978, Michel *et al.* 1980). The potential for generating multimers during ligation of vector and foreign DNA is limited. If the ratio of foreign to vector DNA is elevated in order to increase the proportion of recombinant molecules generated, the yield of transformants will decrease rapidly due to competition between vector–vector and vector–foreign DNA ligations. However, if transformants are obtained, the chances are that the plasmids will replicate as high-molecular-weight multimers. Apparently, insertion of foreign DNA into the commonly-used vectors can impair normal replication, leading to multimer formation (Gruss & Ehrlich 1988). Although this feature is of no value in the initial selection of recombinants it should facilitate any subsequent transformation steps.

Transformation by plasmid rescue

An alternative strategy for transforming *B. subtilis* has been suggested by Gryczan *et al.* (1980a). If plasmid DNA is linearized by restriction endonuclease cleavage, no transformation of *B. subtilis* results. However, if the recipient carries a homologous plasmid and if the restriction cut occurs within a homologous marker, then this same marker transforms efficiently. Since this rescue of donor plasmid markers by a homologous resident plasmid requires the *B. subtilis recE* gene product, it must be due to recombination between the linear donor DNA and the resident plasmid. Since DNA linearized by restriction endonuclease cleavage at a unique site is monomeric this rescue system (*plasmid rescue*) bypasses the requirement for a multimeric vector. The model presented by de Vos *et al.* (1981) to explain the requirement for oligomers (see previous section) can be adapted to account for transformation by monomers by means of plasmid rescue. In practice, foreign DNA is ligated to monomeric vector DNA and the *in vitro* recombinants used to transform *B. subtilis* cells carrying a homologous plasmid. Using such a 'plasmid rescue' system, Gryczan *et al.* (1980a) were able to clone various genes from *B. licheniformis* in *B. subtilis*.

Haima *et al.* (1990) have analysed the parameters affecting the usefulness and efficiency of plasmid rescue and used these to design an efficient *Bacillus* cloning system. In the absence of direct selection for cloned inserts, an important factor is the co-rescue frequency of the selective vector marker and the insert. The amount of DNA separating these two elements and which is non-homologous to the rescue plasmid affects the co-rescue frequency. The longer the non-homologous region, the lower is the frequency. Consequently, the vector system developed by Haima *et al.* (1990) is one in which a selective marker and the site of foreign DNA insertion are located close to one another. Another feature of their system is the ability to directly detect the presence of cloned DNA since the site of insertion in the vector is a *lacZα* gene derived from the pUC vectors (p. 67).

One disadvantage of the plasmid rescue method is that transformants contain both the recombinant molecule and the resident plasmid. Incompatibility will result in segregation of the two plasmids. This may require several subculture steps although Haima *et al.* (1990) observed very rapid segregation. Alternatively, the recombinant plasmids can be transformed into plasmid-free cells.

Transformation of protoplasts

A third method for plasmid DNA transformation in *B. subtilis* involves polyethylene glycol (PEG) induction of DNA uptake in protoplasts and subsequent regeneration of the bacterial cell wall (Chang & Cohen 1979). The procedure is highly efficient and yields up to 80% transformants, making the method suitable for the introduction even of cryptic plasmids. In addition to its much higher yield of plasmid-containing transformants, the protoplast transformation system differs in two respects from the 'traditional' system using physiologically competent cells. First, linear plasmid DNA and non-supercoiled circular plasmid DNA molecules constructed by ligation *in vitro* can be introduced at high efficiency into *B. subtilis* by the protoplast transformation system, albeit at a frequency 10–1000 lower than the frequency observed for CCC plasmid DNA. However, the efficiency of shotgun cloning is much lower with protoplasts than competent cells (Haima *et al.* 1988). Second, while competent cells can be transformed easily for genetic determinants located on the *B. subtilis* chromosome, no detectable transformation with chromosomal DNA is seen using the protoplast assay. Until recently a disadvantage of the protoplast system was that the regeneration medium was nutritionally complex, necessitating a two-step selection procedure for auxotrophic

Table 12.2 Comparison of the different methods of transforming *B. subtilis*

System	Efficiency (transformants/µg DNA)		Advantages	Disadvantages
Competent cells	Unfractionated plasmid	2×10^4	Competent cells readily prepared. Transformants can be selected readily on any medium. Recipient can be Rec$^-$	Requires plasmid oligomers or internally duplicated plasmids which makes shotgun experiments difficult unless high DNA concentrations and high vector/donor DNA ratios are used. Not possible to use phosphatase-treated vector
	Linear	0		
	CCC monomer	4×10^4		
	CCC dimer	8×10^3		
	CCC multimer	2.6×10^5		
Plasmid rescue	Unfractionated plasmid	2×10^6	Oligomers not required. Can transform with linear DNA. Transformants can be selected on any medium	Transformants contain resident plasmid and incoming plasmid and these have to be separated by segregation or retransformation. Recipient must be Rec$^+$
Protoplasts	Unfractionated plasmid	3.8×10^6	Most efficient system. Gives up to 80% transformants. Does not require competent cells. Can transform with linear DNA and can use phosphatase-treated vector	Efficiency lower with molecules which have been cut and re-ligated. Efficiency also very size-dependent, and declines steeply as size increases
	Linear	2×10^4		
	CCC monomer	3×10^6		
	CCC dimer	2×10^6		
	CCC multimer	2×10^6		

markers. Recently details have been presented of a defined regeneration medium (Puyet *et al*. 1987).

The advantages and disadvantages of the three transformation systems are summarized in Table 12.2. It should be noted that although shotgun cloning of chromosomal DNA is difficult, it is relatively easy to isolate recombinants containing fragments of bacteriophage DNA inserted into plasmid DNA. This is simply because of the small size of these genomes. Another point of interest is that in competent cells of *B. subtilis*, transforming DNA is taken up in a single-stranded form. A consequence of this is that it is not possible to clone recombinant molecules freshly constructed *in vitro* using homopolymer tailing. The reason is that recombinant molecules prepared in this way contain single-stranded regions because the homopolymer tails on the two components are generally of different lengths. For similar reasons it is not possible to use phosphatase-treated vectors when cloning directly into competent cells. These problems do not arise if protoplasts are transformed, because the DNA is taken up in a double-stranded form.

Plasmid vectors

The development of *B. subtilis* cloning systems was hindered by the absence of suitable vector replicons; *E. coli* plasmids do not replicate efficiently in *B. subtilis* nor are their markers expressed. A large number of cryptic plasmids had been identified in *B. subtilis* (Le Hegarat & Anagnostopoulos 1977) but in the absence of a detectable phenotype they could not be used for cloning. Ehrlich (1977) showed that plasmids isolated from *S. aureus*, such as pC194 which encodes chloramphenicol resistance, can be transformed into *B. subtilis* where they replicate and express antibiotic resistance normally. As well as pC194, a number of other plasmids

Table 12.3 Properties of some *S. aureus* plasmids used as vectors in *B. subtilis*

Plasmid	Phenotype conferred on host cell	Size	Copy no.	Other comments
pC194	Chloramphenicol resistance	2906 bp	15	Generates large amount of high molecular weight DNA when carrying heterologous inserts
pE194	Erythromycin resistance	3728 bp	10	*cop*-6 derivative has copy number of 100. Plasmid is naturally temperature-sensitive for replication
pUB110	Kanamycin resistance	4548 bp	50	Virtually the complete sequence is involved in replication maintenance, site-specific plasmid recombination or conjugal transfer

such as pUB110, pE194 and pT127 have been used in fundamental studies and for the development of cloning vectors in *B. subtilis*.

As can be seen from Table 12.3, none of the natural *S. aureus* plasmids carries more than one selectable marker and so improved vectors have been constructed by gene manipulation, e.g. pHV11 is pC194 carrying the Tc^R gene of pT127 (Ehrlich 1978). In general, these plasmids are stable in *B. subtilis* but segregative stability is greatly reduced following insertion of exogenous DNA (Bron & Luxen 1985). A major disadvantage of these plasmids is that very low efficiencies of cloning are obtained when selection for recombinants is made directly in *B. subtilis*.

E. coli–B. subtilis shuttle plasmids

Because of the difficulties experienced in direct cloning in *B. subtilis*, hybrid plasmids were constructed which can replicate in both *E. coli* and *B. subtilis*. Originally most of these were constructed as fusions between pBR322 and pC194 or pUB110. With such plasmids, *E. coli* can be used as an efficient intermediate host for cloning. Plasmid preparations extracted from *E. coli* clones are subsequently used to transform competent *B. subtilis* cells. Such preparations contain sufficient amounts of multimeric plasmid molecules to be efficient in *B. subtilis* competent cell transformation (see p. 218). The use of *E. coli* as an intermediate host has a number of disadvantages. First, many genes from Gram-positive bacteria are difficult to clone in *E. coli*, e.g. genes encoding sporulation or competence. Second, complementation studies are not always possible. Finally, use of an intermediate host is time-consuming. Fortunately, improved vectors (see below) have been developed for direct cloning in *B. subtilis* making superfluous the use of *E. coli* as an intermediate host. However, shuttle vectors still are useful for comparative studies between *E. coli* and *B. subtilis*.

Table 12.4 lists some of the commonly used shuttle plasmids. Note that some of them carry some of the features described earlier for *E. coli* plasmids, e.g. the *E. coli lacZα*-complementation fragment, multiple cloning sites (MCS, see p. 67) and the phage f1 origin for subsequent production of single-stranded DNA in a suitable *E. coli* host (see p. 87).

Table 12.4 *B. subtilis–E. coli* shuttle plasmids

Plasmid	Size (kbp)	Replicon E. coli	Replicon B. subtilis	Markers E. coli	Markers B. subtilis	Comments
pHV14	4.6	pBR322	pC194	Ap,Cm	Cm	pBR322/pC194 fusion. Sites: *Pst*I, *Bam*HI, *Sal*I, *Nco*I (Ehrlich 1978)
pHV15	4.6	pBR322	pC194	Ap,Cm	Cm	pHV14, reversed orientation of pC194 relative to pBR322
pHV33	4.6	pBR322	pC194	Ap,Tc,Cm	Cm	Revertant of pHV14 (Primrose & Ehrlich 1981)
pEB10	8.9	pBR322	pUB110	Ap,Km	Km	pBR322/pUB110 fusion (Bron *et al.* 1988)
pLB5	5.8	pBR322	pUB110	Ap,Cm,Km	Cm,Km	Deletion of pBR322/pUB110 fusion, *Cm*^{R} gene of pC194 Segregationally unstable (Bron & Luxen 1985). Sites: *Bam*HI, *Eco*RI, *Bgl*III (in *Km*^{R} gene), *Nco*I (in *Cm*^{R} gene)
pHP3	4.8	pBR322	pTA1060	Em,Cm	Em,Cm	Segregationally stable pTA1060 replicon (Peeters *et al.* 1988). Copy number, ca. 5. Sites: *Nco*I (*Cm*^{R} gene), *Bcl*I and *Hpa*I (both *Em*^{R} gene)
pHP3Ff	5.3	pBR322	pTA1060	Em,Cm	Em,Cm	Like pHP3; phage f1 replication origin and packaging signal
pGPA14	5.8	pBR322	pTA1060	Em	Em	Stable pTA1060 replicon. Copy number, ca. 5. α-Amylase-based selection vector for protein export functions (Smith *et al.* 1987). MCS of M13*mp*11 in *lacZ*α
pGPB14	5.7	pBR322	pTA1060	Em	Em	As pGPA14, probe gene TEM-β lactamase
pHP13	4.9	pBR322	pTA1060	Em,Cm	Cm,Cm	Stable pTA1060 replicon. Copy number, ca. 5. Efficient (shotgun) cloning vector (Haima *et al.* 1987). MCS of M13*mp*9 in *lacZ*α LacZα not expressed in *B. subtilis*. Additional sites: *Bcl*I and *Hpa*I (both *Cm*^{R} gene)
pHV1431	10.9	pBR322	μAMβ1	Ap,Tc,Cm	Cm	Efficient cloning vector based on segregationally stable pAMβ1 (Jannière *et al.* 1990). Copy number, ca. 200. Sites: *Bam*HI, *Sal*I, *Pst*I, *Nco*I. Structurally unstable in *E. coli*
pHV1432	8.8	pBR322	pAMβ1	Ap,Tc,Cm	Cm	pHV1431 lacking stability fragment orfH. Structurally stable in *E. coli*
pHV1436	8.2	pBR322	pTB19	Ap,Tc,Cm	Cm	Low-copy-number cloning vector (Jannière *et al.* 1990) Structurally stable

Improved vectors for cloning in *B. subtilis*

pTA1060 derivatives

Until recently, the use of plasmids for direct cloning in *B. subtilis* was problematical mainly because of structural and segregational instability. Segregational instability appears to be associated with a lack of stability functions and a decrease in the copy number of plasmids containing inserts. Bron and Luxen (1985) found that the segregational instability of a series of plasmids was proportional to their copy number and copy number decreased as the insert size increased. Reasoning that stable host–vector systems in *B. subtilis* are more likely if endogenous plasmids are used, Bron and colleagues have developed the cryptic *Bacillus* plasmid pTA1060 as a vector (Haima *et al.* 1987, Bron *et al.* 1989). For example, plasmid pBB2 is pTA1060 carrying the *Cm*^{R} and *Km*^{R} genes from pC194 and pUB110, respectively (Fig. 12.4). The copy number of pTA1060 and pBB2 is five per chromosome but both plasmids are segregationally more stable than pUB110 derivatives. A useful small derivative of pTA1060 is

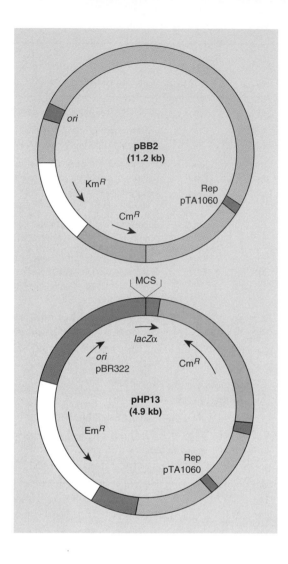

Fig. 12.4 The structure of pTA1060 derivatives pBB2 and pHP13 (not to scale). Sequences derived from pTA1060 are shown in pink and sequences from pUC9 are shown in red. MCS, multiple cloning site.

pHP13 (Haima *et al.* 1987). It contains approximately 1.4 kbp of pTA1060 sequences comprising the primary replication functions, the *Em*^R gene from pE194, the *Cm*^R gene from pC194, and the *lacZα* gene plus replication functions from pUC9 (p. 67). Both antibiotic resistance markers are expressed in *E. coli* and *B. subtilis* and the *lacZα* gene is expressed in *E. coli*. This 4.9-kbp *E. coli–B. subtilis* shuttle plasmid has copy numbers of five per chromosome in *B. subtilis* and about 200 in *E. coli*. Restriction sites enabling clone selection by marker inactivation are *Bcl*I (*Em*^R gene) and *Nco*I (*Cm*^R gene) in *B. subtilis*. In addition, the polylinker site from pUC9 (in *lacZα*) enables direct clone selection in *E. coli*. In restriction-deficient *B. subtilis* hosts, this plasmid is highly efficient for cloning. Clones can be obtained with a high frequency (>30% of the transformants), and large inserts are relatively frequent.

Early in the development of *B. subtilis* cloning vectors it was noted that only short DNA fragments could be efficiently cloned (Michel *et al.* 1980) and that longer DNA segments often undergo rearrangements (Ehrlich *et al.* 1986). This structural instability is independent of the host recombination systems for it still occurs in Rec⁻ strains (Peijnenburg *et al.* 1987).

A major contributing factor to structural instability of recombinant DNA in *B. subtilis* appears to be the mode of replication of the plasmid vector (Gruss & Ehrlich 1989, Jannière *et al.* 1990). All the commonly used vectors replicated by a rolling-circle mechanism (see Box 12.2). Nearly every step in the process does or could digress from its usual function thus effecting rearrangements. Also, single-stranded DNA is known to be a reactive intermediate in every recombination process and single-stranded DNA is generated during rolling-circle replication.

More direct evidence comes from the following observations. First, directly repeated sequences of varying lengths (9–4000 bp) recombined up to 1000 times more frequently when carried on pC194 than when inserted in the chromosome (Niaudet *et al.* 1984, Jannière & Ehrlich 1987). Second, recombination between chromosomally carried long repeats was enhanced over 100-fold when a rolling-circle replicon was inserted in the chromosome close to the repeats (Noirot *et al.* 1987, Young & Ehrlich 1989). Finally, induction of rolling-circle replication strongly stimulated recombination between short direct repeats in an *E. coli* plasmid (Brunier *et al.* 1989).

If structural instability is a consequence of rolling-circle replication then vectors which replicate by an alternative mechanism could be more stable. Jannière *et al.* (1990) have studied two potentially useful plasmids, pAMβ1 and pTB19, which are large (26.5 kb) natural plasmids derived from *Streptococcus (Enterococcus) faecalis* and *B. subtilis* respectively. Replication of these plasmids does not lead to accumulation of detectable amounts of single-stranded DNA whereas the rolling-circle mode of replication does. Also, the replication regions of these two large plasmids share no sequence homology with the corresponding highly conserved regions of the rolling-circle-type plasmids. At least one of them, pAMβ1, undergoes theta replication (Bruand *et al.* 1991).

Vectors derived from both plasmids permitted efficient cloning and stable maintenance of long DNA segments (up to 33 kb) and exhibited a lower recombination frequency of repeated sequences. Despite the improved structural stability of recombinants constructed with these vectors, deletions still can occur, albeit at a much lower frequency. Such deletions may be better adapted to their new hosts and enriched during culture (Leonhardt & Alonso 1991).

Swinfield *et al.* (1991) have identified a region of pAMβ1 which enhances in *B. subtilis* the segregational stability of pAMβ1-derived cloning vectors. This DNA segment encodes a protein with substantial homology to bacterial site-specific recombinases (resolvases). Improved segregational stability is associated with a reduction in plasmid polymerization.

Box 12.2 The two modes of replication of circular DNA molecules

There are two modes of replication of circular DNA molecules: via theta-like structures or by a rolling-circle type of mechanism. Visualization by electron microscopy of the replicating intermediates of many circular DNA molecules reveals that they retain a ring structure throughout replication. They always possess a theta-like shape that comes into existence by the initiation of a replicating bubble at the origin of replication (Fig. B12.1). Replication can proceed either uni- or bidirectionally. As long as each chain remains intact, even minor untwisting of a section of the circular double helix results in the creation of positive supercoils in the other direction. This supercoiling is relaxed by the action of topoisomerases (see Fig. 4.7) which create single-stranded breaks (relaxed molecules) and then re-seal them.

An alternative way to replicate circular DNA is the rolling circle mechanism (Fig. B12.2). DNA synthesis starts with a cut in one strand at the origin of replication. The 5′-end of the cut strand is displaced from the duplex permitting the addition of deoxyribonucleotides at the free 3′-end. As replication proceeds, the 5′-end of the cut strand is rolled out as a free tail of increasing length. When a full-length tail is produced, the replicating machinery cuts it off and ligates the two ends together. The double-stranded progeny can re-initiate replication whereas the single-stranded progeny must first be converted to a double-stranded form. Gruss and

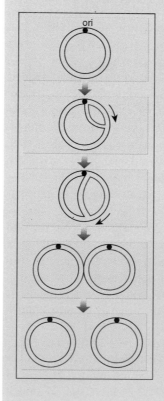

Fig. B12.1 Theta replication of a circular DNA molecule. The original DNA is shown in black and the newly-synthesized DNA in red. • represents the origin of replication and the arrow shows the direction of replication.

continued

Box 12.2 *continued*

Ehrlich (1989) have suggested how deletants and defective molecules can be produced at each step in the rolling-circle process.

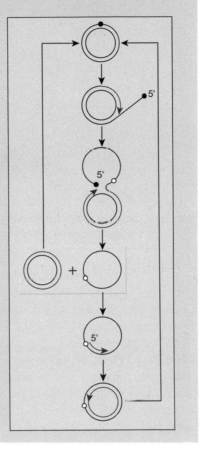

Fig. B12.2 Rolling circle replication of a circular DNA molecule. The original DNA is shown in black and the newly-synthesized DNA in red. The solid and open circles represent the positions of the replication origins of the outer (+) and inner (−) circles respectively.

Presumably the enzyme promotes stability by maintaining the plasmid population in a monomeric state thereby reducing the probability of plasmidfree segregants arising.

Direct clone selection

Until recently, there was no system for direct clone selection in *B. subtilis* and the presence of inserts had to be detected by marker inactivation. Haima *et al*. (1990) now have developed a suitable system based on the *E. coli* β-galactosidase α-complementation system. The plasmid component is pHPS9 which is a derivative of pHP13 (Fig. 12.5). It contains the Cm^R and Em^R markers and a constitutively expressed translational fusion of the *B. pumilis cat-86* gene and the pUC9-derived *LacZα* gene. The pUC9 multiple

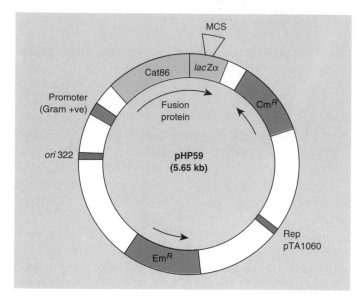

Fig. 12.5 The structure of pHPS9 which is used for direct clone selection in *B. subtilis*. The Gram-positive promoter shown is responsible for the constitutive synthesis of the *cat86 – lacZα* fusion protein.

cloning site, present in the *lacZα* gene, permits the ready insertion of restriction fragments. The host component of the direct selection system is *B. subtilis* strain 6GM15. This strain carries the *lacZΔM15* gene which has been fused to a Gram-positive promoter and translational initiation region. This gene, together with a selectable Km^R marker gene, was stably integrated as a single copy into the chromosome of strain 6GM by a double crossover event. The presence of the Km^R gene adjacent to the *lacZM15* gene facilitates the transfer of the latter to other strains by transformation.

Cells of strain 6GM15 carrying pHPS9 yield blue colonies on XGal containing media. Cells containing inserts in the multiple cloning site of pHPS9 can be selected by virtue of the white colonies which they form on XGal. This system is very efficient in shotgun cloning, in particular when multimeric plasmid molecules are formed by sticky end ligation.

The fate of cloned DNA

When DNA from an organism which has little or no homology with *B. subtilis*, e.g. *B. licheniformis*, is cloned in *B. subtilis* the chimaeric plasmid appears to be stable. However, if the cloned insert has homology with *B. subtilis* the outcome is quite different. The homologous region may be excized and incorporated into the chromosome by the normal recombination process, i.e. substitution occurs via a double crossover event. However, if only a single crossover occurs the entire recombinant plasmid is integrated into the chromosome. It is possible to favour integration by using vectors based on pBR322 which cannot replicate autonomously in *B. subtilis* (Michel *et al.* 1983). Amplification of the cloned gene is possible following integration if an antibiotic-resistance gene has been integrated simultaneously. Although amplified genes are more stable on the chromosome compared with the plasmid, full stability is not ensured (Young & Ehrlich 1989). Integration into the chromosome also can occur with pE194 if selection for Em^R is made at 42°C, a temperature at which it fails to replicate autonomously. A derivative of pE194, called pE194Ts, has been isolated which is unable to grow at temperatures above 37°C (Villafane *et al.* 1987) and gives much tighter selection of integration or plasmid loss. One application of this mutant plasmid is as a delivery vehicle for transposon Tn*917* and improved derivatives of Tn*917* (for review, see Youngman 1990).

Genes cloned into a plasmid and flanked by regions homologous to the chromosome also can integrate without tandem duplication of the chromosomal segments. In this case, the plasmid DNA is linearized before transformation. This technique, which was used in the construction of strain 6GM15 described above, is shown in Fig. 12.6. The same technique can be used to generate deletions. The gene of interest is cloned, a portion of the gene replaced *in vitro* with a fragment bearing an antibiotic marker, and the linearized plasmid transformed into *B. subtilis* with selection made for antibiotic resistance.

Once a recombinant plasmid has integrated into the chromosome it is relatively easy to clone adjacent sequences. Suppose, for example, that a vector carrying *B. subtilis* DNA in the *Bam*HI site (Fig. 12.7) has recombined into the chromosome. If the recombinant plasmid has no *Bgl*II sites it can be recovered by digesting the chromosomal DNA with *Bgl*II, ligating the resulting fragments and transforming *E. coli* to Ap^R. However, the plasmid which is isolated will be larger than the original one because DNA flanking the site of insertion will also have been cloned. In this way Niaudet *et al.* (1982) used a plasmid carrying a portion of the *B. subtilis* *ilvA* gene to clone the adjacent *thyA* gene.

Expression of cloned genes

E. coli is promiscuous in its ability to recognize transcription and translation signals from a wide variety of organisms, e.g. the expression of drug-resistance genes from Gram-positive bacteria (Cohen *et al.* 1972, Ehrlich 1978) and amino acid biosynthetic genes from yeast (Struhl *et al.* 1976). By contrast, *B. subtilis* is restricted in its ability to express genes from other genera (Ehrlich 1978a,b). With one exception (Rubin *et al.* 1980), the only foreign genes expressed in *B. subtilis* are from other Gram-positive bacteria, e.g. drug-resistance genes from *S. aureus* (Ehrlich 1977) and *Streptococcus* (Yagi *et al.* 1978). The lack of expression of *E. coli* genes in *B. subtilis* is not due to structural alterations, for these genes can be isolated from *B. subtilis* and are still functional when re-introduced into *E. coli*. What, then, are the barriers to the expression of *E. coli* genes in *B. subtilis*?

The first step in gene expression is the transcription of DNA by RNA

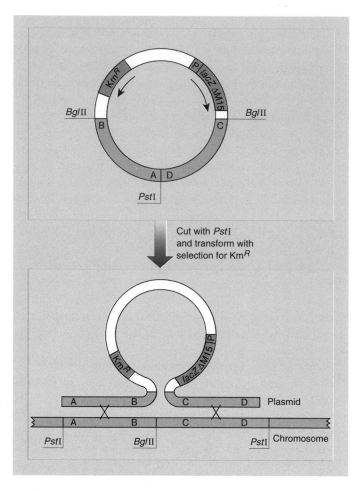

Fig. 12.6 Insertion of plasmid DNA into the chromosome by a double crossover event. The *B. subtilis* DNA is shown in grey and the letters A to D represent different chromosomal sequences. Vector DNA is shown in white and other vector-borne genes in pink.

polymerase. A comparison of the RNA polymerases from the two organisms has not revealed gross differences in the core enzyme but the vegetative sigma factors are of very different sizes: 43 kDa in *B. subtilis* and 70 kDa in *E. coli*. The net result is that the template specificity of the two polymerase differs in a very interesting way. The *E. coli* enzyme can transcribe equally well DNAs of *E. coli* bacteriophage T4 and of *B. subtilis* bacteriophage Φe. By contrast, the *B. subtilis* polymerase is much less active with the *E. coli* than with the *B. subtilis* phage DNAs (Shorenstein & Losick 1973). This behaviour is consistent with the observation that *B. subtilis* genes can function in *E. coli* but the *E. coli* genes cannot function in *B. subtilis*. Moran *et al.* (1982) determined the nucleotide sequence of two *B. subtilis* promoters and found no obvious differences with those of *E. coli*. Since then over 50 *B. subtilis* promoters, recognized by σ factors associated

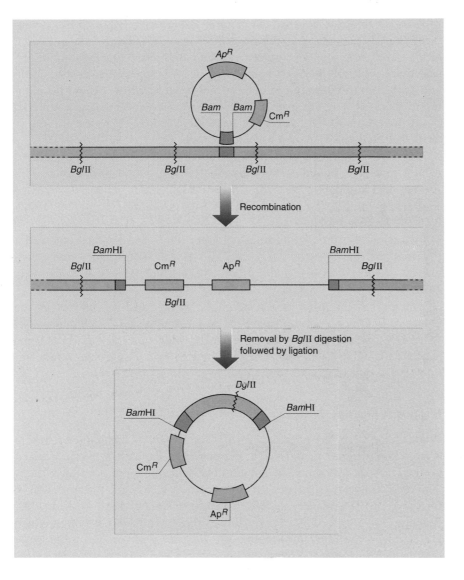

Fig. 12.7 Cloning DNA sequences flanking the site of insertion. The pink bar on the plasmid represents a *Bam*HI fragment of *B. subtilis* chromosomal DNA. Note that the plasmid has no *Bgl*II sites. (See text for details.)

with vegetative growth, have been sequenced and the consensus −35 and −10 sequences are identical to those from *E. coli*. From detailed studies it would appear that recognition of a promoter in *B. subtilis* involves other structural features, but what these features are is not clear.

The translation apparatus of *B. subtilis* differs significantly from that of *E. coli* (for review, see Vellanoweth 1993). This is demonstrated by the observation that *E. coli* ribosomes can support protein synthesis when directed by mRNA from a range of Gram-positive and Gram-negative organisms whereas ribosomes from *B. subtilis* recognize only homologous mRNA (Stallcup *et al.* 1974). The explanation for the selectivity of *B. subtilis* ribosomes is that they lack a counterpart to the largest *E. coli* ribosomal protein, S1 (Higo *et al.* 1982, Roberts & Rabinowitz 1989). Other Gram-positive bacteria such as *Staphylococcus*, *Streptococcus*, *Clostridium* and *Lactobacillus* also lack an S1-equivalent protein and they too exhibit mRNA selectivity. The role of S1 is believed to be to bind RNA non-specifically and bring it to the decoding site of the 30S subunit, where proper positioning of the S–D sequence (see p. 139) and initiation codon signals can take place. Interestingly, *E. coli* ribosomes depleted of S1 preferentially translated *B. subtilis* mRNAs over homologous messages.

The S–D sequences of *B. subtilis* and other Gram-positive organisms exhibit more complementarity with the 16S RNA than is observed with *E. coli* (McLaughlin *et al.* 1981, Moran *et al.* 1982). Thus the average free energy of binding of a *B. subtilis* S–D sequence to the 16S rRNA is -16.7 kcal/mol while that of the average *E. coli* S–D sequence is -11.7 kcal/mol. More significantly, while *E. coli* S–D sequences showed a distribution from -2 to -22 kcal/mol, no *B. subtilis* sequence had a free energy of formation less stable than -12 kcal/mol (Hager & Rabinowitz 1985).

Another major difference between *B. subtilis* and *E. coli* lies in the choice of initiation codon. While 91% of the sequenced *E. coli* genes start with AUG, nearly 30% of the genes of *B. subtilis* and related organisms start with UUG or GUG (Hager & Rabinowitz 1985). CUG also can function as a start codon (Ambulos *et al.* 1990).

Plasmid expression vectors

The development of expression vectors for *B. subtilis* has suffered from a lack of knowledge of controllable promoters in *Bacillus* sp. Vectors based on the *E. coli lac* system or the temperature-sensitive repressor from phages lambda (*E. coli*) or Φ105 (*Bacillus*) have been used to regulate expression of foreign genes (Osburne *et al.* 1984, Yansura & Henner 1984, LeGrice 1990, Breitling *et al.* 1990). Other expression vectors have been developed which are based on the regulatory region of a *B. subtilis* sucrose-inducible gene, *sac*B, which encodes the extracellular enzyme levansucrase (Wong 1989). The regulatory region of *sac*B has been well characterized. There is a terminator-like structure between the *sac*B promoter and its ribosome-binding site. A positive regulatory gene, *sac*Y, which encodes an anti-terminator protein, is responsible for mediating the sucrose-induced expression (Crutz *et al.* 1990). Wu *et al.* (1991) have constructed an expression cassette consisting of *sac*Y coupled to a strong constitutive promoter upstream of the *sac*B promoter (Fig. 12.8). When induced with sucrose expression of the cloned gene increased at least 18-fold.

Plasmid secretion systems

Because of the high levels of industrial enzymes secreted by *Bacillus* sp., members of the genus have been regarded as attractive hosts for the production of both homologous and heterologous secretory proteins. The

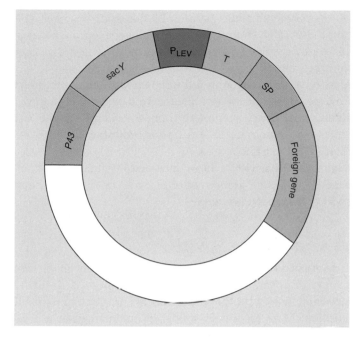

Fig. 12.8 Schematic representation of an inducible expression and secretion cassette for use in *Bacillus subtilis*. P43 is strong constitutive promoter. *SacY* encodes an antitermination molecule which acts on T. P_{LEV} is a promoter inducible with sucrose.

export mechanism in *Bacillus* resembles that of *E. coli* (for review, see Simonen & Palva 1993). However, there are differences in the signal peptides of Gram-positive bacteria compared with those found in *E. coli* or eukaryotes. For example, the NH_2-termini are more positively charged. The signal peptides are also larger and the extra length is distributed among all three regions of the signal peptide (as described on p. 145). Also the wall structure of Gram-positive bacteria is simpler than that of Gram-negative bacteria and proteins secreted through the cytoplasmic membrane appear in the growth medium.

Genes for secreted proteins which originate in Gram-positive bacteria are usually expressed in *B. subtilis* with their own promoters and signal sequences and special secretion vectors are not required. These secreted proteins usually are resistant to the many proteases which the host secretes into the growth medium. Genes for secreted proteins which are derived from Gram-negative bacteria are usually cloned in special secretion vectors. These utilize promoters and ribosome binding sites from *Bacillus* exoprotein genes since the corresponding structures from Gram-negative organisms are often non-functional in *Bacillus*. The join between the vector and the foreign gene insert is usually made close to the signal sequence cleavage site. In this way the functions of the promoter and ribosome-binding site are unaffected structurally. The yield of secreted protein depends on the efficiency of the expression cassette and the sensitivity of the protein to proteolytic degradation. To this end, Wu *et al.* (1991) have constructed a *B. subtilis* strain which lacks all six exoproteinases produced by the wild-type organism and which prolongs the half-life of a secreted β-lactamase from 1.5 to 85 hours.

Problems have been encountered frequently when attempts have been made to over-express and secrete proteins of eukaryotic origin. Many of these proteins are poorly exported from *B. subtilis* despite being true secretory proteins in their normal eukaryotic producer. This may be because *B. subtilis* lacks suitable chaperons.

Simonen and Palva (1993) have included in their review of protein secretion in *Bacillus* spp. details of the levels of different heterologous proteins which have been secreted.

Bacteriophage vectors for use in *B. subtilis*

Although plasmid vectors have been used extensively in *B. subtilis* they do have some disadvantages. Not the least of these is the instability of plasmids bearing inserts, although the factors contributing to instability are being identified. Many of the plasmids are maintained at a greater copy number than the host chromosome, and with some cloned genes the resultant gene-dosage effects can be deleterious for the cell (Banner *et al.* 1983). Finally, homologous inserts of *B. subtilis* DNA cannot be maintained on plasmids except in recombination-deficient hosts. Such Rec⁻ hosts are difficult to construct and maintain and, for those studying sporulation, have the additional disadvantage that they do not sporulate well. Vectors based on bacteriophage Φ105 appear to be free from these defects.

Bacteriophage Φ105 is temperate and thus has a chromosomal attachment site analogous to that of λ in *E. coli*. Thermo-inducible mutants are available. The wild-type genome is 39.2 kb long but, like λ, it can package larger genomes. The upper size limit for efficient packaging is 40.2 kb which is only 1 kb larger than the wild-type genome (Errington & Pughe 1987). However, because some phage DNA can be deleted it is possible to isolate vectors which permit the cloning of up to 5 kb of foreign DNA (Jones & Errington 1987).

Bacteriophage vectors are used in a manner analogous to plasmid vectors (Errington 1984) as shown in Fig. 12.9. DNA is isolated from a suitable vector phage such as Φ105J9, which has unique sites for *Bam*HI and *Xba*I. As with λ, the phage DNA has cohesive ends and multimers are obtained by incubating in the presence of DNA ligase. Treatment of the multimers generates linear molecules containing an entire phage genome flanked by *Bam*HI ends. After ligation with chromosomal DNA fragments, circularized recombinants are selected by transfection of protoplasts. The resultant 'transducing' phages can infect and stably lysogenize *B. subtilis*.

Gibson and Errington (1992) have developed a novel expression vector based on Φ105 and employed it for the production of β-lactamase in *B. subtilis*. Expression of the β-lactamase-encoding gene was low when cloned into the prophage under the control of its own promoter. However,

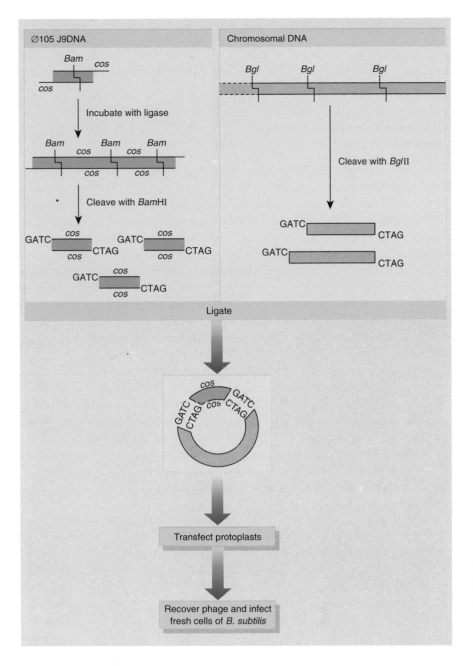

Fig. 12.9 The use of the phage vector Φ105J9 for the construction of genomic libraries in *B. subtilis*. (See text for details.) The cohesive ends of the phage DNA are indicated by *cos*.

expression was considerably elevated when the gene was inserted into the phage genome in the same orientation as phage transcription. A defective Φ105 vector was constructed with a deletion removing a region needed for cell lysis, and with a mutation in the immunity repressor, rendering it temperature sensitive. Production of β-lactamase was induced by a shift in temperature. Since the cells did not lyse, purification of the β-lactamase from the supernatant fluid was very easy.

Review of cloning techniques

An extensive and detailed review of the theory and practice of cloning and gene manipulation in *B. subtilis* has been provided by Harwood and Cutting (1990).

Cloning in Gram-positive bacteria other than *B. subtilis*

The only Gram-positive organisms other than *B. subtilis* in which there has been extensive *in vitro* gene manipulation are the streptomycetes and lactic streptococci. Much of the interest in cloning in these organisms is because of their industrial applications. Lactic streptococci are used in the production of fermented milks, e.g. yoghurt and cheeses, and the requisite metabolic functions are plasmid borne. The techniques of gene manipulation for dairy organisms such as *Streptococcus lactis* are little different from those used with *B. subtilis* (for reviews see Kondo & McKay 1985 and Gasson & Davies 1985). The only technique not described so far is plasmid transfer by the use of protoplast fusion between related species (Okamoto *et al.* 1985) or unrelated species such as *S. lactis* and *B. subtilis* (Baigori *et al.* 1988).

Over 60% of known naturally-occurring antibiotics come from *Streptomyces*, making the genus the major taxonomic class sought by pharmaceutical companies in screening programmes for new naturally-occurring isolates. Recombinant DNA technology is being used to generate novel antibiotics and to improve the titres of naturally-occurring antibiotics.

These topics are covered in a later chapter (see p. 387). As with the lactic streptococci, the methods for handing recombinant DNA in *Streptomyces* are similar to those already described and will not be detailed here (for review see Hopwood *et al.* 1987).

One feature of *Streptomyces* does merit further discussion. Many species spontaneously amplify specific chromosomal DNA sequences, giving rise to several hundred tandem copies which can account for 10% of total DNA. Altenbuchner and Cullum (1987) linked a thiostrepton resistance gene and a tyrosinase gene to an amplifiable sequence from *S. lividans*. This construct was transformed into *S. lividans* where it integrated into the chromosome by homologous recombination. When DNA amplification was induced the cloned genes were co-amplified to give stable high-copy-number clones.

13 Cloning in *Saccharomyces cerevisiae* and other microbial eukaryotes

Introduction

The analysis of eukaryotic DNA sequences has been facilitated by the ease with which DNA from eukaryotes can be cloned in prokaryotes using the vectors described in previous chapters. Such cloned sequences can be obtained easily in large amounts and can be altered *in vivo* by bacterial genetic techniques and *in vitro* by specific enzyme modifications. To determine the effects of these experimentally-induced changes on the function and expression of eukaryotic genes, the rearranged sequences must be taken out of the bacteria in which they were cloned and reintroduced into a eukaryotic organism. Despite the overall unity of biochemistry there are many functions common to eukaryotic cells which are absent from prokaryotes, e.g. localization of ATP-generating systems to mitochondria, association of DNA with histones, mitosis and meiosis, and obligate differentiation of cells. The genetic control of such functions must be assessed in a eukaryotic environment.

Ideally these eukaryotic genes should be re-introduced into the organism from which they were obtained. In this chapter we will discuss the potential for cloning these genes in *Saccharomyces cerevisiae* and other fungi and in later chapters will consider methods for cloning in animal and plant cells. It should be borne in mind that yeast cells are much easier to grow and manipulate than plant and animal cells. Fortunately, the cellular biochemistry and regulation of yeast is very like that of higher eukaryotes. For example, signal transduction and transcription regulation by mammalian steroid receptors can be mimicked in strains of *S. cerevisiae* expressing receptor sequences (Metzger *et al.* 1988, Schena & Yamamoto 1988). There are many yeast homologues of human genes, e.g. those involved in cell division. Thus yeast can be a very good surrogate host for studying the structure and function of eukaryotic gene products.

[237]

Cloning in *Saccharomyces cerevisiae*

Transformation

Transformation of a yeast was first achieved by Hinnen *et al.* (1978) who fused *S. cerevisiae* spheroplasts (i.e. wall-less yeast cells) with polyethylene glycol in the presence of DNA and $CaCl_2$ and then allowed the spheroplasts to regenerate walls in a stabilizing medium containing 3% agar. The transforming DNA used was plasmid pYeLeu 10 which is a hybrid composed of the *E. coli* plasmid ColE1 and a segment of yeast DNA containing the $LEU2^+$ gene. Spheroplasts from a stable $Leu2^-$ auxotroph were transformed to prototrophy by this DNA at a frequency of 1×10^{-7}. Untreated spheroplasts reverted with a frequency of $<1 \times 10^{-10}$. When 42 Leu^+ transformants were checked by hybridization, 35 of them contained ColE1 DNA sequences. Genetic analysis of the remaining seven transformants indicated that there had been reciprocal recombination between the incoming $LEU2^+$ and the recipient $Leu2^-$ alleles.

Of the 35 transformants containing ColE1 DNA sequences, genetic analysis showed that in 30 of them the $LEU2^+$ allele was closely linked to the original $Leu2^-$ allele whereas in the remaining 5, the $LEU2^+$ allele was located on another chromosome. These results can be confirmed by restriction endonuclease analysis since pYeLeu 10 contains no cleavage sites for *Hin*dIII. When DNA from the $Leu2^-$ parent was digested with endonuclease *Hin*dIII and electrophoresed in agarose, multiple DNA fragments were observed but only one of these hybridized with DNA from pYeLeu 10. With the 30 transformants in which the $Leu2^-$ and $LEU2^+$ alleles were linked, only a single fragment of DNA hybridized to pYeLeu 10 but this had an increased size consistent with the insertion of a complete pYeLeu 10 molecule into the original fragment. These data are consistent with there being a tandem duplication of the Leu2 region of the chromosome (Fig. 13.1). With the remaining five transformants, two DNA fragments which hybridized to pYeLeu 10 could be found on electrophoresis. One fragment corresponded to the fragment seen with DNA from the recipient cells, the other to the plasmid genome which had been inserted in another chromosome (see Fig. 13.1). These results represented the first unambiguous demonstration that foreign DNA, in this case cloned ColE1 DNA, can integrate into the genome of a eukaryote. A plasmid such as pYeLeu 10 which can do this is known as a YIp-*yeast integrating* plasmid.

During transformation, the integration of exogenous DNA can occur by recombination with an homologous or an unrelated sequence. In most cases, non-homologous integration is more common than homologous recombination (Fincham 1989) but this is not so in *S. cerevisiae* (Schiestl & Petes 1991). In the experiments of Hinnen *et al.* (1978) described above, sequences of the yeast retrotransposon Ty2 were probably responsible for the integration of the plasmid in novel locations of the genome, i.e. the 'illegitimate' recombinants were the result of homologous crossovers within

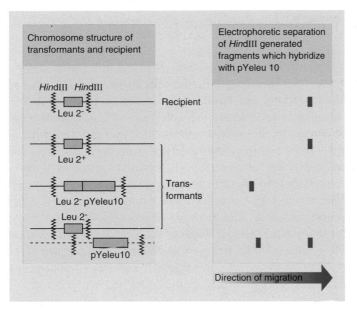

Fig. 13.1 Analysis of yeast transformants. (See text for details.)

a repeated element (Kudla & Nicolas 1992). Based on a similar principle, a novel vector has been constructed by Kudla and Nicolas (1991) which allows integration of a cloned DNA sequence at different sites in the genome. This feature is provided by the inclusion in the vector of a repeated yeast *sigma* sequence present in approximately 20–30 copies per genome and spread over most or all of the 16 chromosomes. Schiestl and Petes (1991) have forced illegitimate recombination by transforming yeast with *Bam*HI-generated fragments in the presence of the *Bam*HI enzyme. The transformants which were obtained had the exogenous DNA integrated into genomic *Bam*HI sites.

Electroporation (see p. 21) now is replacing the use of yeast spheroplasts. It is a much easier and quicker method. In addition, spheroplasts have to regenerate cell walls in a solid matrix (agar) making subsequent retrieval of cells inconvenient. By contrast, cells transformed by electroporation can be selected on the surface of solid media thus facilitating subsequent manipulation.

The development of vectors

Common principles

Since the first demonstration of transformation in yeast a number of different kinds of yeast vector have been constructed. All of them have features in common. First, all of them contain unique target sites for a number of restriction endonucleases. Second, all of them can replicate in

E. coli, often at high copy number. This is important because for many experiments it is necessary to amplify the vector DNA in *E. coli* before transformation of the ultimate yeast recipient. Finally, all of them employ markers that can be selected readily in yeast and which often will complement the corresponding mutations in *E. coli* as well. The four most widely-used markers are *His3*, *Leu2*, *Trp1* and *Ura3*. Mutations in the cognate chromosomal markers are recessive and non-reverting mutants are available. Two yeast selectable markers, *Ura3* and *Lys2*, have the advantage of offering both positive and negative selection. Positive selection is for complementation of auxotrophy. Negative selection is for ability to grow on medium containing a compound that inhibits the growth of cells expressing the wild-type function. In the case of *Ura3* it is 5-fluoroorotic acid (Boeke *et al.* 1984) and for *Lys2* it is α-aminoadipate (Chatoo *et al.* 1979). These inhibitors permit the ready selection of those rare cells which have undergone a recombination or loss event to remove the plasmid DNA sequences. The *Lys2* gene is not utilized frequently because it is large and contains sites within the coding sequence for many of the commonly-used restriction sites.

Yeast episomal plasmids (YEp)

The first YEps were constructed by Beggs (1978) using a naturally-occurring yeast plasmid. The properties of this plasmid have been reviewed by Murray (1987) and can be summarized as follows. The plasmid which is 2 μm long (6.3 kb) is found in many strains of *S. cerevisiae* and has no known function. There are 50–100 copies of the plasmid per haploid cell which represents 2–4% of the total yeast genome. The plasmid is divided by perfect inverted repeats of 599 base-pairs into two unique regions, each with a pair of genes transcribed from a divergent promoter (Fig. 13.2). One of these genes, called *FLP* for its 'flipping' activity, encodes a site-specific recombinase which catalyses genetic crossover between sequences in the inverted repeats. When this occurs the orientation of the non-repeated sequences is inverted and hence the 2 μm plasmid exists as two isomeric forms. Beggs (1978) constructed chimaeric plasmids containing this 2 μm yeast plasmid, fragments of yeast nuclear DNA and the *E. coli* vector pMB9. These chimaeras were able to replicate in both *E. coli* and yeast, transformed yeast with high frequency, and some were able to complement auxotrophic mutations in yeast.

The chimaeric plasmids were constructed in two stages. First, plasmid pMB9 (Fig. 13.2) and the 2 μm plasmid were joined by ligation of the DNA fragments produced by *Eco*RI endonuclease digestion. *Tc*^{R} clones were selected after transforming *E. coli* with the ligated DNAs and these were screened for a hybrid plasmid large enough to carry the complete yeast 2 μm plasmid sequence. Yeast–pMB9 hybrid plasmids theoretically have eight possible configurations. These are determined by the orientation of the insertion into pMB9, which of the *Eco*RI sites on the yeast plasmid is used for insertion and whether the yeast sequence is in the A or B con-

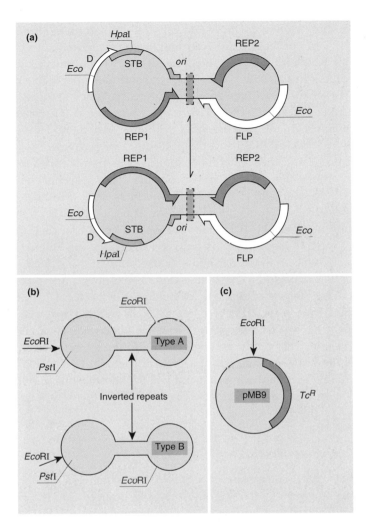

Fig. 13.2 Molecular (a) and schematic (b) structures of the yeast 2-μm plasmid. Only a limited number of restriction sites are shown. REP1 and REP2 are proteins involved in replication of the plasmid and STB is a partition locus. Ori is the origin of replication, *FLP* the gene for a site-specific recombinase, and D a locus which promotes plasmid amplification. The inverted repeats which contain the *FLP* recognition sequences are shown boxed. The structure of pMB9 is shown in (c).

figuration. Five of the eight possible configurations, which can be distinguished by restriction mapping, were found among seven complete hybrid plasmids examined. The exact configuration of the yeast 2 μm plasmid–pMB9 hybrids is probably not important.

For the second stage Beggs (1978) sheared nuclear DNA isolated from *S. cerevisiae* and linked it by means of poly(dA–dT) tails to a mixture of pMB9–yeast 2 μm plasmid hybrids which had been linearized by digestion with *Pst*I. Tc^R transformants were selected in *E. coli* and of the 21 000 obtained, two were found which complemented a Leu⁻ mutant. The plasmids from these clones, pJDB219 and pJDB248, transformed a *leuB* mutant of *E. coli* to Leu$^+$ or tetracycline resistance at the same frequency.

Both pJDB219 and pJDB248 transformed Leu2$^-$ yeast mutants to LEU2$^+$ with a frequency of 5×10^{-4} to 3×10^{-3} transformants per viable cell. This is several orders of magnitude greater than the frequency of transformation obtained by Hinnen *et al.* (1978). However, chromosomal integration of the transforming fragment was essential with the system of Hinnen *et al.* (1978). By contrast pJDB219 and pJDB248 could be recovered as plasmids from yeast cells and their inheritance in yeast was non-Mendelian.

Many other YEp vectors based on the 2 μm plasmid have been developed, e.g. Gerbaud *et al.* 1979, Storms *et al.* 1979, Struhl *et al.* 1979. They possess the same advantages as those of Beggs (1978), viz. they have a high copy number (25–100 copies per cell) in yeast and they transform yeast very well. Indeed, the transformation frequency can be increased 10- to 20-fold if single-stranded plasmids are used (Singh *et al.* 1982).

Chinery and Hinchliffe (1989) have described a novel variant of the YEp type of vector which completely eliminates the bacterial moiety upon introduction into yeast. The bacterial DNA is inserted between a short direct repeat of the *FLP* recognition sequence which has been engineered into the vector. This produces a substrate for site-specific recombination which results in the looping out of a circle of bacterial DNA. This bacterial DNA is unable to replicate in yeast and will be lost from the population. A similar system using a plasmid from *Zygosaccharomyces cerevisiae* as a vector has been described by Awane *et al.* (1992).

Yeast replicating plasmids (YRp)

Struhl *et al.* (1979) constructed a useful vector which consists of a 1.4-kb yeast DNA fragment containing the *trp1* gene inserted into the *Eco*RI site of pBR322. This vector transformed *trp1* yeast protoplasts to TRP1$^+$ at high frequency and transforming sequences were always detected as CCC DNA molecules in yeast. No transformants were found in which the vector had integrated into the chromosomal DNA. Since pBR322 alone cannot replicate in yeast cells (Beggs 1978) a yeast chromosomal sequence must permit the vector to replicate autonomously and to express yeast structural genes in the absence of recombination with host chromosomal sequences. A similar vector based on pBR313 was developed by Kingsman *et al.* (1979). In both cases the yeast gene was linked to a centromere and initially it was thought that this was important. Since then it has been shown that a centromere is not essential; rather, the vector carries an *a*utonomously *r*eplicating *s*equence (*ars*) derived from the chromosome.

Although plasmids containing an *ars* transform yeast very efficiently the resulting transformants are exceedingly unstable. For unknown reasons, YRp plasmids tend to remain associated with the mother cell and are not efficiently distributed to the daughter cell. (Note: *S. cerevisiae* does not undergo binary fission but buds off daughter cells instead.) Occasional stable transformants are found and these appear to be cases in which the entire YRp has integrated into a homologous region on a chromosome in a manner identical to that of YIps (Stinchcomb *et al.* 1979, Nasmyth & Reed 1980).

Yeast centromere plasmids (YCp)

Using a YRp vector Clarke and Carbon (1980) isolated a number of hybrid plasmids containing DNA segments from around the centromere-linked *leu2*, *cdc*10 and *pgk* loci on chromosome III of yeast. As expected for plasmids carrying an *ars* most of the recombinants were unstable in yeast. However, one of them was maintained stably through mitosis and meiosis. The stability segment was confined to a 1.6-kb region lying between the *leu2* and *cdc*10 loci and its presence on plasmids carrying either of two *ars* tested resulted in those plasmids behaving like minichromosomes (Clarke & Carbon 1980, Hsiao & Carbon 1981). Genetic markers on the minichromosomes acted as linked markers segregating in the first meiotic division as centromere-linked genes and were unlinked to genes on other chromosomes. Stinchcomb *et al.* (1982) and Fitzgerald-Hayes *et al.* (1982) have isolated the centromeres from chromosomes IV and XI of yeast and found that they confer on plasmids similar properties to the centromere of chromosome III.

Since then, centromeres from another nine chromosomes have been cloned and found to have a similar structure and function. Structurally, plasmid-borne centromere sequences have the same distinctive chromatin structure that occurs in the centromere region of yeast chromosomes (Bloom & Carbon 1982). Functionally YCps exhibit three characteristics of chromosomes in yeast cells. First, they are mitotically stable in the absence of selective pressure. Second, they segregate during meiosis in a Mendelian manner. Finally, they are found at low copy number in the host cell.

The low copy number of YCp vectors is not altered if they include the yeast 2-μm plasmid amplification system that normally drives plasmids to high copy number (Tschumper & Carbon 1983). This suggests that the centromere is dominant over the 2-μm plasmid amplification system. However, if the *FLP* gene product is synthesized, high-copy-number plasmids are generated *in vivo* in which the centromere sequence has been deleted or inactivated.

By placing an inducible promoter adjacent to a cloned centromere it is possible to modulate centromere function by controlling promoter activity. Such conditional constructs have been used to study centromere structure and function relationships (Hill & Bloom 1987), loss and breakage of dicentric chromosomes (Futcher & Carbon 1986), and to produce regulated copy number plasmids (Chlebowicz-Sledziewska & Sledziewski 1985). When transcription is induced plasmid stability is decreased, unless a second centromere is present, and copy number is increased (Apostol & Greer 1988).

When wild-type yeast cells are forced to maintain multiple YCps bearing independently selectable markers, the cells grow slowly and cell viability is decreased, indicating a toxic effect from the presence of excess centromeres (Futcher & Carbon 1986). A plasmid system has been constructed which permits the selection of cells which tolerate high-copy-number YCps. The plasmids carry a gene conferring resistance to antibiotic G418 but which has a weak promoter ensuring low-level resistance. Selection is made for cells which can grow in the presence of high levels of antibiotic. Three different events can give rise to high level resistance (Tschumper & Carbon 1987). In some cells the centromere sequence has been inactivated and plasmid replication is under the control of the 2-µm plasmid. In other cells mutants have been selected in which the strength of the $G418^R$ gene promoter has been increased. The most interesting class of high-level resistance mutants is that in which chromosomal mutants have arisen which enable yeast cells to tolerate high-copy-number YCps.

Yeast artificial chromosomes (YACs)

All three autonomous plasmid vectors are maintained in yeast as circular DNA molecules — even the YCp vectors which possess yeast centromeres. Thus none of these vectors resembles the normal yeast chromosomes, which have a linear structure. The ends of all yeast chromosomes, like those of all other linear eukaryotic chromosomes, have unique structures that are called *telomeres*. Telomeres have a unique structure (see later) that has evolved as a device to preserve the integrity of the ends of DNA molecules which often cannot be finished by the conventional mechanisms of DNA replication (see Watson 1972 for detailed discussion). Szostak and Blackburn (1982) were able to clone yeast telomeres by developing a linear yeast vector.

The linear yeast vector was constructed from two components. The first was a YRp, pSZ213 (Fig. 13.3) which has no *Bam*HI sites and a single *Bgl*II site. The second component was fragments of an unusual, linear, rRNA-encoding plasmid found in the protozoan *Tetrahymena*. This plasmid DNA was cleaved with *Bam*HI and the end fragments were ligated to *Bgl*II-cut pSZ213. Since *Bam*HI-cut and *Bgl*II-cut ends of DNA are complementary they can be joined by ligase but the product of ligation is not a substrate for either enzyme. Consequently both *Bam*HI and *Bgl*II endonucleases were present in the ligation mixture to cut circularized or dimerized vector molecules and dimerized end fragments. In this way the desired linear vector carrying *Tetrahymena* telomeres accumulated in the reaction mixture and it was purified by agarose gel electrophoresis.

The linear plasmid containing the *Tetrahymena* telomeres retains the $LEU2^+$ marker and this was used for selection of transformants in yeast. Since YRp vectors are capable of integration, those transformants in which the linear plasmid was replicating autonomously were detected by their mitotic instability. Restriction mapping showed that the plasmid in the unstable transformants is a linear molecule identical in structure to the linear molecule constructed *in vitro* and used in the transformation. Furthermore, the ends of *Tetrahymena* rDNA have three unusual structural

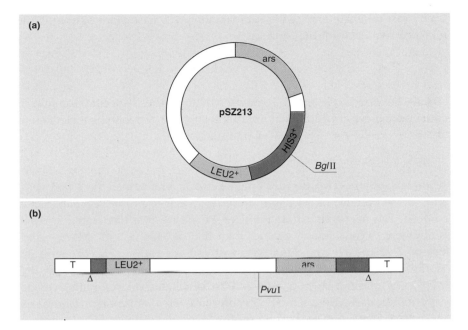

Fig. 13.3 (a) Simplified map of plasmid pSZ213 used for cloning yeast telomeres. The plasmid has no *Bam*HI sites and insertion of telomeric DNA at the unique *Bgl*II site inactivates the *his3*+ gene. (b) Structure of a linear plasmid constructed from pSZ213 showing the asymmetric location of the unique *Pvu*I site. The open triangles indicate the location of *Bam/Bgl* joints created when the telomeric DNA (indicated with a T) was added to *Bgl*II-digested pSZ213.

features: a variable number of short, 5'-CCCCAA-3' repeat units, specific single-strand interruptions within the repeated sequences and a cross-linked terminus. All three structural features were maintained when the *Tetrahymena* sequences were cloned in yeast.

The linear plasmid carrying the *Tetrahymena* telomeres has a single, asymmetrically placed target site for endonuclease *Pvu*I. This enzyme cuts yeast DNA into approximately 2000 fragments and since yeast has 17 chromosomes, 34 of these fragments should contain telomeres. After ligating *Pvu*I-digested chromosomal and plasmid DNA unstable LEU+ transformants were selected. Many of the plasmids generated in this way would have multiple yeast *Pvu*I fragments ligated onto the vector followed by either a yeast end or an end derived from the vector. These plasmids would be as large as, or larger than, the original vector. In contrast, the ligation of a single *Pvu*I telomere fragment from yeast, smaller than the *Tetrahymena*-containing fragment being replaced, would yield a linear plasmid smaller than at the start. After size screening of the plasmids from the transformants, three plasmids carrying a yeast telomere at one end were identified. Analysis of these plasmids shows that yeast telomeres have at least some of the structural features of *Tetrahymena* telomeres.

Each yeast telomere ends with approximately 100 base-pairs of irregularly repeated sequences of the form

$$5'-CCCA\ldots\ldots\ldots$$
$$3'-GGGT\ldots\ldots\ldots$$

attached to specific X and Y sequences. Within these specific terminal X and Y sequences are additional internal sections of the repetitive elements.

Stability of yeast cloning vectors

As noted earlier, YRp vectors are not stably maintained by yeast cells and, in the absence of selection, are quickly lost from the population. By contrast, the segregational stability of YCp vectors which carry a yeast centromere is considerably greater than that of YRp vectors. Murray and Szostak (1983a) have found that YRp vectors have a strong bias to segregate to the mother cell at mitosis. This segregation bias explains how the fraction of plasmid-bearing cells can be small despite the high average copy number of YRp vectors. The presence of a centromere eliminates segregation bias, thus accounting for the increased stability of YCp vectors relative to YRp vectors. YEp vectors are stably maintained in yeast cells provided the strain contains endogenous intact 2 μm circles. In the absence of endogenous 2 μm circles YEp vectors show maternal segregation bias.

Despite the fact that YCp vectors are relatively stable, they are still 1000 times less stable than *bona fide* yeast chromosomes. However, YCp vectors and YRp vectors are circular molecules whereas chromosomes are linear molecules. A linear plasmid vector carrying a centromere would be much more representative of a yeast chromosome. Dani and Zakian (1983) constructed linear yeast plasmids but found that stability was reduced relative to the circular plasmid. However, Murray and Szostak (1983b) found that stability of linear yeast plasmids was related to size. Thus artificial yeast chromosomes which were 55-kb long and contained cloned genes, *ars*, centromeres and telomeres had many of the properties of natural yeast chromosomes. When the artificial chromosomes were less than 20 kb in size, centromere function was impaired.

Whereas the stability of linear YCp vectors is dependent on size, this is not true of YRp vectors. Murray and Szostak (1983a) constructed a linear YRp vector and found that it did not exhibit maternal segregational bias. The model used by them to explain these results is as follows. During DNA replication there is an association between *ars* elements and fixed nuclear sites. Replicated molecules remain attached to this site which is destined to segregate to the mother cell. The replicated plasmids will exist initially as catenated dimers which subsequently are resolved. If the dimers are attached to the putative segregation site by only one of their constituent monomers, their resolution by topoisomerase activity would release one monomer from each dimer and this monomer would be free to segregate at random to either the mother or the daughter cell. With linear plasmids, replication will produce two linear molecules which are not interconnected. Thus, prior to mitosis there will be at least one freely segregating molecule and this explains the increased stability of linear YRp vectors.

Retrovirus-like vectors

The genome of *S. cerevisiae* contains 30–40 copies of a 5.9-kb mobile genetic element called Ty (for review see Fulton *et al.* 1987). This transposable element shares many structural and functional features with retroviruses (see p. 325) and the copia element of *Drosophila* (see p. 362). Ty consists of a central region containing two long, open reading frames (ORF) flanked by two identical terminal 334 base-pair repeats called *delta* (Fig. 13.4). Each delta element contains a promoter as well as sequences recognized by the transposing enzyme. New copies of the transposon arise by a replicative process in which the Ty transcript is converted to a progeny DNA molecule by a Ty-encoded reverse transcriptase. The complementary DNA can transpose to many sites in the host DNA.

The Ty element has been modified *in vitro* by replacing its delta promoter sequence with promoters derived from the phosphoglycerate kinase or galactose-utilization genes (Mellor *et al.* 1985, Garfinkel *et al.* 1985). When such constraints are introduced into yeast on high-copy-number vectors the Ty element is over-expressed. This results in the formation of large numbers of virus-like particles (VLPs) which accumulate in the cytoplasm (Fig. 13.5). The particles, which have a diameter of 60–80 nm, have reverse transcriptase activity. The major structural components of VLPs are proteins produced by proteolysis of the primary translation product of ORF 1. Adams *et al.* (1987) have shown that fusion proteins can be produced in cells by inserting part of a gene from human immunodeficiency virus (HIV) into ORF 1. Such fusion proteins formed hybrid HIV:Ty-VLPs.

The Ty element also can be subjugated as a vector for transposing genes to new sites in the genome. The gene to be transposed is placed between the 3' end of ORF 2 and the 3' delta sequence (Fig. 13.6). Providing the inserted gene lacks transcription termination signals, transcription of

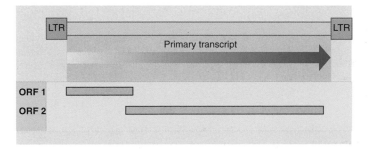

Fig. 13.4 Structure of a typical Ty element. ORF 1 and ORF 2 represent the two open reading frames. The delta sequences are indicated by LTR (long-terminal repeats).

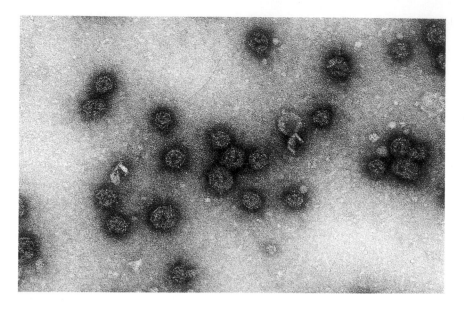

Fig. 13.5 Ty virus-like particles (magnification 80 000) carrying the entire HIV1 TAT coding region. (Photograph courtesy of Dr S. Kingsman.)

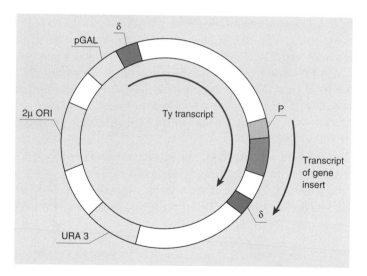

Fig. 13.6 Structure of the multicopy plasmid used for inserting a modified Ty element, carrying a cloned gene, into the yeast chromosome. pGAL and P are yeast promoters, δ represents the long-terminal repeats (delta sequences) and the grey region represents the cloned gene. (See text for details.)

the 3′ delta sequence will occur, which is a prerequisite for transposition. Such constructs act as amplification cassettes for once introduced into yeast, transposition of the new gene occurs to multiple sites in the genome (Boeke *et al*. 1988, Jacobs *et al*. 1988).

Choice of vector for cloning

There are three reasons for cloning genes in yeast. The first of these relates to the potential use of yeast as a cloning host for the over-production of proteins of commercial value. Yeast offers a number of advantages such as the ability to glycosylate proteins during secretion (see p. 265) and the absence of pyrogenic toxins. Commercial production demands over-production and the factors affecting expression of genes in yeast are discussed in a later section (see p. 254). Yeast also is used in the production of food and beverages. The ability to clone in yeast without the introduction of bacterial sequences by using vectors like those of Chinery and Hinchliffe (1989) (see p. 242) is particularly beneficial.

A second reason for cloning genes in yeast is the ability to clone large pieces of DNA. Although there is no theoretical limit to the size of DNA which can be cloned in a bacterial plasmid, large recombinant plasmids exhibit structural and segregative instability. In the case of bacteriophage λ vectors the size of the insert is governed by packaging constraints. Many DNA sequences of interest are much larger than this. For example the gene for blood Factor VIII covers about 190 kbp or about 0.1% of the human X chromosome, and the Duchenne muscular dystrophy gene spans more than a megabase. Long sequences of cloned DNA will also facilitate efforts to sequence the human genome. Yeast artificial chromosomes offer a convenient way to clone large DNA fragments, since there is no practical size limitation to YACs (see p. 244). Indeed, the availability of YACs with large inserts is an essential prerequisite for any genome sequencing project and they were used in the sequencing of the entire *S. cerevisiae* chromosome III (Oliver *et al.* 1992). The method for cloning large DNA sequences developed by Burke *et al.* (1987) is shown in Fig. 13.7.

For many biologists the primary purpose of cloning is to understand what particular genes do *in vivo*. Thus most of the applications of yeast vectors have been in the surrogate genetics of yeast. One advantage of cloned genes is that they can be analysed easily, particularly with the advent of DNA sequencing methods. Thus nucleotide sequencing analysis can reveal many of the elements which control expression of a gene as well as identifying the sequence of the gene product. In the case of the yeast actin gene (Gallwitz & Sures 1980, Ng & Abelson 1980) and some yeast tRNA genes (Peebles *et al.* 1979, Olson 1981) this kind of analysis revealed the presence within these genes of non-coding sequences which are copied into primary transcripts. These *introns* subsequently are eliminated by a process known as *splicing*. (See Box 13.1.) Nucleotide sequence analysis also can reveal the direction of transcription of a gene although this can be determined *in vivo* by other methods. For example, if the yeast gene is expressed in *E. coli* using bacterial transcription signals, the direction of reading can be deduced by observing the orientation of a cloned fragment required to permit expression. Finally, if a single transcribed yeast gene is present on a vector the chimaera can be used as a probe for quantitative solution hybridization analysis of transcription of the gene.

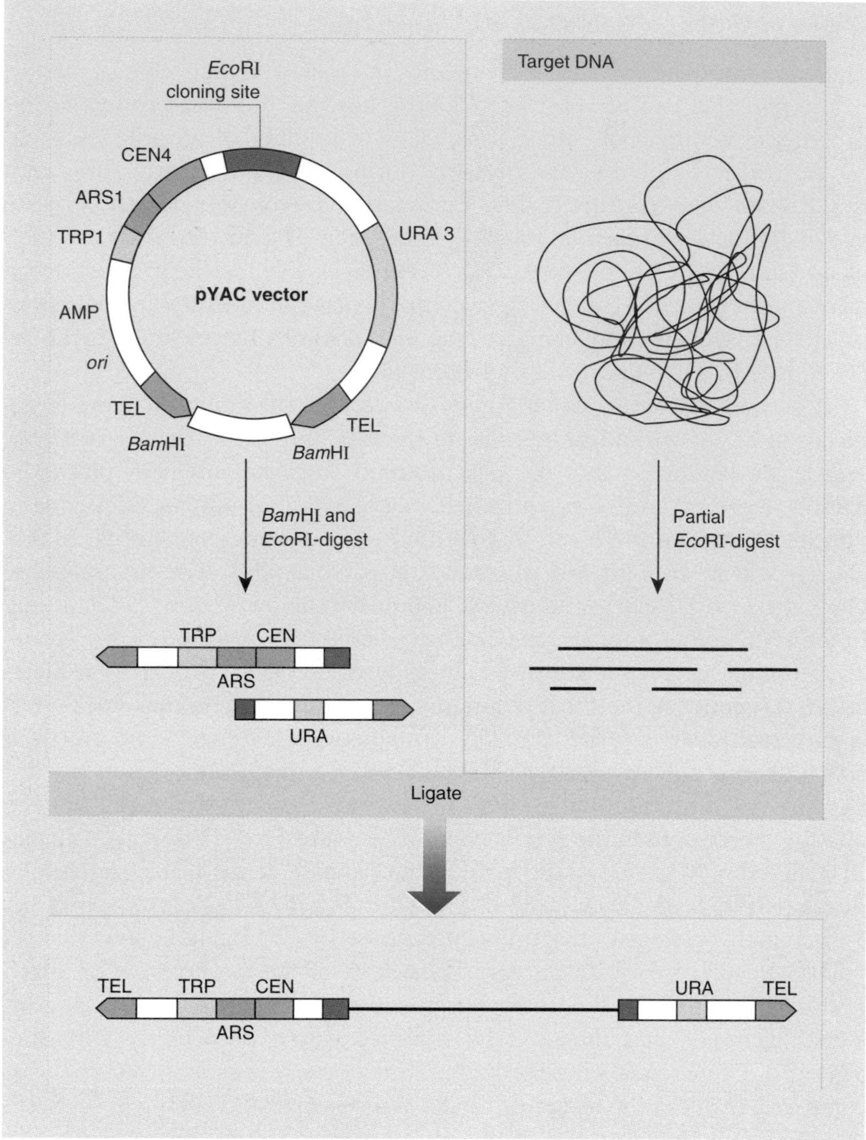

Fig. 13.7 Construction of a yeast artificial chromosome containing large pieces of cloned DNA. Key regions of the pYAC vector are as follows: TEL, yeast telomeres; ARS 1, autonomously replicating sequence; CEN 4, centromere from chromosome 4; *URA3* and *TRP1*, yeast marker genes; Amp, ampicillin-resistance determinant of pBR322; ori, origin of replication of pBR322.

The availability of different kinds of vectors with different properties (see Table 13.1) enables yeast geneticists to perform manipulations in yeast like those long available to *E. coli* geneticists with their sex factors and transducing phages. Thus cloned genes can be used in conventional genetic analysis by means of recombination using YIp vectors or linearized YRp vectors (Orr-Weaver *et al.* 1981). Complementation can be carried out using YEp, YRp, YCp or YAC vectors but there are a number of factors which make YCps the vectors of choice (Rose *et al.* 1987). For example, YEps and YRps exist at high copy number in yeast and this can prevent the isolation of genes whose products are toxic when over-expressed, e.g. the genes for actin and tubulin. In other cases the over-expression of genes other than the gene of interest can suppress the mutation used for selection (Kuo &

Table 13.1 Properties of the different yeast vectors

Vector	Transformation frequency	Copy no./ cell	Loss in non-selective medium	Disadvantages	Advantages
YIp	10^2 transformants per µg DNA	1	Much less than 1% per generation	1 Low transformation frequency 2 Can only be recovered from yeast by cutting chromosomal DNA with restriction endonuclease which does not cleave original vector containing cloned gene	1 Of all vectors, this kind give most stable maintenance of cloned genes 2 An integrated YIp plasmid behaves as an ordinary genetic marker, e.g. a diploid heterozygous for an integrated plasmid segregates the plasmid in a Mendelian fashion 3 Most useful for surrogate genetics of yeast, e.g. can be used to introduce deletions, inversions and transpositions (see Botstein & Davis 1982)
YEp	10^3–10^5 transformants per µg DNA	25–200	1% per generation	Novel recombinants generated *in vivo* by recombination with endogenous 2-µm plasmid	1 Readily recovered from yeast 2 High copy number 3 High transformation frequency 4 Very useful for complementation studies
YRp	10^4 transformants per µg DNA	1–20	Much greater than 1% per generation but can get chromosomal integration	Instability of transformants	1 Readily recovered from yeast 2 High copy number. Note that the copy number is usually less than that of YEp vectors but this may be useful if cloning gene whose product is deleterious to the cell if produced in excess 3 High transformation frequency 4 Very useful for complementation studies 5 Can integrate into the chromosome
YCp	10^4 transformants per µg DNA	1–2	Less than 1% per generation	Low copy number makes recovery from yeast more difficult than that of YEp or YRp vectors	1 Low copy number is useful if product of cloned gene is deleterious to cell 2 High transformation frequency 3 Very useful for complementation studies 4 At meiosis generally shows Mendelian segregation
YAC		1–2	Depends on length: the longer the YAC the more stable it is	Difficult to map by standard techniques	1 High-capacity cloning system permitting DNA molecules greater than 40 kb to be cloned 2 Can amplify large DNA molecules in a simple genetic background
Ty	Depends on vector used to introduce Ty into cell	~20	Stable, since integrated into chromosome	Needs to be introduced into cell in another vector	Get amplification following chromosomal integration

Box 13.1 Excision of introns by yeast

Many eukaryotic genes contain non-coding regions called introns. Introns have two common structure features: their sequences begin with the dinucleotide 5′-GT-3′ and end with the dinucleotide 5′-AG-3′. Besides these invariant nucleotides there is limited structural similarity at and around the intron–exon junction and consensus sequences have been derived from comparison of more than 100 junctions. Because of the similarity of these

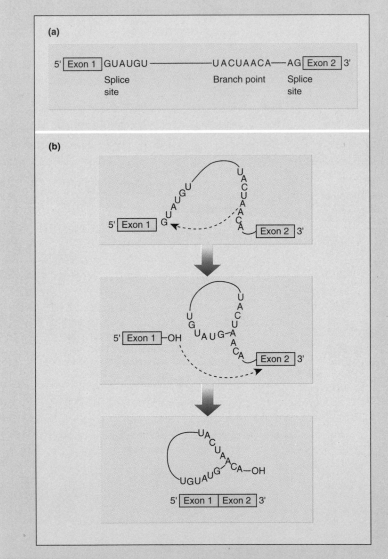

Fig. B13.1 Intron splicing in yeast. (a) The structure of yeast pre-mRNA showing the essential splice site and branch-point sequences. (b) The two *trans*-esterification reactions involved in intron removal.

continued

Box 13.1 *continued*

junction sequences in widely different species, e.g. yeast and man, it was thought that the mechanism of RNA processing to remove introns might be universal. Support for this idea came from the observation that monkey cells can splice out introns from mouse and rat genes, and mouse cells correctly splice transcripts of rabbit and chicken genes. This raises the question whether yeast cells can remove introns from genes of higher eukaryotes. If so this would be of great practical value, for the presence of introns prevents shotgun cloning of functional eukaryotic genes in *E. coli*.

To test the ability of yeast cells to excise introns from foreign genes Beggs *et al.* (1980) transformed *S. cerevisiae* with a hybrid plasmid containing a cloned rabbit chromosomal DNA segment including a complete β-globin gene with two intervening sequences and extended flanking regions. Yeast cells transformed with this chimaera produced β-globin-specific mRNA. However, these globin transcripts were about 20–40 nucleotides shorter at the 5′-end than normal globin mRNA, contained one intron and extended only as far as the first half of the second intron. This result could be taken to indicate that the splicing mechanisms in yeast and rabbit differ, but it could be argued that a complete transcript is a prerequisite for RNA splicing and that the prematurely terminated globin RNA was not a substrate for the yeast splicing enzyme(s). Consequently, Langford *et al.* (1983) inserted into the intron-containing yeast actin gene an intron-containing fragment from either *Acanthamoeba* or duck. In both instances yeast cells removed the natural yeast intron but not the foreign intron from the chimaeric transcript.

Why are introns in foreign genes not removed by yeast cells? The explanation is provided by an understanding of the mechanism of splicing in yeast compared with that in higher eukaryotes (for review, see Ruby & Abelson, 1991 and Woolford & Peebles, 1992). Three short sequences, the 5′ and 3′ splice sites and the branch-point region, in the pre-mRNA intron are essential for the intron to be accurately and efficiently spliced (Fig. B13.1). These sequences are very similar in both yeast and mammalian introns. The branch-point sequence, UACUAAC, is usually located 15–40 nucleotides upstream of the 3′ splice site. However, this sequence is strongly conserved in *S. cerevisiae*, which has very few introns, but is very degenerate in mammals. Mutations within the splice site and branch-point sequences in *S. cerevisiae* usually inactivate the intron, whereas cryptic splice sites are often activated in mammalian systems. The heterologous genes containing introns described above which were not spliced correctly in yeast most likely lacked the precise sequences recognized by *S. cerevisiae*.

Campbell 1983). All the yeast vectors can be used to create partial diploids or partial polyploids and the extra gene sequences can be integrated or extra-chromosomal. Deletions, point mutations and frame shift mutations can be introduced *in vitro* into cloned genes and the altered genes returned to yeast and used to replace the wild-type allele. Excellent reviews of these techniques have been presented by Botstein and Davis (1982), Hicks *et al.* (1982), Struhl (1983) and Stearns *et al.* (1990).

Plasmid construction by homologous recombination in yeast

During the process of analysing a particular cloned gene it often is necessary to change the plasmid's selective marker. Alternatively it may be desired to move the cloned gene to a different plasmid, e.g. from a YCp to a YEp. Again, genetic analysis may require many different alleles of a cloned gene to be introduced to a particular plasmid for subsequent functional studies. All these objectives can be achieved by standard *in vitro* techniques, but Ma *et al.* (1987) have shown that methods based on recombination *in vivo* are much quicker. The underlying principle is that linearized plasmids are efficiently repaired during yeast transformation by recombination with a homologous DNA restriction fragment.

Suppose we wish to move the HIS3 gene from pBR328, which cannot replicate in yeast, to YEp420 (see Fig. 13.8). Plasmid pRB328 is cut with *Pvu*I and *Pvu*II and the HIS3 fragment selected. The HIS3 fragment is mixed with YEp420 which has been linearized with *Eco*RI and the mixture transformed into yeast. Two crossover events occurring between homologous regions flanking the *Eco*RI site of YEp420 will result in the generation of a recombinant YEp containing both the HIS3 and URA3 genes. The HIS3 gene can be selected directly. If this were not possible, selection could be made for the URA3 gene, for a very high proportion of the clones will also carry the HIS3 gene.

Many other variations of the above method have been described by Ma *et al.* (1987), to whom the interested reader is referred for details.

Expression of cloned genes in yeast

As might be expected, most cloned yeast genes are expressed when re-introduced into yeast. More surprising, some bacterial genes are also expressed in yeast (Cohen *et al.* 1980, Jimenez & Davies 1980) and in one instance expression was dependent on a bacterial promoter (Breunïg *et al.* 1982). Since Struhl and Davis (1980) showed that a yeast promoter is functional in *E. coli* it might be thought that transcription signals such as promoters can be active in prokaryotes and eukaryotes. However, this clearly is not the case, for a number of workers failed to get expression of foreign genes in yeast. Thus Rose *et al.* (1981) obtained expression of β-galactosidase in *E. coli* when it was under the control of either the *E. coli* Tc^R promoter or the yeast URA3 promoter but achieved expression in yeast only with the latter promoter. When Beggs *et al.* (1980) introduced the

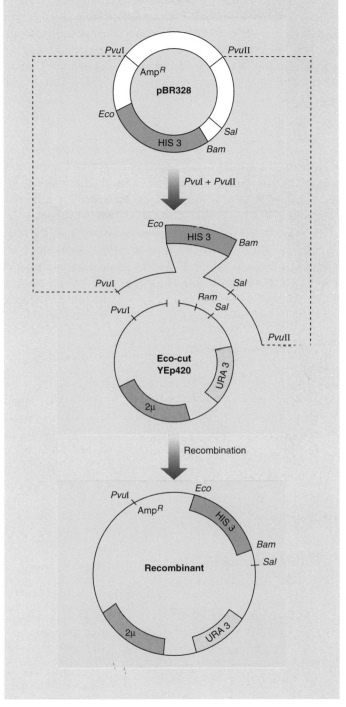

Fig. 13.8 Plasmid construction by homologous recombination in yeast. pRB328 is digested with *Pvu*I and *Pvu*II and the HIS3-containing fragment transformed into yeast along with the *Eco*RI-cut YEp420. Homologous recombination occurs between pBR322 sequences, shown as thin lines, to generate a new plasmid carrying both HIS3 and URA3.

rabbit β-globin gene into yeast, β-globin-specific transcripts were obtained but transcription started at a position downstream from the usual initiation site. Finally, even though a *Drosophila* gene corresponding to the yeast ADE8 locus has been identified by complementation, *Drosophila* genes complementing mutants at other yeast loci have not been obtained (Henikoff *et al.* 1981).

Use of yeast promoters

Because of difficulties in obtaining heterologous gene expression in yeast a number of groups have turned to the use of yeast promoters and translation initiation signals. Thus expression of the *E. coli* β-galactosidase gene was obtained by fusing it to the *N*-terminus of the URA3, CYC1 and ARG3 genes (Guarente & Ptashne 1981, Rose *et al.* 1981, Crabeel *et al.* 1983). Expression in yeast of an interferon-α gene was obtained by fusing it to either the PGK or TRP1 genes (Tuite *et al.* 1982, Dobson *et al.* 1983). In many instances, expression of a mature protein rather than a fusion protein is required. To achieve this Hitzeman *et al.* (1981) started with a

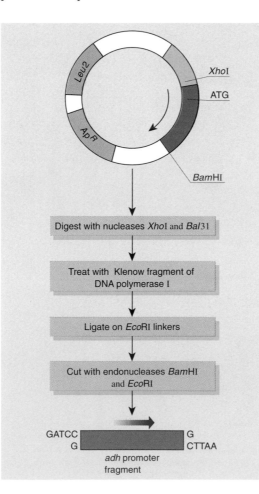

Fig. 13.9 The procedure used to generate *adh* promoter fragments. The *adh* promoter region in the starting plasmid is indicated by the dark red area. The grey area represents another gene fused to the *N*-terminus of the *adh* gene. The distance between the unique *Xho*I site and the initiating ATG codon is 17 bases. The arrow inside the plasmid circle (top) and above the *adh* promoter fragment (bottom) indicates the direction of transcription.

plasmid carrying the promoter and part of the coding sequence of the ADH1 gene (Fig. 13.9). This plasmid was cut with endonuclease *Xho*I for which there is a unique cleavage site downstream from the initiating ATG codon. The linearized vector was digested with nuclease *Bal*31 to remove 30–70 base-pairs of DNA from each end of the molecule. The DNA then was incubated with the Klenow fragment of DNA polymerase and deoxynucleoside triphosphates to fill in the ends and synthetic *Eco*RI linkers added. After cleavage with endonucleases *Eco*RI and *Bam*HI the assorted fragments containing variously deleted ADH1 promoter sequences were isolated by preparative electrophoresis. In this way six different promoter fragments were isolated, joined to an interferon-α gene and used to direct the synthesis of mature interferon in yeast. In a similar fashion synthesis of mature hepatitis B virus surface antigen was achieved from the ADH1 promoter (Valenzuela *et al.* 1982) and mature interferon-γ from the PGK promoter (Derynck *et al.* 1983).

The structure of yeast promoters

When the experiments described above were carried out little was known about the structure of yeast promoters. The general failure of bacterial and higher eukaryotic genes to be expressed in yeast when using their own promoters would suggest that yeast promoters have a unique structure. This is indeed the case (Guarante 1987). Four structural elements can be recognized in the average yeast promoter (Fig. 13.10). First, several consensus sequences are found at the transcription initiation site. Two of these sequences, TC(G/A)A and PuPuPyPuPu, account for more than half of the known yeast initiation sites (Hahn *et al.* 1985, Rudolph & Hinnen 1987). These sequences are not found at transcription initiation sites in higher eukaryotes which implies a mechanistic difference in their transcription machinery compared with yeast.

The second motif in the yeast promoter is the TATA box (Dobson *et al.* 1982). This is an AT-rich region with the canonical sequence TATAT/AAT/A located 60–120 nucleotides before the initiation site. Functionally it can

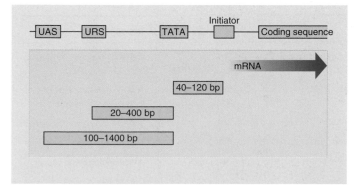

Fig. 13.10 Structure of typical yeast promoters. (See text for details.)

be considered equivalent to the Pribnow box of *E. coli* promoters (see p. 152).

The third and fourth structural elements are upstream activating sequences (UASs) and upstream repressing sequences (URSs). These are found in genes whose transcription is regulated. Binding of positive-control proteins to UASs turns up the rate of transcription and deletion of the UASs abolishes transcription. An important structural feature of UASs is the presence of one or more regions of dyad symmetry (Rudolph & Hinnen 1987). Binding of negative control proteins to URSs turns down the transcription rate of those genes that need to be negatively regulated.

The level of transcription can be affected by sequences located within the gene itself and which are referred to as downstream activating sequences (DAS). Chen *et al.* (1984) noted that using the phosphoglycerate kinase (*PGK*) promoter several heterologous proteins accumulate to 1–2% of total cell protein whereas phosphoglycerate kinase itself accumulates to over 50%. These disappointing amounts of heterologous protein reflect the levels of mRNA which were due to a lower level of initiation rather than a reduced mRNA half-life (Mellor *et al.* 1987). Addition of downstream *PGK* sequences restored the rate of mRNA transcription indicating the presence of a DAS. Evidence for these DASs has been found in a number of other genes.

S. cerevisiae **promoter systems**

The first *S. cerevisiae* promoters used were from genes encoding abundant glycolytic enzymes, e.g. alcohol dehydrogenase (*ADH1*), phosphoglycerate kinase (*PGK*) or glyceraldehyde-3-phosphate dehydrogenase (*GAP*). These are strong promoters and mRNA transcribed from them can accumulate up to 5% of total. They were at first thought to be constitutive but later were shown to be induced by glucose (Tuite *et al.* 1982). Now there is a large variety of native and engineered promoters available (Table 13.2) differing in strength, regulation and induction ratio. These have been reviewed in detail by Romanos *et al.* (1992).

Table 13.2 *S. cerevisiae* promoter systems

Promoter	Strength	Regulation
Phosphoglycerate kinase (*PGK*)	4+ (5% of mRNA)	<20-fold induction by glucose
Galactokinase (*GAL1*)	3+ (1% of mRNA)	1000-fold induction by galactose but subject to glucose repression
Acid phosphatase (*PHO5*)	2+	200-fold repression by inorganic phosphate
Alcohol dehydrogenase II (*ADH2*)	2+	100-fold repression by glucose
Copper metallothionein (*CUP1*)	+	20-fold induction by copper
Mating factor-α_1 (*MFα1*)	+	Constitutive in wild-type α cells but 10^5-fold induction by temperature shift in *sir3*[ts] cells

Box 13.2 Galactose metabolism and its control in *Saccharomyces cerevisiae*

Galactose is metabolized to glucose-6-phosphate in yeast by an identical pathway to that operating in other organisms (Fig. B13.2). The key enzymes and their corresponding genes are a kinase (*GAL1*), a transferase (*GAL7*), an epimerase (*GAL10*) and a mutase (*GAL5*). Melibiose (galactosyl-glucose) is metabolized by the same enzymes after cleavage by an α-galactosidase encoded by the *MEL1* gene. Galactose uptake by yeast cells is via a permease encoded by the *GAL2* gene. The *GAL5* gene is constitutively expressed. All the others are induced by growth on galactose and repressed during growth on glucose.

The *GAL1*, *GAL7* and *GAL10* genes are clustered on chromosome II but transcribed separately from individual promoters. The *GAL2* and *MEL1* genes are on other chromosomes. The *GAL4* gene encodes a protein that activates transcription of the catabolic genes by binding to UAS 5′ to each gene. The *GAL80* gene encodes a repressor that binds directly to *GAL4* gene product thus preventing it from activating transcription. The *GAL3* gene product catalyses the conversion of galactose to an inducer which combines with the *GAL80* gene product preventing it from inhibiting the *GAL4* protein from binding to DNA (Fig. B13.3).

The expression of the *GAL* genes is repressed during growth on glucose. The regulatory circuit responsible for this phenomenon, termed catabolite repression, is superimposed upon the circuit responsible for induction of *GAL* gene expression. Very little is known about its mechanism.

For a review of galactose metabolism in *S. cerevisiae* the reader should consult Johnston (1987).

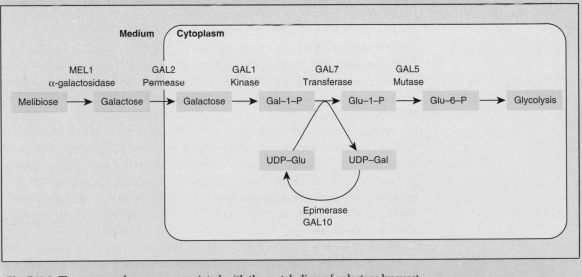

Fig. B13.2 The genes and enzymes associated with the metabolism of galactose by yeast.

continued

Box 13.2 *continued*

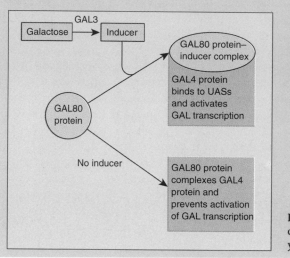

Fig. B13.3 The regulation of transcription of the yeast galactose genes.

The ideal promoter is one which is tightly regulated so that the growth phase can be separated from the induction phase. This minimizes the selection of non-expressing cells and can permit the expression of proteins normally toxic to the cell. The ideal promoter also will have a high induction ratio. One promoter which has these characteristics and which is now the most widely-used is that from the *GAL1* gene. Galactose-regulation in yeast now is extremely well studied and has become a model system for eukaryotic transcriptional regulation (see Box 13.2).

Following addition of galactose, *GAL1* mRNA is rapidly induced over 1000-fold and can reach 1% of total mRNA. However, the promoter is strongly repressed by glucose and so in glucose-grown cultures this induction only occurs following depletion of glucose. To facilitate galactose induction in the presence of glucose, mutants have been isolated which are insensitive to glucose repression (Matsumoto *et al.* 1983, Horland *et al.* 1989). The *trans*-activator *GAL4* protein is present in only one or two molecules per cell and so *GAL1* transcription is limited. With multicopy expression vectors *GAL4* limitation is exacerbated. However, *GAL4* expression can be made autocatalytic by fusing the *GAL4* gene to a *GAL10* promoter (Schultz *et al.* 1987), i.e. *GAL4* expression is now regulated (induced) by galactose.

In an attempt to combine the high activity of glycolytic promoters with the tight regulation of the *GAL* promoters, hybrid promoters have been constructed where a glycolytic UAS is replaced by a *GAL* UAS. The published results (Bitter & Egan 1988, Cousens *et al.* 1990) do not suggest that these hybrid promoters are any more efficient than the *GAL1* promoter.

[261]
CHAPTER 13
Cloning in S. cerevisiae

Detecting protein–protein interactions

Chien *et al.* (1991) have made use of the properties of the GAL4 protein to develop a method for detecting protein–protein interactions. The GAL4 protein has separate domains for the binding to UAS DNA and for transcriptional activation. Plasmids were constructed which encode two hybrid proteins. The first consisted of the DNA-binding domain (residues 1–147) of the GAL4 protein fused to a test protein. The second consisted of residues 768–881 of the GAL4 protein, representing the activation domain, fused to protein sequences encoded by a library of yeast genomic DNA fragments. Interaction between the test protein and a protein encoded by one of the library plasmids led to transcriptional activation of a reporter gene (Fig. 13.11). This method is known as the two-hybrid system. It has been used successfully by a number of groups to detect a variety of protein–protein interactions. For example, Durfee *et al.* (1993) used it to show that the retinoblastoma protein associates with the protein phosphatase type 1 catalytic subunit 1.

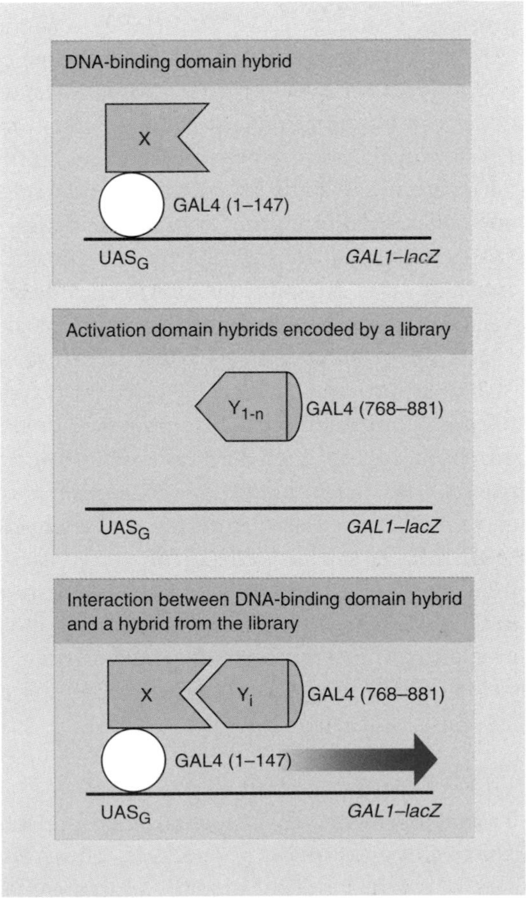

Fig. 13.11 Strategy to detect interacting proteins using the two hybrid system. UAS_G is the upstream activating sequence for the yeast *GAL* genes, which binds the GAL4 protein. Interaction is detected by expression of a *GAL1–lacZ* gene fusion. (Reproduced courtesy of Dr S. Fields and the National Academy of Sciences.)

The impact of mRNA sequence

The sequence of a mRNA molecule can affect its structure and this in turn affects the rate of translation and the stability (half-life) of the mRNA. Translational efficiency is thought to be controlled primarily by the rate of initiation and this is affected by the 5'-untranslated leader sequence. For example, Kniskern *et al.* (1986) found that hepatitis B core antigen accumulated to 0.05% of total soluble protein when viral 5' and 3' sequences were present. Deletion of the 5' viral sequences increased the yield to 26% while deletion of both the 5' and 3' sequences raised it to 41%. In both cases there was no alteration to the mRNA levels.

The half-lives of yeast mRNAs range from 1 to 100 minutes and therefore can have a profound effect on the steady-state level of mRNA. However, little is known about the factors affecting mRNA stability. One parameter which can influence mRNA half-life is the sequence of the 3'-untranslated region. Demolder *et al.* (1992) obtained 10-fold higher yields of murine interleukin-2 by deleting a large part of the mammalian 3'-untranslated region which appeared to be responsible for rapid turnover of the corresponding mRNA.

Theoretically, codon usage will rarely prevent high level gene expression. Bennetzen and Hall (1982), and more recently Sharp and Cowe (1991), have shown that there is an extreme codon bias in highly expressed yeast genes yet high levels of expression have been obtained from foreign genes containing rare yeast codons (Cousens *et al.* 1987, Chen & Hitzeman 1987). Nevertheless, there now are a number of examples of increased yields from synthetic genes with codon selection matched to the *S. cerevisiae* bias. For example, Kotula and Curtis (1991) obtained a 50-fold increase in an immunoglobulin by using a synthetic, codon-optimized gene. However, it is difficult to attribute an observable effect on product yield to a specific feature such as codon usage; changing the gene sequence also affects the primary and secondary structure of the mRNA. A good example of this is provided by studies on the expression of the gene encoding tetanus toxin fragment C (Romanos *et al.* 1991). The levels of the protein increased several thousand-fold when the native gene sequence was replaced with a synthetic one. However, the native gene is AT rich and contained at least six fortuitous polyadenylation sites giving rise to truncated mRNA species. The synthetic gene had an increase in GC content from 29 to 47% and this led to an increase in synthesis of full-length mRNA and toxin fragment C. Romanos *et al.* (1992) have suggested that premature transcriptional termination may be a common reason for low yields of heterologous proteins in *S. cerevisiae*.

Protein stability

By analogy with *E. coli*, a number of factors can control the level of expression of genes in *S. cerevisiae*. These are: (a) the copy number of the gene of interest and the stability of that copy number, (b) the nature of the promoter used and (c) the sequence of the mRNA. These topics have been covered in earlier sections (for review, see Buckholz 1993). A fourth factor is the stability of the protein after synthesis. Some protein products are degraded rapidly either during or shortly after synthesis. Others are lost during cell breakage and subsequent purification. In yeast the vacuole contains at least two endoproteinases, two carboxypeptidases and two aminopeptidases (Jones 1990). These enzymes gain access to heterologous gene products expressed in the cytoplasm when the cells are harvested and lysed. A number of yeast strains have been constructed which lack these proteases and their use has resulted in increased yields of products that initially were lost during their purification.

Where proteins are degraded rapidly in the cytoplasm after their synthesis, one solution is to secrete them into the medium where only low levels of proteases have been detected. However, many proteins whose normal habitat is the cytoplasm cannot be induced to be secreted even when appropriate secretion signals are incorporated in the corresponding gene. Such proteins may be subject to ubiquitin-dependent proteolysis (Wilkinson 1990). The identity of the amino-terminal residue of the protein can greatly affect its half-life in S. cerevisiae (Bachmair et al. 1986). Thus the half-life of β-galactosidase was increased from 2 minutes to 20 hours by replacing the amino-terminal arginine, a target for ubiquitin conjugation, with a variety of other amino acids (Met, Ser, Ala, Thr, Val, Gly).

Secretion of proteins by yeast

In yeast, proteins destined for the cell surface or for export from the cell are synthesized on, and translocated into, the endoplasmic reticulum. From there they are transported to the Golgi body for processing and packaging into secretory vesicles. Fusion of the secretory vesicles with the plasma membrane then occurs constitutively or in response to an external signal (reviewed by Rothman & Orci 1992). Of the proteins naturally synthesized and secreted by yeast, only a few end up in the growth medium, e.g. the mating pheromone α-factor and the killer toxin. The remainder, such as invertase and acid phosphatase, cross the plasma membrane but remain within the periplasmic space or become associated with the cell wall.

Polypeptides destined for secretion have a hydrophobic amino-terminal extension which is responsible for translocation to the endoplasmic reticulum (Blobel & Dobberstein 1975). The extension is usually composed of about 20 amino acids and is cleaved from the mature protein within the endoplasmic reticulum. Such signal sequences precede the mature yeast invertase and acid phosphatase sequences. Rather longer leader sequences precede the mature forms of the α-mating factor and the killer toxin (Kurjan & Herskowitz 1982, Bostian et al. 1984). The initial 20 amino acids or so are similar to the conventional hydrophobic signal sequences but cleavage does not occur in the endoplasmic reticulum. In the case of α factor, which has an 89 amino acid leader sequence, the first cleavage occurs after a Lys-Arg sequence at positions 84 and 85 and happens in the Golgi body (Julius *et al.* 1983, 1984).

To date, a large number of non-yeast polypeptides have been secreted from yeast cells containing the appropriate recombinant plasmid but the rules governing secretion still are not clear. For example, Hitzeman *et al.* (1983) obtained secretion of interferon when the construct carried the endogenous mammalian signal sequence. Sequencing of interferon purified from the growth medium showed it to have the same amino-terminus as natural mature interferon, indicating that the yeast cells had recognized and correctly processed the mammalian signal sequence. Similar results have been obtained with many other mammalian and plant proteins. By contrast, no secretion into the medium was observed with calf prochymosin (Mellor *et al.* 1983) or human α_1-antitrypsin (Cabezon *et al.* 1984) with their natural signal sequences. In the latter case, when the normal signal sequence was replaced with that from the yeast invertase gene, secretion of α_1-antitrypsin was observed (Moir & Dumais 1987).

Interpreting the above results is difficult. First, a whole series of different gene constructs has been used and in many instances little attention has been paid to the level of expression. If a cloned gene is expressed poorly, secretion might not be detectable, particularly if the medium contains proteases. Secondly, most workers have failed to accurately quantitate the levels of protein inside the cell, in the periplasmic space and in the growth medium. Nor have they measured the leakage of intracellular or periplasmic proteins into the medium. Without this information it is difficult to measure the efficiency of secretion. Suffice it to say that a large number of heterologous peptides and proteins have been successfully secreted from *S. cerevisiae* using natural yeast signal peptides. These proteins vary in size from 14 amino acids (somatostatin) to the Epstein–Barr virus envelope glycoprotein which is over 800 residues long (for review, see Romanos *et al.* 1992).

S. cerevisiae cells secrete relatively low levels of protein into the extracellular medium by comparison with filamentous fungi (1% versus 10% of total cellular protein). One possible explanation is that in *S. cerevisiae* one or more proteins involved in secretion are rate-limiting. Consequently, a number of groups have developed methods for identifying host cells with enhanced protein secretion. A general method for use with any protein has been described by Sleep *et al.* (1991). It is a plate assay based on the precipitation of product-specific antibodies. The size of the haloes formed around the colonies is indicative of the secretory capability of the strains. Several rounds of mutagenesis and selection resulted in the isolation of mutants which showed a sixfold increase in human serum albumin secretion and also produced higher levels of intracellularly expressed proteins, e.g. α_1-antitrypsin and human plasminogen activator inhibitor 2. A different method was used by Chow *et al.* (1992). Their method involved the scoring of clearing zones on a lawn of sensitive cells resulting from killer toxin secretion. In this way they identified a gene conferring fourfold-enhanced levels of total protein secreted per cell when re-introduced into the parental strain on a multicopy vector. The range of proteins secreted in greater quantities includes killer toxin, α-factor and acid phosphatase. This is the first example of protein expression being increased by over-expressing a single gene.

Targeting of proteins to the nucleus

The yeast nucleus contains a discrete set of proteins which are synthesized in the cytoplasm. In order to elucidate the mechanism governing protein localization Hall *et al*. (1984) constructed a set of hybrid genes by fusing the yeast *MATα* gene, encoding a presumptive nuclear protein, and the *E. coli lacZ* gene. A segment of the *MAT* gene product which was 13 amino acids long was sufficient to localize β-galactosidase activity in the nucleus. The nuclear location of the β-galactosidase was confirmed by immuno-fluorescence. Similar sequences have been found in other proteins and are termed *nuclear localization signals* (NLS). In the current model of nuclear protein import, proteins containing an NLS are recognized by specific receptors in the cytoplasm. The NLS–receptor complex moves to the nuclear pore which opens by an ATP-mediated reaction and permits protein entry (Bossie & Silver 1992).

Glycosylation of proteins

Many eukaryotic proteins are glycosylated. The mechanism for glyco-sylation resides in the endoplasmic reticulum and Golgi apparatus so glycosylation can occur only if the proteins are directed to these sites via the secretory pathway. In both yeast and mammalian cells an inner core of sugars is added to the protein in the endoplasmic reticulum. This involves the addition of a dolichol-linked oligosaccharide to asparagine residues which are part of the sequence Asn-X-Ser/Thr. Subsequently the protein is transferred to the Golgi apparatus where an outer core of sugars is assembled. In yeast this occurs by stepwise addition of mannose residues. In mammalian cells the outer core can be complex and branched, with sugars other than mannose included in the chain after trimming of the outer core (Dunphy & Rothman 1985).

From the foregoing discussion it should come as no surprise that many of the foreign proteins secreted by yeast are glycosylated. Where a detailed analysis of the secreted proteins has been made glycosylation has occurred at the expected sites (Moir & Dumais 1987, Penttila *et al*. 1988, Yoshizumi & Ashikari 1987) but the composition and sequence of the attached oligosaccharide differs from that occurring naturally. In most instances the unusual glycosylation has no influence on the properties of the protein *in vitro*. However, for proteins which are to be used therapeutically, the difference in glycosylation could affect stability, immunogenicity, and tissue distribution *in vitro*.

The α,3-linked mannose units which occur in large numbers in the outer chain are particularly immunogenic. One solution is to mutate the glycosylation sites as was done with urinary plasminogen activator (Melnick *et al.* 1990). Alternatively, yeast glycosylation mutants can be used. The *mnn*9 mutation prevents the addition of the outer chain of mannose units although some mannose residues are still present (Kukuruzinska *et al.* 1987). The addition of the *mnn*1 mutation totally eliminates mannose residues from the secreted protein. Some 'supersecreting' mutants (see p. 264) are also glycosylation deficient (Rudolph *et al.* 1989).

Cloning in yeasts other than *S. cerevisiae*

Recently there has been considerable interest in the development of cloning and expression systems in yeasts other than *S. cerevisiae*. Most of these alternative systems are based on commercially-important yeasts that have been selected for their favourable growth characteristics at an industrial scale, e.g. *Kluyveromyces lactis*, *Pichia pastoris* and *Hansenula polymorpha*. High levels of expression have been obtained in these yeasts (Table 13.3) and full details of these and other constructs are given in the review by Romanos *et al.* (1992).

Cloning in filamentous fungi

There is an interest in cloning in filamentous fungi for a variety of reasons. For example, to understand the biochemical mechanisms involved in conidiation in fungi (Timberlake & Marshall 1988), to understand the basis of plant pathogenicity or because the organisms are used industrially (Upshall 1986). Some of them have an exceptional secretion capacity, e.g. over 35 g/l hydrolytic enzymes from *Trichoderma reesei* (Durand *et al.* 1988).

An essential step in applying recombinant DNA technology to a species is to develop a DNA-mediated transformation system and this now has been achieved with a wide range of fungi. Efficient transformation occurs with protoplasts and the transforming DNA becomes integrated into the chromosome by homologous recombination. For *Aspergillus* there is no

Table 13.3 High-level heterologous expression in yeasts

Yeast	Protein	Promoter	Level	Location	Reference
Pichia pastoris	Pertactin	Alcohol oxidase	3 g/l	Intracellular	Romanos *et al.* (1991)
Pichia pastoris	Tetanus toxin fragment C	Alcohol oxidase	12 g/l	Intracellular	Clare *et al.* (1991)
Hansenula polymorpha	α-galactosidase	Methanol oxidase	2 g/l	Secreted	Veale *et al.* (1992)
Kluyveromyces lactis	Human serum albumin	Various	3 g/l	Secreted	Fleer *et al.* (1991)

evidence for replicating extra-chromosomal plasmids either in the nucleus or mitochondria, but freely-replicating plasmids have been identified in *Neurospora*. Genes of interest can be cloned by complementation of mutations or by making use of positive selection. For example, a plasmid containing the *amdS* gene of *Aspergillus nidulans* may be used to transform *amdS*$^+$ strains by selecting for increased utilization of acetamide as sole nitrogen source. Analysis of transformants (Kelly & Hynes 1987) has shown that multiple tandem copies of the plasmid can be integrated into the chromosome, commonly at sites other than the *amdS* locus. Such transformants are relatively stable through meiosis and mitosis. Finally, reporter genes can be fused to regulatory sequences and used to assay the effects of promoter mutations made *in vitro* (Hamer & Timberlake 1987) and secretion of heterologous proteins obtained using host signal sequences (Saloheimo & Niku-Paavola 1991, Ward *et al.* 1992, Nyyssonen *et al.* 1993). Heterologous protein production and secretion at levels as high as 3 g/l have been obtained in *Aspergillus oryzae* (Christiensen *et al.* 1988). These results suggest that the techniques needed for cloning in fungi are little different from those for *S. cerevisiae*.

14 Gene transfer to plants

Introduction

The past decade has seen a revolution in our ability to manipulate the genomes of plants. Before the early 1980s there had been no well-substantiated report of expression of a foreign gene in a genetically engineered plant. The situation has changed dramatically since then. Now it is possible to transfer any gene into a plant as a routine procedure. This has come about through very intensive research into vector systems based on the bacterium *Agrobacterium tumefaciens* and alternative strategies based on transfection of plant protoplasts (i.e. plant cells from which the cell wall has been removed), or based on transfection by biolistic devices. This change means that plant molecular biology can advance rapidly and has immense biotechnological implications in the creation of plants with useful genetically engineered characteristics such as herbicide resistance (Chapter 17).

The range of species available for routine transgenic plant production has been limited. This is for two reasons. Firstly, *Agrobacterium*-based gene transfer is efficient only with dicotyledonous plants ('dicots'). Secondly, transgenic plant production by either *Agrobacterium*-based systems or protoplast transfection requires the ability to regenerate plants from single (or a small number of) isolated transfected cells. This regeneration technology

[268]

has been developed to an advanced state in a small number of 'favourite' experimental species; tobacco, tomato and petunia are the most commonly used. Some agriculturally important dicots are not such well-developed experimental systems in this regard – the legumes are an example. With monocots there is the problem that *Agrobacterium* does not interact productively with them under natural circumstances. With the cereal crop species, members of the Graminae, there has been little success until recently. This situation has changed dramatically.

In order to explain these developments this chapter begins with an account of plant callus and protoplast culture. We then describe *Agrobacterium*-based, protoplast, and biolistic transfection systems; and important developments with monocot species. Finally, we briefly discuss the roles of plant viruses in vector systems.

Plant callus culture

Tissue culture is the process whereby small pieces of living tissue (*explants*) are isolated from an organism and grown aseptically for indefinite periods on a nutrient medium. For successful plant tissue culture it is best to start with an explant rich in undetermined cells, e.g. those from the cortex or meristem, because such cells are capable of rapid proliferation. The usual explants are buds, root tips, nodal segments or germinating seeds, and these are placed on suitable culture media where they grow into an undifferentiated mass known as a *callus* (Fig. 14.1). Because the nutrient media used can support the growth of microorganisms, the explant is first washed with a disinfectant such as sodium hypochlorite, hydrogen peroxide or

Fig. 14.1 Close-up view of a callus culture.

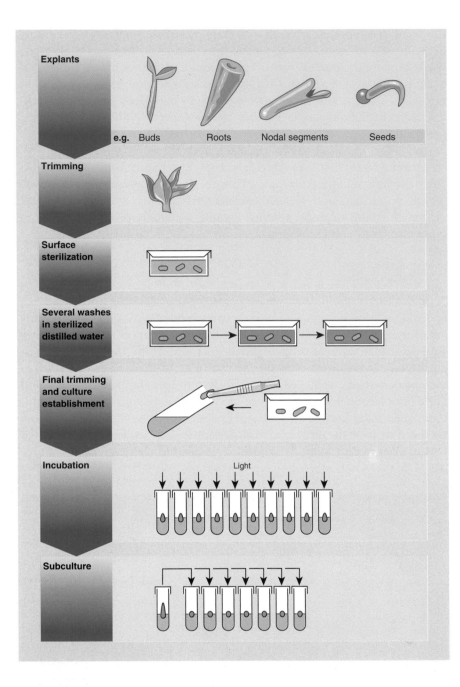

Fig. 14.2 Basic procedure for establishing and maintaining a culture of plant tissue.

mercuric chloride. Once established, the callus can be propagated indefinitely by subdivision (Fig. 14.2).

For plant cells to develop into a callus it is essential that the nutrient medium contain plant hormones, phytohormones, i.e. an auxin, a cytokinin and a gibberellin (Fig. 14.3). The absolute amounts of these which are required vary for different tissue explants from different parts of the same

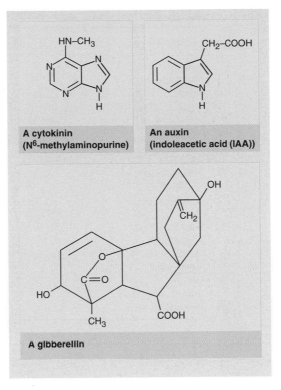

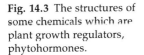

Fig. 14.3 The structures of some chemicals which are plant growth regulators, phytohormones.

plant and for the same explant from different genera of plants. Thus there is no 'ideal' medium. Most of the media in common use consist of inorganic salts, trace metals, vitamins, organic nitrogen sources (glycine), inositol, sucrose and growth regulators. Organic nutrients such as casein hydrolysate or yeast extract and a gelling agent are optional extras. The composition of a typical plant growth medium is shown in Table 14.1.

Plant cell culture and protoplasts

When a callus is transferred to a liquid medium and agitated, the cell mass breaks up to give a suspension of isolated cells, small clusters of cells and much larger aggregates. Such suspensions can be maintained indefinitely by subculture but, by virtue of the presence of aggregates, are extremely heterogeneous. A high degree of genetic instability adds to this heterogeneity. Some plants such as *Nicotiana tabacum* (tobacco) and *Glycine max* (soybean) yield very friable calluses and cell lines obtained from these species are much more homogeneous and can be cultivated both batchwise and continuously.

When placed in a suitable medium, isolated single cells from suspension cultures are capable of division. As with animal cells, for proliferation to occur conditioned medium may be necessary. Conditioned medium is prepared by culturing high densities of cells of the same or different species in fresh medium for a few days and then removing the cells by

Table 14.1 Composition of Murashige and Skoog (MS) culture medium

Ingredient	Amount (mg/l)
Sucrose	30 000
$(NH_4)NO_3$	1 650
KNO_3	1 900
$CaCl_2.2H_2O$	440
$MgSO_4.7H_2O$	370
KH_2PO_4	170
$FeSO_4.7H_2O$	27.8
Na_2EDTA	37.3
$MnSO_4.4H_2O$	22.3
$ZnSO_4.7H_2O$	8.6
H_3BO_3	6.2
KI	0.83
$Na_2MoO_4.2H_2O$	0.25
$CoCl_2.6H_2O$	0.025
$CuSO_4.5H_2O$	0.025
Myo-inositol	100
Glycine	2.0
Kinetin (a cytokinin)	0.04–10.0
Indoleacetic acid	1.0–30.0

filter sterilization. Media conditioned in this way contain essential amino acids such as glutamine and serine as well as growth regulators like cytokinins. Provided conditioned medium is used, single cells can be plated out on solid media in exactly the same way as microorganisms; instead of forming a colony as do microbes, plant cells proliferate to give a callus.

Protoplasts

Protoplasts are cells minus their cell walls. They are very useful materials for plant cell manipulations because under certain conditions those from similar and contrasting cell types can be fused to yield somatic hybrids, a process known as *protoplast fusion*. Protoplasts can be produced from suspension cultures, callus tissue or intact tissues, e.g. leaves, by mechanical disruption or, preferably, by treatment with cellulolytic and pectinolytic enzymes. Pectinase is necessary to break up cell aggregates into individual cells and the cellulase to remove the cell wall proper. After enzyme treatment, protoplast suspensions are collected by centrifugation, washed in medium without enzyme, and separated from intact cells and cell debris by flotation on a cushion of sucrose (Fig. 14.4). When plated onto nutrient medium protoplasts will in 5–10 days synthesize new cell walls and then initiate cell division.

As indicated earlier, the formation and maintenance of callus cultures require the presence of a cytokinin and an auxin, whereas only a cytokinin is required for shoot culture and only an auxin for root culture. Therefore it is no surprise that increasing the level of cytokinin to a callus induces

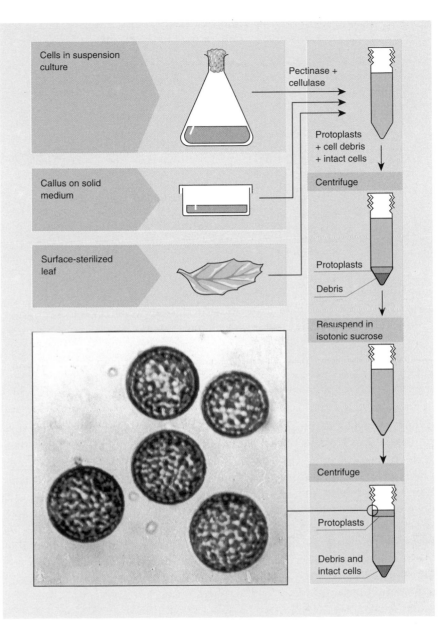

Fig. 14.4 Schematic outline of the enzymatic procedure used to isolate plant protoplasts. The inset shows a photomicrograph of protoplasts.

shoot formation and increasing the auxin level promotes root formation. Ultimately plantlets arise through development of adventitious roots on the shoot buds formed, or through development of shoot buds from tissues formed by proliferation at the base of rootlets (Fig. 14.5). The formation of roots and shoots on callus tissue is known as *organogenesis*. The cultural conditions required to achieve organogenesis vary from species to species, and have not been determined for every type of callus.

Under certain cultural conditions, calluses can be induced to undergo a

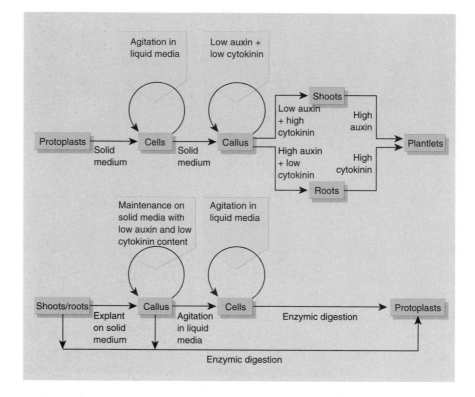

Fig. 14.5 Summary of the different cultural manipulations possible with plant cells, tissues and organs.

different development process known as *somatic embryogenesis*. In this process, the callus cells undergo a pattern of differentiation, similar to that seen in zygotes after fertilization, to produce *embryoids*. Such cells are embryo-like but differ from normal embryos in being produced from *somatic* cells and not from the fusion of two germ cells. These embryoids can develop into fully functional plants without the need to induce root and shoot formation on artificial media. The embryogenic response leading to embryoid formation is stimulated when calluses which were established in medium in which 2,4-dichlorophenoxyacetic acid (2,4-D) was the auxin are transferred to 2,4-D-free medium containing reduced sources of nitrogen, e.g. ammonium salts.

Agrobacterium and genetic engineering in plants

Crown gall disease

Crown gall is a plant tumour, a lump of undifferentiated tissue, which can be induced in a wide variety of gymnosperms and dicotyledonous angiosperms by inoculation of wound sites with the Gram-negative soil bacterium *Agrobacterium tumefaciens* (Fig. 14.6). The disease was first described long ago, and the involvement of bacteria was recognized as early as 1907 (Smith & Townsend 1907). It was subsequently shown that the crown gall tissue represents true oncogenic transformation; callus tissue

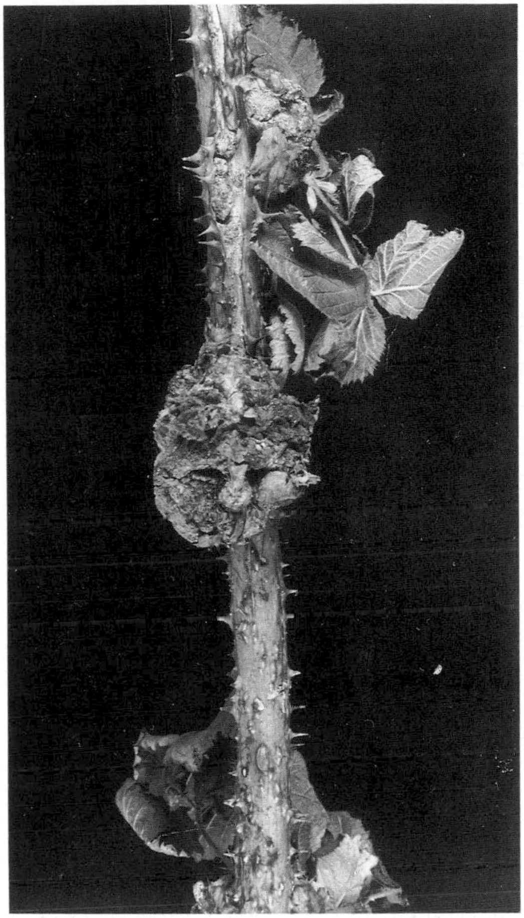

Fig. 14.6 Crown gall on blackberry cane. (Photograph courtesy of Dr C.M.E. Garrett, East Malling Research Station.)

can be cultivated *in vitro* in the absence of the bacterium and yet retain its tumorous properties (Fig. 14.7). These properties include the ability to form an overgrowth when grafted onto a healthy plant, the capacity for unlimited growth as a callus in tissue culture in media devoid of the plant hormones necessary for *in vitro* growth of normal plant cells, and the synthesis of *opines*, which are unusual amino acid derivatives not found in normal tissue (Fig. 14.8). The most common of these opines are octopine and nopaline. In addition, agropine or the agrocinopine family of opines may be present. Crown gall tumour cells continue to synthesize opines in tissue culture, and shoots or whole plants regenerated from tumour cell lines may also continue to synthesize opine (Braun & Wood 1976, Schell & Van Montagu 1977, Wullems *et al.* 1981a,b).

The metabolism of opines is a central feature of crown gall disease. Opine synthesis is a property conferred upon the plant cell when it is transformed by *A. tumefaciens*. The type of opine produced is determined not by the host plant but by the bacterial strain. In general, the bacterium induces the synthesis of an opine which it can catabolize and use as its sole

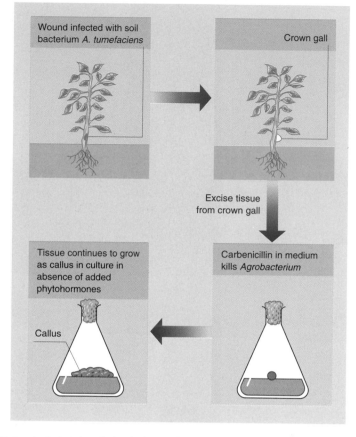

Fig. 14.7 *Agrobacterium tumefaciens* induces plant tumours.

energy, carbon and/or nitrogen source. Thus, bacteria that utilize octopine induce tumours that synthesize octopine, and those that utilize nopaline induce tumours that synthesize nopaline (Bomhoff *et al.* 1976, Montaya *et al.* 1977).

Investigation of the molecular biology of crown gall disease has revealed that *A. tumefaciens* has evolved a natural system for genetically engineering plant cells so as to subvert them for its own ends. Recent research, mainly using tobacco, tomato and petunia as the experimental plants, has enabled gene manipulators to exploit this natural system.

The tumour-inducing principle and the Ti-plasmid

As it was clear that the continued presence of bacteria is not required for transformation of the plant cells, attention was focused on the nature of the 'tumour-inducing principle', the name given to the putative genetic element that must be transferred from the bacterium to the plant at the wound site. For a long time it was believed, correctly, that DNA is transferred from the bacterium to the plant cell. Attempts to detect such

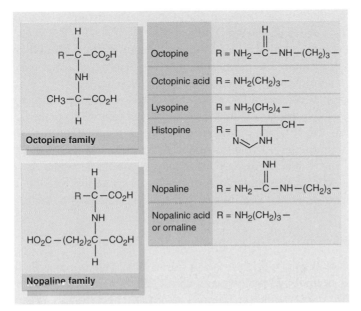

Fig. 14.8 Structures of some opines.

bacterial DNA in the tumour cells failed, simply because the techniques used were not sufficiently sensitive, until plasmids were detected in *Agrobacterium*. Zaenen *et al.* (1974) first noted that tumour-forming (i.e. virulent) strains of *A. tumefaciens* harbour large plasmids (140–235 kb), and it is now clear that the virulence trait is plasmid-borne. Virulence is lost when the bacteria are cured of the plasmid and with at least one strain this can be achieved by growing the cells at 37°C instead of 28°C. Cured strains also lose the capacity to utilize octopine or nopaline (Van Larbeke *et al.* 1974, Watson *et al.* 1975). Virulence is acquired by avirulent strains when a virulence plasmid is reintroduced by conjugation (Bomhoff *et al.* 1976, Gordon *et al.* 1979). If the plasmid from a nopaline strain is transferred to an avirulent derivative of a previously octopine strain, the avirulent strain then acquires the ability to induce nopaline tumours and catabolize nopaline. The virulence plasmid can also be transferred to the legume symbiont *Rhizobium trifolii* which becomes oncogenic and acquires the ability to utilize either octopine or nopaline, depending on the donor.

From the above information it is clear that plasmids are essential for virulence and for this reason they are referred to as Ti (tumour-inducing) plasmids. Furthermore, the genetic information specifying bacterial utilization of opines, and their synthesis by plants, is also plasmid-borne. It should be remembered, however, that the presence of a plasmid in *A. tumefaciens* does not mean that the strain is tumorigenic. Many strains contain very large cryptic plasmids that do not confer virulence, and in some natural isolates a cryptic plasmid is present together with a Ti-plasmid.

Plasmids in the octopine group have been shown to be closely related to each other while those in the nopaline group are considerably more diverse (Currier & Nester 1976, Sciaky *et al.* 1978). Between these two groups there is little DNA homology, except for four limited regions, one of which includes the genes directly responsible for crown gall formation: T-DNA (Drummond & Chilton 1978, Engler *et al.* 1981).

A related organism, *A. rhizogenes*, incites a disease, hairy root disease, in dicotyledonous plants in a manner very similar to *A. tumefaciens*. A large plasmid, the Ri (root inducing) plasmid, is responsible for pathogenicity and the induction of opine synthesis (White & Nester 1980, Chilton *et al.* 1982). The Ri plasmids share little homology with Ti-plasmids. They are of interest because tissue transformed by *A. rhizogenes* readily regenerates into plantlets which continue to synthesize opine (see p. 289).

Incorporation of T-DNA into the nuclear DNA of plant cells

Complete Ti-plasmid DNA is not found in plant tumour cells but a small, specific segment of the plasmid, about 23 kb in size, is found integrated in the plant nuclear DNA (Chilton *et al.* 1977). This DNA segment is called T-DNA (transferred DNA; Fig. 14.9). The structure and organization of the integrated T-DNA in tumour cells has been studied in detail. The main conclusions of these studies (Thomashow *et al.* 1980, Zambryski *et al.* 1980) are listed as follows.

●　Integration of the T-DNA can occur at many different, apparently random, sites in the plant nuclear DNA.

●　The organization of the integrated nopaline T-DNA is simple. It occurs as a single integrated segment.

●　Integrated octopine T-DNA usually exists as two segments. The left segment, TL, includes genes necessary for tumour formation. The right segment, TR, is not necessary for tumour maintenance but codes for enzymes for agropine biosynthesis (Saloman *et al.* 1984). Usually only one copy of TL-DNA is present per cell but as many as ten copies have been found. TR may be present in high copy numbers. The significance of these complications in integrated octopine T-DNA structure is not clear. It is

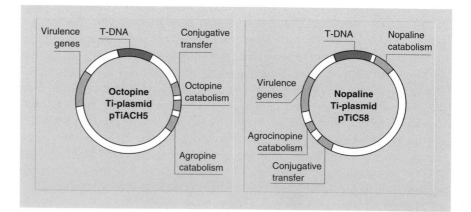

Fig. 14.9 Ti-plasmid gene maps.

likely that rearrangement, amplifications and deletions of T-DNA may follow the initial integration.

As we have noted, the integrated nopaline DNA occurs as a single, 23-kb, segment. Regions including the junctions of nopaline T-DNA with plant DNA have been cloned in phage λ vectors and the DNA sequences at the junctions determined. The right-hand junction with plant DNA appears to be rather precise, whereas the left-hand junction can vary by about 100 nucleotides (Yadav *et al*. 1982, Zambryski *et al*. 1982). On the Ti-plasmid itself the T-region is flanked by two almost perfect, 25 base-pair, direct repeats or *border sequences*. The repeats are not transferred intact to the plant genome, as the T-DNA end-points lie within, or immediately internal to, them. These repeats are conserved between nopaline and octopine Ti-plasmids while the sequences surrounding are not. This strongly suggested that the repeats are important in the integration mechanism (Simpson *et al*. 1982, Yadav *et al*. 1982). Deletions that remove the T-region right repeat abolish tumour formation. But, perhaps surprisingly, the left repeat can be removed without affecting tumour formation. Thus only the right repeat is essential, and in addition it has been shown that sequences further to the right and adjacent to the actual repeat sequence are necessary for full activity in transfer and integration into the plant genome (Shaw *et al*. 1984, Peralta *et al*. 1986). The left repeat has little or no transfer activity alone (Jen & Chilton 1986).

The mechanism by which the T-DNA region of the Ti-plasmid becomes integrated into the plant genome is only partially understood. It appears to closely resemble bacterial conjugation. T-DNA transfer and processing require products of the *vir* genes (*virA, -B, -G, -C, -D* and *-E*), which are located outside the T-DNA region. The expression of *virB, -C, -D* and *-E* is positively regulated at the transcriptional level by plant signal molecules. For tobacco, these signal molecules are known. They are the phenolic compounds acetosyringone and α-hydroxysyringone (Fig. 14.10) (Stachel *et al*. 1985), which are exuded from wounded or actively growing cells. Components of lignin or its precursors also act as signals, and certain

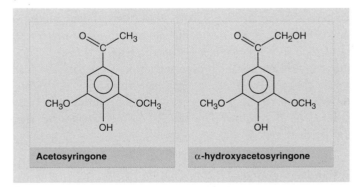

Fig. 14.10 Structures of signal molecules, produced by wounded plant tissue, which activate T-DNA transfer by *Agrobacterium tumefaciens*.

monosaccharides synergize the action of the phenolic signals (Shimoda *et al*. 1990).

The *virA* and *virG* genes are constitutively expressed and control the plant-induced activation of the other *vir* genes. The VirA protein spans the inner bacterial membrane and acts as a receptor for the phenolic and monosaccharide signals. The signal detected by VirA is transduced to the VirG protein, probably by phosphorylation. VirG is a DNA binding protein which activates the transcription of the other *vir* genes. The VirA and VirG proteins show similarities to other two-component regulatory systems common in bacteria (Stachel & Zambryski 1986). In addition to these plasmid-borne genes, genes located on the bacterial chromosome have also been identified as essential for T-DNA transfer.

In response to the activating signal molecule, an endonuclease encoded by *virD* makes two nicks in the T-DNA in the same strand at its border sequences in the Ti-plasmid. The T-DNA single strand is transferred with the right border leading (Yanofsky *et al*. 1986, Stachel *et al*. 1986). This probably accounts for the observation that only the right repeat sequence is essential for transfer. The left repeat probably defines the other end of the transferred DNA by virtue of the nick. If the left repeat is absent, random breakage may define the amount of DNA transferred. At some point in the transfer process the single-stranded DNA must be converted into double-stranded form. In bacterial conjugation this conversion occurs in the recipient cell so if the resemblance with T-DNA transfer holds good, we would predict that the T-DNA enters the plant cell in single-stranded form and that complementary strand is synthesized by plant cell enzymes. The VirE2 protein is a single-stranded DNA binding protein. It covers free T-DNA within the bacterium. The VirD2 protein attaches to the 5'-end of the processed T-DNA, probably covalently. The VirD2 protein has been proposed to protect the T-DNA against nucleases, to target the DNA to the plant cell nucleus, and to integrate it into the plant genome. The protein has two distinct nuclear localization signals, active in plant cells, which may be important in targeting the T-DNA to the nucleus (Tinland *et al*. 1992).

The integration of the T-DNA is not precise to the nucleotide but, as we have noted, it is more precise at the right border than at the left. The significance of this is unclear, as is the observation that separate T-region circles are formed in *Agrobacterium* induced for gene transfer. These circles contain one copy of the border repeat sequences (Koukolikova-Nicola *et al*. 1985). Clearly the ability of *Agrobacterium* to transfer DNA right into the nucleus of the plant cell requires a very intimate association between the plant cell and the bacterium. The details of the fascinating relationship have yet to be discovered.

Gene maps and expression of T-DNA

Genetic maps of Ti-plasmids and T-DNA have been obtained by the study of spontaneous deletions and by transposon mutagenesis (Koekman *et al*.

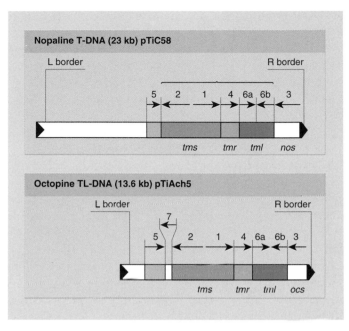

Fig. 14.11 Structure and transcription of T-DNA. The T-regions of nopaline and octopine Ti-plasmids have been aligned to indicate the DNA sequences that are common to both T-regions. The size and orientation of each transcript (numbered) is indicated by arrows. Genetic loci, as defined by deletion and transposon mutagenesis, as follows: *nos*, nopaline synthase; *ocs*, octopine synthase; *tms*, shooty tumour; *tmr*, rooty tumour.

1979, Garfinkel & Nester 1980, Holsters *et al.* 1980, Ooms *et al.* 1980, De Greve *et al.* 1981, Ooms *et al.* 1981). Such studies first revealed the large *vir* region mapping outside the T-DNA. As expected regions of the T-DNA itself were found to affect tumour morphology (Fig. 14.11). The loci affecting tumour morphology are designated 'large' (*tml*), 'shooty' (*tms*) and 'rooty' (*tmr*) to indicate the phenotype of the callus obtained when the loci are inactivated. Ooms *et al.* (1981) proposed that certain mutations in T-DNA appear to affect plant hormone concentrations in the resulting tumours so as to produce rooty and shooty types. The basis for this proposal was the observation that although the tumour grows in media lacking added auxins and cytokinins, the callus actually contains high concentrations of these hormones. Indeed, uncloned primary tumours contain some normal, untransformed cells growing as undifferentiated callus owing to the presence of these endogenous hormones (Gordon 1980). The observations suggested the idea that genes within the T-DNA actually encoded enzymes involved in phytohormone biosynthesis. As we shall see, this idea has been proved correct.

Figure 14.11 shows a transcript map of nopaline and octopine T-DNAs (Willmitzer *et al.* 1982, Willmitzer *et al.* 1983, Winter *et al.* 1984). The transcript encoding octopine synthase (also called lysopine dehydrogenase) has been identified and its gene located at the right end of TL-DNA.

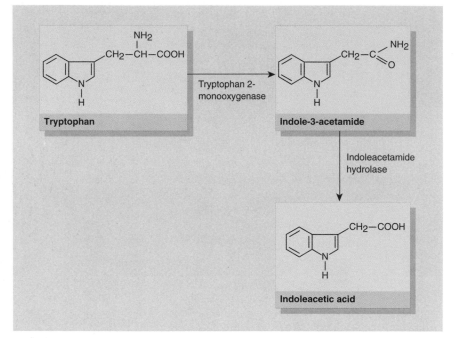

Fig. 14.12 Biosynthesis of the auxin, indoleacetic acid, by the indole-3-acetamide pathway.

Nucleotide sequencing has shown that this gene, *OCS*, has promoter elements which are eukaryotic in character, that it lacks introns, and that it contains a eukaryotic polyadenylation signal near its 3' end (De Greve *et al.* 1982b). Essentially similar results have been obtained for the nopaline synthase gene, *nos* (Depicker *et al.* 1982, Bevan *et al.* 1983a). It is possible that the genes of T-DNA are eukaryotic in origin and have been 'captured' by the Ti plasmid during its evolution. The *ocs* and *nos* gene promoters function in a wide variety of plant cells and have been widely used in constructing plant vectors (see below).

Two of the transcripts, numbered 1 and 2, map to genes (*tms*) involved in auxin production. Transcript 4 maps to a gene (*tmr*) involved in cytokinin production. Transcripts 5 and 6 encode products which appear to suppress differentiation only in cells in which they are expressed, i.e. in a non-hormonal manner.

It has been demonstrated directly (Schroder *et al.* 1984, Akiyoshi *et al.* 1984, Klee *et al.* 1984, Weiler & Schroder 1987) that in fact transcripts 1 and 2 from the auxin genes encode two enzymes, tryptophan 2-monooxygenase and indoleacetamide hydrolase, which convert tryptophan to indole acetic acid (Fig. 14.12). Transcript 4, from the cytokinin gene, encodes isopentenyl transferase, an enzyme that catalyses synthesis of zeatin-type cytokinins (Fig. 14.13).

It is notable that many other microorganisms produce plant hormones (reviewed by Morris 1986). The survival of many plant pathogens must often involve their ability to modify plant physiology by such means. The bakanae fungal disease of rice involves gibberellin production by the

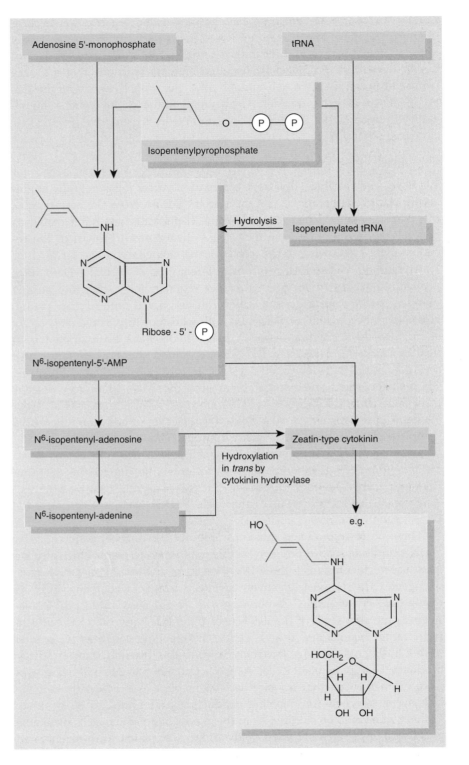

Fig. 14.13 Biosynthesis of cytokinins. Two pathways for the biosynthesis have been described. Isopentenyl pyrophosphate can combine with 5'-AMP, followed by further steps, or it can combine with adenine residues in tRNA followed by hydrolytic release of isopentenyl-5'-AMP.

fungus, and *Corynebacterium fascians* alters meristem organization by producing cytokinin. The bacterium *Pseudomonas syringae* subsp. *savastanoi* induces galls on olive and oleander by the secretion of auxins and cytokinins, which are produced by bacterial enzyme reactions that are very similar to those of crown gall tumour cells. But with *P. savastanoi* there is no evidence of gene transfer. Crown gall induction may be a highly evolved refinement of this type of system (Weiler & Schroder 1987).

Disarmed Ti-plasmid derivatives as plant vectors

We have seen that the Ti-plasmid is a natural vector for genetically engineering plant cells because it can transfer its T-DNA from the bacterium to the plant genome. However, wild-type Ti-plasmids are not suitable as general gene vectors because they cause disorganized growth of the recipient plant cells owing to the effects of the oncogenes in the T-DNA.

The tumour cells which result from integration of normal T-DNA have proven recalcitrant to attempts to induce regeneration, either into normal plantlets, or into normal tissue which can be grafted onto healthy plants. However, tobacco callus transformed with wild-type Ti-plasmid does rarely spontaneously regenerate shoots. These shoots have been grafted onto healthy plants for further analysis. Some grafted shoots were fertile and produced seed that developed into apparently normal plants; however, these plants lacked opine and all or most of the T-DNA had been deleted from them (Braun & Wood 1976, Turgeon *et al.* 1976, Lemmers *et al.* 1980, Wullems *et al.* 1981a,b). A single case has been reported of opine-positive complete plants regenerated from a tumour. This tumour had originally been induced by a shooty mutant of octopine T-DNA (Leemans *et al.* 1982a). In breeding experiments these plants transmitted the octopine trait in a simple Mendelian fashion, giving apparently healthy progeny. Analysis of the T-DNA revealed that little T-DNA was present except for the *ocs* gene (De Greve *et al.* 1982a).

These observations, and our knowledge of the oncogenic functions in T-DNA, indicate that in order to be able to regenerate plants efficiently we must use vectors in which the T-DNA has been *disarmed* by making it non-oncogenic. This is most effectively achieved simply by deleting all of its oncogenes. For example, Zambryski *et al.* (1983) substituted pBR322 sequences for almost all of the T-DNA of pTiC58 leaving only the left and right border regions and the *nos* gene. The resulting construct was called pGV3850 (Fig. 14.14). *Agrobacterium* carrying this plasmid transferred the modified T-DNA to plant cells. As expected, no tumour cells were produced, but the fact that transfer had taken place was evident when the cells were screened for nopaline production and found to be positive. Callus tissue could be cultured from these nopaline-positive cells if suitable phytohormones were provided, and fertile adult plants were regenerated by hormone induction of plantlets. The creation of disarmed T-DNA was an important step forward. But the fact that oncogenic transformation does not now occur means that phytohormone-independent callus production

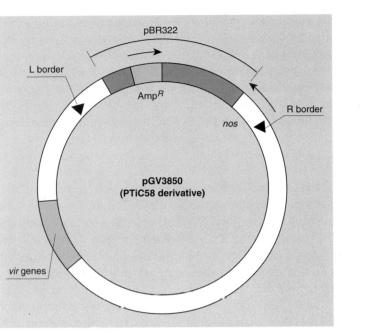

Fig. 14.14 Structure of the Ti-plasmid pGV3850, in which the T-DNA has been disarmed.

cannot be used to self-select recipient plant cells. As we have seen, opine production can be used as a marker but a dominant-acting selectable marker would be much more convenient.

Selectable markers for inclusion in T-DNA

In order to provide suitable markers, chimaeric genes were constructed in which *nos* and *ocs* promoters and polyadenylation sites were exploited to provide expression signals for bacterial neomycin phosphotransferase, which confers resistance to the aminoglycoside antibiotics kanamycin and G418 when expressed in plant cells (compare p. 309), bacterial hygromycin resistance (Shimamoto *et al.* 1989), or bacterial dihydrofolate reductase, which confers resistance to methotrexate or trimethoprim (compare p. 307) (Bevan *et al.* 1983b, Fraley *et al.* 1983, Herrera-Estrella *et al.* 1983a,b, Horsch *et al.* 1984). Inclusion of such selectable markers in T-DNA opened the way for *Agrobacterium*-mediated transfer of T-DNA becoming an immensely powerful gene delivery system. The *bar* gene is discussed below (p. 295).

Insertion of foreign DNA into T-DNA

Wild-type Ti-plasmids are not suitable as experimental gene vectors because their large size means that it is not possible to find adequate unique restriction sites in the T-region. Their size also makes other procedures cumbersome. Intermediate vectors (abbreviated IV) have therefore been developed in which T-DNA has been subcloned into conventional small, pBR322-based, plasmid vectors of *E. coli* (Matzke & Chilton 1981). Standard procedures can then be used to insert any desired DNA into the T-region of such an IV.

The IV, containing foreign DNA in the T-region, can be transferred to *A. tumefaciens* by conjugation. Since the IVs are conjugation-deficient, conjugation must be mediated by the presence in the donor *E. coli* of a helper, conjugation-proficient plasmid which can mobilize the IV. Suitable helper plasmids are pRK2013 (Ditta *et al.* 1980) which consists of the transfer genes from the naturally-occurring plasmid pRK2 cloned onto a Col E1 replicon, or pRN3 (Shaw *et al.* 1983). Neither of these helper plasmids is capable of replicating in *Agrobacterium*. These transfers are conveniently brought about by 'triparental' matings. In these matings three bacterial strains are mixed together. These are: (a) the *E. coli* carrying the conjugation-proficient helper plasmid, (b) the *E. coli* strain carrying the recombinant IV, and (c) the recipient *Agrobacterium*. During the course of the incubation the helper plasmid transfers to the *E. coli* strain carrying the recombinant IV which is then mobilized and transferred to *Agrobacterium*. The *Agrobacterium* recipient frequently receives both the IV and the helper plasmid.

Once the IV has been introduced into *Agrobacterium*, *in vivo* homologous recombination is exploited to insert the IV into a resident non-recombinant Ti-plasmid.

A single recombination event between two circular plasmids will produce a *co-integrate*. This is illustrated in Fig. 14.15 where it is apparent that a Ti-plasmid such as pGV3850 is particularly useful as an acceptor of the IV because the pBR322 sequences in its T-DNA region are homologous with most pBR322-based IV plasmids. In the example shown the IV includes a *neo* marker for selection of recombinant T-DNA in plant cells, and a *kan*^R determinant for selection of cointegrates in the *Agrobacterium*. (Neither the IV nor the helper plasmid can replicate autonomously in *Agrobacterium*.)

Binary Ti-vectors

The *in vivo* recombination technique of co-integrate formation just described has been used widely. The procedure reconstructs a very large, recombinant, disarmed Ti-plasmid. In fact this is unnecessary. We saw earlier that the T-region is distinct from the *vir* region whose functions are responsible for transfer and integration of T-DNA into the plant genome. It is possible to take advantage of this by providing the manipulated disarmed T-DNA carrying foreign DNA and the *vir* functions on separate plasmids (de Framond *et al.* 1983). This is the principle of *binary* vectors (Hoekma *et al.* 1983, Bevan 1984). Thus a modified T-DNA region carrying foreign DNA is constructed in a small plasmid which replicates in *E. coli* such as pRK252. This plasmid, called *mini*-Ti or *micro*-Ti can then be transferred

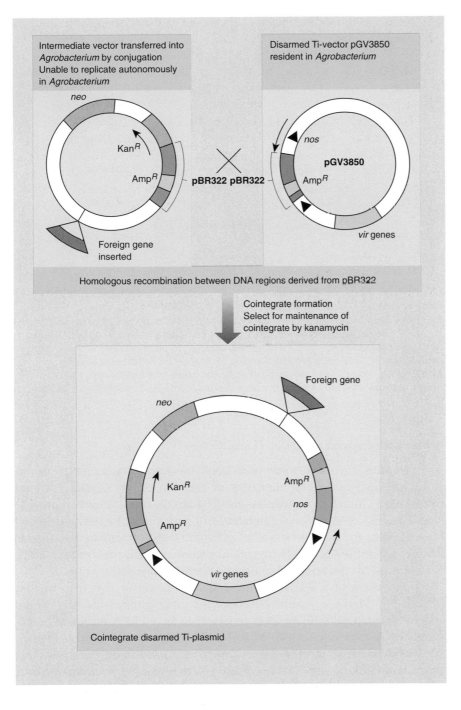

Fig. 14.15 Production of recombinant disarmed Ti-plasmid by cointegrate formation.

conjugatively in a tri-parental mating into *A. tumefaciens* which contains a compatible plasmid-carrying virulence gene. The *vir* functions are supplied in *trans*, causing transfer of the recombinant T-DNA into the plant genome. This binary system simplifies the transfer of foreign genes to plant cells (Fig. 14.16).

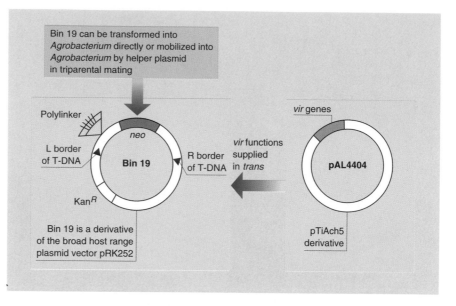

Fig. 14.16 Binary Ti-vector system. The binary vector (Hoekma *et al.* 1983, Bevan 1984) illustrated, Bin 19, is based upon the incP group, broad host-range plasmid pPK252. It contains a *neo* marker for selection of transferred DNA in plant cells and includes a *lacα* peptide gene with inserted polylinker region for blue/white detection of insertion of foreign DNA. *Vir* functions are supplied in *Agrobacterium* by pAL4404, a pTiAch5 derivative from which the T-DNA has been deleted.

Binary vector systems without Ti sequences

In discussing the mechanism of T-DNA transfer from *Agrobacterium* to plant cells we noted the similarity of the process to normal bacterial conjugation, in which double-stranded plasmid DNA is nicked in one strand by the action of *mob* (mobilization) proteins at the *ori*T (origin of transfer) site, and a single strand of plasmid DNA is transferred in a polar manner from this site into the recipient bacterium. The apparent similarity is reinforced by the remarkable observations of Buchanan-Wollaston *et al.* (1987). In place of a T-DNA-containing mini-Ti vector they constructed pJP181 (Fig. 14.17), a plasmid based on the incQ group, wide host range, natural plasmid pRSF1010. A *neo* gene was included for selection of transferred DNA in plant cells. pJP181 was conjugated into *Agrobacterium* which contained pAL4404 and the *Agrobacterium* then used to infect tobacco leaf discs. It was found that kanamycin (or G418) resistant transformant plant cells were produced and that these contained pJP181 DNA integrated into the plant genome. These experiments showed that the bacterial *mob* and *ori*T functions required for normal bacterial conjugation can also promote plasmid transfer to plants and could form the basis of a variety of new binary vectors. It appears therefore that *Agrobacterium* infection provides a means by which bacteria become closely associated with plant cells, and that the actual DNA transfer mechanism to plant cells has evolved from

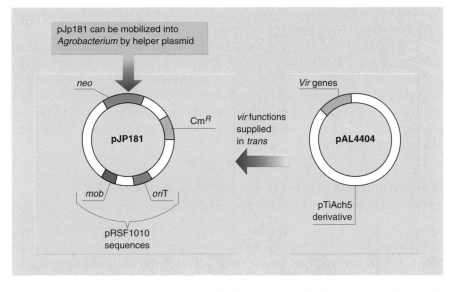

Fig. 14.17 Binary vector system without T-DNA sequences. The binary vector illustrated pJP181 (Buchanan-Wollaston *et al.* 1987) is based upon the incQ group broad host-range plasmid pRSF1010, and includes the *mob* and *oriT* sequences. pJP181 includes a *neo* marker for selection in plant cells. *Vir* functions are supplied in *Agrobacterium* by pAL4404.

the familiar bacterial conjugation mechanism. An important evolutionary implication of these findings is that plant genomes have access to the gene pool of Gram-negative bacteria.

A. rhizogenes and Ri plasmids

We noted previously that Ri plasmids induce hairy root disease in plants. The Ri plasmids are analogous to Ti plasmids. The Ri T-DNA includes genes homologous to the *iaaM* (tryptophan 2-monooxygenase) and *iaaH* (indoleacetamide hydrolase) genes of *A. tumefaciens*. Four other genes present in the Ri T-DNA are named *rol* for *root locus*. Two of these, *rolB* and *rolC*, encode β-glucosidases able to hydrolyse indole- and cytokine-*N*-glucosides. *A. rhizogenes* therefore appears to alter plant physiology by releasing free hormones from inactive or less active conjugated forms (Estruch *et al.* 1991a,b).

Ri plasmids are of interest from the point of view of vector development because opine-producing root tissue induced by Ri-plasmids in a variety of dicots can be regenerated into whole plants by manipulation of phytohormones in the culture medium. Ri T-DNA is transmitted sexually by these plants and affects a variety of morphological and physiological traits but does not in general appear deleterious. The Ri plasmids therefore appear to be already equivalent to disarmed Ti-plasmids (Tepfer 1984). Many of the principles explained in the context of disarmed Ti-plasmids are applicable to Ri-plasmids. An intermediate vector co-integrate system has been developed (Jensen *et al.* 1986) and applied to the study of nodulation in transgenic legumes, *Lotus corniculatus* (bird's-foot trefoil).

Van Sluys *et al.* (1987) have exploited the fact that *Agrobacterium* containing both an Ri plasmid and a disarmed Ti plasmid can frequently co-transfer both plasmids. The Ri plasmid induced hairy root disease in recipient *Arabidopsis* and carrot cells, serving as a transformation marker for the co-transferred recombinant T-DNA, and allowing regeneration of intact plants. No drug resistance marker on the T-DNA was necessary with this plasmid combination.

A simple and general experimental procedure for transferring genes into plants with *Agrobacterium* Ti-plasmid vectors

Once the principle of selectable, disarmed T-DNA vector regions was established, there followed an explosion of experiments in which foreign DNA was transferred into regenerated fertile plants. A precise upper size limit for foreign DNA acceptable by T-DNA has not been determined. It is greater than 50 kb (Herrera-Estrella *et al.* 1983a).

The simple general protocol of Horsch *et al.* (1985) has been widely adopted (Fig. 14.18). First small discs (a few millimetres diameter) are punched from leaves of petunia, tobacco, tomato, or other dicot plant, surface-sterilized, and inoculated in a medium containing *A. tumefaciens* carrying the recombinant disarmed T-DNA (as co-integrate or binary vector) in which the foreign DNA is accompanied by a chimaeric *neo* gene conferring kanamycin resistance on plant cells. The discs are cultured for 2 days and transferred to medium containing kanamycin to select the transferred *neo* gene, and carbenicillin to kill *Agrobacterium*. After 2–4 weeks shoots develop which are excised from the callus and transplanted to root-inducing medium. Rooted plantlets can subsequently be transplanted to soil, about 4–7 weeks after the inoculation step.

This method has the advantage of being simple and relatively rapid. It is superior to previous methods in which transformed plants were regenerated from callus which had itself been derived from protoplasts that had been transformed by co-cultivation with the *Agrobacterium* (De Block *et al.* 1984, Horsch *et al.* 1984). Interestingly, in one application of the protoplast co-cultivation method it was found that transmission of the foreign DNA in regenerated tobacco plants was maternal; it was not transmitted through pollen. Southern blotting chloroplast DNA showed directly that the foreign DNA had become integrated into the chloroplast genome. This was the first demonstration of the, presumably relatively rare, introduction of foreign DNA into the chloroplast DNA by Ti-plasmid vectors (De Block *et al.* 1985).

Agrobacterium and monocots

Agrobacterium normally causes crown gall disease on dicots. In the laboratory, tumours have been induced on the monocot *Asparagus officinalis*.

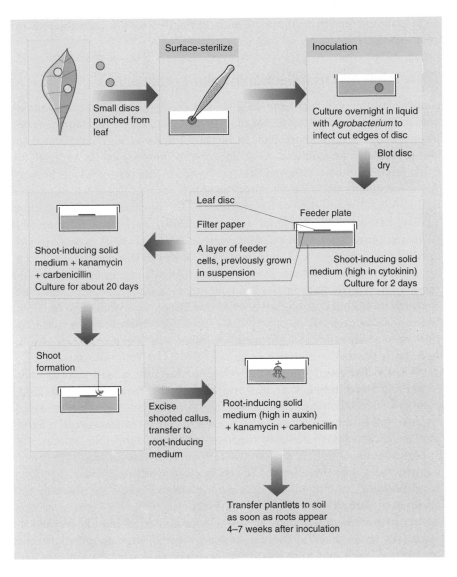

Fig. 14.18 Leaf-disc transformation by Ti-plasmid vectors.

Hooykaas-Van Slogteren *et al.* (1984) showed that *Agrobacterium* T-DNA can be transferred into the genomes of the monocots *Chlorophytum* and *Narcissus*. No tumours were induced but the plant cells synthesized opines. The T-DNA may therefore be naturally disarmed for these monocots.

It has been argued that *Agrobacterium* infection of monocots is inefficient because of the lack of production of *vir* gene activating substances by the wounded monocot tissue. By pretreating *Agrobacterium* with wound exudate from tubers of the potato, a susceptible dicot, Schafer *et al.* (1987) were able to induce crown gall tumours in the important monocot crop plant *Dioscorea bulbifera*, the yam. T-DNA was detected in tumour cell DNA. The experiments demonstrated the usefulness of pre-inducing the *Agrobacterium*, but little progress has been reported with the agriculturally important cereal crop plants.

DNA-mediated transfection of plant protoplasts

Removal of the wall from plant cells makes the resulting protoplasts amenable to transformation by DNA. The process has much in common with animal cell transformation. First, a method for actually getting DNA into the protoplasts is required. A variety of procedures have been used for this: electroporation (see p. 21) has become the favoured technique. Also, selectable genes are required to which foreign DNA can be ligated and which allow selection of transformants. The development of a selection system based on the bacterial genes conferring resistance to neomycin (G418, kanamycin), driven by plant expression signals such as those of *nos*, *ocs* and the CaMV gene VI, has been very useful in this context.

Transformants obtained by such procedures are placed on nutrient medium in which the protoplasts regenerate new cell walls and initiate cell division. Manipulation of the culture conditions then makes it possible, with a wide range of dicotyledonous plants, to induce shoot and root formation. Ultimately transformed plantlets may be grown into fertile transgenic plants.

In early demonstrations of the power of the protoplast transformation technique, Paszkowski, Potrykus and co-workers (Paszkowski *et al.* 1984, Hain *et al.* 1985, Potrykus *et al.* 1985a) showed that selectable foreign DNA became stably integrated into the nuclear genome of transgenic tobacco plants and was transmitted to progeny plants in a Mendelian fashion.

An example of this technology was provided by Meyer *et al.* (1987) who constructed a plasmid, in an *E. coli* vector, which contained a cDNA corresponding to the maize enzyme dihydroquercetin 4-reductase (an enzyme of anthocyanin pigment biosynthesis) linked to promoter and transcript termination signals of the 35S promoter of cauliflower mosaic virus (see later, p. 297). The plasmid also contained a *neo* gene, flanked by the *nos* promoter and the *ocs* polyadenylation site. Protoplasts of a mutant, white coloured, *Petunia* strain were transformed with the plasmid DNA using the mitotic-arrested-cell electroporation technique (p. 21). Kanamycin-resistant microcalli were derived from transfected cells and were induced to regenerate whole plants. The plants produced flowers of a new brick red coloration, owing to expression of the maize cDNA. Other work demonstrated that foreign DNA can be stably introduced into protoplasts from graminaceous plants, including wheat (Lorz *et al.* 1985) and Italian ryegrass, *Lolium multiflorum* (Potrykus *et al.* 1985b). However, by contrast with dicotyledonous plants, cereal species present a problem; with such cereals it has not proven feasible to regenerate whole plants from cultured cells of cereal species. (Rice is exceptional in having been shown to regenerate from cultured cells; Uchimaya *et al.* 1986, Shimamoto *et al.* 1989, Datta *et al.* 1990.) This aspect of plant cell culture is receiving much attention. At present it remains to be seen how amenable graminaceous plants will become to this technology.

Microprojectiles for transfecting living cells: biolistics

A completely novel way of delivering nucleic acids into plant cells involves the use of high velocity microprojectiles, which carry the RNA or DNA, and are literally shot into the cell (Klein *et al.* 1987). Figure 14.19 shows a gun for firing the microprojectiles. A gunpowder charge fires spherical tungsten particles (average diameter 4 μm) into intact cells. In the test system used by Klein *et al.*, intact epidermal cells of the onion, *Allium cepa*, were bombarded with the particles. Most of the cells in the target area of about 1 cm² were found to contain the microprojectiles following bombardment.

To demonstrate the technique, genomic RNA of tobacco mosaic virus was adsorbed onto the tungsten particles before firing. Three days after bombardment it was evident that 30–40% of the *A. cepa* cells which contained the particles also showed signs of virus replication. It was also shown that the gun could deliver DNA into cells. The test DNA consisted of a *CAT* gene construct driven by the CaMV 35S promoter. Extracts of the *A. cepa* epidermal tissue which had been bombarded with microprojectiles coated with the DNA were found to transiently express very high levels of chloramphenicol acetyltransferase.

In place of the original gunpowder method, various devices have been constructed for firing gold or tungsten particles into living cells. These include electric discharge-based (Christou 1988) and pneumatic devices (Sanford *et al.* 1991). The method has also been applied to mammalian cells and organs (Yang *et al.* 1990, Williams *et al.* 1991). The biolistic technique is used to stably transfect nuclear DNA in a variety of cell types in dicotyledonous and monocotyledonous species. [Chloroplast transformation has also been achieved (Svab *et al.* 1990, Staub & Maliga 1992), as has mitochondrial transformation in yeast (Johnston *et al.* 1988).] Transgenic

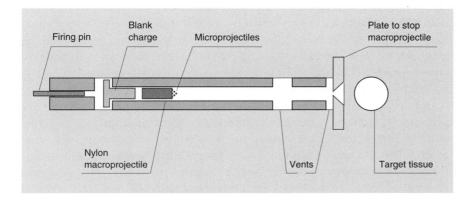

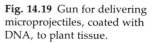

Fig. 14.19 Gun for delivering microprojectiles, coated with DNA, to plant tissue.

plants from an impressively wide variety of species have been recovered by bombardment of a variety of cell types, including: immature zygotic embryos (rice, Christou *et al.* 1991), embryogenic suspension cultures (cotton, Finer & McMullen 1990; maize, Gordon-Kamm *et al.* 1990), embryogenic callus (wheat, Vasil *et al.* 1992; white spruce, Ellis *et al.* 1993), stem sections (cranberry, Serres *et al.* 1992), root sections (*Arabidopsis thaliana*, Seki *et al.* 1991) (for reviews, see Christou 1992, 1993).

Regulated tissue-specific expression of transferred genes in plants

We have already seen how expression signals from constitutively expressed plant genes have been exploited in constructing generally useful selectable markers based on bacterial aminoglycoside phosphotransferases. The expression signals of the *nos*, *ocs* genes and CaMV gene VI have been most widely used.

Transcription directed by these strong constitutive promoters is useful biotechnologically for the suppression of gene expression by anti-sense (or sense) RNA. For a review of this burgeoning field of *trans*-inactivation of gene expression in plants, see Kooter & Mol (1993).

In addition to constitutive expression, it is necessary to be able to direct expression of manipulated genes in particular tissues of a plant. This ability is essential for many actual or foreseen applications of recombinant DNA technology in agriculture. Evidence from animal systems confirms that signals for differentiated expression lie associated with the promoter region and that these signals, together with tissue-specific enhancer and silencer elements, confer tissue specificity of transcription. Fortunately, in plants it also has proven feasible to identify promoter sequences and associated regulatory sequences which normally confer tissue specificity upon their natural, linked gene sequence or upon foreign genes linked to them artificially.

In order to identify tissue-specific control regions, constructs are made in which tissue-specific transcription elements are linked to a reporter. The most widely used reporter for such experiments in plants is β-glucuronidase, GUS (compare reporters for animal cells, Chapter 15). The *GUS* gene used in these constructs is derived from *E. coli* (Jefferson *et al.* 1986, 1987). The tissue-specific expression in transgenic plants can be followed by histochemical staining with the chromogenic substrate 5-bromo-4-chloro-3-indoyl-β-D-glucuronide (X-gluc) (compare X-gal).

Where several genes are involved in a multi-enzyme pathway, it may be possible to activate them all by the constitutive expression of an appropriate transcription factor. An impressive indicator that this might be feasible comes from experiments on regulatory genes, R and C1, that control the pattern of accumulation of anthocyanin in maize tissues. Ubiquitous expression of the maize regulatory genes, in transgenic *Arabidopsis*, led to activation of a multi-enzyme pathway in tissues where it is not normally expressed (Lloyd *et al.* 1992). Furthermore, as this example shows, transcriptional regulators from one species can activate genes in a distantly-related species.

Summary

The *Agrobacterium*-based gene-transfer system has been developed into a relatively simple and efficient technology with wide applicability to many dicotyledonous species. This is an impressive scientific achievement. The conceptually simpler protoplast transformation, and biolistic, methods can do many of the things that the *Agrobacterium*-based system can do, but most importantly, have extended the range of species which can be made transgenic. Transgenic cereals can now be made reproducibly (critically reviewed by Morrish & Fromm 1992), in contrast with several previous irreproducible technologies such as:

- injecting DNA with a syringe into developing floral tillers of rye (de la Pena *et al*. 1987);
- transformation of maize by inoculating the shoot apex with *Agrobacterium* (Gould *et al*. 1991);
- incubating the cut surface of the style (of rice) with DNA, after the pollen tube tip has passed through (Luo & Wu 1988).

The *bar* gene, originally from the bacterium *Streptomyces hygroscopicus*, seems to be emerging as the selectable marker of choice for transgenic cereals, as well as being widely used for dicotyledonous plants (Hamptman *et al*. 1988, Gordon-Kramm *et al*. 1990). The *bar* gene encodes the enzyme phosphinothricin acetyltransferase, which detoxifies the herbicidal compounds phosphinothricin (PPT) (Fig. 14.20), glufosinate (the ammonium salt of PPT), and bialaphos (a compound containing PPT) (De Block *et al*. 1987, Thompson *et al*. 1987). PPT inhibits glutamine synthetase, causing rapid accumulation of ammonia and cell death (Tachibana *et al*. 1986).

An important application of plant transformation is transposon tagging, i.e. random insertion of transferred DNA into the plant genome. This often leads to insertional inactivation of plant genes. When bred into homozygous condition, any gene whose inactivation gives an interesting phenotype can be isolated by virtue of being tagged with the inserted DNA sequence (compare Chapter 15). T-DNA is being used very successfully for such tagging (Walden *et al*. 1991), and has given a large number of tagged *Arabidopsis* lines (Feldmann 1991). The small genome of *Arabidopsis thaliana* is so intensively tagged, that virtually any gene for which a selection procedure can be devised, is now amenable to isolation by this route. In addition, transposons derived from maize, such as the enhancer–inhibitor system (Aarts *et al*. 1993), the activator–dissociation system (Ac/Ds), the *Spm* and *Mu* elements, or derived from *Antirrhinum* (the *Tam* element), all have been investigated for tagging in heterologous species (for review, see Bhatt & Dean 1992).

Current transgenic plant technology leads to integration of transferred DNA at essentially random sites in the genome. As we shall discuss for animals, there is a need to target transferred DNA to particular sites in the

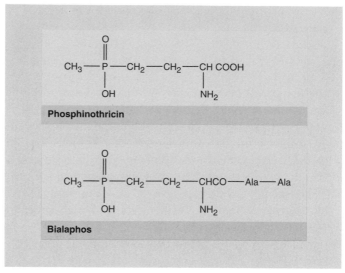

Fig. 14.20 Phosphinothricin (PPT) is a glutamate analogue, and is an inhibitor of glutamine synthetase in both plants and bacteria. Bialaphos is a tripeptide which is produced by *Streptomyces hygroscopicus*. It has little or no inhibitory activity, but is activated, in both plants and bacteria, by intracellular peptidases which remove the alanine residues. The *bar* gene encodes phosphinothricin acetyltransferase, which is an acetyl-coenzyme A-dependent enzyme that acetylates the amino group of PPT and prevents toxicity. In *Streptomyces hygroscopicus* the *bar* gene prevents autotoxicity. For agricultural use as herbicides, PPT is synthesized chemically (Basta, Hoechst AG), and bialaphos is produced by fermentation of *S. hygroscopicus* (Herbiace, Meiji Seika Ltd). The *bar* gene is a useful selective marker for transgenic experiments, and transgenic crops expressing the gene are selectively resistant to these broad-spectrum herbicides.

plant genome (Paszkowski *et al.* 1988), and methods developed in animal cells can possibly be adapted for future use in plants.

Plant viruses as vectors

With the advent of biolistic techniques, many plant species that previously were not amenable to transgenic manipulation, are now no longer so problematical. This progress has removed some of the incentive for developing vectors based upon plant viruses, particularly viral vectors for cereals. Despite this, the development of plant viral vectors is useful in plant virology itself, and for future biotechnological applications.

Plant viruses are attractive as vectors for several reasons.
- Viruses absorb to and infect cells of intact plants.
- Relatively large amounts of virus can be produced from infected plants, leading to the prospect of large amounts of foreign protein being expressed from recombinant viruses. This is an aspect of the inherent gene amplification which accompanies virus replication.
- Some virus infections are *systemic*. They are spread throughout the whole plant. In some cases intact virions are transported through the vascular system of the plant.

• Viruses are known which infect plants for which current alternative technology is limited.

It is possible to envisage the use of a plant virus vector in two distinct ways. One would be a massive infection which is deleterious to the plant and would perhaps ultimately kill it, but which meanwhile leads to the expression of a foreign gene product. The alternative would be a less harmful systemic infection which leads to more-or-less healthy plants in which foreign protein is expressed either to alter the plant to make it a better crop plant, or to express a foreign protein which is a desirable product when purified from the plant tissues. Crop plants can produce biomass cheaply without high technology. Ultimately the production of proteins for therapeutic or other uses may be undertaken not in expensive fermentors with microorganisms such as yeast or bacteria as the host, but in fields of genetically engineered plants. These plants could be engineered by the gene-transfer systems discussed in previous sections, or using virus vectors.

At present, virus vectors have not been developed to the stage where they are widely used as vectors, but progress has been made with the only two groups of plant viruses which have DNA genomes: the caulimoviruses and geminiviruses. Most plant viruses have RNA genomes. Plant RNA viruses which have great potential as prototype vectors are Brome mosaic virus and tobacco mosaic virus.

Cauliflower mosaic virus (CaMV)

CaMV is the best studied of the caulimovirus group. This is a group of spherical viruses (Fig. 14.21) which contain a circular double-stranded

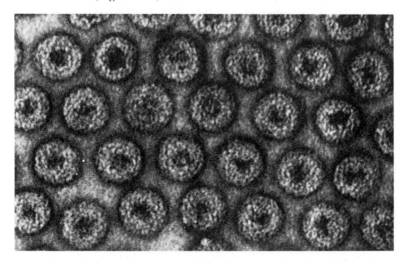

Fig. 14.21 Semi-crystalline array of cauliflower mosaic virus purified from turnip (approx. magnification ×200 000). The dark spots in the centres of the particles are typical and are the result of the outer protein shell being sucked into the hollow core during preparation for electron microscopy. (Photograph courtesy of M. Webb, National Vegetable Research Station.)

DNA genome of about 8 kb. The caulimoviruses are responsible for a number of economically important diseases of cultivated crops. They have restricted host ranges individually but as a group infect a range of dicots.

One feature of CaMV which makes it attractive as a vector is that infection becomes systemic. In order for CaMV to be transmitted through the vascular system of the plant the DNA must be assembled within virions. In infected cells, refractile, round inclusions form which consist of many virus particles embedded in a protein matrix. The matrix protein is virus-encoded and can account for up to 5% of the protein in infected cells. In nature the virus is transmitted on the stylets of aphids. The transmission depends upon a transmission factor which is also a virus-encoded protein present in infected cells, and is not part of the virion (Woolston *et al.* 1983).

CaMV DNA

Several different isolates of CaMV have been sequenced (Franck *et al.* 1980, Gardner *et al.* 1981, Balazs *et al.* 1982). The DNA of CaMV has an unusual structure. There are three discontinuities in the duplex, two in one strand, one in the other (Fig. 14.22). These are regions of sequence overlap (Frank *et al.* 1980, Richards *et al.* 1981).

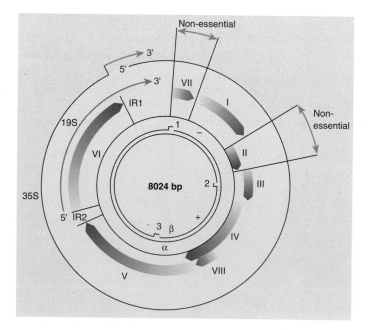

Fig. 14.22 Map of the cauliflower mosaic virus genome. The eight coding regions are shown by coloured boxes, and the different reading frames are indicated by the radial positions of the boxes. The thin lines in the centre indicate the (plus and minus) DNA strands with the three discontinuities. The major transcripts, 19S and 35S, are shown around the outside.

CaMV DNA is infectious. Infections can be initiated by inoculating abraded surfaces of the host plant with DNA. The DNA has a single *Sal*I site, and *Sal*I-linearized DNA is infectious even in the absence of re-ligation *in vitro*. Recircularization occurs in the plant cell. CaMV DNA has been cloned into an *E. coli* vector and propagated in *E. coli*. When the CaMV was released from the vector as a linear CaMV DNA molecule it was found to be infectious, despite the lack of the single-strand discontinuities in the inoculating DNA (Howell *et al.* 1980).

Another unusual feature of CaMV DNA is the presence of ribonucleotides covalently attached to the 5'-termini of the discontinuities. These and other observations have led Hull and Covey (1983a,b) and Pfeiffer and Hohn (1983) to suggest that CaMV replication involves reverse transcription with an RNA genomic intermediate. The replication cycle resembles that of retroviruses and hepatitis B virus (Hohn *et al.* 1985).

The sequence of CaMV DNA reveals eight closely-packed reading frames. There are only two small intergenic regions, and the only non-essential genes are the two small genes II and VII. Thus, most attempts to alter the DNA by insertion or deletion have caused loss of infectivity (Delseny & Hull 1983). The absence of substantial non-essential DNA limits scope for substituting sequences with foreign DNA.

CaMV as a vector

The CaMV envelope does not allow the DNA which it encapsidates to be substantially larger than normal. This was evident when foreign DNA was inserted at the unique *Xho*I site which lies in the non-essential gene II (Gronenborn *et al.* 1981). If DNA longer than a few hundred nucleotides was inserted the infectivity was destroyed. This packaging limitation and the absence of long non-essential sequences which can be deleted in the genome severely limit the capacity for foreign DNA.

In Chapter 15 we demonstrate how similar constraints can be overcome in SV40 by exploiting a helper virus to complement a defective recombinant virus. This does not appear feasible in CaMV because homologous recombination between the helper and recombinant virus DNAs readily expels the foreign DNA with production of wild-type virus.

It appears therefore that capacity is a major limitation of CaMV vectors. But a small foreign DNA, comprising the bacterial *dhfr* gene, replacing the gene II coding sequence, has been successfully expressed in plants (Brisson *et al.* 1984).

As a vector, CaMV has been overshadowed by the success of the *Agrobacterium*-based, and DNA-mediated transfection, systems. Knowledge of CaMV has been of value in providing very strong promoters for driving expression of other genes in plants. We met an example of the use of the 35S promoter (p. 292, Meyer *et al.* 1987). Other constructs employ the promoter of gene VI (Paszkowski *et al.* 1984, Balazs *et al.* 1985). This gene encodes the abundant matrix protein of the inclusion body and has a very strong promoter.

Agroinfection with a geminivirus

Geminiviruses are attractive for vector development because members of this group infect a wide range of monocot and dicot crop plants. The virus particles are twinned (geminate) and contain circular single-stranded DNA which replicates via double-stranded intermediates in the plant nucleus. All monocot-infecting geminiviruses, such as maize streak virus (MSV), are transmitted by leaf-hoppers. Their genomes are very small, 2.7 kb. The dicot-infecting geminiviruses, such as cassava latent virus (CLV), are transmitted by the whitefly *Bemisia tabaci*, and have a two-component genome consisting of two circular single-stranded DNAs which have only a small region of about 200 nucleotides in common. The following experiments with MSV are interesting from several points of view.

The MSV genome has never been introduced successfully into plants as native or cloned DNA. Grimsley *et al.* (1987) were able to introduce the DNA with the aid of *Agrobacterium* as follows. They constructed a plasmid in which MSV cloned DNA was inserted as a tandem dimer of the intact genome. These were inserted into an *Agrobacterium* T-DNA vector, and maize plants were infected with *A. tumefaciens* containing this recombinant T-DNA. Viral symptoms appeared within two weeks of plant inoculation. This process has been termed 'agroinfection'. It has in fact been demonstrated for a number of viruses and viroids that if the T-DNA contains partially or completely duplicated genomes, single copies of the genome can escape and initiate infections. Agroinfection is a very sensitive assay for transfer to the plant cell because of the inherent amplification of virus infection and resulting visible symptoms.

The successful agroinfection of maize plants with MSV leads to two important conclusions. Firstly, *Agrobacterium* can transfer DNA to maize, even if inefficiently by comparison with dicot plants. Secondly, cloned MSV DNA is biologically active and amenable to manipulation and vector development.

Replacement-type vector derivatives of geminiviruses have been investigated. The coat protein gene of CLV is required for insect transmission, but is not essential for replication and infectivity of the viral DNA. It has a strong promoter. Coat protein replacement constructs have been produced, but have yet to find application (Ward *et al.* 1988, Davies & Stanley 1989).

Brome mosaic virus as a vector

Plant viruses with DNA genomes are in the minority; the great majority of plant viruses have RNA genomes. The range of potential vectors is very great. In addition, many of these RNA viruses have filamentous morphology and so it is expected that the length of the virus particle is determined by the length of the viral nucleic acid. Thus there should be no strict size limitation on the RNA to be packaged.

Another common feature of plant viruses is that they may be multi-component. An example of this is Brome mosaic virus. The virus infects a number of Graminae including barley. There are three separate RNA components of the genome, each of which is packaged into a separate particle. Each of these RNAs is also an mRNA and during infection a fourth mRNA is produced which is a subgenomic derivative of RNA 3. This RNA 4 is the mRNA encoding the very abundant viral coat protein. The cloned cDNAs corresponding to RNAs 1–3 can be transcribed *in vitro* to produce transcripts which are infectious when mixed and introduced into barley protoplasts. Only RNAs 1 and 2 are necessary for replication and expression of the genome, so that the coat protein sequence on RNA 3 is available for manipulation, although in the absence of coat protein no virus particles are produced (Ahlquist *et al.* 1987).

The amenability of this system has been demonstrated by French *et al.* (1986) who inserted a bacterial CAT coding sequence into the coat protein gene sequence of RNA 3.

When this construct was transcribed *in vitro*, and introduced into barley protoplasts together with RNAs 1 and 2, high levels of CAT activity were expressed. It is therefore likely that further manipulation of this system is possible, and that the coat protein expression signals may be exploited for high level expression of foreign proteins.

Tobacco mosaic virus as a vector

The approach of using a cDNA copy of an RNA virus genome has also been employed with the tobacco mosaic virus (TMV) genome. The TMV genome is a single-component RNA which is also a messenger RNA. The genome encodes at least four polypeptides. The 130 kDa and 180 kDa proteins are translated directly from the same initiation codon on genomic RNA. The other two proteins, 30 kDa and coat protein, are translated from processed subgenomic RNAs. The 130 kDa and 180 kDa proteins are involved in viral replication, whereas the 30 kDa protein is necessary for cell-to-cell movement of virus. These three proteins are therefore probably essential for TMV propagation in whole plants. The coat protein is not essential for viral multiplication, but is necessary for long-distance spread of infection in the plant. Since the coat protein is synthesized in large amounts (up to several milligrams per gram of infected tobacco leaf), and since it is non-essential, it is an attractive target site for introducing and expressing a foreign gene.

Takamatsu *et al.* (1987) have modified a full-length TMV cDNA clone from which infectious TMV RNA can be transcribed *in vitro*. A bacterial *CAT* gene sequence was placed just downstream of the initiation codon of the coat protein gene. When *in vitro* transcripts of the recombinant TMV cDNA were inoculated into tobacco plants, CAT activity was observed in the inoculated leaves, although the infection was unable to spread systemically throughout the plant.

15 Introducing genes into animal cells

Introduction

Analysis of the regulation of gene expression in animal cells is one of the central themes of current molecular biology. For this and other reasons procedures for introducing manipulated genes into animal cells, where their regulation can be assayed, have been the subject of intensive research. There has been a great diversity of approaches. By direct micro-injection of DNA into fertilized eggs of *Xenopus*, or the mouse, or into early *Drosophila* embryos, it has been possible to incorporate foreign DNA into the genomes of the resulting animals (see Chapter 16). Here we shall consider methods for introducing foreign DNA into animal cells in culture. Most of this research has made use of mammalian cell lines. Insect cells are also important as hosts for baculovirus vectors.

Animal cells

Integration of DNA into the genome of mammalian cells

The ability of mammalian cells to take up exogenously added DNA, and to express genes included in that DNA, has been known for many years. Szybalska and Szybalski (1962) were the first to report DNA-mediated transfer. They transfected* HGPRT$^-$ mutant human cells to HGPRT$^+$ using

---

*The term 'transformation' commonly has two different meanings in the context of this chapter: (i) an inherited change in genotype due to the uptake of foreign DNA, analogous to bacterial transformation; (ii) a change in properties of an animal cell possessing normal growth characteristics to one with many of the characteristics of a cancer cell, i.e. growth transformation. In order to avoid confusion, in this chapter the term *transfection* will be used for the first meaning although originally transfection applied to the uptake of viral DNA. The term *transformation* will be used only with meaning (ii).

[302]

total, uncloned, human nuclear nuclear DNA as the source of the wild-type gene. The rare HGPRT$^+$ transformants were selected by means of the HAT selection system which they had devised (Fig. 15.1). Much later, it was appreciated that successful DNA transfer in these experiments was dependent upon the formation of a co-precipitate of the DNA with calcium phosphate, which is insoluble, and must be formed freshly in the presence of the DNA when the transfection mixture is assembled. Apparently the calcium phosphate granules are phagocytosed by the cells and in a small proportion of the recipients some of the DNA becomes stably integrated into the nuclear genome. The technique became generally accepted after its application by Graham and Van der Eb (1973) to the analysis of infectivity of viral DNA.

Other transformation techniques

The calcium phosphate co-precipitate provides a general method for introducing any DNA into mammalian cells. It can be applied to relatively large numbers of cells in a culture dish but, as originally described, it is limited by the variable and usually rather low (1–2%) proportion of cells that take up exogenous DNA. Only a subfraction of these cells will be stably transfected. Procedures have been designed to increase to about 20% the proportion of cells that take up the DNA (Chu & Sharp 1981). A related procedure involves fusion of cultured cells with bacterial protoplasts containing the exogenous DNA (Schaffner 1980, Rassoulzadegan *et al.* 1982). Here, the proportion of cells taking up exogenous DNA can approach 100%. An alternative approach is to micro-inject DNA into the nucleus of cultured cells (see, for example, Kondoh *et al.* 1983), a procedure which has the advantage that 'hits' are almost certain but which cannot be applied to large numbers of cells.

Recently other general and convenient methods for introducing DNA into cultured mammalian and other vertebrate cells have gained popularity. One problem with the calcium phosphate co-precipitate is that many cell lines do not like having the solid precipitate adhering to them and the surface of their culture vessel. An alternative transfection system employs DEAE-dextran (diethylaminoethyl-dextran) in the transfection medium, in which the DNA is also present. The DEAE-dextran is soluble and no precipitate is involved. It is polycationic and probably acts by mediating, in some unknown way, the productive interaction between negatively-charged DNA and components of the cell surface in endocytosis. The DEAE-dextran procedure is particularly convenient for transient assays in COS cells (see below). It does not appear to be efficient for the production of *stable* transfectants (Sussman & Milman 1984).

Electroporation is very widely used for the production of both transient and stable transfectants. Electroporation uses a short electric impulse to create pores in the plasma membrane, allowing exchange of molecules, including DNA, across the membrane (Chapter 2). The main advantages of electroporation over other transfection methods are its ease of use, and its

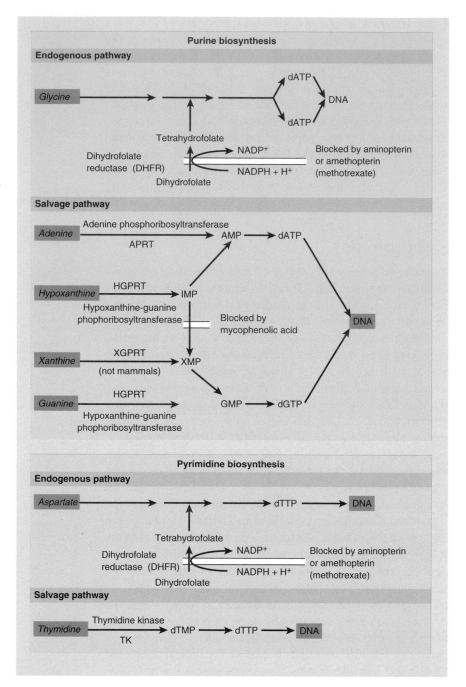

Fig. 15.1 Commonly-used mutants and inhibitors in cell culture. dATP, dGTP and dTTP have two synthetic pathways. A loss of either pathway for any one nucleotide, therefore, is not lethal, so mutants can be isolated which lack one of these pathways. APRT⁻ cells can be isolated because they are *resistant* to toxic base analogues (2,6-diaminopurine, 2-fluoroadenine) but are *killed* in medium containing adenine and hypoxanthine plus azaserine, whereas wild-type cells survive. Azaserine inhibits several reactions in the endogenous pathway of purine biosynthesis. HGPRT⁻ cells are *resistant* to toxic base analogues (thioguanine, azaguanine), but are *killed* in medium containing aminopterin and hypoxanthine in which wild-type cells survive. TK⁻ cells are *resistant* to the thymidine analogue BUdR, but are *killed* in medium containing aminopterin and thymidine. HAT medium contains hypoxanthine, aminopterin and thymidine and is commonly used to select against TK⁻ and HGPRT⁻ cells.

XGPRT is an enzyme that is found in *E. coli* but not in mammalian cells.

applicability to a very broad range of cell types (Toneguzzo *et al.* 1986, Chu *et al.* 1987).

[305]
CHAPTER 15
*Introducing genes
into animal cells*

*Direct application of calcium phosphate-mediated transfection:
transient assays without eukaryotic vector sequences*

An example of the direct application of calcium phosphate-mediated transfection is the work of Rutter's group on DNA sequences controlling cell-specific expression of insulin and chymotrypsin genes (Walker *et al.* 1983). This work illustrates how the activity of manipulated gene sequences can be assayed after they have been introduced into mammalian cells. No mammalian vector is involved – the activity of the introduced sequences was assayed after only a short time period following transfection.

The genes coding for human and rat insulin, and the rat chymotrypsin gene had been cloned and characterized. These genes are expressed at a high level only in the pancreas. Each is expressed in clearly distinct cell types: insulin is synthesized in endocrine β-cells and chymotrypsin in exocrine cells. DNA sequences containing the promoter and 5'-flanking sequences of these genes were linked to the coding sequence of bacterial chloramphenicol acetyltransferase (CAT). This enzyme activity can be assayed very sensitively (see p. 312) and was used here as a 'reporter enzyme' whose activity was taken to be a measure of the transcription of the insulin or chymotrypsin genes. Recombinant plasmids containing such genes were constructed and grown in *E. coli*. These plasmids were not designed to be replicating eukaryotic vectors. Although it is possible that some exogenous DNA replication may occur when they are introduced into mammalian cells (this was not tested), they were expected to persist just long enough in the cell for their *transient* expression to be assayed. Therefore, plasmid DNA was introduced into either pancreatic endocrine or pancreatic exocrine cell lines in culture, and after a subsequent 44-hour incubation cell extracts were assayed for CAT activity. It was found that the constructs retained their preferential expression in the appropriate cell type. The insulin 5'-flanking DNA conferred a high level of CAT expression in the endocrine, but not the exocrine cell line, with the converse being the case for the chymotrypsin 5'-flanking DNA.

The analysis was extended by creating deletions in the 5'-flanking sequences and testing their effects on expression. From such experiments it could be concluded that there are sequences located upstream of the promoter, between 150 and 300 base-pairs of the transcription start site, which are essential for appropriate cell-specific gene transcription.

Co-transfection (co-transformation)

Following the general acceptance of the calcium phosphate method, subsequent experiments showed that the thymidine kinase gene of herpes simplex virus was effective in transfecting Tk$^-$ mammalian cells (see Fig. 15.1) to a stable Tk$^+$ phenotype which can be selected in HAT medium

(Wigler *et al.* 1977). However, the isolation of cells transfected by other genes which do not encode selectable markers remained problematic. A breakthrough was made when it was discovered that cells can be simultaneously co-transfected by a mixture of two *physically unlinked* DNAs in the calcium phosphate precipitate (Wigler *et al.* 1979). To obtain co-transfectants, cultured cells were exposed to the thymidine kinase gene in the presence of a vast excess of a well-defined DNA, such as pBR322 or bacteriophage ΦX174 DNA, for which hybridization probes could be readily prepared. In order to achieve a suitably high DNA concentration for the formation of an effective co-precipitate, 'carrier' DNA was also included. This often consisted of total cellular DNA isolated from salmon sperm. Tk$^+$ transfectants were selected and scored by molecular hybridization for the co-transfer of the *unselected*, pBR322 or ΦX174, DNA.

Wigler *et al.* (1979) demonstrated the co-transfection of mouse Tk$^-$ cells with pBR322, bacteriophage ΦX174 DNA and rabbit β-globin gene sequences and we shall use their ΦX experiments for illustrative purposes. ΦX replicative form DNA was cleaved with *Pst*I which recognizes a single site in the circular genome. The purified thymidine kinase gene (500 pg) was mixed with 10 μg of *Pst*I-cleaved ΦX replicative form DNA. This DNA mixture was added to Tk$^-$ mouse cells and 25 Tk$^+$ transfectants were observed per 10^6 cells after 2 weeks in HAT medium. To determine if these Tk$^+$ transfectants also contained ΦX DNA sequences, high-molecular-weight DNA from the transfectants was cleaved with *Eco*RI which recognizes no sites in the ΦX genome. The cleaved DNA was fractionated by agarose gel electrophoresis, transferred to nitrocellulose filters by 'Southern blotting' and probed with labelled ΦX DNA. These annealing experiments demonstrated that 14 out of 16 Tk$^+$ transfectants had acquired one or more ΦX sequences. Subsequent studies of the integrated DNA tell us something about the molecular events taking place during co-transfection. Although the added DNAs in the calcium phosphate co-precipitate are not physically linked, Southern blot analysis of the DNA integrated into the host genome reveals large concatemeric structures, up to 2000 kb long. These concatemers include copies of the selectable marker, the co-transfecting DNA and fragments of carrier DNA (Perucho *et al.* 1980b). Therefore at some stage during the co-transfection process the DNAs must be physically ligated. The foreign DNA can probably integrate virtually anywhere in the genome (Robbins *et al.* 1981, Scangos *et al.* 1981).

The co-transfection phenomenon allows the stable introduction into cultured mammalian cells of any cloned gene. There is no requirement for a vector capable of replication in the host cell. Originally, co-transfection was discovered using two unlinked DNAs. However, this is an unimportant feature; analogous results can be obtained if the selected and unselected genes are ligated in prior manipulations. Originally, herpes simplex virus DNA provided pure *tk* DNA. Later this *tk* gene was cloned in *E. coli* plasmids, hence providing a very convenient source.

It has been found subsequently that the HSV *tk* gene fragment cloned in *E. coli* plasmids as a 3.4 kb *Bam*HI fragment bears a promoter which is rather weak in mammalian cells. It requires the addition of an enhancer for full activity. Since it is now apparent that efficient expression of selectable markers is required for high transfection efficiencies, the weakness of the *tk* promoter may explain the rather low transfection frequencies obtained in early experiments.

The requirement for a tk⁻ recipient cell line was a serious limitation of the use of *tk* selection. This was overcome by the development of so-called *dominant* selectable markers which could be used with non-mutant cell lines.

Selectable markers other than tk

These markers include:
- dihydrofolate reductase and associated methotrexate resistance;
- rodent CAD;
- bacterial XGPRT; and
- bacterial neomycin phosphotransferase.

Dihydrofolate reductase: methotrexate resistance: amplicons

Dihydrofolate reductase (Fig. 15.1) is sensitive to the inhibitor methotrexate (Mtx). Cultured wild-type cells are sensitive to concentrations of the drug at about 0.1 µg/ml. Mtx-resistant cell lines have been selected and have been found to fall into three categories:
- cells with decreased cellular uptake of Mtx;
- cells over-producing DHFR;
- cells having structural alterations in DHFR, lowering its affinity for Mtx.

It has been found that cells over-producing DHFR contain increased copy numbers of the gene (Schimke *et al.* 1978). The DNA sequence that is amplified can be large, about 100 kb, and it can be amplified to a copy number of up to 1000. Often it is present as small extrachromosomal elements called double minute chromosomes; alternatively the amplified DNA can remain at its original chromosomal site or be transposed elsewhere (Schimke 1984).

The Chinese hamster ovary 'A29' cell line is notable because it is extremely resistant to Mtx and has been shown to synthesize *increased amounts* of an *altered* DHFR (Flintoff *et al.* 1976). Wigler *et al.* (1980) have used genomic DNA of the A29 cell line as a donor of the DHFR gene in co-transfection experiments with Mtx-sensitive cells. This system has the advantage that the high gene copy number in donor DNA gives efficient transfection. Additionally, the selection system is powerful and can be applied to *non-mutant* Mtx-sensitive cell lines.

There is one further advantage of the Mtx system: highly resistant variants of the transfected cell line can be selected so as to give concomitant amplification of the unselected DNA. In order to explore this possibility, Wigler *et al.* (1980) first showed that *Sal*I digestion of A29 DNA did not destroy its ability to transfect cells to Mtx-resistance. The *Sal*I-cleaved A29 DNA was then ligated to *Sal*I-linearized pBR322 and used to transfect a Mtx-sensitive mouse cell line. Resistant colonies were picked, grown, and exposed to increasing concentrations of the drug so as to select highly resistant variants. DNAs of certain variants resistant to 40 µg/ml were analysed by Southern blot hybridization with pBR322 DNA as the probe. This analysis showed that the pBR322 sequence had undergone a substantial amplification of at least 50-fold.

This illustrates the fact that the unit of amplification, i.e. the *amplicon*, can be much larger than the selected *DHFR* gene. DNA that is covalently linked to the *DHFR* gene is co-amplified. This has important biotechnological implications because increasing the copy number of a foreign gene creates the opportunity for obtaining expression of the gene product at a high level (see p. 344).

There are several variations in the use of amplicons. One approach is simply to use co-transfection to integrate and link the foreign gene to a *DHFR* gene. For example, Christman *et al.* (1982) used a procedure very similar to that already described, co-transfecting a cloned hepatitis B virus genome with A29 genomic DNA into mouse 3T3 cells and selecting a cell line expressing HBV surface antigen. Other approaches involve pSV-dhfr (see p. 310) or related vectors expressing mouse dhfr cDNA in cotransfection with genes encoding useful products such as tissue plasminogen activator (Kaufman *et al.* 1985), human interferon-γ (Scahill *et al.* 1983) or hepatitis B virus surface antigen (Patzer *et al.* 1986). The mouse *DHFR* gene is rather long, 31 kb (Crouse *et al.* 1982) so that *DHFR* cDNA is conveniently linked to expression cassettes in such constructs (Kaufman & Sharp 1982).

The principle of amplification can be extended to other selectable markers, several of which are described in more detail below. Table 15.1 lists a range of amplifiable markers that have been used in transfection studies. A variation on the use of DHFR depends upon the fact that bacterial DHFR is intrinsically resistant to Mtx (although sensitive to trimethoprim). O'Hare *et al.* (1981) have constructed plasmids in which the bacterial gene is transcribed from an SV40 promoter. Such plasmids transfected mouse cells to a Mtx-resistant phenotype by integration into the mouse genome.

CAD: PALA resistance

The CAD protein is a multifunctional enzyme catalysing the first three steps of *de novo* uridine biosynthesis: carbamyl phosphate synthetase; aspartate transcarbamylase; and dihydroorotase (Swyryd *et al.* 1974). One of these activities, aspartate transcarbamylase, is inhibited by *N*-phosphonacetyl-L-aspartate (PALA). PALA-resistant mammalian cells over-produce CAD from highly-amplified copies of the *CAD* gene, in a manner analogous to Mtx resistance (Wahl *et al.* 1984). The *CAD* gene of the Syrian hamster has been cloned on cosmid vectors in *E. coli* and has

Table 15.1 Amplifiable selectable markers for animal cells

Selection	Gene	References*
Methotrexate	Dihydrololate reductase	Kaufman *et al.* 1985
Cadmium	Metallothionein	Beach & Palmiter 1981
PALA	Aspartate transcarbamoylase	Wahl *et al.* 1984
Adenosine, alanosine, and 2'-deoxycoformycin	Adenosine deaminase	Kaufman *et al.* 1986
Adenine, azaserine, and coformycin	Adenylate deaminase	Debatisse *et al.* 1981
6-Azauridine or pyrazofuran	UMP synthetase	Kanalas & Suttle 1984
Mycophenolic acid	Xanthine-guanine phosphoribosyltransferase	Chapman *et al.* 1983
Multiple drugs	P-glycoprotein 170	Kane *et al.* 1988, 1989
Methionine sulphoximine	Glutamine synthetase	Cockett *et al.* 1990
β-Aspartyl hydroxamate or Albizziin	Asparagine synthetase	Cartier *et al.* 1987
α-Difluoromethylornithine	Ornithine decarboxylase	Chiang & McConlogue 1988

* For a review, see Kaufman (1990).

been shown to provide a dominant, amplifiable genetic marker that can be selected in non-mutant cells on the basis of resistance to high concentrations of PALA (de Saint Vincent *et al.* 1981, Wahl *et al.* 1984).

XGPRT: mycophenolic acid resistance

The *E. coli* enzyme xanthine–guanine phosphoribosyltransferase, XGPRT, is a bacterial analogue of mammalian HGPRT. However, by contrast with HGPRT, it has the additional ability to convert xanthine to XMP and hence ultimately to GMP (Fig. 15.1). Only hypoxanthine and guanine are substrates of HGPRT. Mulligan and Berg (1980, 1981a,b) have cloned the *XGPRT* gene and incorporated it into a variety of vectors in which transcription of the bacterial gene is directed by a SV40 promoter (see p. 310). Such constructs are capable of transfecting HGPRT$^-$ cells to HGPRT$^+$, but, more importantly, provide a selectable marker for non-mutant cells in medium containing adenine, mycophenolic acid and xanthine (Fig. 15.1). This selection can be made more effective by adding aminopterin, which blocks endogenous purine biosynthesis.

Neomycin phosphotransferase: G418 resistance

Bacterial transposons Tn5 and Tn601 encode distinct neomycin phosphotransferases, whose expression confers resistance to aminoglycoside antibiotics (kanamycin, neomycin and G418), which are protein synthesis inhibitors, active in bacterial or eukaryotic cells. Berg (1981) incorporated a neomycin phosphotransferase gene into constructs analogous to those containing *XGPRT*. Other constructs have linked the neomycin phospho-transferase gene to the herpes simplex virus tk promoter in an *E. coli* plasmid (Colbère-Garapin *et al.* 1981). Transfectants of non-mutant mammalian cells which contain such constructs can be selected by antibiotic resistance. Colbère-Garapin *et al.* (1981) demonstrated the application of their construct to the co-transfection of a variety of cell lines from different mammalian species. Grosveld *et al.* (1982) have also constructed cosmid cloning vectors which include selective markers for growth in the host bacterium (β-lactamase) and animal cells. The markers for animal cells were neomycin phosphotransferase, HSV thymidine kinase, or XGPRT. Such cosmids can be used to construct libraries of eukaryotic genes from which a particular recombinant can be isolated. The recombinant cosmid DNA can then be transfected into animal cells at high efficiency where transfectants in which the DNA has integrated into the nuclear genome can be readily selected. The power of aminoglycoside antibiotic resistance as a selective system in eukaryotes is now very evident. It has application in yeast (Jimenez & Davies 1980; see Chapter 13) and plants (Chapter 14).

The pSV and pRSV plasmids

In previous sections we have mentioned the use of mouse DHFR cDNA, and the bacterial genes encoding XGPRT and neomycin phosphotrans-ferase, in order to provide dominant selectable markers. We stated that expression of these genes was obtained from an SV40 promoter. Here we explain these constructions in a little more detail.

In a subsequent section we shall describe the molecular biology of SV40, but for our purposes here it is only necessary to note that the small *Pvu*II–*Hin*dIII fragment of SV40 DNA (Fig. 15.2) contains the promoter for early transcription and the transcriptional start point but no ATG transla-tional initiation codon. The plasmid pSV contains an expression cassette consisting of this promoter fragment, followed by a sequence containing the small t intron, followed by the transcript polyadenylation signal. These SV40 sequences are inserted into the pBR322 vector. The genes for neomycin phosphotransferase, or XGPRT or mouse DHFR were inserted as *Hin*dII–*Bgl*II fragments into the expression site to produce pSV2-neo (Southern & Berg 1982), pSV2-gpt (Mulligan & Berg 1980, 1981a,b) and pSV2-dhfr (Subramani *et al.* 1981). The SV40 expression signals function in a wide variety of mammalian cells, hence providing a generally useful set of dominant selectable markers. However, it has been observed that the promoter from the retrovirus Rous sarcoma virus (RSV) is more powerful than the SV40 promoter in various cell types (Gorman *et al.* 1982a). The promoter elements of retroviruses are located in their long terminal repeat (LTR) sequences (see p. 339). A 524 nucleotide pair fragment of one LTR of RSV has been isolated and incorporated in place of the *Pvu*II–*Hin*dIII SV40 promoter fragment in the pSV series of plasmids. This has given rise to the pRSV series of expression plasmids, including pRSV-neo and pRSV-gpt

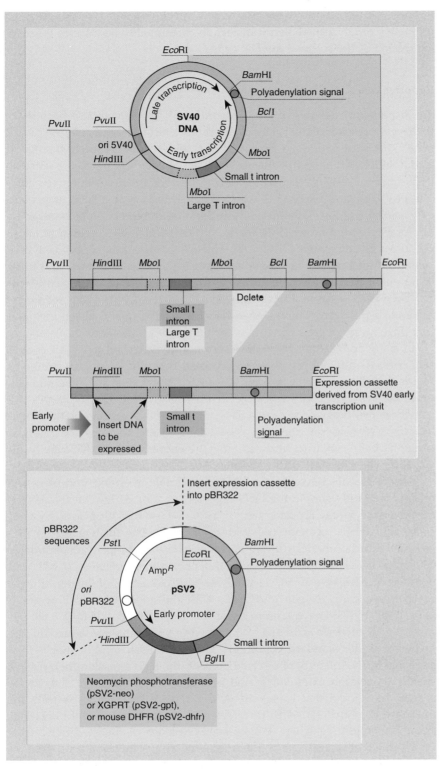

Fig. 15.2 Construction of pSV2-neo, pSV2-gpt and pSV2-dhfr.

(Gorman *et al*. 1983). These plasmids appear to give a relatively high frequency of DNA-mediated transfection, presumably because of efficient expression of the selected marker gene.

Reporter genes

The expression cassettes of the pSV and pRSV types have also been used to drive expression of the bacterial chloramphenicol acetyltransferase (*CAT*) gene. This gene is derived from Tn*9* of *E. coli*, where it confers resistance to the antibiotic chloramphenicol. We have already discussed a particular application of *CAT* as a reporter enzyme earlier in this chapter (p. 305). The pSV-CAT and pRSV-CAT constructs (Gorman *et al*. 1982b) have been very widely used as tools for analysing transient expression of transfected DNAs where a promoter (SV40 early, or RSV) which is expressed in many cell types is convenient. The CAT enzyme activity can be assayed very sensitively and rapidly in simple homogenates of cells or tissues, and there is no endogenous eukaryotic activity (Fig. 15.3). The pSV-CAT construct has been the source of a DNA fragment bearing a CAT coding sequence, linked to a polyadenylation signal, which has subsequently been included in a huge range of gene constructs for particular applications.

In many applications a 'minimal' promoter is attached to a CAT coding sequence. A minimal promoter will consist, typically, of a short promoter region including a TATA box and a transcriptional start site. The minimal promoter will give only a low level of 'background' transcription. The activity of cell type-specific enhancers, or other DNA sequences that regulate transcription, can then be assessed by fusing them to the minimal promoter/CAT construct and assaying the expression of CAT activity, in whatever cell types are under investigation.

Reporter genes have proven widely applicable in animal and plant cell biology (see for example the GUS reporter, p. 294). Firefly luciferase was introduced as a new reporter gene in 1986, for use in both plant (Ow *et al*. 1986) and animal systems (de Wet *et al*. 1987). The luciferase from the common North American firefly *Photinus pyralis* is a single polypeptide of 550 amino acids (compare the application of bacterial *lux* genes, p. 373, and Szittner & Meighen 1990). It catalyses the oxidation of luciferin, in a reaction requiring oxygen and ATP, with the emission of yellow-green light (Fig. 15.4). The quantum yield of the reaction is high, 0.88 per molecule of luciferin oxidized. When excess substrates are added to the luciferase under standard conditions, there is a flash of light that is proportional to the quantity of enzyme present; the light emission then decays rapidly to give an extended period of low-intensity emission. Recent research has shown that coenzyme A can replace ATP as a substrate for luciferase (Wood 1991). With this substrate, the emission of light does not decay so rapidly as when ATP is used. Because of the sensitivity with which the light can be detected in luminometers, or by using a scintillation counter as a luminometer, or even using photographic film (Wood & de

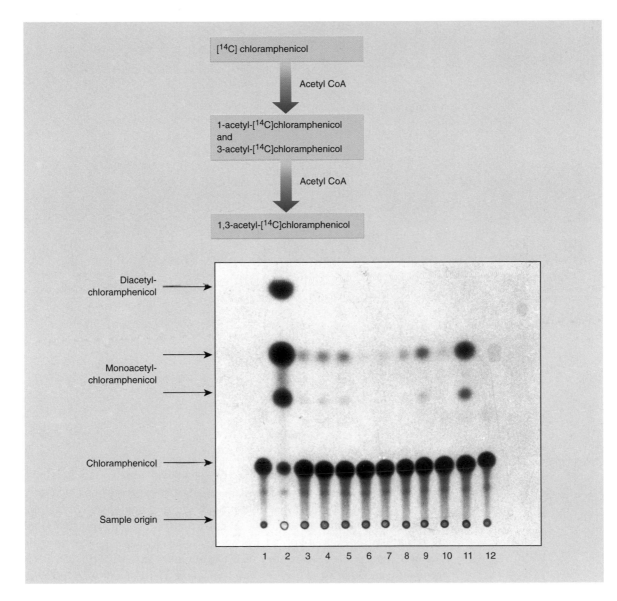

Fig. 15.3 The enzyme chloramphenicol acetyltransferase (CAT) catalyses the transfer of acetyl groups from acetyl-CoA to chloramphenicol. A mixture of monoacetylated forms of chloramphenicol is the first product of the reaction. Further reaction produces 1,3-acetyl-chloramphenicol. The activity is conveniently assayed with ^{14}C-labelled chloramphenicol as substrate. Thin-layer chromatography separates the products, which can be detected by autoradiographing the chromatogram. An autoradiograph of a typical set of assays is shown. Sample 1 is a negative control with no enzyme activity. Sample 2 is a positive control showing extensive activity. Samples 3–12 exhibit intermediate CAT activity.

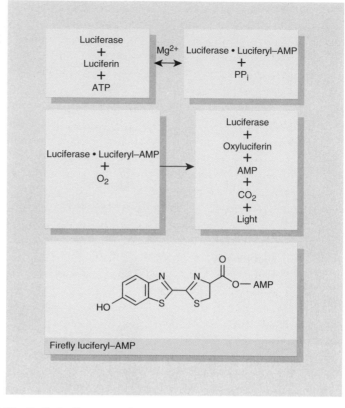

Fig. 15.4 The firefly luciferase reaction.

Luca 1987), luciferase assays can be very sensitive, much more sensitive than CAT assays (de Wet *et al.* 1987). Assaying the expression of luciferase in cell extracts, as a reporter, can therefore be a rapid and sensitive method for monitoring promoter activity. A particularly striking application of the luciferase reporter involved transgenic tobacco plants expressing the luciferase gene under the control of the cauliflower mosaic virus 35S RNA promoter (see Chapter 14). When the transgenic plants were watered with luciferin solution, light was emitted from intact organs (leaves, stems, roots) (Ow *et al.* 1986).

Future development of the luciferase system may take advantage of luciferases from other insects. For example, tropical click beetle luciferases have been cloned which elicit bioluminescence of different colours (Wood *et al.* 1989).

The fate of transfecting DNA

In discussing the fate of transfecting DNA it is necessary to distinguish between procedures in which a high concentration of high-molecular-weight 'carrier DNA' (usually total genomic DNA from a cheap, convenient source such as salmon or herring sperm, or chicken red blood cells) has been used, and those where it has not been used. Carrier DNA was used in the co-transfection procedure of Wigler *et al*. (1979) discussed above. We noted there that the selectable marker DNA, the unselected DNA and the carrier DNA are eventually found in large concatemeric structures. It appears that during the transformation the DNAs become fragmented and then joined in random combinations. Recombination events also take place (Miller & Temin 1983). The large concatemer of selected and non-selected DNA interspersed with carrier DNA is called a *transgenome*. It forms early in the transformation process and at a later stage integrates at an apparently 'random site' into the genome, and thereupon the cells become stably transformed. In stable transformations there is usually a single transgenome (Robbins *et al*. 1981). The transgenome is susceptible to partial or more-or-less complete deletion (studied for example by reversion from tk^+ to a tk^- phenotype) at a relatively high frequency (Perucho *et al*. 1980b).

The apparently random integration of the transfecting DNA is turned to advantage in gene isolation procedures which exploit the 'tagging' approach described in the following section.

Where stable transfection has taken place with pure plasmids in the absence of carrier DNA using the calcium phosphate technique, the plasmids integrate into the genome as single (or, rarely, a few) copies found at one to five separate chromosomal locations. This is the case, for example, with pSV2-neo (Southern & Berg 1982) or pSV2-gpt (Mulligan & Berg 1981a). Using transfection with pure plasmid by the electroporation technique a similar small number of copies are integrated. Where the DNA is micro-injected into the nucleus, the DNA also becomes integrated at random sites. Multiple head-to-tail concatemers are integrated if a large amount of DNA is micro-injected; single copies are integrated upon micro-injection of smaller amounts. Whether the transfection technique employs calcium phosphate precipitation, electroporation or micro-injection, linearization of the plasmid seems to increase the efficiency of integration and the possibility of concatemer formation (Huttner *et al*. 1981, Folger *et al*. 1982, Potter *et al*. 1984).

Isolation of genes transferred to animal cells in culture

Methods for transferring DNA into cultured animal cells have been described. Techniques have been devised for screening libraries specifically for the transferred sequences. Thus any gene which can be selected, or recognized by its phenotypic effects, in tissue culture can be isolated. There are several variants of this approach. The simplest is 'plasmid rescue' (Hanahan *et al*. 1980, Perucho *et al*. 1980a). In the example shown in Fig. 15.5, plasmid rescue is used to isolate a chicken tk^+ gene. First *Hind*III fragments of non-mutant, total nuclear chicken DNA are ligated into pBR322. This recombinant DNA is used directly to transform chicken tk^- cells, from which tk^+ transformants are selected in HAT medium. DNA from these transformants is prepared and used in a second round of

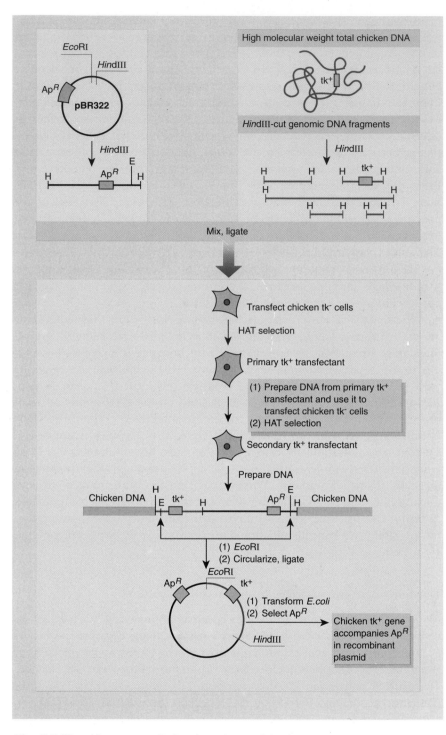

Fig. 15.5 Plasmid rescue, applied to the isolation of the chicken thymidine kinase gene (Perucho *et al*. 1980a). (See text for details.)

transformation. This second round eliminates most of the non-selected recombinant plasmid DNA which becomes integrated in the first round. DNA from these secondary tk$^+$ transformants is cut with an appropriate restriction enzyme, circularized and selected in *E. coli* on the basis of plasmid-borne drug resistance. The tk$^+$ gene accompanies the drug resistance marker in the plasmid which has been rescued. Thus in the plasmid rescue approach the transferred DNA is 'tagged' with the drug resistance marker, which can be selected in *E. coli*. In a related approach the transferred DNA is tagged with an amber suppressor gene. This is illustrated in Fig. 15.6, in which the cloning of a human oncogene is used as an example (Goldfarb *et al.* 1982).

In other variants of this approach, the transferred DNA is 'tagged' with DNA sequences that are not selected genetically but which can be detected by molecular hybridization probes. Figure 15.7 shows the procedure of Lowy *et al.* (1980) in which this tag is simply the plasmid pBR322 to which the transferred DNA has been ligated. There is, therefore, a formal resemblance to the plasmid rescue approach, but an advantage of this procedure is the incorporation of the phage λ cloning step. As we saw in Chapter 6, genomic libraries can be prepared in phage λ very efficiently when combined with *in vitro* packaging. In an alternative variant, the 'tag' for the transferred DNA is not provided artificially, but is provided by 'Alu' sequences (Rubin *et al.* 1983). These are members of a highly repetitive family of sequences. Copies are dispersed throughout the human genome such that any substantial fragment of human DNA several kilobases in size is likely to contain at least one copy of an Alu sequence. It has been demonstrated that under stringent hybridization conditions the presence of Alu sequences can be used to establish the presence of human DNA in mouse or hamster cells (Gusella *et al.* 1980, Murray *et al.* 1981, Perucho *et al.* 1981).

Gene targeting in animal cells: homologous recombination and allele replacement

The armoury of techniques available to the gene manipulator allows any cloned gene sequence to be altered as desired *in vitro*. It would be a great advance if such alterations could be engineered into copies of a chosen gene *in situ* within the chromosomes of a living animal cell. The strategy for achieving this desirable aim is to bring about the change in the endogenous gene through homologous recombination between it and incoming mutated copies of the gene introduced by a DNA transfection procedure. If this capability were available for mouse cells it would, for example, be possible to introduce a mutation into a chosen gene within embryo-derived stem (ES) cells in culture. These cells can then be incorporated into mouse embryos at the blastocyst stage.

ES cells have pluripotential developmental capacity, and can give rise to many cell types in the resulting adult mouse, including germ cells. Careful selection of the ES cell line is necessary in order to maximize the

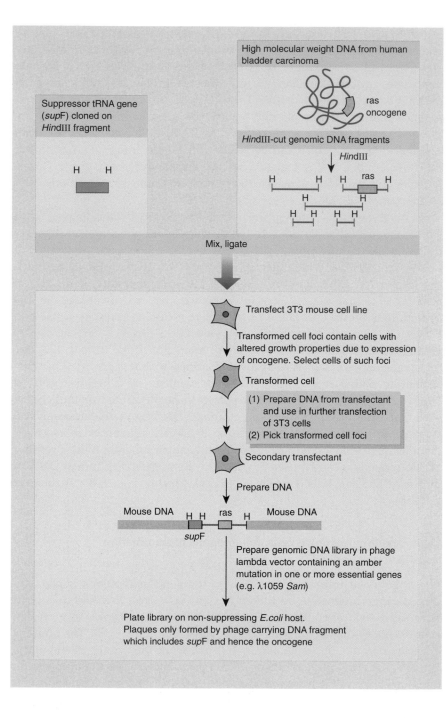

Fig. 15.6 Suppressor rescue, applied to the isolation of the human, *ras* oncogene in the T24 bladder carcinoma (Goldfarb *et al.* 1981).

production of chimaeras. By injecting vigorous ES cells into inbred host embryos, the colonization of the host germ line can be substantial. In addition, by using XY ES cells in random, XX or XY, host embryos, the high contribution of XY ES cells results in a large sex bias towards maleness in the resulting chimaeras (McMahon & Bradley 1990). Interbreeding

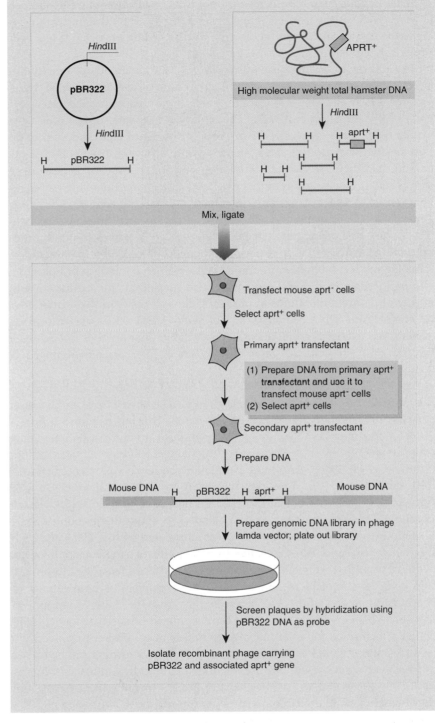

Fig. 15.7 Tag identification by hybridization, applied to the isolation of the hamster adenine phosphoribosyl transferase (*aprt*) gene (Lowy *et al*. 1980).

of heterozygous sibling progeny produced from such adults yields animals homozygous for the desired mutation (Fig. 15.8). Thus the functioning of any gene can be studied in whole animals, providing only that it has been previously cloned.

In Chapter 13 we discussed how, in yeast, homologous recombination is relatively easily detected. Homologous recombination between transforming DNA and the endogenous gene is stimulated by the free DNA ends of the transforming DNA. The situation in animal cells seems to be similar to that in yeast, with the important difference that random integration of the transfecting DNA occurs relatively very frequently, hence making it difficult to detect homologous recombination (or gene conversion) events. One solution to this problem in studying homologous recombination is to use test systems in which selectable inactivation of a test gene (*HGPRT*) or activation (Thomas *et al.* 1986) of a mutated test gene (*neo*) occurs by homologous recombination (or gene conversion) but not by random integration. These experiments have defined some parameters of homologous recombination in animal cells such as the length of homology required (Hasty *et al.* 1991), and have also revealed evidence for stimulation of unplanned mutations in the test gene as a result of the introduction of homologous DNA (Thomas & Cappechi 1986). However, these test systems are artificial in that a suitable selectable phenotype is not available for the great majority of genes for which allele replacement might be desired.

An alternative and general approach to the problem of detecting rare homologous recombination events is simply to use brute force and examine very large numbers of individual transfectants. This has been the approach used for studying homologous recombination in the human β-globin locus of a human fibroblast × mouse erythroleukaemia hybrid somatic cell line (Smithies *et al.* 1985). A modified β-globin sequence was 'tagged' with a *supF* gene (Fig. 15.9). This tag was used to isolate cloned DNA from transformed cells into which the plasmid had been introduced by electroporation. These experiments were designed to study integration of a test plasmid by a single recombination event stimulated by free DNA ends, i.e. the whole plasmid DNA became integrated by a single recombination event as though it were circular, although it had been linearized to provide free ends to stimulate recombination. The frequency of homologous recombination was found to be between 10^{-2} and 10^{-3} that of random integration. Other workers have used the polymerase chain reaction (Chapter 10) as a technique for examining large numbers of different transfectants (see Hogan & Lyons 1988, Zimmer & Gruss 1989). Screening can be carried out relatively quickly without the need for cloning. In summary, these experiments show that with either electroporation or nuclear micro-injection the frequency of homologous recombination can be in the range of 10^{-2}–10^{-3} transformants or injected cells.

Note that as shown in Fig. 15.9 a single homologous recombination event leads to duplication of sequences and not allele replacement: this requires either a gene conversion event, or two separate stages of recom-

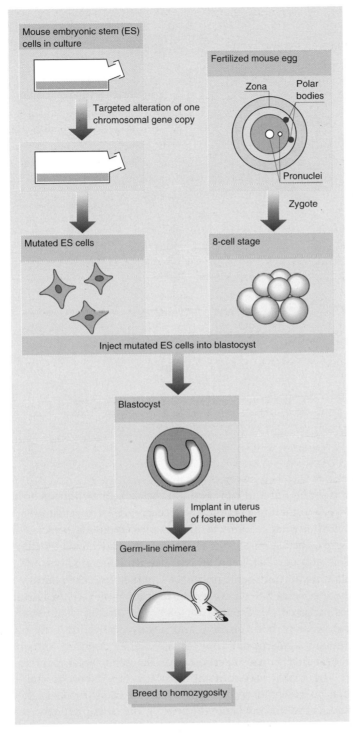

Fig. 15.8 Gene targeting in ES cells to give homozygous mutant mice.

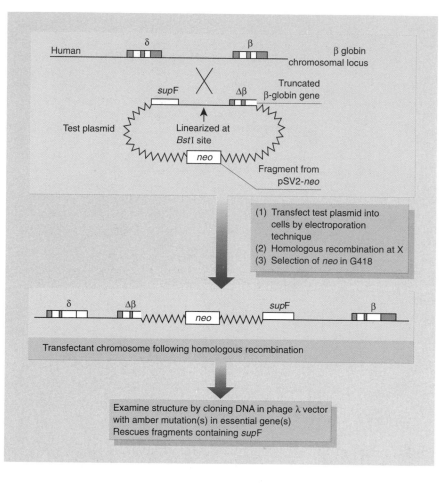

Fig. 15.9 Homologous recombination between test plasmid and human β-globin locus.

bination (see the 'hit-and-run' method, below), or two more-or-less simultaneous recombination events. The occurrence of two 'simultaneous' recombination events, leading to allele replacement, is selected for in the following practicable strategy for disrupting, and hence inactivating, any gene. This strategy has been widely used for 'gene-knockouts', creating null mutations, in transgenic mice (see Chapter 16). The strategy exploits a generally applicable selection system for enriching cells in which the rare event has occurred (Mansour *et al.* 1988). The strategy is illustrated in Fig. 15.10 and involves both positive and negative selection. The transfecting DNA contains a disrupted copy of any gene, depicted as gene *A*. The disruption is due to the insertion of a *neo* gene which confers G418 resistance. The DNA also contains the *tk* gene of herpes simplex virus. When two homologous recombination events occur between the targeting vector and a chromosomal copy of gene *A*, the disrupted gene replaces the chromosomal copy without including the *HSV–tk* gene. Random integration events elsewhere in the genome will usually lead to insertion of the entire targeting vector including the *HSV–tk* gene. Transfection by both kinds of process is positively selected by virtue of G418 resistance, but

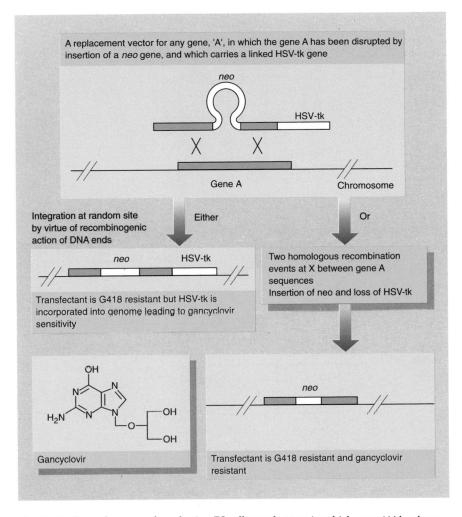

Fig. 15.10 General strategy for selecting ES cell transfectants in which gene 'A' has been disrupted by insertion of a *neo* gene. Ganciclovir is a nucleoside analogue which selectively kills cells expressing HSV tk because the substrate requirement of HSV tk is less stringent than that of cellular tk.

random integration is excluded by selection for the absence of the *HSV−tk* gene. (*HSV−tk* confers sensitivity to the nucleoside analogues gancyclovir and FIAU; see Figs 15.10 and 15.11.) This combination of positive and negative selection leads to a 2000-fold enrichment for ES cells that contain a targeted mutation. Although this strategy is expected to be general in that gene *A* could be any gene, in its present form it is limited to disruption of targeted genes by *neo*. This is quite adequate for the envisaged application of disrupting genes in ES cells with a view to creating mutant mice. But an important application of homologous recombination is in the field of gene therapy for human inherited diseases. Here the requirement is to introduce functional, wild-type alleles into cells with a mutant allele.

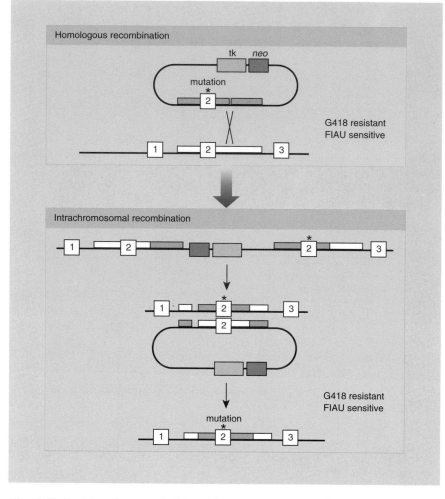

Fig. 15.11 The hit-and-run method for allele replacement in ES cells. An initial homologous recombination event involves a single reciprocal exchange between the target gene in the chromosome, and the transfecting DNA. The transfecting DNA consists of the linearized vector, which carries the HSV *tk* gene, and a *neo* gene, as well as the altered version of the target sequence. This exchange creates a duplication of sequences. Because of the *tk* and *neo* genes, the cell becomes G418-resistant and FIAU-sensitive. FIAU, 1-(2-deoxy-2-fluoro-β-D-arabinofuranosyl)-5-iodouridine, is a toxic nucleoside analogue (Borrelli *et al.* 1988) in cells expressing HSV tk. Intrachromosomal recombination between duplicated sequences leads to loss of vector sequences, the *tk* and *neo* genes, and in the event illustrated leads to loss of the original target sequence. The cell becomes FIAU-resistant, hence providing selection for intrachromosomal recombination. FIAU-resistant cells can be screened for the replacement allele by PCR, or by other methods (Hasty *et al.* 1991).

There are two reasons for the desire to introduce the functional gene at the precise chromosomal location of the mutated gene. First, random integration may have deleterious effects due to disruption of important gene functions at the unplanned site of integration. Second, it is clear that transferred genes are often expressed much less efficiently than endogenous genes, and hence insufficient gene product may not correct the deficiency. In the case of transfected globin genes, sequences which normally flank the β-globin locus at considerable distances have been found to have an important effect, possibly by influencing chromatin structure so as to place it in an 'open' configuration. Inclusion of these sequences with transfected β-globin genes allows normal, relatively high level expression in transgenic mice (Grosveld *et al.* 1987).

In order to introduce subtle changes into genes, the hit-and-run gene targeting method has been developed (Fig. 15.11). This generates ES cells, and thus transgenic animals, having specific changes in the targeted gene, with no selectable marker present (Hasty *et al.* 1991).

Viral vectors

Many animal viruses have been subjugated as vectors. Virtually every virus that has been studied in any detail and that has a DNA genome or a DNA stage in its replication cycle has been manipulated in this way. As we have seen in bacterial systems, viruses provide an efficient way of introducing foreign gene sequences into a cell: the gene manipulator can exploit their natural ability to adsorb to cells and infect them. Also as in bacterial systems, animal viruses often contain powerful promoters which can be of general applicability in driving gene expression, and in many cases they have the ability to replicate their genomes to high intracellular copy numbers, hence providing a route to high level expression of foreign genes. Certain animal viruses, of which retroviruses are the most prominent example, naturally integrate their DNA into the host chromosome as part of their replication cycle: another feature open to exploitation by the gene manipulator. Viral vectors can provide alternative and distinctive means for transferring genes into animal cells in addition to the transfection approaches discussed so far.

Examples of vectors derived from several types of animal virus will follow in this chapter, but we begin with a close look at some early SV40 manipulation. Development of SV40 derivatives preceded others because SV40 molecular biology was the first to be worked out in detail. Its small genome was the first animal virus genome to be completely sequenced. Ramifications of SV40 'vectorology' are to be found widespread in animal cell molecular biology.

SV40

Basic properties of SV40

The virus particle contains the circular duplex DNA genome, of about 5.2 kb (Fig. 15.12), associated with the four histones H4, H2a, H2b and H3 in a mini-chromosome. H1 is absent. The capsid is composed of 420 subunits of the 47 000 kDa polypeptide VP1. Two minor polypeptides VP2

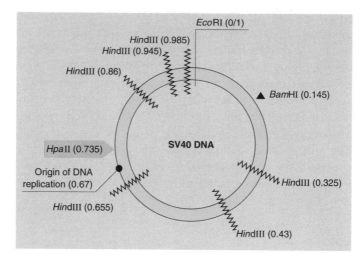

Fig. 15.12 Restriction endonuclease sites on SV40 DNA. The map coordinates of each site are shown in parentheses.

and VP3, which consist largely of identical amino acid sequences, are also present.

SV40 can enter two types of life cycle depending upon the host cell. In *permissive* cells, which are usually permanent cell lines derived from the African green monkey, virus replication occurs in a normal infection. In *non-permissive* cells, usually mouse or hamster cell lines, there is no lytic infection because the virus is unable to complete DNA replication. However, growth transformation of non-permissive cells can occur. In such cells SV40 DNA sequences are integrated into the host genome. The SV40 sequences are often amplified and rearranged in such transformed cell DNA. Because of the unpredictability in this integration and the subsequent rearrangements, recombinant SV40 virus vectors are not commonly used as integrative vectors.

The lytic infection of monkey cells by SV40 can be divided into three distinct phases. During the first 8 hours the virus particles are uncoated and the DNA moves to the host cell nucleus. In the following 4 hours, the *early* phase, synthesis of early mRNA and early protein occur and there is a virus-induced stimulation of host cell DNA synthesis. The *late* phase occupies the next 36 hours and during this period there is synthesis of viral DNA, late mRNA, and late protein, and the phase culminates in virus assembly and cellular disintegration.

A functional map of the SV40 genome is given in Fig. 15.13. The region of about 400 base-pairs around the origin of DNA replication is extremely interesting. Closely associated with the origin are control signals regulating the initiation of early and late primary transcripts. The early transcription is stimulated by two tandem, 72 base-pairs, enhancer sequences located within this region.

An important feature of SV40 gene expression is the complex pattern of

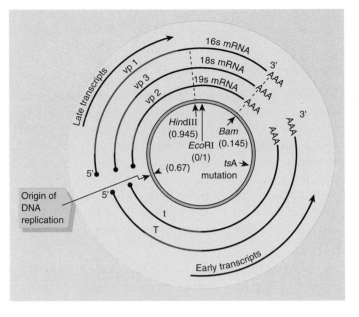

Fig. 15.13 Transcripts and transcript processing of SV40. Intron sequences which are spliced out of the transcripts are shown by red lines.

RNA splicing (Fig. 15.13). Alternative splicing pathways process the early primary transcript into two different early mRNAs. The late primary transcript can be processed into three different late mRNAs.

The two early mRNAs encode the large T and small t proteins (*Tumour* proteins or antigens). The late mRNAs, designated according to their sedimentation coefficients as the 16S, 18S, and 19S mRNAs, encode VP1, VP3 and VP2, respectively. The 18S and 19S mRNAs have coding sequences in common and direct the synthesis of VP3 and VP2 such that they contain amino acid sequences in common. More striking, however, is the finding that the VP1 coding region overlaps VP2 and VP3 in a different translational reading frame.

SV40 vectors: strategies

The assembly of SV40 virions imposes a strict size limitation on the amount of recombinant DNA which can be packaged. In view of this, two strategies have been employed in developing SV40 vectors. The first is to replace a region of the viral genome with an equivalently sized fragment of foreign DNA, and hence produce a recombinant DNA that can replicate and be packaged into virions in permissive cells. In order to supply genetic functions lost by replacement of virus sequences, a helper SV40 virus must genetically complement the recombinant.

In the alternative strategy, the size limitation of the virion is avoided. Recombinants are constructed which are never packaged into virions and give no lytic infection. These are maintained in host cells transiently as high copy number, unintegrated, plasmid-like DNA molecules.

First experiments with late region replacement

SV40 lacking the entire late region functions can be propagated in *mixed infections* with a temperature-sensitive helper virus that can complement the late region defect. The mixed infection is maintained since the *ts* defect in the helper's early region must be complemented. A suitable helper is a *ts*A mutant, which produces a temperature-sensitive T protein (Mertz & Berg 1974).

Based on this observation Goff and Berg (1976) prepared an SV40 vector by excising virtually the entire late region of the viral DNA by cleavage with *Hpa*II and *Bam*HI restriction endonucleases (Fig. 15.14). Cleavage with

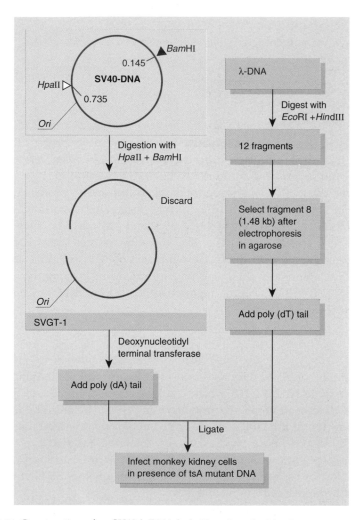

Fig. 15.14 Construction of an SV40-λ DNA hybrid as described in the text.

these two enzymes produces fragments approximately 0.6 and 0.4 of the genome length. These two fragments were separated by electrophoresis and the smaller fragment discarded. The large fragment, called SVGT-1, was then modified by the addition of poly(dA) tails using deoxynucleotidyl terminal transferase.

For an insert into this vector, a test fragment of DNA was chosen that was the same size as the discarded late region. This test fragment was produced by digesting phage λ DNA (Fig. 15.14). It was tailed with dA and annealed with the vector. The annealed DNA was then used to transfect monkey kidney cells in the presence or absence of *ts*A helper virus DNA. At the restrictive temperature (41°C) *ts*A DNA alone gave no plaques, the annealed DNA alone gave no plaques, but transfection with the two DNAs together produced 2.5 × 10³ p.f.u./μg of annealed DNA.

Construction of an improved SV40 vector

From the information presented above on transcription of SV40 it is clear that the ideal vector would retain all the regions implicated in transcriptional initiation and termination splicing and polyadenylation. Inspection of the SV40 map (Fig. 15.13) reveals two restriction endonuclease sites that could be used to generate a suitable vector. First, there is a *Hind*III site at map position 0.945 which is six nucleotides *proximal* to the initiation codon for VP1 (Fig. 15.15) and 50 nucleotides *distal* to the site at which the leader

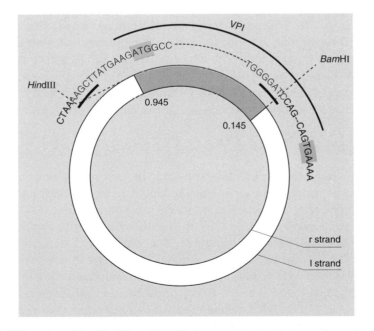

Fig. 15.15 Location of the *Hind*III and *Bam*HI cleavage sites in relation to the coding sequence for VP1. The triplets enclosed in boxes show the initiation and termination signals for translation of VP1. The sequences underlined are the recognition sites for the *Hind*III and *Bam*HI restriction endonucleases.

sequence is joined to the body of 16S mRNA. Second, the *Bam*HI site at map position 0.145 is 50 nucleotides proximal to the termination codon for VP1 translation and 150 nucleotides before the poly A sequence at the 3′ end of 16S RNA (Fig. 15.15). If the DNA between coordinates 0.945 and 0.145 were removed, the remaining molecule could be used as a vector for it would retain:

* the origin of replication;
* the regions at which splicing and polyadenylation occur;
* the entire early region, and hence could be complemented by a *ts*A mutant.

Such a vector (SVGT-5) has been constructed and used successfully to clone the rabbit β-globin gene in monkey kidney cells, to create SVGT-5-RaβG (Mulligan *et al.* 1979). In SVGT-5-RaβG the coding sequence of VP1 has been precisely replaced by a cDNA copy of rabbit β-globin mRNA. Following infection of monkey cells with the mixed recombinant/helper virus stock, the recombinant directed the synthesis of authentic rabbit β–globin.

Application of late-region replacement vectors

Vectors of this type have been effective in obtaining expression of many foreign genes in monkey cells: mouse β-globin (Hamer & Leder 1979a); rat preproinsulin (Gruss & Khoury 1981, Gruss *et al.* 1981a); the p21 protein of Harvey murine sarcoma virus (Gruss *et al.* 1981b); influenza virus hae-magglutinin (Gething & Sambrook 1981; Sveda & Lai 1981), and hepatitis B virus surface antigen (Moriarty *et al.* 1981). Late-region replacement vectors have been used to study post-translational RNA processing and stability. Some of these studies have demonstrated a requirement for an intron in the late-region primary transcript if a cytoplasmic mRNA is to be produced (Hamer & Leder 1979b). However, this has not proved consistently to be the case for all constructs (Gething & Sambrook 1981, Gruss *et al.* 1981a, Sveda & Lai 1981) so that the reasons for the requirement remain unclear.

Early region replacement vectors; COS cells

When the early region of SV40 is replaced, the essential T function is lost and must be provided by complementation. A major breakthrough in the development of SV40 vectors came when it was found that the T protein can be complemented in the COS cell line (this must not be confused with *cos*, the cohesive end site of phage λ). This cell line is a derivative of the permissive CV-1 monkey cell line. When CV-1 cells are infected with SV40, the normal lytic cycle ensues and no growth transformants are recovered. Gluzman (1981) found that CV-1 cells could be transformed by a segment of SV40 early-region DNA (cloned in an *E. coli* plasmid vector) in which the SV40 origin of DNA replication had been inactivated by a 6 bp deletion. Had the origin of replication been functional, the plasmid DNA would have replicated many times under the influence of T protein and made the host cell inviable. The resulting COS cell line (CV-1, origin of *SV40*) expresses T protein from integrated SV40 sequences and does so in a cellular background which is permissive for SV40 DNA replication. Infection of these cells with SV40 lacking a functional early region therefore leads to a normal lytic cycle. Since mixed infections are not necessary there is no contaminating helper virus. Therefore SV40 recombinants in which early DNA sequences have been replaced with foreign DNA can be constructed and propagated. Gething & Sambrook (1981) have made recombinants of this type which express influenza virus haemagglutinin in COS cells.

Overview of recombinant SV40 viruses

Both late- and early-region replacement vectors have the advantage of producing recombinant virion particles which introduce the foreign DNA into cells without the need for DNA-mediated transfection. However, advances in transfection technology are reducing the importance of this advantage. The replication of the recombinant DNA to a high copy number is a considerable advantage. For many biotechnological production pro cesses a limitation on the use of recombinant SV40 viruses is the fact that expression of foreign genes is transient in infected cells; this is often not as convenient as a stable transfected cell line. The maximum capacity of about 2.5 kb of foreign DNA is another potential limitation.

COS cells and SV40 replicons: transient expression — the SV40 enhancer

In COS cells *any* circular DNA, with no definite size limitation, containing a functional SV40 origin of replication should be replicated independently of the cellular DNA as a plasmid-like episome. Many laboratories have constructed vectors based on this principle (Myers & Tjian 1980, Lusky & Botchan 1981, Mellon *et al.* 1981). In general, these vectors consist of a small SV40 DNA fragment containing the SV40 origin cloned in an *E. coli* plasmid vector. Recombinant derivatives can be constructed and grown in *E. coli*. These are then transfected into COS cells where very high copy numbers of the recombinants are obtained. It is important that the *E. coli* plasmid sequence does not contain the so-called 'poison' sequence. This has been identified as a small region of pBR322 which in simian cells causes inefficient replication of any DNA in which it is included (Lusky & Botchan 1981). Plasmid pAT153 (see Chapter 4) and other related deletion derivatives of pBR322 have lost this site and are, therefore, suitable for constructing shuttle vectors for mammalian cells.

Permanent cell lines are not established when transient vectors are transfected into COS cells because the massive vector replication makes the cells inviable, but even though only a low proportion of cells are transfected, the high copy number is compensatory and the *transient* expression of cloned genes can be analysed. The major application of these vectors is in providing a rapid means of screening the effects of *in vitro* manipulations upon transcriptional and post-transcriptional control sequences. In such studies the important transcriptional stimulation by the SV40 enhancer sequences was discovered (Banerji *et al*. 1981, Khoury & Gruss 1983).

The enhancer sequences are a pair of 72-nucleotide, tandemly repeated sequences located near the origin of DNA replication and which are necessary for efficient early transcription. By linking the enhancer sequences to other genes it was found that they stimulate or *enhance* transcription from virtually any promoter placed near them. This effect operates on both sides of the enhancer, i.e. it is independent of enhancer orientation and extends to a promoter placed several kilobases away.

Subsequent to the discovery of the SV40 enhancer, virtually every cellular gene transcribed by RNA polymerase II has been found to have one or more associated enhancers. Some enhancers are cell-type specific. It is not entirely clear how enhancers work; it is possible that several mechanisms are involved. Enhancers and their relationship to transcription factors and other proteins which influence transcription are the subject of the most intensive research, which is beyond the scope of this book (Maniatis *et al*. 1987, Wasylyk 1988). The SV40 enhancer sequences are important to gene manipulators because of their stimulatory effect on the SV40 early transcription unit to which foreign genes may be fused and hence expressed.

Alternatively, transcription from *cellular* promoters may be enhanced by the enhancer. The SV40 enhancer has the advantageous property of stimulating transcription in a wide variety of cell types. It is very complex. Detailed analysis has revealed many distinct sequence motifs within the enhancer region, with different cellular specificities, which in combination give the rather non-specific generalized enhancement.

In the original COS cell lines the SV40 T protein was expressed constitutively from integrated SV40 sequences. Subsequent COS-type cell lines have been produced in which the SV40 T protein activity is regulated. This has been achieved either by using a temperature-sensitive T protein (Rio *et al*. 1985) or by placing the T protein gene under the control of a human metallothionein gene promoter which is inducible by heavy metals (Gerard & Gluzman 1985). Using such cell lines it is possible to establish *stable* transfectants in which the SV40 replicon is maintained episomally at a low copy number. The T protein activity can be increased when expression of foreign genes carried on the vector is required.

SV40 and polyoma virus mini-replicons

A series of vectors has been constructed which comprise an *E. coli* plasmid replicon derived from pBR322, an SV40 origin of DNA replication, a functional early region providing T protein, and an SV40 transcription unit into which foreign genes can be inserted and expressed. These vectors are grown and manipulated in *E. coli* and then transferred into permissive monkey cells by standard transfection procedures (Berg 1981, Mulligan & Berg 1981b).

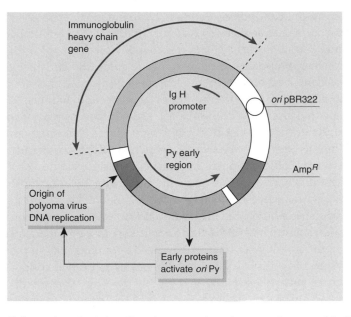

Fig. 15.16 Polyoma-based mini-replicon for expression of a mouse immunoglobulin heavy chain gene in mouse lymphocytes.

Analogous vectors contain a virus origin of DNA replication and early region derived from polyoma virus, whose structural organization closely resembles that of SV40. Mouse cells, but not monkey cells, are permissive for polyoma virus and therefore are the host for such polyoma-derived vectors.

Because of the intact T function, these mini-replicon vectors replicate in permissive cells and attain very high copy numbers over a period of a few days. This eventually kills the transfected host cells, but massive transient expression of a foreign gene is possible. If appropriate, the foreign gene can be expressed from its own promoter rather than a viral one. Deans *et al.* (1984) used such a polyoma-based mini-replicon vector to express an immunoglobulin heavy-chain gene in transfected mouse lymphocytes (Fig. 15.16).

Bovine papillomavirus (BPV) DNA

Bovine papillomavirus causes warts in cattle. It is a member of a group of viruses which induce warts and papillomas in a range of mammals. Certain of the human papillomaviruses are implicated in causing cervical cancer. Like other papillomaviruses, BPV normally infects terminally differentiated squamous epithelial cells, but it can cause growth transformation in dividing cultured mouse cell lines. In such transformed mouse cells the viral DNA is found as a multicopy plasmid. This property has been the impetus for developing vector molecules (reviewed in Campo 1985) based on the BPV replicon.

BPV has a circular double-stranded DNA genome of about 7.9 kb, which has been completely sequenced (Chen *et al.* 1982). The molecular biology of the virus (Giri & Danos 1986) is considerably more complex than that of SV40, to which it appears to be distantly related. The early region of the genome is contained on a large subgenomic fragment of about 5.5 kb. This fragment makes up 69% of the entire genome and has been found to contain all the functions necessary for growth transformation of mouse cells, including the origin of BPV DNA replication. This fragment is called BPV$_{69T}$. Simple vectors have been constructed which consist of this fragment cloned in pBR322. Such a vector has been used to express a rat preproinsulin genomic DNA in mouse cells, using growth transformation as the basis for selecting transfected cells (Sarver *et al.* 1981a,b). The pBR322 sequences appeared to be inhibitory and so were excised from the plasmid before DNA-mediated transfection into the mouse cells. The remaining BPT$_{69T}$ rat gene fragment cyclized spontaneously in the host cell and replicated as a plasmid at about 100 copies per cell. Expression of the rat gene from its own promoter was readily detected. Other workers have constructed similar vectors and found that the *E. coli* plasmid sequence could be retained. For example, a recombinant carrying human β-globin gene sequences was maintained at 10–30 copies per cell (Di Maio *et al.* 1982). In similar experiments with a simple BPV$_{69T}$-based vector, transformed cells contained as many as 200 copies of the recombinant DNA (Ostrowski *et al.* 1983). There are examples of transfectants that have maintained the recombinant plasmid as an episome, without integration into the chromosome, for long periods (Fukunaga *et al.* 1984), but in other cases the DNA has been found to integrate into the host genome (Ostrowski *et al.* 1983, Sambrook *et al.* 1985). The simple vectors use growth transformation as the means of selecting transfectants. This is not always convenient and limits the range of mouse cell lines available as hosts. By inserting a neomycin phosphotransferase gene into the vector plasmids, transfectants can then be selected for resistance of G418 (Law *et al.* 1983).

BPV-derived vectors are useful because *permanent* cell lines can be obtained carrying the recombinant DNA either episomally or integrated at relatively high copy numbers. An additional advantage is the ability to carry large DNA inserts.

Recombinant vaccinia viruses

Vaccinia virus is closely related to variola virus, which causes smallpox, and inoculation with vaccinia virus provides a high degree of immunity to smallpox. There has been a proposal to construct vaccinia virus recombinants which express antigens of unrelated pathogens and use them as live vaccines against those pathogens. Smith *et al.* (1983b,c) adopted a clever strategy for expressing the hepatitis B virus surface antigen (HBsAg) in vaccinia which took into account:

- the large size (187 kb) of vaccinia DNA;

- the lack of infectivity of isolated viral DNA;
- the packaging of viral enzymes necessary for transcription within the virion; and
- the probability that vaccinia virus has evolved its own transcriptional regulatory sequences operative in the cytoplasm where viral transcription and replication occur.

Briefly, fragments of vaccinia DNA were cloned in an *E. coli* plasmid vector that contained a non-functional vaccinia thymidine kinase gene. This gene had been rendered inactive owing to the insertion of a vaccinia DNA fragment containing a promoter derived from another early vaccinia gene. The HBsAg gene was inserted next to the vaccinia promoter in the correct orientation. This chimaeric HBsAg gene was then inserted into vaccinia DNA by homologous recombination as follows. Monkey cells were infected with wild-type vaccinia and simultaneously transfected with the recombinant *E. coli* plasmid. Homologous recombination could then replace the functional thymidine kinase gene of the wild-type virus with the non-functional tk gene sequence which included the HBsAg chimaeric gene. Such virus would be TK$^-$ and would be selectable on the basis of resistance to BUdR (Fig. 15.1). When cells were infected with such TK$^-$ virus, they were found to synthesize HBsAg and secrete it into the culture medium. Vaccinated rabbits rapidly produced high-titre antibodies to HBsAg. A similar strategy (Fig. 15.17) was then used to construct an infectious vaccinia virus recombinant which expresses the influenza haemagglutinin gene and induces resistance to influenza virus infection in hamsters (Smith *et al.* 1983a).

Subsequently recombinant vaccinia viruses expressing other important genes have been constructed, including an AIDS virus envelope gene (Hu *et al.* 1986), HTLV-III envelope gene (Chakrabarti *et al.* 1986) and hepatitis B virus surface antigen gene (Moss *et al.* 1984, Smith *et al.* 1983a).

These experiments showing the potential of vaccinia vectors for immunization have been followed by the actual immunization of wild foxes against rabies. A recombinant vaccinia virus that expresses the rabies antigen was administered to the wild population of foxes in north-eastern France by providing 'bait' consisting of chicken heads spiked with the virus. A substantial proportion of the wild foxes were shown to have acquired immunity to rabies by this means (Blancou *et al.* 1986). See Chapter 17.

Baculovirus vectors for insect cells and insects

Baculoviruses infect insects. They have large double-stranded, circular DNA genomes within a rod-shaped capsid. The baculoviruses which have been exploited as vectors are the *Autographa california* nuclear polyhedrosis virus (AcMNPV) (Smith *et al.* 1983a, Miller *et al.* 1987) and the silkworm virus, *Bombyx mori* nuclear polyhedrosis virus (BmNPV) (Maeda *et al.* 1985). During normal infection the viruses produce nuclear inclusion bodies which consist of virus particles embedded in a protein matrix, the major

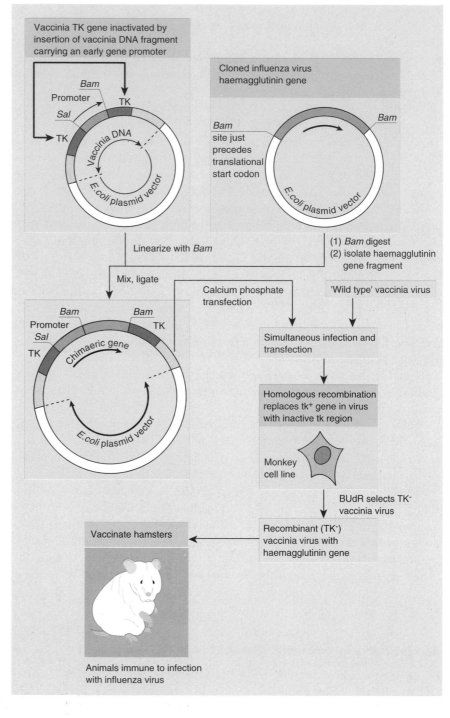

Fig. 15.17 Construction of an infectious vaccinia virus recombinant expressing influenza virus haemagglutinin.

component of which is a virus-encoded protein called polyhedrin. Large amounts of virus and polyhedrin are produced. Transcription of the polyhedrin gene is driven by an extremely active promoter, which is therefore ideally suited as a promoter for driving expression of foreign genes. This is all the more attractive because the polyhedrin gene product is not essential for viral replication. Construction of expression vectors has therefore consisted of inserting a foreign coding sequence just downstream of the polyhedrin promoter. This cannot be achieved directly because the large size (about 130 kb) of the viral DNA precludes simple *in vitro* manipulation. The strategy for inserting the foreign DNA into the virus has many similarities to that already described for vaccinia virus. The strategy involves a small, recombinant *transfer vector*, and homologous recombination to generate the recombinant viral genome. A range of such transfer vectors is available for the production of nonfused and polyhedrin-fused proteins. (For extensive reviews, see Webb & Summers 1990, King & Possee 1992, O'Reilly *et al.* 1992.)

Consider the insertion of a human interferon (IFN)-α gene into BmNPV (Maeda *et al.* 1985) as an example. Essentially, the same strategy has been used for expressing foreign genes in AcMNPV. First, a fragment of DNA from the polyhedrin gene region was cloned into the *E. coli* vector pUC9 so as to place a polylinker site just downstream from the start point of transcription directed by the polyhedrin promoter. On the other side of the polylinker, opposite from the promoter fragment, was inserted a DNA fragment containing downstream sequences of the polyhedrin gene. A human IFN-α coding sequence was then ligated into the polylinker site so that it was placed between the polyhedrin promoter and downstream sequences in the correct orientation so as to give an expression construct. Homologous recombination *in vivo* was then used to replace the wild-type polyhedrin gene in viral DNA with the disrupted polyhedrin gene of the expression construct. This was achieved by co-transfecting *Bombyx mori* cultured cells with expression plasmid DNA and BmNPV DNA. Recombinant viruses gave characteristic plaques because of their failure to form inclusion bodies, and so could be isolated. These recombinant viruses replicated in silkworm caterpillars, producing up to 50 μg per larva.

The silkworm host has advantages as a production system employing recombinant BmNPV because silkworms can be cultured easily and at low cost, as in the long-established silk industry. Expression of recombinant AcMNPV is usually carried out by infecting cultured cells of the insect *Spodoptera frugiperda* (Fig. 15.18). The foreign gene is expressed during the infection and very high yields of protein can be achieved by the time the cells lyse, about 3 days post-infection (e.g. Miyamoto *et al.* 1985, Kuroda *et al.* 1986).

Retroviruses

Retroviruses have useful properties for exploitation as vectors: (a) they cover a wide host range including avian, mammalian and other animal

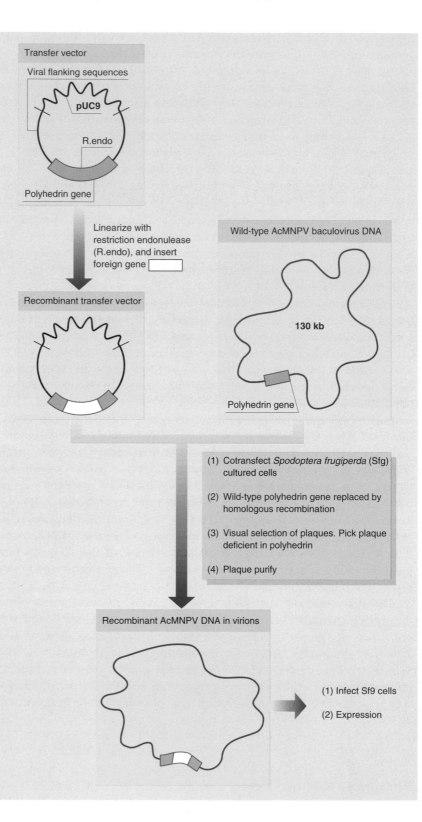

Fig. 15.18 Generating
recombinant baculovirus,
AcMNPV.

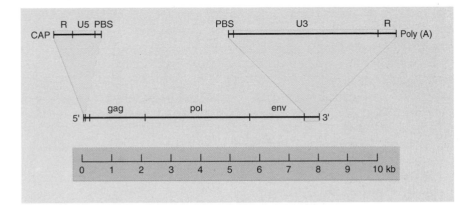

Fig. 15.19 The genome of murine leukaemia virus.

hosts, (b) infection does not lead to cell death, infected cells produce virus over an indefinite period, (c) viral gene expression is driven by strong promoters – these can be harnessed to foreign genes, and (d) in the case of murine mammary tumour virus the promoter function can be switched on and off experimentally. Transcription is induced by glucocorticoid hormones (Lee *et al.* 1981, Scheidereit *et al.* 1983).

Retroviruses contain RNA genomes. The virus particle actually contains two copies of viral RNA. Each RNA genome has many features similar to a eukaryotic mRNA in that it has a poly(A) sequence of approximately 200 residues at the 3′ terminus and a typical cap structure at the 5′ terminus. Figure 15.19 shows a simple map of a typical retrovirus, Moloney-MuLV.

A very abbreviated scheme of the retroviral replication cycle is given in Fig. 15.20. This illustrates the main points necessary for an appreciation of vector development. (For a complete account of retroviral biology see Weiss *et al.* 1985.) When the viral RNA enters the cell it is accompanied by reverse transcriptase and integrase which are packaged into the virion. The reverse transcriptase (and associated RNase H activity) then engages in a complex series of cDNA synthesis reactions which lead to the production of a double-stranded DNA copy of the viral RNA. This DNA copy, which is called the proviral DNA, is slightly longer than the RNA from which it was derived because terminal sequences are duplicated in the process of converting it to double-stranded form. The proviral DNA circularizes and, through the action of the integrase protein, inserts into the host genome. There is usually only a single copy (or a small number of copies) of proviral DNA integrated per cell, the site of integration in the genome is 'random' (there may be a preference for transcriptionally active regions) and the proviral DNA integrates such that it is bounded by the long terminal repeat, LTR, sequences which include a strong promoter for RNA polymerase II. The proviral genome contains three genes: *gag*, *pol* and *env*. These are transcribed and translated into precursor proteins which are subject to proteolytic cleavages to produce mature proteins. Full-length transcripts of the proviral DNA constitute the RNA genome which is packaged into virus particles. The *psi* site in full-length RNA is important

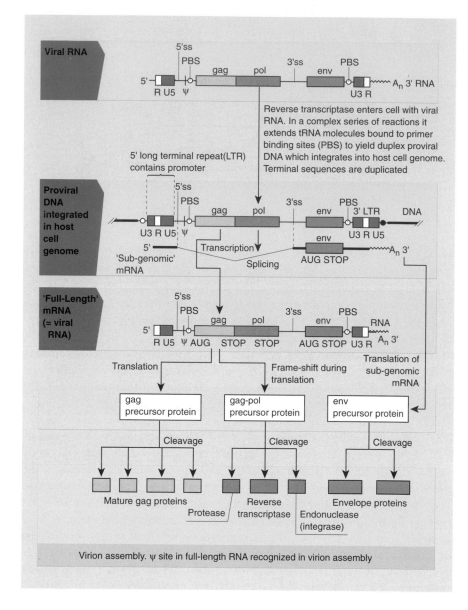

Fig. 15.20 Abbreviated
scheme of retroviral
replication cycle. (See text for
details.)

for the interaction of viral RNA with proteins in the assembly of virions.
Assembly takes place at the cell membrane with virus budding off from the
cell. There is no cell lysis. Figure 15.21 is a schematic representation of the
structural proteins within the virion.

Strategies of vector construction

Certain retroviruses are acutely oncogenic. Most of these have genome
structures similar to that of Moloney MuLV shown in Fig. 15.19 but differ
in that they contain an oncogene sequence. These viral oncogenes are

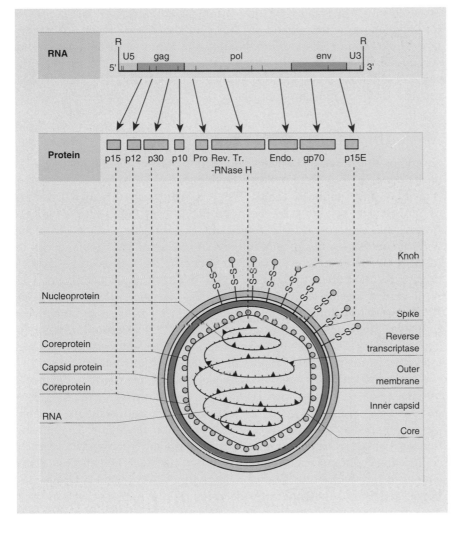

Fig. 15.21 Gene map, polypeptide products and schematized virion particle of a typical retrovirus, murine leukaemia virus.

derived from cellular genes and are often the result of obvious gene fusions with viral genes. As a result such oncogenic viruses have lost essential viral gene functions. These viruses are therefore defective. They can replicate in a mixed infection with a helper virus which provides the lost functions. Such oncogenic retroviruses demonstrate a natural ability of retroviruses to act as vectors.

For the deliberate construction of retrovirus vectors the starting point has been cloned DNA of the retrovirus. It is necessary to know which *cis*-acting sequences in the genome are essential for its replication and packaging. The LTRs are essential for efficient transcription of proviral DNA and for generating the 3'-end of the full-length transcripts. The LTRs are also essential for integration of the proviral DNA into the host cell DNA. Other sequences important for viral replication are located between the 5' LTR and the *gag* gene, and between the *env* gene and the 3' LTR.

These include a minus-strand tRNA primer-binding site (PBS) (necessary for initiation of the first strand of DNA synthesis during reverse transcription), a plus-strand tRNA primer-binding site (required for initiation of the second strand of DNA during reverse transcription), and the *psi* site. The *gag*, *pol* and *env* genes are dispensable because their functions can be provided from another proviral DNA.

Generation of infectious recombinant retroviruses

Several groups of researchers developed retroviral vectors in which regions of the proviral DNA were replaced by foreign DNA (Tabin *et al.* 1982, Wei *et al.* 1981, Shimotohno & Temin 1981, 1982). Subsequently a great variety of replacement vectors have been produced (see Weiss *et al.* 1985 for review). It is common to insert a selectable marker gene, such as *neo*, along with the non-selectable gene of interest. A variety of arrangements can be used to obtain expression of inserted genes. If the LTR is to be used to drive expression then the inserted gene is most simply expressed if its ATG is in a similar location to that of the *gag* gene which it replaces. Alternatively, the foreign gene may be driven by its own accompanying promoter. There are examples where selection for expression of one gene appears to interfere with high expression of the second inserted gene in integrated retrovirus recombinants (Emerman & Temin 1986).

The construction of such recombinants takes place in *E. coli* using proviral DNA sequences carried by an *E. coli* plasmid vector. Infectious recombinant virus can then be generated by using standard calcium phosphate-mediated transfection of a suitable animal cell line. The missing functions of the recombinant retrovirus are most simply provided either by co-infecting the transfected cell line with non-defective helper virus, or by co-transfection with non-defective helper virus DNA. In the latter case the transfected DNAs integrate into the genome in a non-specific way with respect to their own sequences. Transcription of intact, non-permuted copies leads to virus expression. The mixed virus preparation can then be used to infect cells in the normal way. These infected cells will continue to produce virus indefinitely. Since the packaging of the recombinant genome is dependent upon the helper virus, it will have the host-range properties of the helper virus (and the helper will also provide the proteins necessary for the initial stages of infection, e.g. reverse transcriptase). The helper virus is said to *pseudotype* the recombinant.

The use of the helper virus just described leads to the production of a mixed virus preparation containing defective recombinant virus and non-defective helper. An important alternative strategy is to use a *helper cell line* to provide the missing functions. Such helper cell lines contain mutant proviruses which have non-functional packaging signals. These mutant proviruses have deletions of the *psi* site. The first of these helper cell lines to be constructed, psi-2, is a NIH 3T3 mouse cell line with a transfected provirus derived from Moloney-MuLV (Mann *et al.* 1983). When such a helper cell line is transfected with defective, recombinant proviral DNA, fully infectious recombinant virus particles are produced. These viruses can be applied to other normal cell lines where recombinant provirus formation, integration, and expression occur efficiently, but where no infectious viruses are produced.

The host range of the recombinant retrovirus produced with the aid of the helper cell line will depend upon its pseudotype. The Moloney-MuLV provirus in psi-2 cells confers a narrow host range. Therefore in order to broaden the range of hosts available, helper cell lines have now been produced which are based upon amphotropic strains of murine leukaemia virus. Amphotropic MuLVs replicate in cells from a wide variety of mammalian and even avian species. These new helper cell lines allow the transfer of retroviral recombinant DNAs to human cells (Cone & Mulligan 1984, Sorge *et al.* 1984, Miller *et al.* 1985, Danos & Mulligan 1988).

In addition to the helper-virus or helper cell line approaches already mentioned, there is a final possible strategy, that is to produce non-defective recombinant virus which contains a complete set of essential genes. This has been achieved with derivatives of the avian virus, Rous sarcoma virus (e.g. Hughes & Kosick 1984). This approach has not been extensively explored for mammalian retroviruses, but in view of the fact that mammalian retroviruses can accommodate an increase in overall genome size it should be possible to include foreign DNA *in addition* to the essential gene sequences. In fact it appears that retroviruses have considerable flexibility in the size of the genome that can be packaged, ranging from considerably smaller than, up to a few kilobases larger than, the normal genome size.

Retroviruses can be used to be reduce the size of gene inserts containing introns. Should the original insert in the DNA construct of a recombinant retrovirus contain introns, then the retroviral life cycle which follows transfection into a susceptible cell leads to the removal of intron sequences by RNA splicing. A shortened, intronless RNA copy is packaged into virions and is later represented in proviral DNA copies.

Adenovirus vectors

Adenoviruses have a linear double-stranded DNA genome of about 36 kb. Recombinant viruses can be created by deleting genes E1A/E1B and replacing them with the foreign DNA (Eloit *et al.* 1990). The deleted genes are transcriptional regulators, and are essential for virus replication, and so the recombinant virus is defective. For propagating the defective recombinant virus, a transfected cell line is used that constitutively expresses the E1A/E1B functions. In addition, the E3 gene may be deleted from the adenovirus genome: this gene downregulates the immune response of the host *in vivo*, but is not necessary for replication *in vitro*. With the space created by these deletions, and the limited flexibility in the size of DNA that can be packaged into virions, the maximum insert size is 6–8 kb. When defective, recombinant virus is applied to normal susceptible cells; the recombinant DNA persists episomally within the cell for relatively long periods, from days up to months in post-mitotic brain cells *in vivo* (La Salle *et al*. 1993, Davidson *et al*. 1993, Akli *et al*. 1993, Bajocchi *et al*. 1993, Neve 1993). Adenoviruses naturally infect the respiratory tract, and for this and other reasons recombinant adenoviruses have been applied to transfecting airway cells for somatic gene therapy in cystic fibrosis (Rosenfeld *et al*. 1992). Recombinant adenoviruses have also been effective for gene transfer to skeletal muscle (Ragot *et al*. 1993). See Chapter 16.

High-level expression of foreign genes in animal cells

Expression of transferred genes in mammalian cells is often a prerequisite for cell and molecular biological research. In addition, in important biotechnological applications, animal cells are necessary hosts for the expression of foreign proteins; this is often the case where the protein product must be glycosylated in a characteristically mammalian way for therapeutic applications.

The following general considerations apply when high-level expression is desired:

• Use a strong enhancer/promoter such as the SV40 enhancer/early promoter, the RSV LTR, or the human cytomegalovirus immediate early promoter.

• Secondary structure within the 5′-untranslated region of the mRNA is likely to prevent efficient translation (Kozak 1984).

• Initiation codons within the 5′-untranslated region are detrimental to downstream initiation at the correct initiation codon, especially when the upstream AUG is not followed by an in-frame stop codon before the correct AUG (Kozak 1984). Because of this, and the preceding point, it is often desirable to shorten the 5′-untranslated region of the mRNA.

• The sequence around the initiation codon should conform to Kozak's rules (Kozak 1986). Of most importance is a purine at the −3 position (the A of the AUG codon is counted as +1), and a G at the +4 position.

• AU-rich sequences can reduce mRNA stability when present in the 3′-untranslated region of mRNA (Shaw & Kamen 1986).

Various expression systems have been discussed in this chapter. The choice of a particular system will depend upon the following criteria.

• The host cell and its culture conditions. For large-scale applications an important consideration is the cost of culture medium (with or without serum supplement).

• Whether a transient or stable expression system is suitable. For therapeutic applications, a stable, permanently-expressing cell line has the advantage of consistency and reproducible quality.

• The transfection system, and the scale of the transfection reaction. This is especially important if transient expression is envisaged.

• The amount of expression required for the application: it may not necessarily be the maximum attainable.

• The possible need for an inducible expression system. For this the metallothionein promoter may be chosen. Transcription is induced by

Table 15.2 High-level expression systems in animal cells

Host cell	Mode of transfer	Applications
Monkey COS cell	DEAE dextran or electroporation. Transient	Rapid characterization of constructs and expressed products. Cloning by expression
Chinese hamster ovary cell lines	Stable transfectants. DHFR/Mtx amplification	Constitutive expression, favoured for therapeutic proteins
Various mammalian	Recombinant vaccinia virus	Transient expression in wide range of cells
Various mammalian	Infection by vaccinia expressing T7 RNA pol*	Very high transient expression
Various mammalian and avian	Recombinant retrovirus	Long-term expression, variety of species
Insect cells	Recombinant baculovirus	Very high transient expression

*The recombinant vaccinia virus expresses T7 RNA polymerase in the host cell. Target genes are constructed by linking to T7 promoter and terminator regions. When cells are infected by the recombinant vaccinia virus, and transfected with plasmid containing the target gene, the target gene is expressed at a very high level (Fuerst *et al.* 1986).

metal ions such as cadmium and zinc. In Chinese hamster ovary cells a 200-fold induction has been attained (McNeall *et al.* 1989). Alternatively a promoter may be chosen that is under the control of nuclear receptors of the steroid/thyroid hormone/retinoic acid receptor superfamily. These receptors are ligand-activated transcription factors, which bind to specific short DNA sequences, the response elements. The response elements are usually located within the promoter region of hormone-responsive genes. The best established example in this context is the mouse mammary tumour virus (MMTV) LTR which is a complex promoter region that includes a glucocorticoid response element. Binding of a steroid, such as dexamethasone, to the glucocorticoid receptor, leads to induction of transcription from the MMTV LTR (Israel & Kaufman 1989). The molecular biology of these receptors has advanced greatly in recent years, and as the receptors themselves can be manipulated (for example by exchanging domains for ligand binding between different receptor types – 'domain-swaps') the scope for exploiting this system is expanding rapidly (Green *et al.* 1993, Smith *et al.* 1993).

Table 15.2 summarizes some animal-cell expression systems.

16 Transferring genes into animal oocytes, eggs, embryos and specific animal tissues

Introduction

The ability to transfer a new or altered gene into the germline of mice, and hence produce families of transgenic mice that inherit the gene, has opened up a new era of experimentation. From the beginning, studies on transgenic mice focused on the cell-type specificity of transgene expression, and the DNA sequences controlling the specificity. Later, the ability to disrupt the normal development, or functioning, of specific tissues through cell-type-specific gene expression has been exploited to illuminate the developmental, or other functional, interactions between cells within the tissues of a living animal.

The technology for replacing alleles precisely, with completely non-functional copies (null alleles), or alleles altered in more subtle ways, or replacing defective genes with functional copies, has endowed the genetic engineer with the almost fabulous power to alter any particular gene sequence within a living mammal, at will. At the simplest level, observing the phenotypic effects of 'knocking out' a gene (replacing it with a null allele), is the most direct way of establishing the gene's normal function.

The technology for altering the germline of mice is well established: the technology for altering somatic cells in particular tissues is less well developed but is being actively researched. This research is driven by the potential for somatic gene therapy in man, i.e. altering the genetic constitution of particular somatic cells for therapeutic purposes, in either inherited or acquired disease.

While progress with mammalian species has been very dramatic, other species such as the frog, *Xenopus laevis*, or the fruit-fly, *Drosophila melanogaster*, provide well-developed experimental systems. *Xenopus* is not readily amenable to the production of truly transgenic adults. But the embryos are

[346]

very suitable for micro-injection of mRNA and DNA, and this is exploited by developmental biologists. The ability to make transgenic flies, and the related technology of *enhancer trapping*, have revolutionized *Drosophila* genetics.

Xenopus

Xenopus oocytes as a heterologous expression system

Since Gurdon *et al.* (1971) first demonstrated that oocytes synthesized large amounts of globin after they had been micro-injected with rabbit globin mRNA, this expression system has been a valuable tool for expressing a very wide range of proteins from plants and animals (Colman 1984). Oocytes can be obtained in large numbers by removal of the ovary of adult female *Xenopus*. Each fully-grown oocyte is a large cell (0.8–1.2 mm diameter) arrested at first meiotic prophase (Box 16.1). This large cell has a correspondingly large nucleus (called the *germinal vesicle*) which is located in the darkly pigmented hemisphere of the oocyte.

Because of the large size of oocytes, mRNA — either natural, or synthesized by transcription *in vitro* using phage T7 RNA polymerase (Melton 1987) — can be introduced into the cytoplasm readily by micro injection (Fig. 16.1). This is achieved using a finely-drawn glass capillary as the injection needle, held in a simple micromanipulator. The nucleus can also be micro-injected. The oocyte nucleus contains a store of the three eukaryotic RNA polymerases, enough to furnish the needs of the developing embryo at least until the 60 000 cell stage (Box 16.1). The RNA polymerases are available for the transcription of injected exogenous DNA.

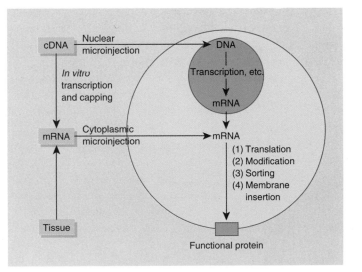

Fig. 16.1 DNA can be micro-injected into the *Xenopus* oocyte nucleus. More commonly, either natural mRNA, or synthetic mRNA, is micro-injected into the cytoplasm. Many foreign proteins are modified and targeted appropriately in oocytes.

Box 16.1 Events in oogenesis and early embryogenesis of *Xenopus*

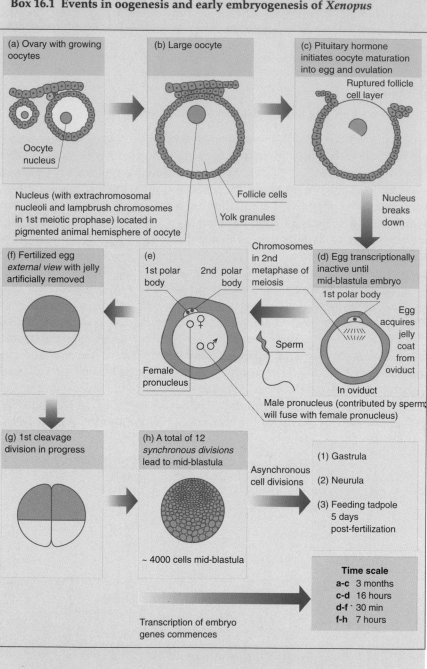

Thus, for example, cDNAs linked to a heat shock promoter, or to mammalian virus promoters, have been transcribed and translated (Fig. 16.1) (Ballivet *et al.* 1988, Ymer *et al.* 1989, Swick *et al.* 1992). In addition, methods exist for expressing cDNA in the cytoplasm of oocytes. Oocytes can either be injected with a recombinant vaccinia virus containing the foreign cDNA, linked to a vaccinia early promoter, or vaccinia virus expressing T7 RNA polymerase can be injected together with a plasmid in which a T7 promoter drives transcription of the foreign cDNA (Yang *et al.* 1991).

An important aspect of the oocyte expression system is that the foreign protein products are often correctly post-translationally modified, and directed to the correct cellular location. For example, oocytes translate a wide variety of mRNAs encoding secretory proteins, modify them and correctly secrete them (Lane *et al.* 1980, Colman *et al.* 1981). Oocytes glycosylate and phosphorylate (Gedamu *et al.* 1978, Matthews *et al.* 1981) foreign proteins, target foreign lysosomal proteins to endogenous lysosomes (Faust *et al.* 1987), and correctly assemble multi-subunit proteins (Ceriotti & Colman, 1990). Foreign plasma-membrane proteins are often correctly targeted to the plasma membrane of the oocyte, where they can be shown to be functional.

The first plasma-membrane protein to be expressed in this system was the acetylcholine receptor from the electric organ of the ray, *Torpedo marmorata* (Sumikawa *et al.* 1981). Injected oocytes translated mRNA extracted from the electric organ, and assembled functional multi-subunit receptor molecules in the plasma membrane (Barnard *et al.* 1982). Following this work, the oocyte has become a standard heterologous expression system for plasma–membrane proteins, including ion channels, carriers and receptors. The variety of successfully expressed plasma membrane proteins is very impressive. However, there are examples of foreign channels and receptors being non-functional in oocytes, either due to lack of coupling to second messenger systems in the oocyte, incorrect post-translational processing, or for unknown reasons (reviewed in Goldin 1991).

Xenopus oocytes as a system for functional expression cloning of plasma membrane proteins

Functional expression cloning using oocytes was first developed by Noma *et al.* (1986), using a strategy outlined in Fig. 16.2. Essentially similar strategies have been adopted for the cloning of a wide variety of plasma membrane proteins (Table 16.1). The following example, the cloning of the substance K receptor is illustrative (Fig. 16.3).

It has been found that oocytes can be made responsive to the mammalian tachykinin neuropeptide, substance K, by injecting a mRNA preparation from bovine stomach into the oocyte cytoplasm. The mRNA preparation contains mRNA encoding the substance K receptor protein, which is evidently expressed and inserted into the oocyte membrane in

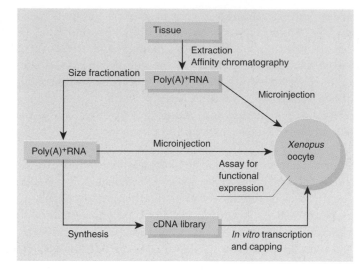

Fig. 16.2 Strategy for functional expression cloning, using *Xenopus* oocytes as a heterologous expression system.

Table 16.1 Examples of cDNAs encoding plasma membrane proteins which have been cloned by functional expression in *Xenopus* oocytes

Protein	Reference
Receptors	
Thyrotropin-releasing hormone receptor (mouse pituitary)	Straub *et al.* 1990
Serotonin 5-HT$_{IC}$ receptor (mouse choroid plexus)	Lubbert *et al.* 1987
Substance K receptor (bovine stomach)	Masu *et al.* 1987
Glutamate receptor (rat brain)	Hollmann *et al.* 1989
Platelet-activating factor receptor (guinea-pig lung)	Honda *et al.* 1991
Carriers	
Na$^+$/glucose co-transporter (rabbit intestine)	Hediger *et al.* 1987
Na$^+$-independent neutral amino acid transporter (rat kidney)	Tate *et al.* 1992
Channels	
K$^+$ channel (rat brain)	Frech *et al.* 1989
Cl$^-$ channel (Madin Darby canine kidney epithelial cell)	Paulmichl *et al.* 1992

functional form. Masu *et al.* (1987) exploited this property of oocytes to isolate a cDNA clone encoding the receptor. The principle was to make a cDNA library from stomach mRNA, using a vector in which the cDNA was flanked by a promoter for the SP6 or T7 RNA polymerase. This allowed *in vitro* synthesis of mRNA from the mixture of cloned cDNAs in the library.

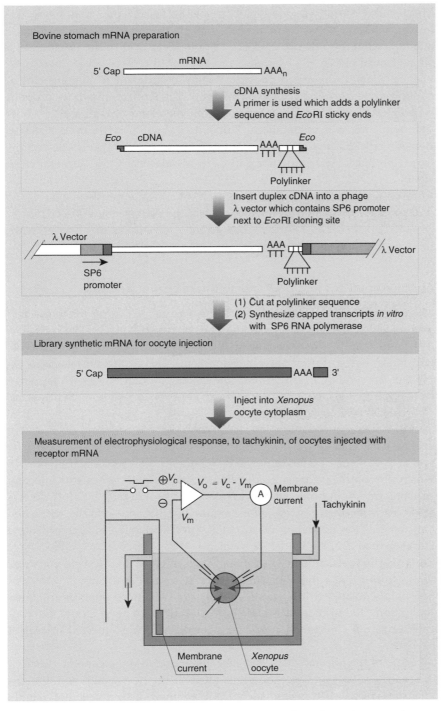

Fig. 16.3 Cloning isolation of bovine substance K receptor through oocyte expression system. (See text for details.)

The receptor clone was identified by testing for receptor expression following injection of synthetic mRNA into the oocyte cytoplasm. Repeated subdivision of the mixture of cDNAs in the library led to the isolation of a single cloned cDNA.

The strategy described above can only be applied to cloning single-subunit proteins, not proteins composed of different subunits, nor proteins whose function in oocytes requires more than one foreign polypeptide. This limitation was overcome by Lubbert *et al.* (1987) who used a hybrid depletion procedure to clone a serotonin receptor cDNA. A related strategy, involving hybrid arrest of translation, was devised by St. Germain *et al.* (1990).

A prerequisite for using the oocyte in functional expression cloning, as described in this section, is a knowledge of the oocyte's own ion channels, carriers and receptors. Endogenous activity may mask, or interfere with, the sought-after function (for a review, see Goldin 1991).

Injection of mRNA and DNA into fertilized eggs of *Xenopus*

Messenger RNA, synthesized and capped *in vitro*, can be micro-injected into de-jellied, fertilized eggs at the one- or two-cell stage. The mRNA is distributed more-or-less evenly in the descendants of the injected cells, and is expressed during early development. This approach has been exploited very widely for examining the developmental effects which result from overexpressing normal or altered gene products (e.g. Smith *et al.* 1993, reviewed in Vize *et al.* 1991).

DNA also may be micro-injected into de-jellied, fertilized eggs, at the one- or two-cell stage. As the embryo develops, exogenous DNA persists, and at least some of it continues to be replicated. In a typical experiment in which a recombinant plasmid carrying *Xenopus* globin genes was injected, the amount of plasmid DNA increased 50- to 100-fold by the gastrula stage. At subsequent stages, the amount of DNA per embryo decreased and most of the persisting DNA co-migrated with high-molecular-weight chromosomal DNA (Bendig & Williams 1983). Etkin *et al.* (1987) have analysed the replication of a variety of DNAs injected into *Xenopus* embryos. It was found that various plasmids increase to different extents. This was not simply related to the size of the plasmid. Some sequences inhibited replication. Replication has also been found to depend upon the concatemerization conformation and number of molecules injected (Marini *et al.* 1989).

Some of the injected DNA becomes integrated into the genome (Rusconi & Schaffner 1981) and may be transmitted to the next generation, but the ability of frogs to transmit such DNA to their progeny is not readily investigated because the time scale of breeding experiments in *Xenopus* (at least 6 months from egg to adult) is long.

Efficient expression of cloned genes, injected into embryos *Xenopus* embryos, depends upon the promoter/enhancer present in the construct. Certain promoters from *Xenopus* itself, that are normally expressed in embryogenesis, are efficiently expressed following injection. Some of these show correct temporal (GS17, a gastrula-specific promoter, Krieg & Melton 1985) or tissue-specific (e.g. Wilson *et al.* 1986, Jonas *et al.* 1989) patterns of expression. However it is commonly found that tissue-specific promoters do not show a good, tissue-specific, restricted, pattern of expression in injected embryos. Another limitation of this type of experiment is that, unlike mRNA which is expressed in all the descendants of injected cells, DNA injected into embryos is expressed in a mosaic pattern. A number of strategies have been investigated to overcome this limitation. These include injecting the DNA into oocytes and then maturing them *in vitro* (with progesterone, which stimulates the oocytes to complete meiosis) before implanting the matured, injected eggs back into a laying female so that they are laid, complete with jelly-coat, and can be fertilized (Holwill *et al.* 1987, Heasman *et al.* 1991). However, no really satisfactory method is yet available. Success in this area would be a great advance in further exploiting the potential of *Xenopus* as an experimental system (Kay & Peng 1991).

Transgenic mammals

The ability to introduce genes into the germline of mammals is one of the greatest technical advances in recent biology. The results of gene manipulation are inherited by the offspring of these animals. All cells of these offspring inherit the introduced gene as part of their genetic make-up. Such animals are said to be *transgenic*. Transgenic mammals have provided a means for studying gene regulation during embryogenesis and in differentiation, for studying the action of oncogenes, and for studying the intricate interactions of cells in the immune system. The whole animal is the ultimate assay system for manipulated genes which direct complex biological processes. In addition, transgenic animals provide exciting possibilities for expressing useful recombinant proteins and for generating precise animal models of human genetic disorders.

Methods for producing transgenic mammals

Several methods for introducing foreign DNA into the germline of mammals have been developed (Fig. 16.4). A fundamental requirement was the availability of techniques for removing fertilized eggs or early embryos, culturing them briefly *in vitro*, and then returning them to foster mothers where further embryogenesis could proceed. This opened the way for the mixing of cells from different embryos, i.e. chimaera production, for introducing pluripotent cells such as ES cells into developing embryos, for micro-injecting DNA, and for infection by retroviruses.

Micro-injection of DNA

In early experiments on micro-injecting DNA into mouse embryos, SV40 DNA was deposited in embryos at the pre-implantation blastocyst (4–30

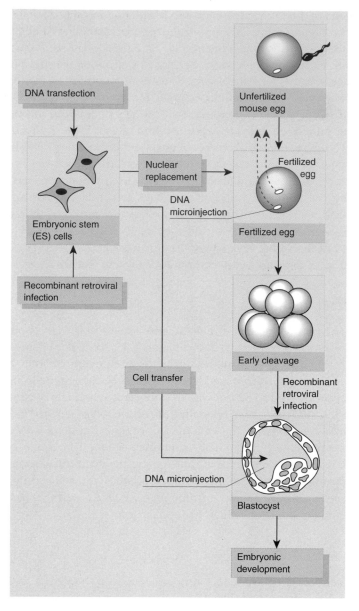

Fig. 16.4 Methods of producing transgenic mice that have been devised or proposed. (After Palmiter & Brinster 1986.)

cell) stage (Jaenisch & Mintz 1974). These embryos were implanted into the uteri of foster mothers and allowed to develop. Some cells of the embryo incorporated DNA into their chromosomes, but the adult animals which resulted were mosaics, with only a proportion of the cells in a tissue containing integrated DNA. However, integration into some germline cells did occur and genetically-defined substrains could be obtained in the next generation. In later experiments viral DNA (cloned proviral Moloney murine leukaemia virus DNA) was injected into the cytoplasm of one-cell embryos (zygotes). Such embryos developed into adults carrying a single inserted copy of the viral DNA in every cell (Harbers et al. 1981).

The procedure which has revolutionized transgenic mouse production is the direct micro-injection of DNA into one of the *pronuclei* of the newly fertilized egg (reviewed by Palmiter & Brinster 1986). The process is shown in Fig. 16.5.

The male pronucleus is contributed by the sperm and, being larger than the female pronucleus, is the one usually chosen for micro-injection. Typically about 2 pl of DNA-containing solution is introduced. The two pronuclei subsequently fuse to form the diploid, zygote nucleus of the fertilized egg. The injected embryos are cultured *in vitro* to morulae or blastocysts and then transferred to pseudopregnant foster mothers (Gordon & Ruddle 1981) (Box 16.2). In practice, between 3 and 40% of the animals developing from these embryos contain copies of the exogenous DNA (Lacy et al. 1983). In such *transgenic* mice the foreign DNA must have been integrated into one of the host chromosomes at an early stage of embryo development. There is usually no mosaicism, and so the foreign DNA is transmitted through the germline. In different transgenic animals the copy number of integrated plasmid sequences differs, ranging from one copy to several hundred in a head-to-tail array, and the chromosomal location differs (Palmiter et al. 1982a, Lacy et al. 1983).

In general, the foreign DNA is stably transmitted for generations.

Other methods: retroviruses, ES cells

As we discussed in Chapter 15, proviral DNA of retroviruses integrates by a precise mechanism into the genome of infected cells. Only a single proviral copy is integrated at a given chromosomal site; the chromosomal site is 'random' but the junctions of the proviral DNA are precise with respect to the viral sequences. Infection of pre-implantation embryos by natural or recombinant retroviruses can lead to germline integration and hence transgenic animals. An advantage of this method is its technical simplicity; eight-cell embryos with the zona pellucida removed are exposed to concentrated virus stock and transferred to foster mothers. However it has the limitations that additional steps are required for construction of recombinant virus, size limitations of foreign DNA, mosaicism of the founder animals because infection occurs after cell division begins, and possible interference of the proviral LTR sequences with the expression of foreign genes. Therefore this approach is promising for some applications but limited in general utility (reviewed by Jaenisch 1988).

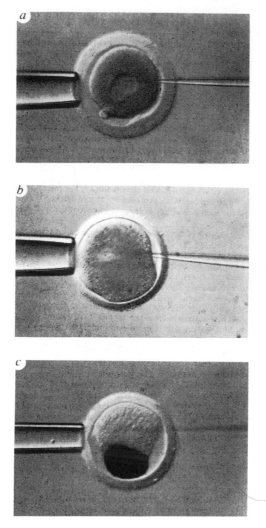

Fig. 16.5 Injection of DNA into the pronucleus of a newly fertilized mouse egg.

A further method for producing transgenic mice was discussed in Chapter 15. This involved the transfection of ES cells, and their subsequent incorporation into mouse embryos. ES cells can also be transduced by recombinant retrovirus, as an alternative to transfection.

Expression of foreign DNA in transgenic mice

The mouse metallothionein gene promoter

The mouse metallothionein-1 (*MMT*) gene encodes a small cysteine-rich polypeptide that binds heavy metals and is thought to be involved in zinc homeostasis and detoxification of heavy metals. The protein is present in many tissues of the mouse, but is most abundant in the liver. Synthesis of

Box 16.2 Production of transgenic mice

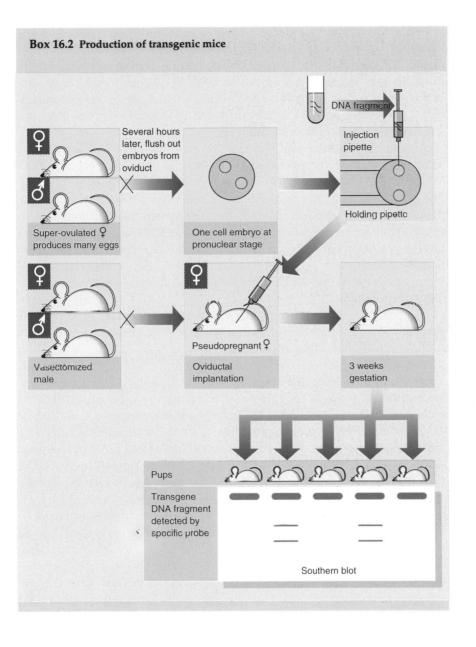

DNA fragment

Injection pipette

Several hours later, flush out embryos from oviduct

Holding pipette

Super-ovulated ♀ produces many eggs

One cell embryo at pronuclear stage

Vasectomized male

Pseudopregnant ♀

Oviductal implantation

3 weeks gestation

Pups

Transgene DNA fragment detected by specific probe

Southern blot

the protein is induced by heavy metals and glucocorticoid hormones. This regulation occurs at the transcriptional level (Durnam & Palmiter 1981).

Brinster *et al*. (1981) constructed plasmids in which the *MMT* gene promoter and upstream sequences had been fused to the coding region of the herpes simplex virus *tk* gene. The thymidine kinase enzyme can be assayed readily and provides a convenient 'reporter' of *MMT* promoter function. The fused *MK* (metallothionein-thymidine kinase) gene was injected into the male pronucleus of newly fertilized eggs which were then incubated *in vitro* in the presence or absence of cadmium ions (Brinster *et al.* 1982). The thymidine kinase activity was found to be induced by the heavy metal. By making a range of deletions of mouse sequences upstream of the *MMT* promoter sequences, the minimum region necessary for inducibility was localized to a stretch of DNA 40–180 nucleotides upstream of the transcription initiation site. Additional sequences that potentiate both basal and induced activities extended to at least 600 base-pairs upstream of the transcription initiation site.

The same *MK* fusion gene was injected into embryos which were raised to transgenic adults (Brinster *et al.* 1981). Most of these mice expressed the *MK* gene and in such mice there were from one to 150 copies of the gene. The reporter activity was inducible by cadmium ions and showed a tissue distribution very similar to that of metallothionein itself (Palmiter *et al.* 1982b). Therefore these experiments showed that DNA sequences necessary for heavy metal induction and tissue specific expression can be functionally dissected in eggs and transgenic mice. For unknown reasons, there was no response to glucocorticoids in either the egg or transgenic mouse experiments.

As expected, the transgenic mice transmitted the *MK* gene to their progeny. The genes were inherited as though they were integrated into a single chromosome. When reporter activity was assayed in these offspring the amount of expression could be very different from that in the parent. Examples of increased, decreased, or even totally extinguished expression were found. In some, but not all, cases the changes in expression correlated with changes in methylation of the gene sequences (Palmiter *et al.* 1982b).

In a dramatic series of experiments, Palmiter *et al.* (1982a) fused the *MMT* promoter to a rat growth hormone genomic DNA. This hybrid gene (*MGH*) was constructed using the same principles as the *MK* fusion. Of 21 mice that developed from micro-injected fertilized eggs, seven carried the *MGH* fusion gene and six of these grew significantly larger than their littermates. The mice were fed zinc to induce transcription of the *MGH* gene, but this did not appear to be absolutely necessary since they showed an accelerated growth rate before being placed on the zinc diet. Mice containing high copy numbers of the *MGH* gene (20–40 copies per cell) had very high concentrations of growth hormone in their serum, some 100–800 times above normal. Such mice grew to almost double the weight of littermates at 74 days old (Fig. 16.6).

These experiments have subsequently been repeated with the *MMT* promoter linked to a human growth hormone gene (Palmiter *et al.* 1983). Synthesis of growth hormone was found to be inducible by heavy metals. The gene was expressed in all tissues examined, but the ratio of human growth hormone mRNA to endogenous metallothionein-1 mRNA varied among different tissues and animals, suggesting that the expression of the fused gene was affected by factors such as cell type and integration site.

These experiments showed the power of the *MMT* promoter in obtaining high levels of expression of any gene to which it is fused. The application of this technology to the production of important polypeptides in farm animals is discussed by Palmiter and his co-authors (1982a). The concen-

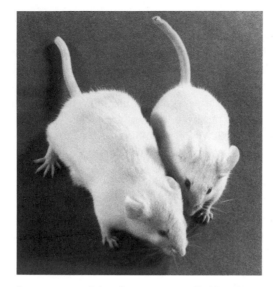

Fig. 16.6 Transgenic mouse containing the mouse metallothionein promoter fused to the rat growth hormone gene. The photograph shows two male mice at about 10 weeks old. The mouse on the left contains the *MGH* gene and weighs 44 g; his sibling without the gene weighs 29 g. In general, mice that express the gene grow 2–3 times as fast as controls and reach a size up to twice the normal. (Photograph by courtesy of Dr R. L. Brinster.)

tration of growth hormone in the transgenic mice was impressively high, much greater than bacterial or cell cultures genetically engineered for growth hormone production. The genetic farming concept is comparable to the practice of raising valuable antisera in animals, except that a single injection of a gene into a fertilized egg would substitute for multiple somatic injections. An added advantage is the heritable nature of genes, but the variability in expression already encountered in the progeny may be problematical here (see Chapter 17).

Gene regulation in transgenic mice

The similarities between the tissue distribution of *MK* expression and normal *MMT* expression encouraged the hope that transgenic mice would provide a general assay for functionally dissecting DNA sequences responsible for tissue-specific or developmental regulation of a variety of genes. But in early experiments correct tissue-specific expression of genes was not often observed. It was soon realised that the presence of bacterial vector sequences is detrimental to correct expression of certain genes including β-globin and α-fetoprotein (Chada *et al.* 1986, Kollias *et al.* 1986, Hammer *et al.* 1987). Other genes appeared to be less sensitive to this effect, and to chromosomal position effects, e.g. immunoglobulin and elastase genes (Storb *et al.* 1984, Swift *et al.* 1984, Davis & MacDonald 1988). Most researchers routinely remove prokaryotic vector sequences to avoid possible adverse effects. A large number of regulated genes have been assayed in transgenic mice. In order to discriminate the products of the injected gene from the endogenous counterpart the gene must be 'marked' in some way such as by fusion to a reporter enzyme or by other modifications. This research allows the dissection of *cis*-acting sequences which are required for correctly regulated expression (Palmiter & Brinster 1986).

In certain cases, variable expression has been observed among offspring in a transgenic line. There are several possible explanations of this, including extreme sensitivity to variables of chromatin structure such as nucleosomal phasing. Variable methylation of the genes is also a likely explanation of differences in expression.

Transgenic mice and cancer

Many oncogene constructs have been introduced into transgenic mice, and have been found to elicit tumorigenic responses. For example, mice that carry the SV40 enhancer and region coding for the large T-antigen, reproducibly develop tumours of the choroid plexus which are derived from the cells lining the ventricles of the brain (Palmiter *et al.* 1985). In similar experiments the *c-myc* oncogene was fused to the LTR of mouse mammary tumour virus. In one transgenic line, females characteristically developed mammary carcinomas during their second or third pregnancy (Stewart *et al.* 1984). In other experiments the *c-myc* oncogene was driven by immunoglobulin enhancers and gave rise to malignant lymphoid tumours in transgenic mice (Adams *et al.* 1985). The evidence from these experiments suggests that the tumours were clonal in origin, i.e. derived from a single transformed cell. In contrast with this, when an oncogenic human *ras* gene was fused to an elastase promoter/enhancer construct, transgenic mice were born with pancreatic neoplasms that appeared to be due to transformation of *all* of the differentiating pancreatic acinar cells (Palmiter & Brinster 1986).

Such transgenic lines as those described, which have a predisposition to specific tumours, are of immense value in the investigation of the events leading to malignant transformation.

Gene 'knock-outs' in transgenic mice

In Chapter 15, ES cell technology was described that enables functional genes to be replaced by non-functional, null, alleles. Experiments of this kind are being performed in ever-increasing numbers as the technology becomes more widely disseminated. The phenotypes of homozygous, null mutant mice provide important clues to the normal function of the gene. Some gene knock-outs have resulted in surprisingly little phenotypic effect, much less severe than might have been expected. For example, *myoD*, whose expression in transfected fibroblasts causes them to differentiate into muscle cells, and which was therefore a good candidate as a key regulator of myogenesis, is not necessary for development of a viable animal (Rudnicki *et al.* 1992). Similarly, the retinoic acid γ-receptor, is not necessary for viable mouse development as defined by the knock-out test

(Lohnes *et al.* 1993), even although this receptor is a necessary component of the pathway for signalling by retinoids, and has a pattern of expression quite distinct from other retinoic acid receptors in embryos. Such observations have prompted speculation that genetic redundancy may be common in development, and may include compensatory upregulation of some members of a gene family when one member is inactivated. An example of this may be the upregulation of *myf-5* in mice lacking *myoD* (Rudnicki *et al.* 1992).

Introduction of cloned genes into the germline of *Drosophila* by micro-injection

P elements of *Drosophila*

P elements are transposable DNA elements which, in certain circumstances, can be highly mobile in the germline of *Drosophila melanogaster*. The subjugation of these sequences as specialized vector molecules in *Drosophila* represents a landmark in modern *Drosophila* genetics. Through the use of P element vectors any DNA sequence can be introduced into the genome of the fly.

P elements are the primary cause of a syndrome of related genetic phenomena called *P–M hybrid dysgenesis* (Bingham *et al.* 1982, Rubin *et al.* 1982). Dysgenesis occurs when males of a P (paternally contributing) strain are mated with females of an M (maternally contributing) strain, but usually not when the reciprocal cross is made. The syndrome is confined mainly to effects of the germline and includes a high rate of mutation, frequent chromosomal aberrations and, in extreme cases, failure to produce any gametes at all.

P strains contain multiple genetic elements, the P elements, which may be dispersed throughout the genome. These P elements do not produce dysgenesis within P strains because transposition is repressed, probably due to the presence of a P-encoded repressor of a P element-specific *transposase* which is also encoded by the P element. However, when a sperm carrying chromosomes harbouring P elements fertilizes an egg of a strain that does not harbour P elements (i.e. an M strain), the P element transposase is temporarily derepressed owing to the absence of repressor. P element transposition occurs at a high frequency and this leads to the dysgenesis syndrome; the high rate of mutation results from the insertion into and consequent disruption of genetic loci.

Several members of the P transposable element family have been cloned and characterized (O'Hare & Rubin 1983). It appears that the prototype is a 2.9-kb element and that other members of the family have arisen by different internal deletion events within this DNA. The elements are characterized by a perfect 31 base-pair inverted terminal repeat. It is likely that this repeat is the site of action of the putative transposase. Three long open-reading frames have been identified in the prototype DNA sequences. One of these open-reading frames encodes the transposase protein. The messenger RNA for this protein is produced by removal of introns from its

Box 16.3 Targeting genes to particular tissues and organs in adult mammals

Although tissue-specific gene expression can be achieved by coupling suitable enhancer/promoter elements to a gene in a transgenic animal, this strategy for targeting gene expression to a particular tissue is not appropriate for gene therapy in man: somatic gene therapy is the preferred option. Much of the research on somatic gene therapy has focused on inherited disorders, but the prospects for applications in infections (notably HIV), and acquired diseases such as cancer and atherosclerosis are exciting.

The first trials did not involve therapy as such, but marking of cells with transferred genes. The cell-marking experiments use retroviral vectors to transfer marker genes (such as *neo*) into cells that will be returned to the patient from whom the cells were obtained. In an example of this, tumour-infiltrating lymphocytes are first removed from patients, marked, grown *in vitro* in large numbers and then returned to the patient. The drug-resistance marker can be used to assess the persistence and homing of the cells after re-infusion. In other marking studies, the marking technique relies on polymer chain reaction (PCR)-based detection of the transferred sequence, rather than selection of a drug-resistance marker.

The first therapeutic gene transfer trials involved ADA (adenosine deaminase) deficiency. Patients have been re-infused with their own T-cells that have been transduced with a retroviral vector that expresses neomycin resistance and human ADA. Following initial success with these trials, a wide variety of other trials have followed (reviewed by Miller 1992).

The following table lists some of the transfection methods that have been investigated for tissue-specific targeting, in man or other mammals.

Cell or tissue	Transfection system
Lymphocytes (T and B cell)	Retroviral
Haematopoietic stem cells	Retroviral
Hepatocytes	Retroviral
Hepatocytes	DNA–asialoglycoprotein complex
Airway epithelium (cotton rats)	Adenovirus (intratracheal instillation)
Brain[1]	Adenovirus (injection *in situ*)
Brain[1]	DNA–liposome complex (injection *in situ*)
Muscle cells	DNA injection, Adenovirus[3]
Arterial wall[2]	Retroviral or DNA–liposome complex

References: (1) reviewed in Neve (1993), (2) Nabel *et al.* (1990), (3) Ragot *et al.* (1993). The remaining examples are reviewed in Miller (1992).

primary transcript. The germline specificity of transposition is due to the fact that splicing out of one of these introns occurs in germ cells but not somatic cells. Laski *et al.* (1986) showed this clearly by making a P element construct in which the intron had been precisely removed. This element, which lacked the intron, showed a high level of somatic transposition activity. Some naturally-occurring short P elements are defective. They cannot encode functional transposase but are transposable in *trans* in the presence of a non-defective P element within the same nucleus.

Spradling & Rubin (1982) devised an approach for introducing the P element DNA into *Drosophila* chromosomes which mimics events taking place during a dysgenic cross. Essentially, a recombinant plasmid which consisted of a 2.9-kb P element together with some flanking *Drosophila* DNA sequences, cloned in the pBR322 vector, was micro-injected into the posterior pole of embryos from an M-type strain. The embryos were injected at the syncytial blastoderm stage. This is a stage of insect development in which the cytoplasm of the multinucleate embryo has not yet become partitioned into individual cells (Fig. 16.7). The posterior pole was chosen because it is the site at which the cytoplasm is first partitioned, resulting in cells that will form the germline. P element DNA introduced in this way became integrated into the genome of one or more posterior pole cells. Because of the multiplicity of such germline precursor cells the integrated P element DNA was expected to be inherited by only some of the progeny of the resulting adult fly. Therefore the progeny of injected embryos were used to set up genetic lines which could be genetically tested for the presence of incorporated P elements.

A substantial proportion of progeny lines were indeed found to contain P elements integrated at a variety of sites in each of the five major chromosomal arms, as revealed by *in situ* hybridization to polytene chromosomes. It may be asked whether integration really does mimic normal P element transposition or whether it is simply some non-specific integration of the micro-injected plasmid. The answer is that integration occurs by a mechanism analogous to transposition. By probing Southern blots of restricted DNA it was found that the integrated P element was not accompanied by the flanking *Drosophila* or pBR322 DNA sequences present in the recombinant plasmid that was micro-injected (Spradling & Rubin 1982). Injected plasmid DNA must presumably have been expressed at some level *before* integration so as to provide transposase activity for integration by the transposition mechanism.

These experiments, therefore, showed that P elements can transpose with a high efficiency from injected plasmid into diverse sites in chromosomes of germline cells. At least one of the integrated P elements in each progeny line remained functional, as evidenced by the hypermutability it caused in subsequent crosses to M strain eggs.

Development of P element as a vector

Rubin and Spradling (1982) exploited their finding that P elements can be artificially introduced in the *Drosophila* genome. A possible strategy for using the P element as a vector would be to attempt to identify a suitable site in the 2.9-kb P element sequence where insertion of foreign DNA could be made without disrupting genes essential for transposition. However, an alternative strategy was favoured. A recombinant plasmid was

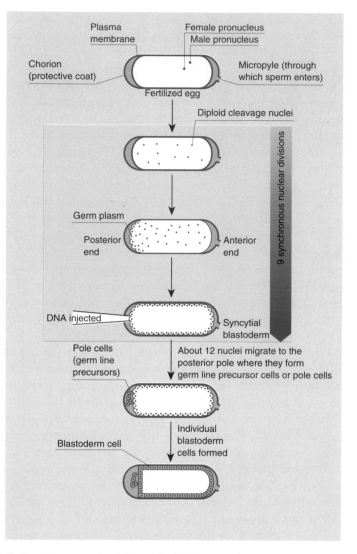

Fig. 16.7 Early embryogenesis of *Drosophila*. DNA injected at the posterior end of the embryo just prior to pole cell formation is incorporated into germline cells.

isolated which comprised a short (1.2 kb), internally deleted member of the P element family together with flanking *Drosophila* sequences, cloned in pBR322. This naturally defective P element cannot encode any of the putative protein products of the 2.9-kb prototype element (O'Hare & Rubin 1983). Target DNA was ligated into the defective P element. The aim was to integrate this recombinant P element into the germline of injected embryos by providing transposase function in *trans*. Two approaches for doing this were tested. In one approach a plasmid carrying the recombinant P element was injected into embryos derived from a P–M dysgenic cross in which transposase activity was therefore expected to be high. This ap-

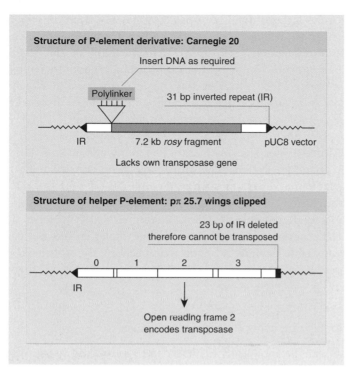

Fig. 16.8 P element derivatives as vector system. (See text for details.)

proach does have the disadvantage that frequent mutations and chromosomal aberrations would also be expected. In the other approach the plasmid carrying recombinant P element was co-injected with a plasmid carrying the non-defective 2.9-kb element.

In the first experiments of this kind, embryos homozygous for a *rosy* mutation were micro-injected with the P element vector containing a wild-type *rosy* gene. Both methods for providing complementing transposase were effective. Rosy$^+$ progeny, recognized by their wild-type eye colour, were obtained from 20 to 50% of injected embryos. The chromosomes of these flies contained one or two copies of the integrated *rosy*$^+$ DNA.

The *rosy* gene is a particularly useful genetic marker. It produces a clearly visible phenotype: Rosy$^-$ flies have brown eyes instead of the characteristic red colour of Rosy$^+$ flies. The *rosy* gene encodes the enzyme xanthine dehydrogenase, which is involved in the production of a precursor of eye pigments. The *rosy* gene is not cell autonomous: expression of *rosy*$^+$ anywhere in the fly, for example in a genetically mosaic fly developing from an injected larva, results in a wild-type eye colour.

A simple modern P element vector is shown in Fig. 16.8 (Rubin & Spradling 1983). It consists of a P element cloned in a bacterial vector, pUC8. Most of the P element has been replaced by a *rosy*$^+$ gene, but the terminal repeats essential for transposition have been retained. The vector includes a polylinker site for inserting foreign sequences. The capacity of such a vector is large: a definite upper limit has not been determined, but inserts of over 40 kb have been successfully introduced into flies by a P element vector (Haenlin *et al.* 1985). Transposition of the recombinant vector into the genome of injected larvae is brought about by co-injecting a helper P element which provides transposase in *trans*, but which cannot transpose itself because of a deletion in one of its terminal inverted repeats. Such an element is referred to as a *wings-clipped* element.

Pπ25.7 wc is an example of a *wings-clipped* helper element (Fig. 16.8) (Karess & Rubin 1984). The *rosy* marker has many advantages. But it is not selectable. Therefore P element vectors have been constructed which include selectable marker genes such as alcohol dehydrogenase, *Adh* (Adh$^+$ flies can be selected on food containing 6% ethanol), or *neo*, which confers resistance to food containing G418 (Goldberg *et al.* 1983, Steller & Pirotta 1985). These markers have the disadvantage that they do not confer an immediately visible phenotype, hence complicating the maintenance of fly stocks. However, they can be combined in a vector with *rosy*$^+$ to give both a visible and a selectable marker.

Transposon tagging using P element derivatives

P elements can be mobilized to insert into genes and cause insertional mutagenesis. The mutated gene can then be cloned by virtue of having been tagged with P element sequences. Systems for controlled transposon mutagenesis have been developed (reviewed by Cooley *et al.* 1988). Two fly strains are involved. One strain contains the 'mutator', a defective P element carrying useful marker genes, which can be mobilized when provided with a source of transposase. A second strain contains a *wings-clipped* type of element which provides transposase in *trans*. These are called 'jumpstarter' elements. During a controlled mutagenesis screen, a single jumpstarter element is crossed into a mutator-containing strain. Transposition occurs. In subsequent generations the mutator element is stabilized when the chromosome bearing the jumpstarter element segregates from the target chromosome into which the mutator element has been inserted.

Recovery of the P element and associated gene sequences from the inactivated gene is most easily accomplished by a plasmid rescue technique (see Chapter 15). For this approach, an *E. coli* plasmid origin of replication must be included in the mutator element.

Enhancer trapping

Enhancer trapping is an important application of P element derivatives. In principle, the method employs a *lacZ* reporter construct in which the reporter gene is transcribed from a weak promoter. Expression from the promoter is weak because the promoter lacks an enhancer to stimulate its transcriptional activity. P element-mediated transposition is used to transpose the construct into many different genomic positions in separate fly lines. In some flies, by chance, the construct is transposed to a position where it comes under the influence of an enhancer that activates transcription from the weak promoter. It is often found in practice, by using a histochemical stain for *lacZ* activity, that the pattern of expression shows cell specificity. Sometimes the pattern of expression is remarkably refined and detailed (O'Kane & Gehring 1987). It is conjectured that the pattern of *lacZ* expression reflects the cell-type specificity of the enhancer. Presumably an endogenous gene, located within range of the enhancer's effect, has the same pattern of expression as the reporter. This conjecture is known to be valid in some cases, but it appears not to be always the case that the reporter expression exactly matches an endogenous gene.

By including a gene such as ampicillin resistance in the transposed construct, and an *E. coli* plasmid origin of replication, the chromosomal region surrounding the construct can be isolated simply, by a plasmid rescue procedure (see Chapter 15). DNA sequences isolated in this way can be the starting point for a chromosome walk leading to isolation of the enhancer, and the endogenous gene.

The main use of enhancer traps is in fact not the cloning of enhancers. Rather, the cell-specific expression that is revealed can be harnessed in other ways. For example, instead of driving expression of a *lacZ* reporter, the principle could be applied to the expression of a toxic gene (such as the toxic B-chains of diphtheria toxin, or ricin), leading to cell death and thus ablation of specific cell lineages in the fly. This has enormous potential in developmental studies, for example in studies of the development of the nervous system (O'Kane & Moffat 1992).

In order to facilitate the use of enhancer trapping as a general method for driving cell-specific expression, a modified strategy has been developed. This depends upon the Gal4 transcription factor, originally from yeast, *Saccharomyces cerevisiae* (see p. 13). The Gal4 transcription factor binds to a response element in the promoter region of responsive genes, and by binding turns transcription on. The enhancer trapping principle in Fig. 16.9 is modified so that the *lacZ* reporter gene is replaced by *gal4*. No *lacZ* reporter is included. However, the pattern of expression of Gal4 can be

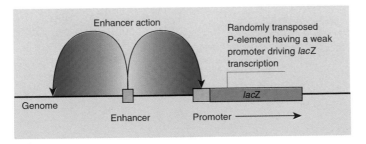

Fig. 16.9 Enhancer trapping. A reporter construct consists of *lacZ* linked to a weak promoter, which requires an enhancer for significant transcriptional activity. P element-mediated transposition is used to insert the construct at random sites in the fly genome. When the promoter is inserted within the active range of an enhancer, expression of *lacZ* can be detected.

revealed by introducing a chromosome containing a *lacZ* gene coupled to a promoter with a Gal4 response element. The chromosome can be introduced by a suitable cross between fly strains. The revealed pattern of *lacZ* expression therefore reflects the pattern of Gal4 expression. The beauty of this system is that a bank of fly stocks with different trapped enhancers can be built up, each with a defined pattern of Gal4 expression. Once the patterns of Gal4 expression are known, crosses can be performed to introduce chromosomes containing constructs in which any desired gene is coupled to a Gal4-dependent promoter, giving a particular cell-specific pattern of expression.

17 The impact of recombinant DNA technology: the generation of novelty

Introduction

Genetics is the fundamental biological science, for without genes there is no life. Thus a full understanding of any biological process can be achieved only when there has been a detailed analysis of gene structure and function. Classically, this analysis was undertaken by making mutants, studying their properties, mapping them and generating hypotheses for further testing. The archetypal example of the success of this approach was the development of the operon concept from studies on the regulation of lactose metabolism. What recombinant DNA technology has done is to add new weapons to the armoury of the geneticist. For example, hypothesizing what has happened at the DNA level is no longer necessary: the genes now can be cloned and sequenced and the location and nature of the mutation identified precisely. More important, the cloned gene can be introduced into exotic hosts to determine the effect it has. In some instances designer changes are made to microbes, plants and animals. In other cases the introduced genes serendipitously prove beneficial. Examples of both are described later.

Mutation still remains an essential tool for the geneticist but instead of seeking mutants with interesting phenotypes, mutations of different types are introduced at will and at pre-determined locations. The net effect is that

Table 17.1 Applications for labelled hybridization probes

Southern blots	Detection of gel-fractionated DNA molecules following transfer to a membrane
Northern blots	Detection of gel-fractionated RNA molecules following transfer to a membrane
Dot blots	Detection of unfractionated DNA or RNA molecules immobilized on a membrane
Colony/plaque blots	Detection of DNA released from lysed bacteria or phage and immobilized on a membrane
S1/RNase mapping	Positional mapping of termini of target molecules
In situ hybridization	Detection of DNA or RNA molecules in cytological preparations

the whole analytical process is speeded up by several orders of magnitude. This in turn has opened up the field of protein engineering where startling progress has been made as described later (p. 382).

Many of the techniques of gene manipulation depend on the hybridization of a nucleic acid probe to a target DNA or RNA sequence. In previous chapters we have described many applications of the technology and some examples are given in Table 17.1. Of equal importance are the diagnostic applications of probe technology and, since these have not been dealt with elsewhere, they will be covered in the discussion that follows. These applications are not just of interest for their commercial value: the principles behind them have much to offer the research scientist.

Overall, gene manipulation has provided novel solutions to experimental problems in biology. These solutions have led to novel products. The development of these products has raised novel problems — and so *ad infinitum*. Some of this novelty is captured in the sections which follow.

Recombinant DNA technology and the new diagnostics

The production of reagents for the diagnosis of biological disorders and infectious disease is a major industry. On a global scale revenue from sales is estimated at many billions of dollars. Traditionally diagnostic reagents have taken two forms. These are biochemical reagents for assaying specific enzymes and antibodies for detecting specific proteins by such techniques as immunofluorescence, radioimmunoassay and enzyme-linked immunosorbent assay. Although the former are used principally in clinical biochemistry, the latter have applications in almost all industries, e.g. health care, food and agriculture. Recently a number of new types of diagnostic reagent have been developed of which the most widely used are nucleic acid probes. Although based on an old technique, nucleic acid hybridization, development of these reagents awaited the availability of milligram quantities of specific genes. The ready provision of such large amounts of gene sequences is easy if the sequences have been cloned in bacteria since a 1 litre culture can yield up to 0.5 mg of plasmid DNA. The development of amplification methods, particularly the polymerase chain reaction (see p. 178), has greatly reduced the amount of target DNA which can be detected thereby increasing the sensitivity of the test. Finally, the availability of robust and sensitive non-radioactive detection methods has widened the appeal of the method (Wolcott 1992).

There are three basic applications of probes. The first of these is the detection of specific nucleic acid sequences as in the identification of microorganisms in clinical specimens. The sequences detected are usually large (i.e. an entire gene or cluster of genes) and considerable mismatching between the target and probe sequences is expected. In the second application, most commonly found in clinical genetics, the objective is to detect changes to specific sequences. These changes can include deletions or rearrangements in the test DNA but the methods used must be precise enough to routinely detect single base changes. The third type of application is a forensic one. Here the need is to detect the size distribution of DNA fragments bearing repetitive DNA sequences. This technique is referred to as *DNA profiling* or *fingerprinting*. In this technique a considerable degree of sequence mismatch is expected.

Detection of sequences at the gross level

There are many applications of hybridization technology where all that is required is to determine if a particular nucleic acid sequence is present. In almost all instances the requirement is to detect a specific microorganism. The greatest potential is seen in clinical microbiology (Tenover 1988) where a single technology (hybridization) could replace a whole variety of disparate test procedures. For example, a patient with a disorder of the gastrointestinal tract could be infected with any one of the infectious agents shown in Table 17.2. In conventional diagnostic practice,

Table 17.2 Pathogens causing infection of the gastrointestinal tract

Bacteria
Salmonella spp.
Shigella spp.
Yersinia enterocolitica
Enterotoxigenic *E. coli*
Clostridium difficile
Clostridium welchii
Campylobacter spp.
Aeromonas spp.
Vibrio spp.

Viruses
Rotaviruses
Enteroviruses
Enteric adenoviruses

Protozoa
Entamoeba histolytica
Giardia lamblia

identification of the causative organism would require cultivation of stool samples on a variety of different media in a variety of different ways, microscopy, animal cell culture and immunoassay. Clearly, hybridization of the test sample with a battery of probes should be much simpler. Although the development of nucleic acid probes now is commonplace, the selection of probes with the correct diagnostic potential is much more difficult (see, for example, Rotbart 1991).

There are other advantages in the use of hybridization for the diagnosis of infectious disease. First, if sufficient of the infectious agent is present then no cultivation of the microbe is necessary. This eliminates hazards and should shorten the time required for identification of the pathogen. The polymerase chain reaction (PCR) is particularly powerful in this respect. For example, Ou *et al.* (1988) were able to detect HIV-1 DNA in peripheral blood mononuclear cells in less than 1 day whereas virus isolation took 3–4 weeks. However, PCR does have some shortcomings. The system is susceptible to contamination with extraneous DNA fragments that could be amplified along with the sample. In the clinical laboratory it is essential that such contamination is eliminated or at least minimized. This invariably means many specialized disinfection procedures (Wolcott 1992) as well as dedicated staff and work areas. Although PCR has been semi-automated the technique still requires a significant amount of labour and the protocols are not yet sufficiently developed and reproducible for many assays.

A second advantage of hybridization technology is that it can be used even when the target organism cannot be cultured. This is often the case with viral infections, and the current diagnostic method is to look for rising titres of antibody in the blood. The method will also work with latent infections when no antibody may be present. A third advantage is that a single probe can identify all of the serotypes of an infectious agent whereas a single antibody may not. Finally, it is easy to clone a gene known to be involved in pathogenicity and to develop a suitable probe. Generating an antibody to the product of that gene can be much more difficult.

An important diagnostic tool is *in situ* hybridization. In most instances *in situ* hybridization is carried out on formalin-fixed, paraffin-embedded tissues. This technique has proven particularly useful for detection of viral pathogens (Brigatti *et al.* 1983). It allows us to examine the tissue first by traditional staining methods, such as haematoxylin and eosin staining, and then to correlate the cytopathology seen with the presence of infectious agents as shown by a non-radioactive probe.

Hybridization as a diagnostic tool is not restricted to clinical microbiology. There are many applications in plant pathology (Hull & Al-Hakim 1988). For example, the identification of viruses is important in the prediction of plant diseases in annual crops, the prevention of infection in planting stock, in monitoring disease control methods and in diagnosing disease in plants held in quarantine. Hybridization also is being used as a tool in the study of microbial ecology (Holben & Tiedje 1988, Stahl *et al.* 1988).

A virus-based diagnostic test for pathogens

As indicated earlier, one of the disadvantages of nucleic acid probes is that they can readily detect the presence of microbial nucleic acids but cannot determine if the source was alive or dead. This is of particular concern to the food industry which wants rapid methods for the detection and enumeration of *viable* microbes in food which may have been heat treated. One such method makes use of the fact that bacteriophages can only replicate in viable bacteria. The bacteriophage genome is modified by the inclusion of *lux* genes encoding bioluminescence (Stewart *et al.* 1990). When such bacteriophages infect cells, the *lux* genes are expressed and light is emitted (Fig. 17.1). The light output can be measured in a simple luminometer and is proportional to the number of bacteria infected. It should be noted that, since multiple copies of the phage genome are present inside infected cells, there is a built-in amplification system.

The detection of genetic disorders

There are several hundred recognized genetic diseases in man which result from single recessive mutations. In some of these a protein product that is defective or absent has been identified but in many others the nature of the mutation is unknown. For many of these genetic diseases there is no definitive treatment although in some cases human gene therapy may become possible (see p. 405). Their prevention is the current strategy in many countries. Primary prevention by identifying heterozygous carriers and dissuading them from reproducing with other carriers is not practicable, and where it has been attempted it has failed. Usually the discovery that both partners are carriers is made only after the birth of an affected child. When a pregnant woman has been identified as at risk, as indicated by a previous birth or other factors, the approach is to offer antenatal diagnosis and the possibility of an abortion.

An essential prerequisite for prenatal diagnosis is the availability of fetal DNA. Fetal cells can be obtained by amniocentesis (Orkin 1982) but this method is not entirely satisfactory for it cannot be carried out early in pregnancy. An alternative approach is to obtain fetal DNA from biopsies of trophoblastic villi in the first trimester of pregnancy (Williamson *et al.* 1981). In this technique some villi of the trophoblast – this is an external part of the human embryo which functions in implantation and becomes part of the placenta – are biopsied with the aid of an endoscope passed through the cervix of the uterus. Up to 100 μg of pure fetal DNA can be obtained between the 6th and 10th weeks of pregnancy (Old *et al.* 1982a). The efficiency of these techniques has been greatly improved with the advent of the polymerase chain reaction. First, no longer is it necessary to grow fetal cells in culture following amniocentesis in order to isolate sufficient DNA for analysis (Bugawan *et al.* 1988). Second, amplification can be done without purification of the cellular DNA (Kogan *et al.* 1987). This

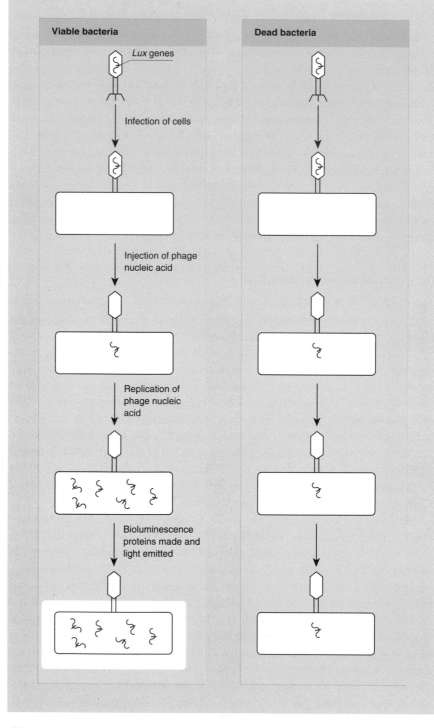

Fig. 17.1 The detection of viable bacteria using bacteriophages genetically-engineered to emit light following infection of susceptible cells. See text for details.

improvement circumvents losses of DNA during purification and leads to more rapid results. In their study Kogan *et al.* (1987) were able to determine the sex of the fetus by using primers surrounding a DNA segment specific to the Y chromosome. Amplified 'male' DNA could be directly visualized on a gel.

Of all genetic diseases the inherited haemoglobin disorders have been the most extensively studied at the DNA level. In what follows, haemoglobinopathies will be taken as examples. Their antenatal diagnosis by recombinant DNA techniques has served as a prototype for other genetic disorders (Weatherall & Old 1983). Clinically the most important haemoglobinopathies are sickle-cell anaemia and the thalassaemias. The latter are a group of disorders in which there is an imbalance in the synthesis of globin chains due to the low, or totally absent, synthesis of one of them. Many α-thalassaemias are caused by gene deletions, although several non-deletion forms have been identified (Weatherall & Clegg 1982). The β-thalassaemias are complex and more than 20 different molecular lesions have been identified. Most are nonsense or frameshift mutations in exons of the β-globin gene. Others are point mutations affecting transcript processing, or point mutations affecting the promoter. Only one form has been identified as due to a major deletion of the β-globin gene (Weatherall & Clegg 1982).

Fetal DNA analysis

Clearly, if a mutation either removes or produces a restriction enzyme site in genomic DNA, this can be used as a marker for the presence or absence of the defect. The mutation from GAG to GTG in sickle-cell anaemia eliminates a restriction site for the enzyme *Dde*I (CTNAG) or the enzyme *Mst*II (CCTNAGG) (Chang & Kan 1981, Orkin *et al.* 1982). The mutation can therefore be detected by digesting mutant and normal DNA with the restriction enzyme and performing a Southern-blot hybridization with a cloned β-globin DNA probe (Fig. 17.2). Such an approach is applicable only to those disorders where there is an alteration in a restriction site, or where a major deletion or rearrangement alters the restriction pattern. It is not applicable to most β-thalassaemias.

There are many polymorphic restriction sites scattered throughout the β-globin gene cluster (Weatherall & Old 1983). These are revealed as restriction fragment-length polymorphisms in Southern-blot experiments. They can be used as linkage markers for antenatal diagnosis, i.e. the close physical linkage with a β-globin gene will mean that the polymorphic site will trace the inheritance of that gene. These polymorphisms can be used in two ways. First, some polymorphisms are linked to specific globin mutations; for example in the USA among the black population the sickle mutation is associated 60% of the time with a polymorphic mutation near the gene that eliminates a *Hpa*I site (Kan & Dozy 1978). The linkage could form the basis of diagnosis, but is inferior to the direct analysis with *Mst*II. Examples of such associations (so-called linkage disequilibrium) are rare. A

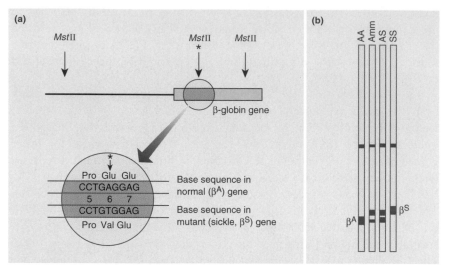

Fig. 17.2 Antenatal detection of sickle cell genes. Normal individuals are homozygous for the β^A allele, while sufferers from sickle-cell anaemia are homozygous for the β^S allele. Heterozygous individuals have the genotype $\beta^A\beta^S$. In sickle-cell anaemia, the 6th amino acid of β-globin is changed from glutamate to valine. (a) Location of recognition sequences for restriction endonuclease *Mst*II in and around the β-globin gene. The change of A → T in codon 6 of the β-globin gene destroys the recognition site (CCTGAGG) for *Mst*II as indicated by the asterisk. (b) Electrophoretic separation of *Mst*II-generated fragments of human control DNAs (AA, AS, SS) and DNA from amniocytes (Amn). After Southern blotting and probing with a cloned β-globin gene, the normal gene and the sickle gene can be clearly distinguished. Examination of the pattern for the amniocyte DNA indicates that the fetus has the genotype $\beta^A\beta^S$, i.e. it is heterozygous.

second approach is required in which it is necessary to establish linkages between polymorphic restriction sites and a particular β-globin gene mutation by carrying out a family study before antenatal diagnosis (Fig. 17.3). This is not always possible, and in any case suitable polymorphic markers may not be present. It has been estimated that this approach is feasible in no more than 50% of β-thalassaemia cases in the UK (Weatherall & Old 1983).

A very powerful and direct approach to analysing point mutations has been devised by Conner *et al.* (1983) who synthesized two 19-mer oligonucleotides, one of which was complementary to the amino-terminal region of the normal β-globin (β^A) gene, and one of which was complementary to the sickle cell β-globin gene (β^S). These oligonucleotides were radiolabelled and used to probe Southern blots. Under appropriate conditions, the probes could distinguish the normal and mutant alleles. The DNA from normal homozygotes only hybridized with the β^A probe, and DNA from sickle-cell homozygotes only hybridized with the β^S probe. DNA of heterozygotes hybridized with both probes. These experiments, therefore, showed that oligonucleotide hybridization probes can discriminate between a fully complementary DNA and one containing a single

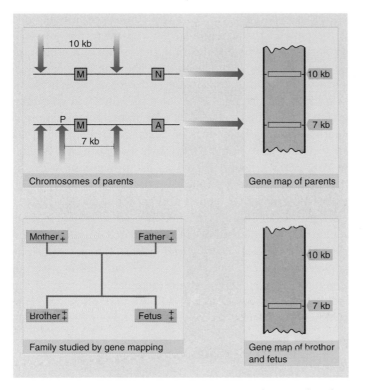

Fig. 17.3 An example of prenatal diagnosis using restriction fragment-length polymorphism (RFLP) linkage analysis. The parents are both carriers for a deleterious gene (A): one of their chromosomes carries this determinant, the other its normal allele (N). One of the parental chromosomes carries a polymorphic restriction enzyme site P which is close enough to A or N so that they will not be separated in successive generations. On the chromosome which does not contain this site (−) a particular restriction enzyme cuts out a piece of DNA 10 kb long which contains another locus (M) for which we have a radioactive probe. On the chromosome containing the polymorphism (+) a single base change produces a new site and hence the DNA fragment containing locus M is now only 7 kb. On gene mapping of the parents' DNA using probe M we see two bands representing either the + or − chromosomes. A previously born child had received the deleterious gene A from both parents and on mapping we find that it has the + + chromosome arrangement, i.e. only a single 7 kb band. Hence the mutation must be on the + chromosome in both parents. To identify the disease in a fetus in subsequent pregnancies we will be looking for an identical pattern, i.e. the 7 kb band only. (Reproduced courtesy of Professor D. Weatherall and Oxford University Press.)

mismatched base. Similar results have subsequently been obtained with a point mutation in the α-antitrypsin gene which is implicated in pulmonary emphysema (Cox *et al.* 1985). The generality of this approach is very impressive. It should be applicable to other genetic disorders provided that the nucleotide sequences around the mutation site, which could be a substitution, insertion or deletion, can be established.

The current state of clinical genetics (see Weatherall 1991) is such that pregnant women and newborn infants can be screened for a wide range of genetic and congenital disorders. This will almost certainly be followed by screening for genetic risk (prognostics). However, the advance of the technology is raising a whole series of social, moral and ethical questions. Where pre-natal screening is practised then abortion is a possible outcome. On its own this is a contentious issue and centres on whether or not a foetus has a 'right to life'. With the recent identification of evidence for a homosexuality locus (Hamer *et al.* 1993) the issues become more complex. What is 'normal' and will eugenics be practised in the future? Pandora's box is now well and truly open. Prognostics also raise many fears particularly now that private health insurance is growing in many countries concomitant with a decline in publicly-funded medical care. Once an individual has been genetically screened and shown to have a high chance of developing a particular disease in later life, what are their prospects of purchasing life or medical insurance? Unfortunately this is not an imaginary scenario: it has happened already and offsets the potential benefits of prognostic screening. Last, but not least, a major issue associated with genetic screening is the inadequate and inequitable delivery of the service within and between different countries. All too often any cost-benefit analysis which is done lacks rigour (Modell & Kuliev 1993).

DNA fingerprinting (DNA profiling)

Much of the human genome cannot vary greatly between individuals because it has an essential coding function. In non-coding regions this requirement does not exist and the DNA sequence can accommodate changes. One change which does occur is the tandem repetition of DNA sequences. At each locus the sequence that is repeated will probably be unique but tends to be GC-rich and 9–40 bp in length. The DNA at such loci often involves thousands of tandem repeats causing an abnormal base composition and hence the term *mini-satellite* is used to define these regions. Mini-satellites often are observed to be *hypervariable*, that is, they differ greatly between unrelated individuals. This polymorphism is due to the variation in the number of repeats arising from the loss or gain of the repeat sequence through mutation. Consequently, an alternative name for these loci is *variable number tandem repeats* or VNTRs.

Many repeated sequences have evolved from a common core ancestral sequence or set of core sequences. Thus each hypervariable locus has some sequence similarity with many other loci. This can be demonstrated by using any such locus as a probe under low-stringency hybridization conditions such that the probe acts generically detecting any sequence with sufficient sequence similarity to permit hybridization (Jeffreys *et al.* 1985a). The multi-locus probes described by Jeffreys and colleagues (Fig. 17.4) were derived from a tandemly repeated sequence within an intron of the myoglobin gene and shown subsequently to hybridize to other autosomal genes. Other suitable probes include the insulin, inter-ζ and a α-globin 3′ HVRs (Fowler *et al.* 1988) and DYXS15 (Simmler *et al.* 1987). Under the right conditions it is possible to use bacteriophage M13 DNA as a probe

	A
Core sequence	G G A G G T G G G C A G G A G G
Probe 33.6	[(A G G G C T G G A G G)₃]₁₈
Probe 33.15	(A G A G G T G G G C A G G T G G)₂₉
Probe 33.5	C
	(G G G A G T G G G C A G G A G G)₁₄

Fig. 17.4 Probes used for
DNA fingerprinting.

(Vassart *et al.* 1987) the effective sequence being two clusters of 15 base repeats within the protein III gene.

Analysis of a DNA sample involves cleavage of total DNA with restriction enzymes which cut on either side of the mini-satellite but not within the repeated sequence. Thus the restriction fragment length will depend on the number of repeats between sites and generate DNA fragment-length polymorphisms. The DNA fragments are separated by electrophoresis and after Southern blotting are detected with one of the probes described above. This results in an individual-specific *DNA fingerprint* or *DNA profile*.

DNA fingerprinting has been used in pedigree analysis in cats and dogs (Jeffreys & Morton 1987) and to monitor the behaviour and breeding success of bird populations (Burke & Bruford 1987). It also has been used to confirm cell line authenticity in animal cell cultures (Devor *et al.* 1988, Stacey *et al.* 1992). In humans it has been used in anthropological studies charting the origins and migration of ancient populations (Cherfas 1991). However, the technique has been most extensively used in forensic science and for other medico-legal applications (Gill *et al.* 1985, 1987, Hill & Jeffreys 1985). Of particular value, suitable high-molecular-weight DNA can be isolated from such stains as blood and semen made on clothing several years previously. Also, sperm nuclei can be separated from the vaginal cellular debris present in semen-contaminated vaginal swabs taken from rape victims.

The first criminal court case to use DNA fingerprinting evidence was in Bristol, UK, in 1987. In this case DNA provided the link between a burglary and a rape. The following year, DNA fingerprinting evidence was used in the USA. DNA evidence is now widely used in Western countries but the UK and the USA and the main proponents. In the UK it also is used in immigration cases (see Box 17.1). It is worth noting that DNA evidence also has been used to prove innocence (see, for example, Gill & Werrett 1987) as well as guilt. Indeed, it is easier to prove innocence than guilt as shown by the recent bitter debate in the USA over interpretation of data (Chakraborty & Kidd 1991, Lewontin & Hartl 1991, Roberts 1991, 1992). Proponents of DNA fingerprinting claim that the probability of two DNA samples matching by chance is very low, somewhere between 10^{-6} and 10^{-15}. Lewontin & Hartl (1991) have challenged such calculations since they assume random mating between individuals. They say this ignores evidence that different ethnic groups actually are made up of multiple subpopulations. It should be noted that the challenge is not to the value of the technique, but the way in which the results are interpreted.

Box 17.1 Use of DNA fingerprinting in an immigration test case

In 1984, a Ghanaian boy was refused entry into Britain because the immigration authorities were not satisfied that the woman claiming him as her son was in fact his mother. Analysis of serum proteins and erythrocyte antigens and enzymes showed that the alleged mother and son were related but could not determine whether the woman was the boy's mother or aunt. To complicate matters, the father was not available for analysis nor was the mother certain of the boy's paternity. DNA fingerprints from blood samples taken from the mother and three children who were undisputedly hers as well as the alleged son were prepared by Southern blot hybridization to two of the mini-satellite probes shown in Fig. B17.1. Although the father was absent, most of his DNA fingerprint could be reconstructed from paternal-specific DNA fragments present in at least one of the three undisputed

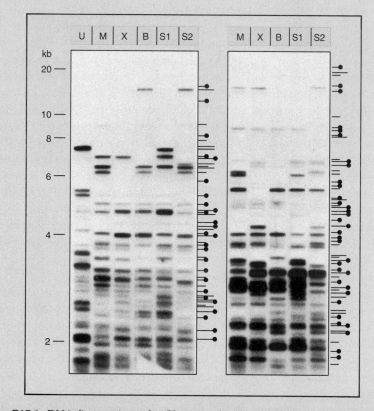

Fig. B17.1 DNA fingerprints of a Ghanaian family involved in an immigration dispute. Fingerprints of blood DNA are shown for the mother (M), the boy in dispute (X), his brother (B), sisters (S1, S2) and an unrelated individual (U). Fragments present in the mother's (M) DNA are indicated by a short horizontal line (to the right of each fingerprint); paternal fragments absent from M but present in at least one of the undisputed siblings (B, S1, S2) are marked with a long line. Maternal and paternal fragments transmitted to X are shown with a dot. (Photo courtesy of Dr A Jeffreys and the editor of *Nature*.)

continued

Box 17.1 *continued*

siblings but absent from the mother. The DNA fingerprint of the alleged son contained 61 scorable fragments, all of which were present in the mother and/or at least one of the siblings. Analysis of the data showed the following.

• The probability that either the mother or the father by chance possess all 61 of the alleged son's bands is 7×10^{-22}. Clearly the alleged son is part of the family.

• There were 25 maternal-specific fragments in the 61 identified in the alleged son and the chance probability of this is 2×10^{-15}. Thus the mother and alleged son are related.

• If the alleged mother of the boy in question is in fact a maternal aunt, the chance of her sharing the 25 maternal-specific fragments with her sister is 6×10^{-6}.

When presented with the above data (Jeffreys et al. 1985b), as well as results from conventional marker analysis, the immigration authorities allowed the boy residence in Britain. In a similar kind of investigation a man originally charged with murder was shown to be innocent (Gill & Werrett 1987).

The variable control region of the mitochondrial genome contains a very high number of variant base substitutions. Coupled with its high copy number, mitochondrial DNA (mt DNA) is suited for the analysis of severely degraded DNA from decomposed remains. Amplified mt DNA was used to confirm that skeletons found in Ekaterinburg, Russia, were the remains of the last Tsar and his family (Gill *et al.* 1994).

Mini-satellite alleles vary not only in the number of copies of the repeat but also in the sequence of the repeat. That is, along any one individual mini-satellite the individual repeats may have different sequences. With one particular hypervariable locus, D1S8, two classes of repeat unit have been identified that differ by a single base substitution which creates or destroys a *Hae*III restriction site. Jeffreys *et al.* (1991) have developed a method for displaying the sequence pattern of these two repeat units along mini-satellite alleles. This produces DNA profiles which can be digitized and stored in computer databases for forensic applications.

DNA profiling now makes much more use of *single-locus probes* under conditions of high stringency (Wong *et al.* 1986, Nakamura *et al.* 1987). Thus these probes detect two DNA bands amongst DNA fragments separated by size, one band corresponding to each allele at that locus (Fig. 17.5). In general, four single-locus probes are used for forensic analysis. The chance of two individuals having the same pattern of eight bands is 10^{-6} and in paternity suits the chance of two males having the same pattern is 10^{-3}. Not surprisingly, the use of single-locus probes has been challenged in the courts and, again, it is easier to prove innocence rather than guilt. The technical and legal issues surrounding the use of these probes have been detailed by Debenham (1992).

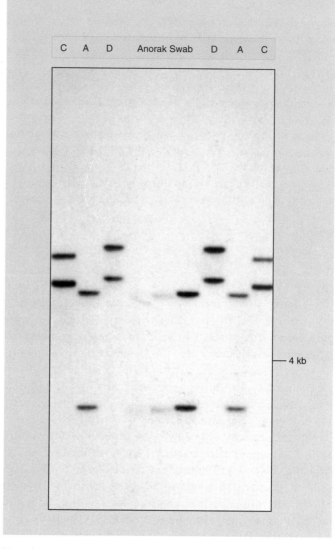

Fig. 17.5 Use of a single locus probe to determine the identity of a rapist. Semen was extracted from an anorak and a vaginal swab. The victim's profile is in track D and that of two suspects in tracks A and C. The profile matches individual A. (Photo courtesy of Dr P. Gill.)

A number of different techniques have been developed for DNA profiling of plants (Rafalski & Tingey 1993). However, the requirements of a diagnostics programme for agricultural genetics are quite different from those for forensic purposes. In forensic applications one needs to have extremely high confidence in the accuracy of each assay but relatively few assays are required. The value of each assay thus is very high. By contrast plant breeding programmes require an assay that is inexpensive and automatable because in a single season 100 000 plants may need to be assayed for multiple loci by each breeder.

The use of DNA analysis in forensics, disease and animal and plant identification has been reviewed by Alford and Caskey (1994).

The generation of novel proteins: protein engineering

One of the most exciting aspects of recombinant DNA technology is that it permits the design, development and isolation of proteins with improved operating characteristics and even completely novel proteins. The simplest example of protein engineering involves site-directed mutagenesis to alter key residues, as originally shown by Winter and colleagues (Winter *et al.* 1982, Wilkinson *et al.* 1984). From a detailed knowledge of the enzyme tyrosyl-tRNA synthetase from *Bacillus stearothermophilus*, including its crystal structure, they were able to predict point mutations in the gene which should increase the enzyme's affinity for the substrate ATP. These changes were introduced and in one case a single amino acid change improved the affinity for ATP by a factor of 100. Using a similar approach the stability of an enzyme can be increased. Thus Perry and Wetzel (1984) were able to increase the thermostability of T4 lysozyme by the introduction of a disulphide bond. However, although new cysteine residues can be introduced at will, they will not necessarily lead to increased thermal stability (Wetzel *et al.* 1988).

Increasing the bioactivity of proteins

Many human proteins are being tested as potential therapeutic agents and a number of them already are commercially available. Protein engineering now is being used to generate second-generation variants with improved pharmacokinetics, structure, stability and bioavailability (Bristow 1993). For example, in the neutral solutions used for therapy, insulin is mostly assembled as zinc-containing hexamers. This self-association may limit absorption. By making single amino acid substitutions Brange *et al.* (1988) were able to generate insulins which are essentially monomeric at pharmaceutical concentrations. Not only have these insulins preserved their biological activity, they are absorbed two to three times faster. Similarly, replacing an asparagine residue with glutamine altered the glycosylation pattern of tissue plasminogen activator. This in turn significantly increased the circulatory half-life which in the native enzyme is only 5 min (Lau *et al.* 1987). Proteins also can be engineered to be resistant to oxidative stress as has been shown with α_1-antitrypsin (see Box 17.2).

Subtilisin: a paradigm for protein engineering

Proof of the power of gene manipulation coupled with the techniques of site-directed mutagenesis is provided by the work on subtilisin (Carter & Wells 1987, Wells & Estell 1988, Carter *et al.* 1991, Arnold 1993). Almost

Box 17.2 Oxidation-resistant variants of α_1-antitrypsin (AAT)

Cumulative damage to lung tissue is thought to be responsible for the development of emphysema, an irreversible lung disease characterized by loss of lung elasticity. The primary defence against elastase damage is AAT, a glycosylated serum protein of 394 amino acids. The function of AAT is known because its genetic deficiency leads to a premature breakdown of connective tissue. In healthy individuals there is an association between AAT and neutrophil elastase followed by cleavage of AAT between methionine residue 358 and serine residue 359 (see Fig. B17.2).

After cleavage, there is negligible dissociation of the complex. Smokers are more prone to emphysema because smoking results in an increased concentration of leukocytes in the lung and consequently increased exposure to neutrophil elastase. In addition, leukocytes liberate oxygen free radicals and these can oxidize methionine-358 to methionine sulphoxide. Since methionine sulphoxide is much bulkier than methionine it does not fit into the active site of elastase. Hence oxidized AAT is a poor inhibitor. By means of site-directed mutagenesis an oxidation-resistant mutant of AAT has been constructed by replacing methionine-358 with valine (Courtney *et al.* 1985). In a laboratory model of inflammation, the modified AAT was an effective inhibitor of elastase and was not inactivated by oxidation. Clinically this could be important since intravenous replacement therapy with plasma concentrates of AAT is already being tested on patients with a genetic deficiency in AAT production.

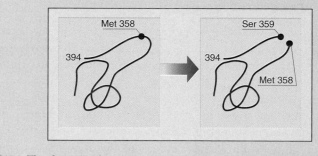

Fig. B17.2 The cleavage of α_1-antitrypsin on binding to neutrophil elastase.

every property of this serine protease has been altered including its rate of catalysis, substrate specificity, pH rate profile and stability to oxidative, thermal and alkaline inactivation. Many of these changes have been introduced in order to improve the industrial uses of subtilisin for hydrolysing proteins, e.g. in detergents. However, serine proteases can be used to synthesize peptides and this approach has a number of advantages over conventional methods (Abrahmsen *et al.* 1991). A problem with the use of subtilisin for peptide synthesis is that hydrolysis is strongly favoured over aminolysis unless the reaction is undertaken in organic solvents. Solvents, in turn, reduce the half-life of subtilisin. Using site-directed mutagenesis a

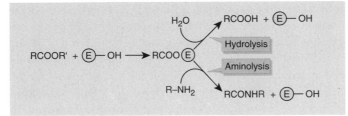

Fig. 17.6 The aminolysis (synthetic) and hydrolysis reactions mediated by an acylated protease.

number of variants of subtilisin have been isolated with greatly enhanced solvent stability (Wong *et al.* 1990, Zhong *et al.* 1991). Changes introduced included the minimization of surface changes to reduce solvation energy, the enhancement of internal polar and hydrophobic interactions, and the introduction of conformational restrictions to reduce the tendency of the protein to denature. Designing these changes requires an extensive knowledge of the enzyme's structure and function. Chen and Arnold (1991, 1993) have provided an alternative solution. They utilized random mutagenesis combined with screening for enhanced proteolysis in the presence of solvent (dimethyl formamide) and substrate (casein).

The engineering of subtilisin now has gone one step further in that it has been modified such that aminolysis (synthesis) is favoured over hydrolysis, even in aqueous solvents. This was achieved by changing a serine residue in the active site to cysteine (Abrahmsen *et al.* 1991). The reasons for this enhancement derive mainly from the increased affinity and reactivity of the acyl intermediate for the amino nucleophile (Fig. 17.6). These engineered 'peptide ligases' are in turn being used to synthesize novel glycopeptides (Wong 1992). A glycosyl amino acid is used in peptide synthesis to form a glycosyl peptide ester which will react with another *C*-protected peptide in the presence of the peptide ligase to form a larger glycosyl peptide.

Successful mutagenesis without knowledge of protein tertiary structure

Successful protein engineering of the kind discussed above depends upon the availability of reliable structural information. Often little or nothing is known about the tertiary structure of the protein of interest. In such cases, successful modification through site-directed mutagenesis may be achieved if a homologous protein of known structure is available. Using this strategy, a neutral protease from *Bacillus stearothermophilus* was stabilized using thermolysin as a model protein (Imanaka *et al.* 1986).

In the absence of a homologous protein of known structure there is little one can do other than to resort to random mutagenesis. Clearly, it helps if there is a good direct selection method for the improved variant as with the solvent-resistant subtilisin described above. A modification of this technique has been used to isolate a nucleotidyl transferase with enhanced thermal stability (Liao *et al.* 1986). The gene encoding the transferase, which specifies kanamycin resistance, was isolated from a mesophile and cloned in a kanamycin-sensitive thermophile. The thermophile was grown at elevated temperatures and when selection was made for kanamycin resistance two naturally-occurring mutants were selected. Both mutants produced enzymes with single amino acid substitutions and these substituted enzymes were more thermostable than the original protein.

Macro-modification of proteins

With many methods of mutagenesis the aim is to replace one, or at most a few, amino acids in a protein sequence to realise some 'improvement' of the protein in question. Much larger modifications can be made to a protein. For example, part of a gene can be deleted by eliminating a restriction fragment or by chemically synthesizing only part of the gene. In this way it was possible to produce the Klenow fragment of DNA polymerase free of the $3' \rightarrow 5'$ exonuclease activity associated with the intact enzyme. Conversely, it is possible to insert additional amino acids into a protein sequence and this has been done to create purification fusions (see below) or to stabilize foreign proteins in *E. coli* (see p. 62).

Although many different purification fusions have been described (Sassenfeld 1990) only the two most elegant will be given in detail here. Most proteins do not have a high affinity for metals but this property can be engineered into them. A high-affinity site can be formed by inserting the sequence His-X_3-His into an α-helix such that it is exposed on the surface (Arnold & Haymore 1991). This surface motif can form a ternary complex with a metal ion such as copper and an anchoring ligand attached to a polymer or solid support. Such metal binding proteins also can have increased resistance to denaturation.

An alternative approach to purification fusions is polyarginine tailing. In this method the gene sequence encoding the desired protein is extended by the inclusion at the 3'-end of a number of codons for arginine. When such genes are expressed the resultant proteins have a polyarginine 'tail' which makes them more basic. Upon ion-exchange chromatography, such proteins are separated from the bulk of the host cell proteins, which are more acidic (Fig. 17.7). The polyarginine tail is then removed with the enzyme carboxypeptidase B, which for convenience can be immobilized. The de-tailed protein is re-chromatographed on an identical ion-exchange resin to separate it from any remaining contaminating proteins which will be more basic (Sassenfeld 1990). The advantage of this latter method is that it produces an authentic protein indistinguishable in amino acid sequence from the natural molecules.

The most dramatic macro-modifications are ones in which one gene is fused with all or part of another to generate completely novel proteins. The best examples of these are provided by the many different variants of antibodies which have been created. For example, single-chain antibodies

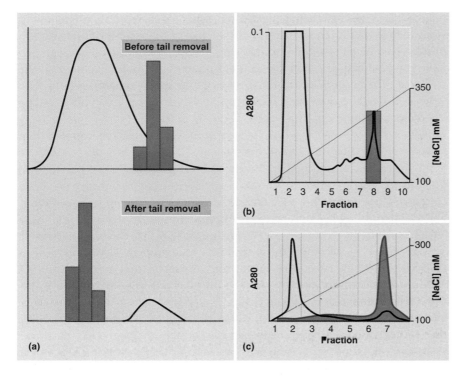

Fig. 17.7 The use of polyarginine-tailing to facilitate protein purification. (a) (*left*) Schematic representation of a hypothetical protein before and after enzymic removal of the C-terminal arginine residues. (b) (*top right*) Separation of polyarginine-tailed urogastrone from the bulk of the proteins in an *E. coli* cell extract. The red line shows the salt gradient and the tinted bar the urogastrone activity. (c) (*bottom right*) Chromatographic behaviour of tailed and untailed urogastrone. The red line shows the salt gradient.

(SCA) are artificial antibodies composed of heavy- and light-chain binding regions linked chemically and produced in a microorganism (Bird *et al*. 1988, Huston *et al*. 1988). To make an SCA the gene sequences for the N-terminal end of the heavy and light chains are joined by a gene sequence encoding a linker peptide. The entire construct then is inserted into an appropriate expression vector.

Novel routes to vaccines

An effective vaccine generates humoral and/or cell-mediated immunity which prevents the development of disease upon exposure to the corresponding pathogen. This is accomplished by presenting pertinent antigenic determinants to the immune system in a fashion which mimics that in natural infections. Conventional viral vaccines consist of inactivated, virulent strains or live, attenuated strains, but they are not without their problems. For example, many viruses have not been adapted to grow to high titre in tissue culture, e.g. hepatitis B virus. There is a danger of vaccine-related disease when using inactivated virus since replication-competent virus may remain in the inoculum. Outbreaks of foot-and-mouth disease in Europe have been attributed to this cause. Finally, attenuated virus strains have the potential to revert to a virulent phenotype upon replication in the vaccinee. This occurs about once or twice in every million people who receive live polio vaccine. Recombinant DNA technology offers some interesting solutions.

Given the ease with which heterologous genes can be expressed in various prokaryotic and eukaryotic systems it is not difficult to produce large quantities of purified immunogenic material for use as a subunit vaccine. A whole series of immunologically pertinent genes have been cloned and expressed but in general the results have been disappointing. For example, of all the polypeptides of foot-and-mouth disease virus, only VP1 has been shown to have immunizing activity. However, polypeptide VP1 produced by recombinant means was an extremely poor immunogen (Kleid *et al.* 1981). Perhaps it is not too surprising that subunit vaccines produced in this way do not generate the desired immune response, for they lack authenticity. The hepatitis B vaccine, which is commercially available (Valenzuela *et al.* 1982), differs in this respect, for expression of the surface antigen in yeast results in the formation of virus-like particles. A similar phenomenon is seen with a yeast Ty vector carrying a gene for HIV coat protein (Adams *et al.* 1987). These subunit vaccines also have another disadvantage. Being inert they do not multiply in the vaccinee and so they do not generate the effective cellular immune response essential for the recovery from infectious disease.

An alternative approach to the development of live vaccines is to start with the food-poisoning organism *Salmonella typhimurium*. This organism can be attenuated by the introduction of lesions in the *aro* genes, which encode enzymes involved in the biosynthesis of aromatic amino acids, *p*-aminobenzoic acid and enterochelins. Whereas doses of 10^4 wild-type *S. typhimurium* reproducibly kill mice, *aro* mutants do not kill mice when fed orally, even when doses as high as 10^{10} organisms are used. However, the mutant strains can establish self-limiting infections in the mice and can be detected in low numbers in organs such as the liver and spleen. Such attenuated strains of *S. typhimurium* are particularly attractive as carriers of heterologous antigens because they can be delivered orally and because they can stimulate humoral, secretory and cellular immune responses in the host (Charles & Dougan 1990). Already a wide range of heterologous antigens have been expressed in such vaccine strains (Hackett 1993).

The use of the BCG vaccine strain as an alternative vector has many advantages including its known safety, low cost, widespread use as a childhood vaccine and ability of a single dose to induce long-lasting protection. Several recombinant BCG strains have been constructed that stably express foreign genes (Stover *et al.* 1991) and preliminary results from animal studies are very encouraging. An alternative vector is the human oral commensal *Streptococcus gordonii* (Fischetti *et al.* 1993).

A different procedure for attenuating a bacterial pathogen to be used as

Table 17.3 Immune response to heterologous antigens expressed by vaccinia virus recombinants

Antigen	Neutralizing antibodies	Cellular immunity	Animal protection
Rabies virus glycoprotein	+	+	+
Vesicular stomatitis glycoprotein	+	+	+
Herpes simplex virus glycoprotein D	+	+	+
Hepatitis B surface antigen	+	+	+
Influenza virus haemagglutinin	+	+	+
Human immunodeficiency virus envelope	+	+	not determined

Reproduced with permission from Tartaglia & Paoletti (1988).

a vaccine has been proposed by Kaper *et al.* (1984). They attenuated a pathogenic strain of *Vibrio cholera* by deletion of DNA sequences encoding the A_1 subunit of the cholera enterotoxin. A restriction endonuclease fragment encoding the A_1, but not the A_2 or B sequences was deleted *in vitro* from cloned cholera toxin genes. The mutation then was recombined into the chromosome of a pathogenic strain. The resulting strain produces the immunogenic but non-toxic B subunit of cholera toxin but is incapable of producing the A subunit. This strain has been found to be safe and immunogenic in carefully-controlled clinical trials in a number of countries (Cryz 1992).

Yet another approach to vaccine preparation is to make use of an animal virus as a vector to express immunologically pertinent proteins. Vaccinia is the strongest candidate because of its successful history as an immunizing agent. Vaccinia virus was the immunogen used to accomplish the global eradication of smallpox. Among the properties which contributed to its success as a live vaccine were its stability in a freeze-dried preparation, its low production cost, and the ability to administer it by simple dermal abrasion. A number of viral proteins have been expressed by vaccinia virus recombinants, basically using the method described on p. 334. These recombinants show great promise for they induce the correct immunological responses in experimental animals (Table 17.3).

One disadvantage of this approach in humans is the unacceptably high rate of adverse reactions associated with the smallpox vaccine. To eliminate this problem the vector was further attenuated by making it thymidine kinase deficient (Cooney *et al.* 1991). This vector was used to deliver HIV gp160 protein to HIV sero-negative males and found to be safe. However, in subjects who had previously been immunized against smallpox it produced only a weak anti-HIV response. This suggests that pre-existing immunity to a vaccine vector may suppress the host's ability to respond to recombinant antigen.

The most successful vaccination campaign to date using a recombinant vaccine is the elimination of rabies with a vaccinia-based vaccine. This has

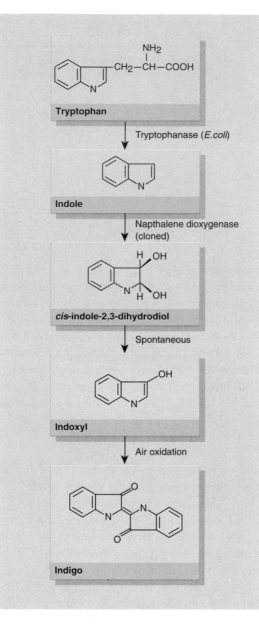

NH2
|
CH2—CH—COOH

Tryptophan

Tryptophanase (*E.coli*)

Indole

Napthalene dioxygenase
(cloned)

H   OH

H   OH

cis-**indole-2,3-dihydrodiol**

Spontaneous

OH

Indoxyl

Air oxidation

O

O

Indigo

Fig. 17.8 Proposed pathway
for indigo biosynthesis by a
strain of *E. coli* carrying a
cloned gene for naphthalene
dioxygenase.

been administered to the wild population of foxes in a large part of central
Europe by providing 'bait' consisting of chicken heads spiked with the
recombinant virus. As few as 12–15 vaccine baits per km^2 are sufficient to
immunize the entire fox population of an area. The epidemiological effect
of vaccination has been most evident in eastern Switzerland. High rabies
prevalence lasting for 18 years came to a sudden end after only three
vaccination campaigns (Anderson 1991, Brochier *et al.* 1991, Flamand *et al.*
1992).

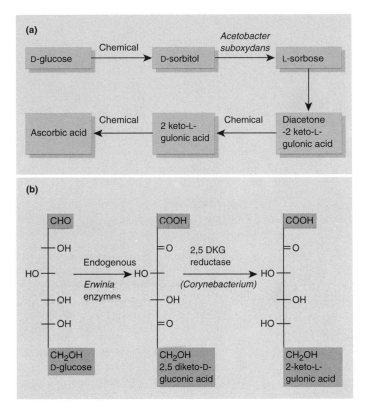

Fig. 17.9 Simplified route to vitamin C (ascorbic acid) developed by cloning in *Erwinia* the *Corynebacterium* gene for 2,5-diketogluconic acid reductase. (a) Classical route to vitamin C. (b) The simplified route to 2-ketogulonic acid, the immediate precursor of vitamin C.

New routes to small molecules: metabolite engineering

Recombinant DNA technology does not just offer novel methods for the generation of proteins, it also provides new ways of making low molecular weight compounds. Good examples are the microbial synthesis of the blue dye indigo (Ensley *et al.* 1983) and the black pigment melanin (Della-Cioppa *et al.* 1990). Neither compound is normally produced by microbes. The cloning of a single gene from *Pseudomonas putida*, that encoding naphthalene dioxygenase, resulted in the generation of an *E. coli* strain able to synthesize indigo in a medium containing tryptophan (Fig. 17.8). Similarly, cloning a tyrosinase gene in *E. coli* led to conversion of tyrosine to dopaquinone which spontaneously converts to melanin, in the presence of air. With both indigo (Murdock *et al.* 1993) and melanin, yields are improved by increasing the levels of cofactors. Thus, increasing the expression of a cofactor-requiring protein as part of a metabolic engineering scheme may require the supply of the cofactor to be enhanced as well.

A slightly different approach to that above has yielded a new route to vitamin C. The conventional process starts with glucose and comprises one

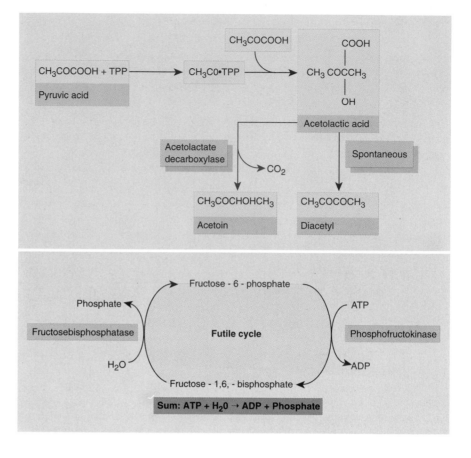

Fig. 17.10 The biochemical steps manipulated in the generation of improved yeast strains.

microbiological and four chemical steps (Fig. 17.9). By cloning in *Erwinia* a single gene, that from *Corynebacterium* encoding 2,5-diketogluconic acid reductase, the process can be simplified to a single microbiological and a single chemical step (Anderson *et al.* 1985). After observations of unexpectedly low yields of 2-ketogulonic acid in the recombinant strain it was found that 2-ketogulonic acid was converted to L-idonic acid by an endogenous 2-ketoaldonate reductase. Cloning, deletion mutagenesis and homologous recombination of the mutated reductase gene into the chromosome were part of several steps taken to develop an organism capable of accumulating large amounts (120 g/litre) of 2-ketogulonic acid (Lazarus *et al.* 1990).

Just as novel proteins can be produced by recombinant DNA techniques so too can novel small molecules. The *Streptomyces coelicolor* gene cluster encoding the biosynthesis of the isochromanequinone antibiotic actinorhodin has been cloned. When the cloned genes were introduced into a variety of other *Streptomyces* spp. producing different isochromanequinones at least three new antibiotics were detected (Hopwood *et al.* 1985b). Clearly, actinorhodin, or one of its precursors, is a novel metabolite in these other *Streptomyces* spp. and is subject to further or different enzymatic modifications (for a review see Hopwood 1993). Other novel antibiotics produced in this way include 2-norethyromycins A, B, C and D (McAlpine *et al.* 1987) and isovaleryl spiramycin (Epp *et al.* 1989). A novel erythromycin derivative was isolated in a different way. The gene *eryF* in *Saccharopolyspora erythrae* encodes the first enzyme in the pathway from 6-deoxyerythronolide B to erythromycin. After the targeted disruption of this gene using an integrative plasmid, 6-deoxyerythronolide was converted to an erythromycin derivative more stable at the low pH of the stomach (Weber *et al.* 1991).

Yeast (*Saccharomyces cerevisiae*) has been used in baking and brewing for several millenia. Yet even these processes can be improved by gene manipulation. For example, during fermentation, α-acetohydroxy acids leak into the medium where spontaneous hydroxylation produces diacetyl. This diacetyl has an undesirable flavour and is removed during a lengthy lagering step. By cloning in yeast a *Klebsiella* gene encoding α-acetolactate decarboxylase, acetoin was formed instead of diacetyl (Fig. 17.10). Unlike diacetyl, acetoin only influences flavour at high concentrations. Pilot brewing studies with the engineered strain showed that it yielded quality beer in only 2 weeks instead of 5 weeks because the lagering step now could be omitted (Suihko *et al.* 1990).

The production of CO_2 by yeast is important for the baking process. During production of CO_2, ATP is produced and this inhibits phosphofructokinase and pyruvate kinase, two key enzymes in glucose catabolism. To overcome this, a futile cycle with phosphofructokinase was created by expressing cloned yeast fructose 1,6-bisphosphatase from a yeast promoter that is induced by glucose (Fig. 17.10). This yeast strain produced 25% more CO_2 than the parent strain (Rogers & Hiller 1990).

The generation of novel plants and foods

It is widely recognized that gene manipulation techniques have revitalized the biotechnology industry to such an extent that genetic engineering and biotechnology are almost synonymous. However, this viewpoint is heavily biased towards the health care industry. Yet, in the ten years since the first expression of heterologous genes in transgenic plants, spectacular progress has been made in developing improved and novel crops. Initially efforts were focused on single genes conferring improved agronomic traits such as herbicide and disease resistance. More recently these efforts have been expanded to include multigenic traits. Next came a second phase whose objective was the development of plants with improved food processing characteristics, e.g. decelerated ripening in tomatoes. Now there is a third phase of work in which plants are being developed as factories for biological and chemical products. Examples of the successes to date are set out below. However, as noted above, plant biotechnology is an extremely fast moving area of research. The reader who wishes to keep up to date is advised to look for the annual review of topics in plant biotechnology which appears in *Current Opinion in Biotechnology*.

Table 17.4 Mode of action of herbicides and method of engineering herbicide-resistant plants

Herbicide	Pathway inhibited	Target enzyme	Basis of engineered resistance to herbicide
Glyphosate	Aromatic amino acid biosynthesis	5-enol-pyruvyl shikimate-3-phosphate (EPSP) synthase	Over-expression of plant EPSP gene or introduction of bacterial glyphosate-resistant *aroA* gene
Sulphonylurea	Branched-chain amino acid biosynthesis	Acetolactate synthase (ALS)	Introduction of resistant *ALS* gene
Imidazolinones	Branched-chain amino acid biosynthesis	Acetolactate synthase (ALS)	Introduction of mutant *ALS* gene
Phosphinothricin	Glutamine biosynthesis	Glutamine synthetase	Over-expression of glutamine synthetase or introduction of the *bar* gene, which detoxifies the herbicide
Atrazine	Photosystem II	Q_B	Introduction of mutant gene for Q_B protein or introduction of gene for glutathione-S-transferase, which can detoxify atrazines
Bromoxynil	Photosynthesis		Introduction of nitrilase gene which detoxifies bromoxynil

Herbicide and disease resistance: single-gene traits

Herbicides generally affect processes that are unique to plants, e.g. photosynthesis or amino acid biosynthesis (see Table 17.4). These processes are shared by both crops and weeds, and developing herbicides which are selective for weeds is very difficult. An alternative approach is to modify crop plants so that they become resistant to broad-spectrum herbicides. Two approaches to engineering herbicide resistance have been adopted. In the first of these the target molecule in the cell either is rendered insensitive or is over-produced. In the second approach a pathway that degrades or detoxifies the herbicide is introduced into the plant.

Glyphosate inhibits 5-enol-pyruvylskikimate-3-phosphate (EPSP) synthase, a key enzyme in the biosynthesis of aromatic amino acids in plants and bacteria. A glyphosate-tolerant *Petunia* cell line obtained after selection for glyphosate resistance was found to over-produce the EPSP synthase as a result of gene amplification. A gene encoding the enzyme was isolated and introduced into petunia plants under the control of a CaMV promoter. Transgenic plants expressed increased levels of EPSP synthase in their chloroplasts and were significantly more tolerant to glyphosate (Shah *et al*. 1986). An alternative approach to glyphosate resistance has been to introduce a gene encoding a mutant EPSP synthase.

Fig. 17.11 Evaluation of phosphinothricin resistance in transgenic tobacco plants under field conditions. (a) Untransformed control plants. (b) Transgenic plants. (Photographs courtesy of Dr J. Botterman and the editor of *Biotechnology*.)

This mutant enzyme retains its specific activity but has decreased affinity for the herbicide. Transgenic tomato plants expressing this gene under the control of an opine promoter were also glyphosate tolerant (Comai *et al*. 1985).

Phosphinothricin (PPT) is an irreversible inhibitor of glutamine synthetase in plants and bacteria. Bialaphos, produced by *Streptomyces hygroscopicus*, consists of PPT and two alanine residues. When these residues are removed by peptidases the herbicidal component PPT is released. To prevent self-inhibition of growth, bialaphos-producing strains of *S. hygroscopicus* produce an acetyltransferase that inactivates PPT by acetylation. The *bar* gene that encodes the acetylase has been introduced into potato, tobacco and tomato cells using *Agrobacterium*-mediated transformation. The resultant plants were resistant to commercial formulations of PPT and bialaphos in the laboratory (De Block *et al*. 1987) and in the field (De Greef *et al*. 1989) (Fig. 17.11). More recently, it has been shown (Uchimiya *et al*. 1993) that bialaphos-resistant transgenic rice plants which were inoculated with the fungi causing sheath blight disease and subsequently treated with the herbicide were completely protected from infection. This agronomically important result depends on the observation that bialaphos is toxic to fungi as well as being a herbicide. Other examples of these approaches to engineering herbicide resistance have been reviewed by Botterman and Leemans (1988).

Major crop losses occur every year as a result of viral infections, e.g. tobacco mosaic virus (TMV) causes losses of tomato plants of over $50 million per annum. Currently, there is little the farmer can do to control such infections and he is dependent on the plant breeder for the production of novel, virus-resistant cultivars. Unfortunately, little is known about the genetics of virus resistance. However, there is a useful phenomenon known as cross-protection in which infection of a plant with one strain of virus protects against superinfection with a second, related strain. The mechanism of cross-protection is not understood but it is believed that the viral coat protein is important. Powell-Abel *et al.* (1986) developed transgenic plants which express the TMV coat protein and which had greatly reduced disease symptoms following virus infection. Since that observation the principle has been extended to many different plants and viruses (for review, see Beachy *et al.* 1990). In the case of resistance to TMV, the coat protein must be expressed in the epidermis and in the vascular tissue through which the virus spreads systemically (Clark *et al.* 1990).

A different method of minimizing the effects of plant virus infection was developed by Gehrlach *et al.* (1987). They developed plants which expressed the satellite RNA of tobacco ringspot virus and such plants were phenotypically resistant to infection with tobacco ringspot virus itself. Another potential method of inducing resistance to viruses is the production of antiviral proteins in transgenic plants. American pokeweed produces an antiviral protein that functions as a ribosome-inactivating protein. The cDNA for this protein has been cloned (Lin *et al.* 1991) and this should facilitate further studies. For a review of the use of a gene manipulation in developing virus-resistant plants the reader should consult Fitchen and Beachy (1993).

Progress also has been made in developing resistance to plant pathogenic fungi. Broglie *et al.* (1991) have shown that expression of a bean chitinase can protect tobacco and oilseed rape from post-emergent damping off caused by *Rhizoctonia solani*. Instead of using an enzyme to provide protection, Hain *et al.* (1993) used the phytoalexin stilbene. Tomato plants expressing the grapevine genes for stilbene synthase demonstrated increased resistance to infection by *Botrytis cinerea*. Similarly, Anzai *et al.* (1989) have used a bacterial gene encoding tabtoxin detoxification to protect tomato plants against *Pseudomonas syringae* infection.

Plants can be made resistant to insects as well as to herbicides and pathogens. Insect-resistant plants are particularly desirable for a number of reasons. First, innate resistance eliminates the need for insecticides which are both costly and time consuming to administer. Second, insecticides are not selective and can kill off desirable as well as undesirable insects. Finally, many insecticides accumulate in the environment and can lead to long-term changes in fauna. Thus expression of the insect-control protein genes of *Bacillus thuringiensis* in plants has been used in an attempt to solve the agricultural challenge of insect control. These toxins are specific for

lepidopteran insects and exhibit no activity against humans, other vertebrates or beneficial insects.

The relevant *B. thuringiensis* genes have been introduced into tomato
(Fischoff *et al.* 1987), tobacco (Vaeck *et al.* 1987) and cotton (Perlak *et al.*
1990) resulting in the production of insecticidally-active protein. However,
field tests of these plants revealed that higher levels of the toxin in the
plant tissue would be required to obtain commercially useful plants
(Delannay *et al.* 1989). Attempts to increase the expression of the toxin
gene in plants by use of different promoters, fusion proteins and leader
sequences were not successful. Examination of the bacterial gene indicated
that it differs significantly from plant genes in a number of ways (Perlak *et
al.* 1991). For example, localized regions of AT richness resembling plant
introns, potential plant polyadenylation signal sequences, ATTTA sequences which can destabilize mRNA and rare plant codons were all
found. Elimination of many of these undesirable sequences resulted in
greatly enhanced expression of the insecticidal toxin and good insect resistance of the transgenic plants in field tests (Koziel *et al.* 1993).

It is well-known that certain plant varieties are more resistant to disease
than others. Some progress has been made in identifying the genes responsible for such resistance. For example, the tomato gene *Mi* which specifies
resistance to the root knot nematode has been subjected to RFLP mapping
(Messeguer *et al.* 1991, Klein-Lankhorst *et al.* 1991). In maize, resistance to
the fungus *Cochliobolus carbonum* is mediated by a reductase which inactivates the phytopathogenic toxin (Meeley & Walton 1991). RFLP mapping
also has identified a tomato gene conferring resistance to the fungus
Fusarium oxysporum (Segal *et al.* 1992) and the construction of a tomato
YAC library has led to the identification of genes conferring resistance to
tobacco mosaic virus and *Pseudomonas syringae* (Martin *et al.* 1992). Considerable progress also is being made in identifying genes that are important for tolerance to environmental stress (cold, drought, salt, etc.) and in
defining stress-responsive gene promoters and signal-transduction pathways (Vierling & Kimpel 1992, Koizumi *et al.* 1993).

Pigmentation in transgenic plants

Plants are widely used for ornamental purposes so it is not surprising that
considerable attempts have been made to develop varieties exhibiting new
colours or pigmentation patterns. Pigmentation of flowers is mainly due
to three classes of compound: the flavonoids, the carotenoids and the
betalains. Of these, the flavonoids are the best characterized with much
information now available concerning the chemistry, biochemistry and
molecular genetics. Several flavonoid genes have been cloned and different
approaches have been used for their introduction and expression in receptor plants, mainly *Petunia* (Forkmann 1993). Although there have been a
few successes, these are the exception and not the rule. The problem is a
phenomenon known as co-suppression.

Co-suppression can be described most easily with the specific example of chalcone synthase (CHS), a key enzyme in anthocyanin biosynthesis. In an attempt to over-produce flower pigments in petunias an additional *CHS* gene was introduced by gene manipulation techniques. Surprisingly, the introduction of the *CHS* transgene was found to suppress both alleles of the endogenous *CHS* genes causing the production of pure white or patterned flowers (Napoli *et al*. 1990). Suppression was restricted to homologous gene expression and no other genes in the biosynthetic pathway were affected. Purple revertants could be isolated and these had normal levels of gene expression.

Co-suppression now has been demonstrated in numerous other systems. It does not appear to be a dosage effect resulting from competition for transcription factors. Nor is it a result of a system that detects specific duplicated plant genes. Rather, it appears to be the result of a homology-dependent interaction between homologous sequences.

The use of antisense RNA

During transcription of a gene only one strand of DNA, the 'sense' strand, is copied into mRNA. The location of the promoter 5' to the gene ensures that the mRNA is produced from the correct strand. If the gene is positioned in the wrong orientation relative to the promoter, the opposite or 'antisense' strand is transcribed and an RNA is made that is complementary to the mRNA. This molecule is termed *antisense RNA*.

Antisense RNA was first engineered into tomatoes to deliberately reduce the levels of polygalacturonase, an enzyme involved in softening and over-ripening (Sheehy *et al*. 1988, Smith *et al*. 1988). The transgenic tomatoes which were produced had an increased shelf-life and increased resistance to bruising. This enables the farmer to retain the tomatoes on the vine longer as well as simplifying the transport of the tomatoes to market. More recently, extended shelf-life of tomatoes has been achieved by using antisense RNA to suppress two key enzymes in the biosynthesis of the plant hormone ethylene (Hamilton *et al*. 1991, Oeller *et al*. 1991). An alternative method of reducing ethylene synthesis depends not on antisense RNA but on the introduction of a bacterial gene encoding an enzyme which degrades a key intermediate (Klee *et al*. 1991). If there is a need for ripening to be accelerated this can be achieved by the administration of exogenous ethylene. The molecular genetics of tomato ripening has been reviewed by Fray and Grierson (1993).

The introduction of antisense constructs of pigmentation genes into a fully-coloured *Petunia* has been used to develop variants with reduced or erratic pigmentation or white flowers (Van der Krol *et al*. 1988). The same antisense constructs were also effective in *Solanum* and *Nicotiana* sp. Antisense RNA constructs also have been used to block the replication of tomato golden mosaic virus in tobacco plants (Day *et al*. 1991). The resulting transgenic plants were much more resistant to infection with the virus than were the controls.

Although antisense RNA has been shown to work in each of the above examples it is not clear exactly how it functions. Proposed explanations (Bejarano & Lichtenstein 1992) are:

- duplex formation with the DNA template to block transcription;
- blocking intron splicing;
- failure of the antisense:mRNA duplex to be transported to the cytoplasm;
- promoting rapid RNA degradation;
- blocking initiation of translation.

In practice more than one of these mechanisms may be operating simultaneously.

Altering the food content of plants

Starch is the major storage carbohydrate in higher plants. It consists of two components: amylose, which is a linear polymer, and amylopectin, which is a branched polymer. The physicochemical, nutritional and textural properties of food are significantly influenced by the nature of the starch. For example, cooked amylose produces fibre-like structures that are resistant to digestion, serve as dietary fibre and require less insulin for metabolism. Amylopectin, by contrast, is waxy and viscous. A wide range of different starches is used by the food and other industries These are obtained by sourcing the starch from different plant varieties coupled with chemical and enzymatic modification. Today, genetic modification of plants offers a new approach to creating novel starches with new functional properties.

The first step in starch biosynthesis in plants is catalysed by the enzyme ADP-glucose pyrophosphorylase (GP). Stark et al. (1992) cloned the gene for an E. coli mutant GP which was deficient in allosteric regulation. When this gene was inserted into potato plants the tubers accumulated higher levels of starch but had normal starch composition and granule size. Since transgenic plants expressing the wild-type E. coli GP had normal levels of starch, the allosteric regulation of GP, and not its absolute level, must influence starch levels. A different method of modulating GP activity has been used by Muller-Rober et al. (1992). They used an antisense construct to reduce the level of plant GP to 2–5% of wild type. These plants had very low levels of starch (2–5% of wild type) but had a very much higher number of tubers. These tubers had six- to eightfold more glucose and sucrose and much lower levels of patatin and other storage proteins.

Several genes involved in the biosynthesis of amylose and amylopectin have been cloned. Modification of these genes should permit the production of novel starches. In the meantime, Visser et al. (1991) have modified the starch composition of potato plants. They demonstrated that potato plants expressing an antisense gene to the granule-bound starch synthase have little or no amylose compared to wild-type potato where it can be as much as 20%.

Whereas some plants accumulate starch as a carbon reserve, others accumulate high levels of triacylglycerols in seeds or mesocarp tissues.

Higher plants produce over 200 kinds of fatty acids some of which are of food value. However, many are likely to have industrial (non-food) uses of higher value than edible fatty acids (Murphy 1992, Kishare & Somerville 1993). Thus there is considerable interest in using gene manipulation techniques to modify the fatty acid content of plants. As a first step towards this goal, a number of genes encoding desaturases have been cloned, e.g. the $\Delta15$ desaturase which converts linoleic acid to α-linolenic acid (Arondel *et al.* 1992).

In most plants the $\Delta9$ stearoyl ACP desaturase catalyses the first desaturation step in seed-oil biosynthesis, converting stearoyl-ACP to oleoyl-ACP. When antisense constructs were used to reduce the levels of the enzyme in developing rapeseed embryos, the seeds had greatly increased stearate levels (Knutzon *et al.* 1992). The significance of this result is that high stearate content is of value in margarine and confectionary fats.

Most oilseed crops accumulate triacylglycerols with C_{16} to C_{22} acyl chains. However, plants which accumulate medium chain triacylglycerols (MCT), with C_8 to C_{12} acyl chains, would be very useful for these lipids are used in detergent synthesis. Some plants such as the California bay accumulate MCT because of the presence of a medium-chain specific thioesterase. When the gene for this thioesterase was cloned in rape (canola) and *Arabidopsis* the transgenic plants accumulated high levels of MCT (Voelker *et al.* 1992).

Phytate is the main storage form of phosphorus in many plant seeds but bound in this form it is a poor nutrient for monogastric animals. For these animals inorganic phosphate has to be added to the fodder. A phytase from *Aspergillus niger* has been developed for use in pig and poultry diets since it functions between pH2 and 5.5 and thus is active in the gastrointestinal tract. Pen *et al.* (1993) have engineered this enzyme into tobacco seeds where it was expressed as 1% of soluble protein. Supplementation of broiler diets with transgenic seeds resulted in improved growth rate comparable to diets supplemented with phosphate or fungal phytase.

Plants as producers of speciality chemicals

Plants are easy to grow and, in contrast to bacteria, their cultivation does not require specialist equipment or chemicals. Where materials accumulate to high levels in plants their production costs very little, e.g. the production of starch from maize costs about 1 cent per kilogram. Thus there is great interest in using plants as 'factories' for materials which normally are expensive to produce. For example, a common storage material of bacteria, polyhydroxybutyrate (PHB), has been produced in plants. Genes from *Alcaligenes eutrophus* which encode two key enzymes, acetyl-CoA reductase and PHB synthase, were introduced into *Arabidopsis* (Poirier *et al.* 1992). The transgenic plants only produced about 0.1 mg/g plant tissue but no effort had been made to maximize expression.

Table 17.5 Major cultivars that have been modified using recombinant DNA technology and which have been field tested. (Modified from Kareiva 1993)

Crop	Modified trait
Alfalfa	Herbicide tolerance, virus resistance
Apple	Insect resistance
Oilseed rape (canola)	Herbicide tolerance, insect resistance, modification of seed oils
Cantaloupe	Virus resistance
Corn	Herbicide tolerance, insect resistance, virus resistance, wheat germ agglutinin
Cotton	Herbicide tolerance, insect resistance
Cucumber	Virus resistance
Melon	Virus resistance
Papaya	Virus resistance
Potato	Herbicide tolerance, virus resistance, insect resistance, starch increase, and modification to make a variety of non-potato products such as chicken lysozyme
Rice	Insect resistance, modified seed protein storage
Soybean	Herbicide tolerance, modified seed protein storage
Squash	Virus resistance
Strawberry	Insect resistance
Sunflower	Modified seed protein storage
Tobacco	Herbicide tolerance, insect resistance, virus resistance
Tomato	Virus resistance, herbicide tolerance, insect resistance, modified ripening, thermal hysteresis
Walnut	Insect resistance

Plants also have been used to produce mammalian proteins, e.g. interferon, enkephalins and human serum albumin. In many cases, the levels of these proteins in the plants have been surprising. For example, Hiatt *et al.* (1989) found that transgenic plants expressing immunoglobulin genes produced antibody light or heavy chains at levels of more than 1% of soluble leaf protein. Sexual crossing of transgenic plants producing either light or heavy chains generated plants which produced both immunoglobulin chains. There is in theory no reason why transgenic plants could not be used in the future as sources of pharmaceutical proteins in the same way that pharmacologically-active chemicals such as aspirin are extracted from plants today.

Regulatory aspects of novel plants and novel foods

Ten years ago a key question was whether plants could be genetically engineered to produce varieties with improved agronomic, environmental or consumer benefits. From the foregoing it is clear that this no longer is in doubt. The key issue now is whether there are any environmental hazards associated with the large-scale open cultivation of transgenic plants and whether food derived from such plants is safe to eat. Commercially, much hangs on resolving these issues (Fuchs & Perlak 1992) for a wide range of transgenic plants currently is being field-tested (Table 17.5).

Environmentalists have been concerned about the ecological risks of a new crop escaping from cultivation and displacing natural vegetation. This spectre is real to them because many 'exotic' species have become pests when introduced into a new environment. This, of course, is not the same as introducing an existing crop carrying a few extra genes but it would help if botanists knew what traits constitute invasiveness. Crawley *et al.* (1993) have reported the first detailed study to determine the invasiveness of a transgenic plant and probably the most comprehensive study ever undertaken in plant ecology. The study was designed to follow the population growth of normal and genetically-engineered oilseed rape plants across a wide range of environments. Experimentally, this involved three climatically distinct sites, four habitats (wet, dry, sunny, shady) and presence or absence of vertebrate grazers, insect herbivores and fungal pathogens and cultivated or uncultivated background vegetation. Under no conditions did the transgenic cultivars exhibit different rates of population growth to those of the wild-type plants. However, this study does not tell us anything about the invasiveness, or lack of it, in genetically-engineered crops in general. For example, in the study of Crawley *et al.* (1993) the markers selected for study were kanamycin resistance and herbicide (Basta) resistance. The outcome might have been different had other markers, such as stress tolerance or insect resistance, been selected (Kareiva 1993). What is unfortunate is that in the USA there have been over 370 field trials of transgenic plants but the regulatory authorities there have insisted on total destruction of the crops on completion of the trials. Thus an agronomic evaluation has been obtained but no ecological data.

A range of experts in the food safety, toxicology, biotechnology and plant breeding from academia and industry has worked under the auspices of the International Food Biotechnology Council to review the safety of foods derived from recombinant plants. The outcome of this is a guide (International Food Biosafety Council 1991) in the form of decision trees to assess potential food safety risks.

Transgenesis: the generation of novel animals

The use of gene manipulation to permanently modify the germ cells of animals ('transgenesis') was described in Chapter 16. The archetypal example of transgenesis in animals is the production of 'supermice' which are extra-large as a result of the over-production of human growth hormone (see Fig. 16.6). Although of great academic interest, such supermice have no commercial value. However, transgenic mice have been produced which carry genetic lesions identical to those existing in certain human inherited diseases (Table 17.6). Such mice can be used as models for the development and evaluation of new pharmaceutical entities. The power of this technique can be illustrated by its application to studies on tumour development (Adams & Cory 1991). Numerous cancer-prone strains of mice have been created by the introduction of candidate tumour-promoting genes into fertilized eggs. Each transgenic strain is predisposed to develop specific types of tumours, but they usually arise stochastically because of the need for spontaneous mutation of genes that collaborate with the

Table 17.6 Human disease equivalents derived from genetic alterations in the mouse

Genetic alteration	Human disease equivalent	Reference
Introduction of mutant collagen gene into wild-type mice	Osteogenesis	Sinn *et al.* 1987
Inactivation of mouse gene coding hypoxanthine-guanine phosphoribosyl transferase (HPRT)	HPRT deficiency	Kuehn *et al.* 1987
Mutation at locus for X-linked muscular dystrophy	X-linked muscular dystrophy	Chamberlain *et al.* 1988
Introduction of activated human *ras* and *c-myc* oncogenes	Induction of malignancy	Sinn *et al.* 1987
Introduction of mutant (Z) allele of human α_1-antitrypsin gene	Neonatal hepatitis	Dycaico *et al.* 1988
Introduction of HIV *tat* gene	Kaposi's sarcoma	Vogel *et al.* 1988
Over-production of atrial natriuretic factor	Chronic hypotension	Field *et al.* 1991
Introduction of rat angiotensinogen gene	Hypertension	Kimura *et al.* 1992
Constitutively active tyrosine kinase	Cardiac hypertrophy	Chow *et al.* 1991
Over-expression of amyloid precursor protein	Alzheimer's disease	Quon *et al.* 1991
Trisomy 16	Alzheimer's disease	Holzman *et al.* 1992
Expression of Simian cholesteryl ester transfer protein	Athorosclerosis	Marotti *et al.* 1993

introduced oncogene. Thus these mice can provide insights into the effects of individual oncogenes on cellular proliferation, differentiation and viability as well as cooperation between different oncogenes. Their predisposed state also imposes sensitivity to chemical and viral carcinogenesis.

In addition to their use as animal models of human disease, transgenic mice can be used for mutagenicity testing. For many years mice have been used in long-term toxicity testing of new chemicals. In these tests the animal is acutely or chronically exposed to the test compound and observed constantly for occurrence of tumours. However, it is difficult to assess the mutagenic effects of a particular chemical from tumour formation alone for carcinogenesis is a multi-step phenomenon. Nor, for time and cost considerations, is such a long-term assay practical for screening large numbers of new compounds. With this in mind Kohler *et al.* (1990) developed a short-term, *in vivo* assay to study the mutagenic effects of chemical exposure. Transgenic mice were generated using a λ shuttle vector containing a *lacZ* target gene. Following exposure to mutagens this target can be rescued efficiently from genomic DNA prepared from tissues of the treated mice. Mutations in the target gene appear as colourless plaques on a background of blue plaques when plated on Lac⁻ *E. coli* on Xgal containing media (Fig. 17.12).

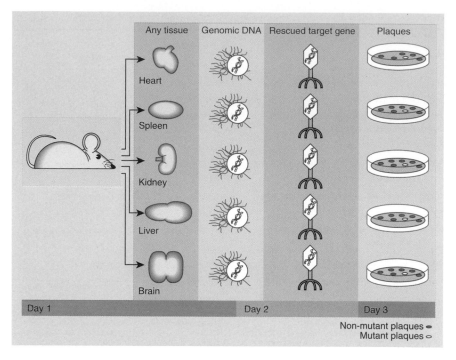

Fig. 17.12 Principle of the *in vivo* assay for detecting mutagens. Mice transgenic for the *E. coli* β-galactosidase gene/phage λ vector are exposed to potential mutagens. At appropriate intervals the mice are sacrificed and genomic DNA extracted from various tissues. The λ vector is recovered by mixing the DNA with an *in vitro* packaging extract which recognizes the *cos* ends of the phage within the genomic DNA. Each packaged phage, which represents a single rescued target gene, is used to infect Lac⁻ *E. coli* growing on Xgal media. The ratio of colourless plaques to blue plaques is a measure of the mutation frequency. (Diagram reproduced courtesy of Dr J. Short.)

Transgenic animals for food production

The most obvious uses for transgenic animals in food production are through direct manipulation of output, either by genetically enhancing existing traits or by programming animals to produce novel products (see next section). Many of the processes that determine the production traits of farm animals, e.g. fertility, growth rate and milk yield, are regulated by protein hormones. Thus it is not surprising that many of the experiments to date have involved supplementing the normal levels of these hormones, particularly with the success of early work leading to the generation of 'supermice'. Thus over-expression of growth hormone has been tried in order to increase the rate of growth of livestock, poultry and fish (Pursel *et al.* 1989, Chen *et al.* 1989) but the results have been far from encouraging.

That success is possible is shown by the generation of transgenic sheep which secrete human proteins (see below) and disease-resistant chickens. In the latter case, a transgenic chicken line was developed which expresses an envelope glycoprotein from avian leukosis virus. The transgenic chickens showed no signs of virus infection up to 40 weeks after inoculation whereas the wild-type birds became viraemic and died of lymphoid leukosis (Salter & Crittenden 1988).

Animal 'pharming'

Based on the work with rat growth hormone, transgenic mice have been generated which over-express foreign proteins and secrete them in their milk. This was achieved by coupling the gene of interest to promoters derived from genes encoding milk proteins, e.g. casein. Proteins synthesized in this way include human tissue plasminogen activator (tPA) (Gordon *et al.* 1987, Pittius *et al.* 1988) and urokinase (Meade *et al.* 1990). Although these proteins are produced at a high concentration, e.g. 50 ng tPA/ ml and are biologically active, mice are unlikely to be acceptable as a pharmaceutical production system. Fortunately, significant progress has been made in producing pharmaceutical proteins in the milk of transgenic farm animals. Apart from the greater regulatory acceptability of farm animals they offer the potential of high volumetric productivity since milk contains tens of grams of proteins per litre.

Wright *et al.* (1991) have generated transgenic sheep which produce α_1-antitrypsin (ATT), a protein used as replacement therapy for genetically-deficient individuals at risk from emphysema (see Box 17.2 on p. 384). Artificially inseminated eggs were micro-injected with a DNA construct containing an AAT gene fused to a β-lactoglobin promoter. These eggs were implanted into surrogate mothers of which 112 gave birth. Four females and one male were found to have incorporated intact copies of the gene and all five developed normally. The four females subsequently gave birth to a mixture of transgenic and non-transgenic lambs and milk from the lactating mothers contained from 1–35 g/litre of AAT. Over the lactation period sheep can produce 250–800 litres of milk so the production potential is significant.

Using similar protocols Ebert *et al.* (1991) have demonstrated the production of a variant of human tPA in goat milk. Of 29 offspring, one male and one female contained the transgene. The transgenic female underwent two pregnancies and one out of five offspring was transgenic. Milk collected over her first lactation contained only a few mg/l of tPA but improved expression constructs have since resulted in an animal generating grams/ litre. Despite the initial low levels of tPA Denman *et al.* (1991) were able to develop a purification protocol which gave 25% recovery of total activity with a purity of 98% as judged by SDS gel electrophoresis.

The potential applications of transgenic animals in medicine have been expanded with the expression of proteins in blood, tissues and animal organs for transplantation to humans (Logan 1993).

Novel therapies for human disease: gene therapy

The characterization of a growing number of hereditable disorders at the gene and protein level has led to the conceptualization of therapies directly

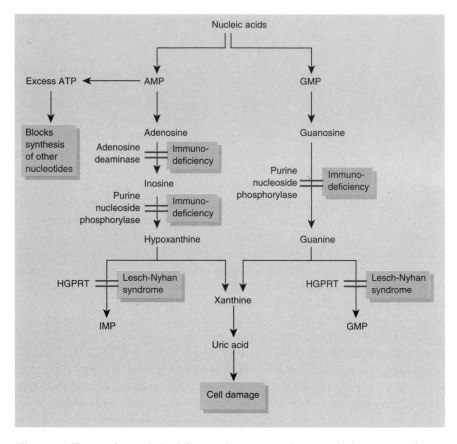

Fig. 17.13 The simple metabolic defects in three genetic diseases which are targets for human gene therapy. The red bars show the metabolic consequences of the defects.

aimed at correcting the molecular defects. However, translating vision into reality has been much more difficult. In addition to the technical problems there are significant safety, social and ethical concerns. Consequently the first human genetic engineering experiment was one of *gene marking*. In this experiment which was undertaken in 1989, tumour infiltrating lymphocytes from patients with advanced cancer were genetically marked with a neomycin-resistance gene and injected back into the same patient. The objectives here were to demonstrate that an exogenous gene could be safely transferred into a patient and that this gene could be detected subsequently in cells removed from the patient at a later date. Both objectives were met (Rosenberg *et al.* 1990).

While experiments on gene marking are continuing (Anderson 1992) experiments on human gene therapy have been initiated. Although major genetic diseases like cystic fibrosis, sickle-cell anaemia and thalassaemia are attractive targets their genetics are too complex or too poorly understood to contemplate therapeutic strategies at this time. For gene therapy it was wiser to focus initially on a disease in which corrected cells might have a selective growth advantage. Three diseases fit this description (Fig. 17.13).

Of these three diseases, ADA deficiency had been cured by bone marrow transplantation, i.e. ADA-normal cells are curative and are able to overgrow the patient's own ADA-deficient cells.

The first human gene therapy was initiated in 1990 and involved a 4-year-old girl suffering from ADA deficiency. Cells from the patient were subjected to leukopheresis and mononuclear cells isolated. These were grown in culture under conditions that stimulated T-lymphocyte activation and growth. These were incubated with a retroviral vector carrying a normal ADA gene as well as the neomycin-resistance gene and then infused into the patient. Both this patient and a second one, who began treatment in early 1991, have shown an improvement in their clinical condition as well as in a battery of *in vitro* and *in vivo* immune function studies (Anderson 1992). Similar experiments are now underway in several European countries.

The first cancer gene-therapy experiments also have been initiated and these are an extension of the early gene-marking experiments. The tumour infiltrating leukocytes now carry a gene for tumour necrosis factor (TNF) in addition to the neomycin-resistance gene. Although TNF is highly toxic to humans at levels as low as $10\,\mu g/kg$ body weight, there have been no side effects from the gene therapy and no apparent organ toxicity from secreted TNF (Hwu *et al.* 1993).

To date, no adverse effects have been noted in any human gene-therapy experiments. Nor has any malignancy been observed as a result of a replication-defective retroviral vector. However, great care is needed for Kolberg (1992) observed malignant T-cell lymphomas in monkeys after a bone-marrow transplantation and gene-transfer protocol. The retroviral vector preparation used was contaminated with helper virus which was probably responsible for the lymphoma (Anderson 1993).

Human gene therapy has only been practised on somatic cells. Ethics apart, could germ cells be modified, i.e. are transgenic humans a possibility in the near future? The answer must be in the negative for three basic reasons. First, micro-injection has a high failure rate even in experienced hands, and this presupposes that sufficient human fertilized eggs could be obtained. Second, micro-injection of eggs can produce deleterious effects because there is no control over where the injected DNA will integrate in the genome. A high level of spontaneous abortion may be tolerated in farm animals but it would not be acceptable in humans. Finally, there is the question of limited usefulness. Most of the serious genetic disorders result in death before puberty, or infertility in homozygous patients. Therefore the worst case that can be imagined is that both parents are heterozygous and, on average, only one-quarter of the offspring will suffer the consequences of homozygosity. The question is, which ova are affected? Currently there is no way to tell and manipulating all ova poses far greater hazards!

Appendix 1: Restriction endonuclease target sites

AatII 5′ . . . GACGT▾C . . . 3′
3′ . . . C▴TGCAG . . . 5′

AccI 5′ . . . GT▾$^{AT}_{CG}$AC . . . 3′

3′ . . . CA$^{AT}_{GC}$▴TG . . . 5′

AflII 5′ . . . C▾TTAAG . . . 3′
3′ . . . GAATT▴C . . . 5′

AhaI 5′ . . . GPu▾CGPyC . . . 3′
3′ . . . CPyGC▴PuG . . . 5′

AluI 5′ . . . AG▾CT . . . 3′
3′ . . . TC▴GA . . . 5′

AlwI 5′ . . . GGATC(N)₄▾ . . . 3′
3′ . . . CCTAG(N)₅▴ . . . 5′

AlwNI 5′ . . . CAGNNN▾CTG . . . 3′
3′ . . . GTC▴NNNGAC . . . 5′

ApaI 5′ . . . GGGCC▾C . . . 3′
3′ . . . C▴CCGGG . . . 5′

ApaLI 5′ . . . G▾TGCAC . . . 3′
3′ . . . CACGT▴G . . . 5′

AseI 5′ . . . AT▾TAAT . . . 3′
3′ . . . TAAT▴TA . . . 5′

AvaI 5′ . . . C▾PyCGPuG . . . 3′
3′ . . . GPuGCPy▴C . . . 5′

AvaII 5′ . . . G▾G$^{A}_{T}$CC . . . 3′

3′ . . . CC$^{T}_{A}$G▴G . . . 5′

AvrII 5′ . . . C▾CTAGG . . . 3′
3′ . . . GGATC▴C . . . 5′

BalI 5′ . . . TGG▾CCA . . . 3′
3′ . . . ACC▴GGT . . . 5′

BamHI 5′ . . . G▾GATCC . . . 3′
3′ . . . CCTAG▴G . . . 5′

BanI 5′ . . . G▾GPyPuCC . . . 3′
3′ . . . CCPuPyG▴G . . . 5′

BanII 5′ . . . GPuGCPy▾C . . . 3′
3′ . . . C▴PyCGPuG . . . 5′

BbvI 5′ . . . GCAGC(N)₈▾ . . . 3′
3′ . . . CGTCG(N)₁₂▴ . . . 5′

BclI 5′ . . . T▾GATCA . . . 3′
3′ . . . ACTAG▴T . . . 5′

BglI 5′ . . . GCCNNNN▾NGGC . . . 3′
3′ . . . CGGN▴NNNNCCG . . . 5′

BglII 5′ . . . A▾GATCT . . . 3′
3′ . . . TCTAG▴A . . . 5′

BsmI 5′ . . . GAATGCN▾ . . . 3′
3′ . . . CTTAC▴GN . . . 5′

BsmAI 5′ . . . GTCTC(N₁)▾ . . . 3′
3′ . . . CAGAG(N₅)▴ . . . 5′

        G C
Bsp1286I 5′ . . . GAGCA▾C . . . 3′
        T T
        C G
3′ . . . C▴TCGTG . . . 5′
        A A

BspHI 5′ . . . T▾CATGA . . . 3′
3′ . . . AGTAC▴T . . . 5′

BspMI 5′ . . . ACCTGC(N)₄▾ . . . 3′
3′ . . . TGGACG(N)₈▴ . . . 5′

BspMII 5′ . . . T▾CCGGA . . . 3′
3′ . . . AGGCC▴T . . . 5′

BsrI 5′ . . . ACTGGN▾ . . . 3′
3′ . . . TGAC▴CN . . . 5′

BssHII 5′ . . . G▾CGCGC . . . 3′
3′ . . . CGCGC▴G . . . 5′

BstI 5′ . . . TT▾CGAA . . . 3′
3′ . . . AAGC▴TT . . . 5′

BstEII 5′ . . . G▾GTNACC . . . 3′
3′ . . . CCANTG▴G . . . 5′

BstNI 5′ . . . CC▾$^{A}_{T}$GG . . . 3′

3′ . . . GG$^{A}_{T}$▴CC . . . 5′

BstUI 5′ . . . CG▾CG . . . 3′
3′ . . . GC▴GC . . . 5′

BstXI 5′ . . . CCANNNNN▾NTGG . . . 3′
3′ . . . GGTN▴NNNNACC . . . 5′

BstYI 5′ . . . Pu▾GATCPy . . . 3′
3′ . . . PyCTAG▴Pu . . . 5′

*Bsu*36I 5′...CC▼TNAGG...3′
3′...GGANT▲CC...5′

*Cla*I 5′...AT▼CGAT...3′
3′...TAGC▲TA...5′

*Dde*I 5′...C▼TNAG...3′
3′...GANT▲C...5′

       CH$_3$
*Dpn*I 5′...GA▼TC...3′
3′...CT▲AG...5′
       CH$_3$

*Dra*I 5′...TTT▼AAA...3′
3′...AAA▲TTT...5′

*Dra*III 5′...CACNNN▼GTG...3′
3′...GTG▲NNNCAC...5′

*Eae*I 5′...Py▼GGCCPu...3′
3′...PuCCGG▲Py...5′

*Eag*I 5′...C▼GGCCG...3′
3′...GCCGG▲C...5′

*Ear*I 5′...CTCTTCN▼NNN...3′
3′...GAGAAGNNNN▲...5′

*Eco*NI 5′...CCTNN▼NNNAGG...3′
3′...GGANN▲NNTCC...5′

*Eco*O 109I 5′...PuG▼GNCCPy...3′
3′...PyCCNG▲GPu...5′

*Eco*RI 5′...G▼AATTC...3′
3′...CTTAA▲G...5′

*Eco*RV 5′...GAT▼ATC...3′
3′...CTA▲TAG...5′

*Fnu*4HI 5′...GC▼NGC...3′
3′...CGN▲CG...5′

*Fok*I 5′...GGATG(N)$_9$▼...3′
3′...CCTAC(N)$_{13}$▲...5′

*Fsp*I 5′...TGC▼GCA...3′
3′...ACG▲CGT...5′

*Hae*II 5′...PuGCGC▼Py...3′
3′...Py▲CGCGPu...5′

*Hae*III 5′...GG▼CC...3′
3′...CC▲GG...5′

*Hga*I 5′...GACGC(N)$_5$▼...3′
3′...CTGCG(N)$_{10}$▲...5′

*Hgi*AI 5′...G$_A^T$GC$_A^T$▼...3′

3′...C$_A^T$CG$_A^T$G...5′

*Hha*I 5′...GCG▼C...3′
3′...C▲GCG...5′

*Hinc*II 5′...GTPy▼PuAC...3′
3′...CAPu▲PyTG...5′

*Hind*III 5′...A▼AGCTT...3′
3′...TTCGA▲A...5′

*Hinf*I 5′...G▼ANTC...3′
3′...CTNA▲G...5′

*Hin*PI 5′...G▼CGC...3′
3′...CGC▲G...5′

*Hpa*I 5′...GTT▼AAC...3′
3′...CAA▲TTG...5′

*Hpa*II 5′...C▼CGG...3′
3′...GGC▲C...5′

*Hph*I 5′...GGTGA(N)$_8$▼...3′
3′...CCACT(N)$_7$▲...5′

*Kpn*I 5′...GGTAC▼C...3′
3′...C▲CATGG...5′

*Mbo*I 5′...▼GATC...3′
3′...CTAG▲...5′

*Mbo*II 5′...GAAGA(N)$_8$▼...3′
3′...CTTCT(N)$_7$▲...5′

*Mlv*I 5′...A▼CGCGT...3′
3′...TGCGC▲A...5′

*Mnl*I 5′...CCTC(N)$_7$▼...3′
3′...GGAG(N)$_7$▲...5′

*Mse*I 5′...T▼TAA...3′
3′...AAT▲T...5′

*Msp*I 5′...C▼CGG...3′
3′...GGC▲C...5′

*Nae*I 5′...GCC▼GGC...3′
3′...CGG▲CCG...5′

*Nar*I 5′...GG▼CGCC...3′
3′...CCGC▲GG...5′

*Nci*I 5′...CC▼$_G^C$GG...3′

3′...GG$_G^C$▲CC...5′

*Nco*I 5′...C▼CATGG...3′
3′...GGTAC▲C...5′

*Nde*I 5′...CA▼TATG...3′
3′...GTAT▲AC...5′

*Nhe*I 5′...G▼CTAGC...3′
3′...CGATC▲G...5′

*Nla*III 5′...CATG▼...3′
3′...▲GTAC...5′

*Nla*IV 5′...GGN▼NCC...3′
3′...CCN▲NGG...5′

APPENDIX 1

*Restriction endonuclease
target sites*

*Not*I 5′ . . . GC▼GGCCGC . . . 3′
3′ . . . CGCCGG▲CG . . . 5′

*Nru*I 5′ . . . TCG▼CGA . . . 3′
3′ . . . AGC▲GCT . . . 5′

*Nsi*I 5′ . . . ATGCA▼T . . . 3′
3′ . . . T▲ACGTA . . . 5′

*Pae*R7I 5′ . . . C▼TCGAG . . . 3′
3′ . . . GAGCT▲C . . . 5′

*Pfl*MI 5′ . . . CCANNNN▼NTGG . . . 3′
3′ . . . GGTN▲NNNNACC . . . 5′

*Ple*I 5′ . . . GAGTC(N₄)▼ . . . 3′
3′ . . . CTCAG(N₅)▲ . . . 5′

*Ppu*MI 5′ . . . PuG▼G$_T^A$CCPy . . . 3′

3′ . . . PyCC$_T^A$G▲Pu . . . 5′

*Pst*I 5′ . . . CTGCA▼G . . . 3′
3′ . . . G▲ACGTC . . . 5′

*Pvu*I 5′ . . . CGAT▼CG . . . 3′
3′ . . . GC▲TAGC . . . 5′

*Pvu*II 5′ . . . CAG▼CTG . . . 3′
3′ . . . GTC▲GAC . . . 5′

*Rsa*I 5′ . . . GT▼AC . . . 3′
3′ . . . CA▲TG . . . 5′

*Rsr*II 5′ . . . CG▼G$_T^A$CCG . . . 3′

3′ . . . GCC$_T^A$G▲GC . . . 5′

*Sac*I 5′ . . . GAGCT▼C . . . 3′
3′ . . . C▼TCGAG . . . 5′

*Sac*II 5′ . . . CCGC▼GG . . . 3′
3′ . . . GG▲CGCC . . . 5′

*Sal*I 5′ . . . G▼TCGAC . . . 3′
3′ . . . CAGCT▲G . . . 5′

*Sau*3AI 5′ . . . ▼GATC . . . 3′
3′ . . . CTAG▲ . . . 5′

*Sau*96I 5′ . . . G▼GNCC . . . 3′
3′ . . . CCNG▲G . . . 5′

*Sca*I 5′ . . . AGT▼ACT . . . 3′
3′ . . . TCA▲TGA . . . 5′

*Scr*FI 5′ . . . CC▼NGG . . . 3′
3′ . . . GGN▲CC . . . 5′

*Sfa*NI 5′ . . . GCATC(N)₅▼ . . . 3′
3′ . . . CGTAG(N)₉▲ . . . 5′

*Sfi*I 5′ . . . GGCCNNNN▼NGGCC . . . 3′
3′ . . . CCGGN▲NNNNCCGG . . . 5′

*Sma*I 5′ . . . CCC▼GGG . . . 3′
3′ . . . GGG▲CCC . . . 5′

*Sna*BI 5′ . . . TAC▼GTA . . . 3′
3′ . . . ATG▲CAT . . . 5′

*Spe*I 5′ . . . A▼CTAGT . . . 3′
3′ . . . TGATC▲A . . . 5′

*Sph*I 5′ . . . GCATG▼C . . . 3′
3′ . . . C▲GTACG . . . 5′

*Ssp*I 5′ . . . AAT▼ATT . . . 3′
3′ . . . TTA▲TAA . . . 5′

*Stu*I 5′ . . . AGG▼CCT . . . 3′
3′ . . . TCC▲GGA . . . 5′

*Sty*I 5′ . . . C▼C$_{TT}^{AA}$GG . . . 3′

3′ . . . GG$_{TT}^{AA}$C▲C . . . 5′

Taq^{α}I 5′ . . . T▼CGA . . . 3′
3′ . . . AGC▲T . . . 5′

*Tth*111I 5′ . . . GACN▼NNGTC . . . 3′
3′ . . . CTGNN▲NCAG . . . 5′

*Xba*I 5′ . . . T▼CTAGA . . . 3′
3′ . . . AGATC▲T . . . 5′

*Xho*I 5′ . . . C▼TCGAG . . . 3′
3′ . . . GAGCT▲C . . . 5′

*Xma*I 5′ . . . C▼CCGGG . . . 3′
3′ . . . GGGCC▲C . . . 5′

*Xmn*I 5′ . . . GAANN▼NNTTC . . . 3′
3′ . . . CTTNN▲NNAAG . . . 5′

Appendix 2:
The genetic code and single-letter amino acid designations

5'-OH terminal base	Middle base				3'-OH terminal base
	U	C	A	G	
U	Phe	Ser	Tyr	Cys	U
	Phe	Ser	Tyr	Cys	C
	Leu	Ser	STOP	STOP	A
	Leu	Ser	STOP	Trp	G
C	Leu	Pro	His	Arg	U
	Leu	Pro	His	Arg	C
	Leu	Pro	Gln	Arg	A
	Leu	Pro	Gln	Arg	G
A	Ile	Thr	Asn	Ser	U
	Ile	Thr	Asn	Ser	C
	Ile	Thr	Lys	Arg	A
	Met	Thr	Lys	Arg	G
G	Val	Ala	Asp	Gly	U
	Val	Ala	Asp	Gly	C
	Val	Ala	Glu	Gly	A
	Val*	Ala	Glu	Gly	G

*Codes for Met if in the initiator position.

Alanine	A	Leucine	L
Arginine	R	Lysine	K
Asparagine	N	Methionine	M
Aspartic Acid	D	Phenylalanine	F
Cysteine	C	Proline	P
Glycine	G	Serine	S
Glutamic Acid	E	Threonine	T
Glutamine	Q	Tryptophan	W
Histidine	H	Tyrosine	Y
Isoleucine	I	Valine	V

References

Aarts M.G.M., Dirkse W., Stiekema W.J. & Pereira A. (1993) Transposon tagging of a male sterility gene in *Arabidopsis*. *Nature* **363**, 715–17.

Aaij C. & Borst P. (1972) The gel electrophoresis of DNA. *Biochim. Biophys. Acta* **269**, 192–200.

Abrahmsen L., Tom J., Burnier J., Butcher K.A., Kossiakof A. & Wells J.A. (1991) Engineering subtilisin and its substrates for efficient ligation of peptide bonds in aqueous solution. *Biochemistry* **30**, 4151–9.

Adams J.M. & Cory S. (1991) Transgenic models of tumor development. *Science* **254**, 1161–7.

Adams J.M., Harris A.W., Pinkert C.A., Corcoran L.M., Alexander W.S., Cory S., Palmiter R.D. & Brinster R.L. (1985) The c-*myc* oncogene driven by immunoglobulin enhancers induces lymphoid malignancy in transgenic mice. *Nature* **318**, 533–8.

Adams S.E., Dawson K.M., Gull K., Kingsman S.M. & Kingsman A.J. (1987) The expression of hybrid HIV: Ty virus-like particles in yeast. *Nature* **329**, 68–70.

Ahlquist P., French R. & Bujarski J.J. (1987) Molecular studies of brome mosaic virus using infectious transcripts from cloned cDNA. *Adv. Virus Res.* **32**, 215–42.

Akiyoshi D.E., Klee H., Amasino R.M., Nester E.W. & Gordon M.P. (1984) T-DNA of *Agrobacterium tumefaciens* encodes an enzyme of cytokinin biosynthesis. *Proc. Natl. Acad. Sci. USA* **81**, 5994–8.

Akli S., Caillaud C., Vigne E., Stratford-Perricaudet L.D., Poenaru L., Perricaudet M., Kahn A. & Peschanski M.R. (1993) Transfer of a foreign gene into the brain using adenovirus vectors. *Nature Genet.* **3**, 224–8.

Alford R.L. & Caskey C.T. (1994) DNA analysis in forensics, disease and animal/plant identification. *Curr. Opinion Biotechnol.* **5**, 29–33.

Altenbuchner J. & Cullum J. (1987) Amplification of cloned genes in *Streptomyces*. *Biotechnology* **5**, 1328–9.

Alwine J.C., Kemp D.J., Parker B.A., Reiser J., Renart J., Stark G.R. & Wahl G.M. (1979) Detection of specific RNAs or specific fragments of DNA by fractionation in gels and transfer to diazobenzyloxymethyl paper. *Methods Enzymol.* **68**, 220–42.

Amann E., Brosius J. & Ptashne M. (1983) Vectors bearing a hybrid *trp–lac* promoter useful for regulated expression of cloned genes in *Escherichia coli*. *Gene* **25**, 167–78.

Ambulos N.P. Jr., Smith T., Mulbry W. & Lovett P.S. (1990) CUG as a mutant start codon for *cat*-86 and *xylE* in *Bacillus subtilis*. *Gene* **94**, 125–8.

Anderson R.M. (1991) Immunization in the field. *Nature* **354**, 502–3.

Anderson W.F. (1992) Human gene therapy. *Science* **256**, 808–13.

Anderson W.F. (1993) What about those monkeys that got T-cell lymphoma [editorial]. *Hum. Gene Ther.* **4**, 1–2.

Anderson S., Gait M.J., Mayol L. & Young I.G. (1980) A short primer for sequencing DNA cloned in the single-stranded phage vector M13mp2. *Nucleic Acids Res.* **8**, 1731–43.

Anderson S., Marks C.B., Lazarus R., Miller J., Stafford K., Seymour J., Light D., Rastetter W. & Estell D. (1985) Production of 2-keto-L-gulonate, an intermediate in L-ascorbate synthesis by a genetically modified *Erwinia herbicola*. *Science* **230**, 144–9.

Anzai H., Yoneyama K. & Yamaguchi I. (1989) Transgenic tobacco resistant to a bacterial disease by the detoxification of a pathogenic toxin. *Mol. Gen. Genetics* **219**, 492–4.

Apostol B. & Greer C.L. (1988) Copy number and stability of yeast 2µ-based plasmids carrying a transcription-conditioned centromere. *Gene* **67**, 59–68.

Arber W. (1965) Host specificity of DNA produced by *Escherichia coli*. V. The role of methionine in the production of host specificity. *J. Mol. Biol.* **11**, 247–56.

Arber W. & Dussoix D. (1962) Host specificity of DNA produced by *Escherichia coli*. I. Host controlled modification of bacteriophage λ. *J. Mol. Biol.* **5**, 18–36.

Arkin A.P. & Youvan D.C. (1992) Optimizing nucleotide mixtures to encode specific subsets of amino acids for semi-random mutagenesis. *Biotechnology* **10**, 297–300.

Arnold F.H. (1993) Protein engineering for unusual environments. *Curr. Opinion Biotechnol.* **4**, 450–55.

Arnold F.H. & Haymore B.L. (1991) Engineered metal-binding proteins: purification to protein folding. *Science* **252**, 1796–7.

Arondel V., Lemieux B., Hwang I., Gibson S., Goodman H. & Somerville C.R. (1992) Map-based cloning of a gene controlling omega-3 fatty acid desaturation in *Arabidopsis*. *Science* **258**, 1353–5.

Awane K., Naito A., Araki H. & Oshima Y. (1992) Automatic elimination of unnecessary bacterial sequences from yeast vectors. *Gene* **121**, 161–5.

Axel R., Fiegelson P. & Schutz G. (1976) Analysis of the complexity and diversity of mRNA from chicken oviduct and liver. *Cell* **11**, 247–54.

Babitzke P. & Kushner S.R. (1991) The ams (altered mRNA stability) protein and ribonuclease E are encoded by the same structural gene of *Escherichia coli*. *Proc. Natl. Acad. Sci. USA* **88**, 1–5.

Bachmair A., Finley D. & Varshavsky A. (1986) *In vivo* half-life of a protein is a function of its amino-terminal residue. *Science* **234**, 179–86.

Backman K. & Boyer H.W. (1983) Tetracycline resistance determined by pBR322 is mediated by one polypeptide. *Gene* **26**, 197–203.

Backman K., Ptashne M. & Gilbert W. (1976) Construction of plasmids carrying the *cI* gene of bacteriophage λ. *Proc. Natl. Acad. Sci. USA* **73**, 4174–8.

Bagdasarian M.M., Amann E., Lurz R., Ruckert B. & Bagdasarian M. (1983) Activity of the hybrid *trp–lac* (*tac*) promoter of *Escherichia coli* in *Pseudomonas putida*. Construction of broad-host-range, controlled-expression vectors. *Gene* **26**, 273–82.

Bagdasarian M., Bagdasarian M.M., Coleman S. & Timmis K.N. (1979) New vector plasmids for gene cloning in *Pseudomonas*. In *Plasmids of Medical, Environmental and Commercial Importance*, eds Timmis K.N. & Pühler A., pp. 411–22. Elsevier/North-Holland Biomedical Press, Amsterdam.

Bagdasarian M., Lurz R., Rückert B., Franklin F.C.H., Bagdasarian M.M., Frey J. & Timmis K.N. (1981) Specific-purpose plasmid cloning vectors. II. Broad host range, high copy number, RSF1010-derived vectors, and a host–vector system for gene cloning in *Pseudomonas*. *Gene* **16**, 237–47.

Baigori M., Sesma R., de Ruiz A.P. & de Mendoza D. (1988) Transfer of plasmids between *Bacillus subtilis* and *Streptococcus lactis*. *Appl. Environ. Microbiol.* **54**, 1309–11.

Bajocchi G., Feldman S.H., Crystal R.G. & Mastrangeli A. (1993) Direct *in vivo* gene transfer to ependymal cells in the central nervous system using recombinant adenovirus vectors. *Nature Genet.* **3**, 229–34.

Balazs E., Bouzoubaa S., Guilley H., Jonasd G., Paszkowski J. & Richards K. (1985) Chimeric vector construction for higher plant transformation. *Gene* **40**, 343–8.

Balazs E., Guilley H., Jonard G. & Richards K. (1982) Nucleotide sequence of DNA from an altered-virulence isolate D/H of the cauliflower mosaic virus. *Gene* **19**, 239–49.

Balbás P., Soberon X., Merino E., Zurita M., Lomeli H., Valle F., Flores N. & Bolivar F. (1986) Plasmid vector pBR322 and its special-purpose derivatives – a review. *Gene* **50**, 3–40.

Ballivet M., Nef P., Coutourier S., Rungger D., Bader C.R., Bertrand D. & Cooper E. (1988) Electrophysiology of a chick neuronal nicotinic acetylcholine receptor expressed in *Xenopus* oocytes after cDNA injection. *Neuron* **1**, 847–52.

Bandyopadhyay P.K., Studier W.F., Hamilton D.L. & Yuan R. (1985) Inhibition of the type I restriction-modification enzymes EcoB and EcoK by the gene 0.3 protein of bacteriophage T7. *J. Mol. Biol.* **182**, 567–78.

Banerji J., Rusconi S. & Schaffner W. (1981) Expression of a β-globin gene is enhanced by remote SV40 DNA sequences. *Cell* **27**, 299–308.

Banner C.D.B., Moran C.P. & Losick R. (1983) Deletion analysis of a complex promoter for a developmentally regulated gene from *Bacillus subtilis*. *J. Mol. Biol.* **168**, 351–65.

Barbour A.G. (1988) Plasmid analysis of *Borrelia burgdorferi*, the Lyme disease agent. *J. Clin. Microbiol.* **26**, 475–8.

Barbour A.G. (1993) Linear DNA of *Borrelia* species and antigenic variation. *Trends in Microbiol.* **1**, 236–9.

Bardwell J.C.A., McGovern K. & Beckwith J. (1991) Identification of a protein required for disulphide bond formation *in vivo*. *Cell* **67**, 581–9.

Barnard E.A., Houghton M., Miledi R., Richards B.M. & Sumikawa K. (1982) Molecular genetics of the acetylcholine receptor and its insertion and organization in the membrane. *Biol. Cell* **45**, 383.

Barnes W.M. (1980) DNA cloning with single-stranded phage vectors. In *Genetic Engineering*, eds Setlow J.K. & Hollaender A., Vol. 2, pp. 185–200. Plenum Press, New York.

Bass S., Greener R. & Wells J.A. (1990) Hormone phage: an enrichment method for variant proteins with altered binding properties. *Proteins* **8**, 309–14.

Bass S., Mulkerin M. & Wells J.A. (1991) A systematic mutational analysis of hormone-binding determinants in the human growth hormone receptor. *Proc. Natl. Acad. Sci. USA* **88**, 4498–502.

Bates P.F. & Swift R.A. (1983) Double *cos* site vectors: simplified cosmid cloning. *Gene* **26**, 137–46.

Beach L.R. & Palmiter R.D. (1981) Amplification of the metallothionein-I gene in cadmium-resistant mouse cells. *Proc. Natl. Acad. Sci. USA* **78**, 2110–14.

Beachy R., Loesch-Fries S. & Tumer N. (1990) Coat-protein mediated resistance against virus infection. *Annu. Rev. Phytopathol* **28**, 451–74.

Beaucage S.L., Miller C.A. & Cohen S.N. (1991) Gyrase-dependent stabilization of pSC101 plasmid inheritance by transcriptionally active promoters. *EMBO J.* **10**, 2583–8.

Beck E. & Bremer E. (1980) Nucleotide sequence of the gene *ompA* encoding the outer membrane protein II of *Escherichia coli* K-12. *Nucleic Acids Res.* **8**, 3011–24.

Beggs J.D. (1978) Transformation of yeast by a replicating hybrid plasmid. *Nature* **275**, 104–9.

Beggs J.D., van den Berg J., van Ooyen A. & Weissmann C. (1980) Abnormal expression of chromosomal rabbit β-globin gene in *Saccharomyces cerevisiae*. *Nature* **283**, 835–40.

Bejarano E.R. & Lichtenstein C.P. (1992) Prospects for engineering virus resistance in plants with antisense RNA. *Trends Biotechnol.* **10**, 383–8.

Bell G.I., Merryweather J.P., Sanchez-Pescador R. *et al.* (1984) Sequence of a cDNA clone encoding human preproinsulin-like growth factor II. *Nature* **310**, 775–77.

Bender W., Spierer P. & Hogness D.S. (1983) Chromosomal walking and jumping to isolate DNA from the *Ace* and *rosy* loci and the bithorax complex in *Drosophila melanogaster*. *J. Mol. Biol.* **168**, 17–33.

Bendig M.M. & Williams J.G. (1983) Replication and expression of *Xenopus laevis* globin genes injected into fertilized *Xenopus* eggs. *Proc. Natl. Acad. Sci. USA* **80**, 6197–201.

Bennetzen J.L. & Hall B.D. (1982) Codon selection in yeast. *J. Biol. Chem.* **257**, 3026–31.

Benson D., Lipmann D.J. & Ostell J. (1993) GenBank. *Nucleic Acids Res.* **21**, 2963–5.

Benton W.D. & Davis R.W. (1977) Screening λgt recombinant clones by hybridization to single plaques *in situ*. *Science* **196**, 180–2.

Berg P. (1981) Dissections and reconstructions of genes and chromosomes. *Science* **213**, 296–303.

Bevan M. (1984) Binary *Agrobacterium* vectors for plant transformation. *Nucleic Acids Res.* **12**, 8711–21.

Bevan M., Barnes W. & Chilton M.D. (1983a) Structure and transcription on the nopaline synthase gene region of T-DNA. *Nucleic Acids Res.* **11**, 369–85.

Bevan M.W., Flavell R.B. & Chilton M.D. (1983b) A chimaeric antibiotic resistance gene as a selectable marker for plant cell transformation. *Nature* **304**, 184–7.

Bhatt A.M. & Dean C. (1992) Development of tagging systems in plants using heterologous transposons. *Curr. Opinion Biotechnol.* **3**, 152–8.

Bickle T.A., Brack C. & Yuan R. (1978) ATP-induced conformational changes in restriction endonuclease from *Escherichia coli* K12. *Proc. Natl. Acad. Sci. USA* **75**, 3099–103.

Bierne H., Ehrlich S.D. & Michel B. (1991) The replication termination signal *ter*B of the *Escherichia coli* chromosome is a deletion hot spot. *EMBO J.* **10**, 2699–705.

Bingham A.H.A. & Busby S.J.W. (1987) Translation of *gal*E and coordination of galactose operon expression in *Escherichia coli*: effects of insertions and deletions in the non-translated leader sequence. *Mol. Microbiol.* **1**, 117–24.

Bingham P.M., Kidwell M.G. & Rubin G.M. (1982) The molecular basis of P-M hybrid dysgenesis: the role of the P element, a P-strain-specific transposon family. *Cell* **29**, 995–1004.

Bird A.P. (1986) CpG-rich islands and the function of DNA methylation. *Nature* **321**, 209–13.

Bird A.P. & Southern E.M. (1978) Use of restriction enzymes to study eukaryotic DNA methylation: I. The methylation pattern in ribosomal DNA from *Xenopus laevis*. *J. Mol. Biol.* **118**, 27–47.

Bird R.E., Hardmann K.D., Jacobson J.W., Johnson S., Kaufmann B.M., Lee S.-M., Lee T., Pope S.H., Riordan G.S. & Whitlow M. (1988) Single-chain antigen binding proteins. *Science* **242**, 423–6.

Bitter G.A. & Egan K.M. (1988) Expression of interferon-gamma from hybrid yeast GPD promoters containing upstream regulatory sequences from the *GAL1–GAL10* intergenic region. *Gene* **69**, 193–207.

Blancou J., Kieny M.P., Lathe R., Lecocq J.P., Pastovet P.P., Soulebot J.P. & Desmettre P. (1986) Oral vaccination of the fox against rabies using a live recombinant vaccinia virus. *Nature* **322**, 373–5.

Blattner F.R., Williams B.G., Blechl A.E., Deniston-Thompson K., Faber H.E., Furlong L.A., Grunwald D.J., Kiefer D.O., Moore D.D., Schumm J. W., Sheldon E.L. & Smithies O. (1977) Charon phages: safer derivatives of bacteriophage lambda for DNA cloning. *Science* **196**, 161–9.

Blobel G. & Dobberstein B. (1975) Transfer of proteins across membranes. I. Presence of proteolytically processed and unprocessed nascent immunoglobulin light chains on membrane-bound ribosomes of murine myeloma. *J. Cell Biol.* **67**, 835–51.

Bloom K.S. & Carbon J. (1982) Yeast centromere DNA is a unique and highly ordered structure in chromosomes and small circular minichromosomes. *Cell* **29**, 305–17.

Blum P., Velligan M., Lin N. & Matin A. (1992) DnaK-mediated alterations in human growth hormone protein inclusion bodies. *Biotechnology* **10**, 301–4.

Boeke J.D., Vovis G.F. & Zinder N.D. (1979) Insertion mutant of bacteriophage fl sensitive to *Eco*RI. *Proc. Natl. Acad. Sci. USA* **76**, 2699–702.

Boeke J.D., Lacroute F. & Fink G.R. (1984) A positive selection for mutants lacking orotidine-5'-phosphate decarboxylase activity in yeast: 5-fluoro-orotic acid resistance. *Mol. Gen. Genetics* **197**, 345–6.

Boeke J.D., Xu H. & Fink G.R. (1988) A general method for the chromosomal amplification of genes in yeast. *Science* **239**, 280–2.

Bolivar F., Rodriguez R.L., Betlach M.C. & Boyer H.W. (1977a) Construction and characterization of new cloning vehicles. I. Ampicillin-resistant derivatives of the plasmid pMB9. *Gene* **2**, 75–93.

Bolivar F., Rodriguez R.L., Greene P.J., Betlach M.V., Heynecker H.L., Boyer H.W., Crosa J.H. & Falkow S. (1977b) Construction and characterization of new cloning vehicles. II. A multipurpose cloning system. *Gene* **2**, 95–113.

Bomhoff G.H., Klapwijk F.M., Kester M.C.M., Schilperoort R.A., Hernalsteens J.P. & Schell J. (1976) Octopine and nopaline synthesis and breakdown genetically controlled by a plasmid of *Agrobacterium tumefaciens*. *Mol. Gen. Genet.* **145**, 177–81.

Borrelli E., Heyman R., Hsi M.A. & Evans R.M. (1988) Targeting of an inducible toxic phenotype in animal cells. *Proc. Natl. Acad. Sci. USA* **85**, 7572–6.

Bossi L. & Smith D.M. (1984) Conformational change in the DNA associated with an unusual promoter mutation in a tRNA operon of *Salmonella*. *Cell* **39**, 643–52.

Bossie M.A. & Silver P.A. (1992) Movement of macromolecules between the cytoplasm and the nucleus in yeast. *Curr. Opinion Genetics Devel.* **2**, 768–74.

Bostian K.A., Ellio O., Bussey H., Burn V., Smith A. & Tipper D.J. (1984) Sequence of the prepro-toxin ds RNA gene of Type 1 killer yeast: multiple processing events produce a two-component toxin. *Cell* **36**, 741–51.

Botstein D. & Davis R.W. (1982) Principles and practice of recombinant DNA research with yeast. In *The Molecular Biology of the Yeast Saccharomyces*, eds Strathern J.N., Jones E.W. & Broach J.R. Cold Spring Harbor Press, Cold Spring Harbor, New York.

Botterman J. & Leemans J. (1988) Engineering herbicide resistance in plants. *Trends Genetics* **4**, 219–22.

Bottger E.C. (1988) High-efficiency generation of plasmid cDNA libraries using electro-transformation. *Biotechniques* **6**, 878–80.

Bouret P. & Belasco J.G. (1992) Control of RNase E-mediated RNA degradation by 5'-terminal base pairing in *E. coli*. *Nature* **360**, 488–91.

Boyko W.L. & Ganschow R.E. (1982) Rapid identification of *Escherichia coli* transformed by pBR322 carrying inserts at the *Pst*I site. *Anal. Biochem.* **122**, 85–8.

Brange J., Ribel U., Hansen J.F., Dodson G., Hansen M.T., Haveland S., Melberg S.G., Norris F., Norris K., Snel L., Sørenson A.R. & Voist H.O. (1988) Monomeric insulins obtained by protein engineering and their medical implications. *Nature* **333**, 679–82.

Braun A.C. & Wood H.N. (1976) Suppression of the neoplastic state with the acquisition of specialised functions in cells, tissues and organs of crown gall teratoma of tobacco. *Proc. Natl. Acad. Sci. USA* **73**, 496–500.

Breitling R., Sorokin A.V., Ellingert & Behnke D. (1990) Controlled gene expression in *Bacillus subtilis* based on the temperature-sensitive λ cI repressor. In *Genetics and Biotechnology of Bacilli*, eds Zukowski M.M., Ganesan A.T. & Hoch J.A., Vol. 3, pp. 3–12. Academic Press, San Diego.

Breter H.J., Knoop M.-T. & Kirchen H. (1987) The mapping of chromosomes in *Saccharomyces cerevisiae*. I. A cosmid vector designed to establish, by cloning *cdc* mutants, numerous start loci for chromosome walking in the yeast genome. *Gene* **53**, 181–90.

Breunig K.D., Mackedonski V. & Hollenberg C.P. (1982) Transcription of the bacterial β-lactamase gene in *Saccharomyces cerevisiae*. *Gene* **20**, 1–10.

Brickman E. & Beckwith J. (1975) Analysis of the regulation of *Escherichia coli* alkaline phosphatase synthesis using deletions and phi-80 transducing phages. *J. Mol. Biol.* **96**, 307–16.

Brigatti D.J., Myerson D., Leary J.J., Spalholz B., Travis S.Z., Fong C.K.Y., Hsuing G.D. & Ward D.C. (1983) Detection of viral genomes in cultured cells and paraffin-embedded tissue sections using biotin-labelled hybridization probes. *Virology* **126**, 32–50.

Brinster R.L., Chen H.Y., Trumbauer M., Senear A.W., Warren R. & Palmiter R.D. (1981) Somatic expression of Herpes thymidine kinase in mice following injection of a fusion gene into eggs. *Cell* **27**, 223–31.

Brinster R.L., Chen H.Y., Warren R., Sarthy A. & Palmiter R.D. (1982) Regulation of metallothionein-thymidine kinase fusion plasmids injected into mouse eggs. *Nature* **296**, 39–42.

Brisson N., Paszkowski J., Penswick J.R., Gronenborn B., Potrykus I. & Hohn T. (1984) Expression of a bacterial gene in plants by using a viral vector. *Nature* **310**, 511–14.

Bristow A.F. (1993) Recombinant-DNA-derived insulin analogues as potentially useful therapeutic agents. *Trends Biotechnol.* **11**, 301–305.

Brochier B., Kieny M.P., Costy F., Coppens P., Bauduin B., Lecocq J.P., Languet B., Chappuis G., Desmettre P., Afiademanyo K., Libois R. & Pastoret P.P. (1991) Large-scale eradication of rabies using recombinant vaccinia-rabies vaccine. *Nature* **354**, 520–2.

Broda P. (1979) *Plasmids*. W.H. Freeman & Co., San Francisco.

Broglie K., Chet I., Holliday M., Cressman R., Biddle P., Knowlton S., Mauvais C.J. & Broglie R. (1991) Transgenic plants with enhanced resistance of the fungal pathogen *Rhizoctonia solanii*. *Science* **254**, 1194–7.

Bron S. & Luxen E. (1985) Segregational instability of pUB110-derived recombinant plasmids in *Bacillus subtilis*. *Plasmid* **14**, 235–44.

Bron S., Bosma P., Van Belkum M. & Luxen E. (1988) Stability function in the *Bacillus subtilis* plasmid pTA1060. *Plasmid* **18**, 8–16.

Bron S., Peijnenburg A., Peeters B., Haima P. & Venema G. (1989) Cloning and plasmid (in)stability in *Bacillus subtilis*. In *Genetic Transformation and Expression*, eds Butler O.O., Harwood C.R. & Moseley B.E.B., pp. 205–19. Intercept Ltd, Andover.

Bruand C., Ehrlich S.D. & Jannière L. (1991) Unidirectional theta replication of the structurally stable *Enterococcus faecalis* plasmid pAMβ1. *EMBO J.* **10**, 2171–7.

Brumbaugh J.A., Middendorf L.R., Grone D. & Ruth J.L. (1988) Continuous, on-line DNA sequencing using oligodeoxynucleotide primers with multiple fluorophores. *Proc. Natl. Acad. Sci. USA* **85**, 5610–14.

Brunier D., Bron S., Peeters B. & Ehrlich S.D. (1989) Mechanisms of recombination between short homologous sequences. *EMBO J.* **8**, 3127–33.

Brunschwig E. & Darzins A. (1992) A two-component T7 system for the overexpression of genes in *Pseudomonas aeruginosa*. *Gene* **111**, 35–41.

Buchanan-Wollaston V., Passiatore J.E. & Channon F. (1987) The *mob* and *ori* T functions of a bacterial plasmid promote its transfer to plants. *Nature* **328**, 172–5.

Buckholz R.G. (1993) Yeast systems for the expression of heterologous gene products. *Curr. Opinion Biotechnol.* **4**, 538–542.

Buell G., Schulz M.-F., Selzer G., Chollet A., Movva N.R., Semon D., Escanez S. & Kawashima E. (1985) Optimizing the expression in *E. coli* of a synthetic gene encoding somatomedin-C (IGF-I). *Nucleic Acids Res.* **13**, 1923–38.

Bugawan T.L., Saiki R.K., Levenson C.H., Watson R.M. & Ehrlich H.A. (1988) The use of non-radioactive oligonucleotide probes to analyze enzymatically amplified DNA for prenatal diagnosis and forensic HLA typing. *Biotechnology* **6**, 943–7.

Burke T. & Bruford M.W. (1987) DNA fingerprinting in birds. *Nature* **327**, 149–52.

Burnette W.N. (1981) Western blotting: electrophoretic transfer of proteins from sodium dodecyl sulphate–polyacrylamide gels to unmodified nitrocellulose and radiographic detection with antibody and radioiodinated protein A. *Anal. Biochem.* **112**, 195–203.

Cabezon T., De Wilde M., Herion P., Lorian R. & Bollen A. (1984) Expression of human alpha$_1$-antitrypsin cDNA in the yeast *Saccharomyces cerevisiae*. *Proc. Natl. Acad. Sci. USA* **81**, 6594–8.

Cameron J.R., Panasenko S.M., Lehman I.R. & Davis R.W. (1975) *In vitro* construction of bacteriophage λ carrying segments of the *Escherichia coli* chromosome: selection of hybrids containing the gene for DNA ligase. *Proc. Natl. Acad. Sci. USA* **72**, 3416–20.

Campo M.S. (1985) Bovine papillomavirus DNA: a eukaryotic cloning vector. In *DNA Cloning: A Practical Approach*, Vol. II, ed. Glover D., pp. 213–39. IRL Press, Oxford.

Canosi U., Iglesias A. & Trautner T.A. (1981) Plasmid transformation in *Bacillus subtilis*: effects of insertion of *Bacillus subtilis* DNA into plasmid pC194. *Mol. Gen. Genetics* **181**, 434–40.

Canosi U., Morelli G. & Trautner T.A. (1978) The relationship between molecular structure and transformation efficiency of some *S. aureus* plasmids isolated from *B. subtilis*. *Mol. Gen. Genetics* **166**, 259–67.

Cariello N., Swenberg J. & Skopek T. (1991) Fidelity of *Thermococcus litoralis* DNA polymerase (Vent^{TM}) in PCR determined by denaturing gradient gel electrophoresis. *Nucleic Acids Res.* **19**, 4193–8.

Carle G.R., Frank M. & Olson M.V. (1986) Electrophoretic separations of large DNA molecules by periodic inversion of the electric field. *Science* **232**, 65–8.

Carle G.R. & Olson M.V. (1984) Separation of chromosomal DNA molecules from yeast by orthogonal-field-alternation gel electrophoresis. *Nucleic Acids Res.* **12**, 5647–64.

Carlton B.C. & Helinski D.R. (1969) Heterogeneous circular DNA elements in vegetative cultures of *Bacillus megaterium*. *Proc. Natl. Acad. Sci. USA* **64**, 592–9.

Carter P., Bedouelle H. & Winter G. (1985) Improved oligonucleotide site-directed mutagenesis using M13 vectors. *Nucleic Acids Res.* **13**, 4431–43.

Carter P., Abrahmsen L. & Wells J.A. (1991) Probing the mechanism and improving the rate of substrate-assisted catalysis in subtilisin BPN. *Biochemistry* **30**, 6142–8.

Carter P. & Wells J.A. (1987) Engineering enzyme specificity by 'substrate-assisted catalysis'. *Science* **237**, 394–9.

Cartier M., Chang M. & Stanners C. (1987) Use of the *Escherichia coli* gene for asparagine synthetase as a selective marker in a shuttle vector capable of dominant transfection and amplification in animal cells. *Mol. Cell Biol.* **7**, 1623–8.

Ceriotti A. & Colman A. (1990) Trimer formation determines the rate of influenza virus hemagglutinin transport in the early stages of secretion in *Xenopus* oocytes. *J. Cell Biol.* **111**, 409–20.

Cesarini G., Muesing M.A. & Polisky B. (1982) Control of Col E1 DNA replication: the *rop* gene product negatively affects transcription from the replication primer promoter. *Proc. Natl. Acad. Sci. USA* **79**, 6313–17.

Cesarini G., Helmer-Citterich M. & Castagnoli L. (1991) Control of Col E1 plasmid replication by antisense RNA. *Trends Genetics* **7**, 230–5.

Chada K., Magram J. & Constantini F. (1986) An embryonic pattern of expression of a human fetal globin gene in transgenic mice. *Nature* **319**, 685–9.

Chakrabarty R. & Kidd K.K. (1991) The utility of DNA typing in forensic work. *Science* **254**, 1735–9.

Chakrabarti S., Robert-Guroff M., Wong-Staal F., Gallo R.C. & Moss B. (1986) Expression of the HTLV-III envelope gene by a recombinant vaccinia virus. *Nature* **320**, 535–7.

Chamberlain J.S., Pearlman J.A., Muzny D.M., Gibbs R.A., Ranier J.E., Reeves A.A. & Caskey C.T. (1988) Expression of murine muscular dystrophy gene in muscle and brain. *Science* **239**, 1416–18.

Chang A.C.Y., Ehrlich H.A., Gunsalus R.P., Nunberg J.H., Kaufman R.J., Schimke R.T. & Cohen S.N. (1980) Initiation of protein synthesis in bacteria at a translational start codon of mammalian cDNA: effects of the preceding nucleotide sequence. *Proc. Natl. Acad. Sci. USA* **77**, 1442–6.

Chang A.C.Y., Nunberg J.H., Kaufman R.K., Ehrlich H.A., Schimke R.T. & Cohen S.N. (1978) Phenotypic expression in *E. coli* of a DNA sequence coding for mouse dihydro-folate reductase. *Nature* **275**, 617–24.

Chang J.C. & Kan Y.W. (1981) Antenatal diagnosis of sickle-cell anaemia by direct analysis of the sickle mutation. *Lancet* **2**, 1127–9.

Chang L.M.S. & Bollum F.J. (1971) Enzymatic synthesis of oligodeoxynucleotides. *Biochemistry* **10**, 536–42.

Chang S. & Cohen S.N. (1979) High-frequency transformation of *Bacillus subtilis* pro-toplasts by plasmid DNA. *Mol. Gen. Genetics* **168**, 111–15.

Chapman A.B., Costello M.A., Lee R. & Ringold G.M. (1983) Amplification and hormone-regulated expression of a mouse mammary tumor virus-Ecogpt fusion plasmid in mouse 3T6 cells *Mol. Cell Biol.* **3**, 1421–9.

Charbit A., Molla A., Saurin W. & Hofnung M. (1988) Versatility of a vector for expressing foreign polypeptides at the surface of Gram-negative bacteria. *Gene* **70**, 181–9.

Charles I. & Dougan G. (1990) Gene expression and the development of live enteric vaccines. *Trends Biotechnol* **8**, 117–21.

Chatoo B.B., Sherman F., Azubalis D.A., Fjellstedt T.A., Mehvert D. & Oghur M. (1979) Selection of *lys2* mutants of the yeast *Saccharomyces cerevisiae* by the utilisation of α-aminoadipate. *Genetics* **93**, 51–65.

Chen C.Y. & Hitzeman R.A. (1987) Human, yeast and hybrid 3-phosphoglycerate kinase gene expression in yeast. *Nucleic Acids Res.* **15**, 643–60.

Chen C.Y., Oppermann H. & Hitzeman R.A. (1984) Homologous versus heterologous gene expression in the yeast *Saccharomyces cerevisiae*. *Nucleic Acids Res.* **12**, 8951–70.

Chen E., Howley P., Levenson A. & Seeburg P. (1982) The primary structure and genetic organisation of the bovine papilloma virus type I genome. *Nature* **299**, 529–34.

Chen H.Y., Garber E.A., Rosenblum C.I., Taylor J.E., Kopchick J.J., Smith R.G., Smith J. & Mills E.O. (1989) Expression of bovine growth hormone gene in transgenic chickens. *J. Cell Biochem.* **13B**, 178.

Chen J.-D. & Morrisson D.A. (1987) Cloning of *Streptococcus pneumoniae* DNA fragments in *Escherichia coli* requires vectors protected by strong transcriptional terminators. *Gene* **55**, 179–87.

Chen K. & Arnold F.H. (1991) Enzyme engineering for non-aqueous solvents: random mutagenesis to enhance activity of subtilisin E in polar organic media. *Biotechnology* **9**, 1073–7.

Chen K. & Arnold F.H. (1993) Tuning the activity of an enzyme for unusual environments: sequential random mutagenesis of subtilisin E for catalysis in dimethylformamide. *Proc. Natl. Acad. Sci. USA* **90**, 5618–22.

Cherfas A. (1991) Ancient DNA: still busy after death. *Science* **253**, 1354–1356.

Chiang T.-R. & McConlogue L. (1988) Amplification of heterologous ornithine decar-boxylase in Chinese hamster ovary cells. *Mol. Cell Biol.* **8**, 764–9.

Chien C.-T., Bartel P.L., Sternglanz R. & Fields S. (1991) The two-hybrid system: a method to identify and clone genes for proteins that interact with a protein of interest. *Proc. Natl. Acad. Sci. USA* **88**, 9578–82.

Chilton M.-D., Drummond M.H., Merlo D.J., Sciaky D., Montoya A.L., Gordon M.P. & Nester E.W. (1977) Stable incorporation of plasmid DNA into higher plant cells: the molecular basis of crown gall tumorigenesis. *Cell* **11**, 263–71.

Chilton M.-D., Tepfer D.A., Petit A., David C., Casse-Delbart F. & Tempe J. (1982) *A. rhizogenes* inserts T-DNA into the genomes of host plant root cells. *Nature* **295**, 432–4.

Chinery S.A. & Hinchliffe E. (1989) A novel class of vector for yeast transformation. *Curr. Genetics* **16**, 21–5.

Chlebowicz-Sledziewska E. & Sledziewski A.Z. (1985) Construction of multicopy yeast plasmids with regulated centromere function. *Gene* **39**, 25–31.

Chow L.H., Yee S.P., Pawson T. & McManus B.M. (1991) Progressive cardiac fibrosis and myocyte injury in V-*fps* transgenic mice. *Lab. Invest.* **64**, 457–62.

Chow T.Y.-K., Ash J.J., Dignard D. & Thomas D.Y. (1992) Screening and identification of a gene, PSE-1, that affects protein secretion in *Saccharomyces cerevisiae*. *J. Cell Sci.* **101**, 709–19.

Christensen T., Woeldike H., Boel E., Mortensen S.B., Hjortschoejk., Thim L. & Hansen M.T. (1988) High level expression of recombinant genes in *Aspergillus oryzae*. *Biotechnology* **6**, 1419–22.

Christou P., McCabe D.E. & Swain W.F. (1988) Stable transformation of soybean callus by DNA-coated gold particles. *Plant Physiol.* **87**, 671–4.

Christman J.K., Gerber M., Price P.M., Flordellis C., Edelman J. & Acs G. (1982) Amplification of expression of hepatitis B surface antigen in 3T3 cells cotransfected with a dominant-acting gene and cloned viral DNA. *Proc. Natl. Acad. Sci. USA* **79**, 1815–19.

Christou P. (1992) Genetic transformation of crop plants using microprojectile bombardment. *Plant J.* **2**, 275–81.

Christou P. (1993) Particle gun mediated transformation. *Curr. Opinion Biotechnol.* **4**, 135–141.

Christou P., McCabe D.E. & Swain W.F. (1988) Stable transformation of soybean callus by DNA-coated gold particles. *Plant Physiol.* **87**, 671–4.

Christou P., Ford T.L. & Kofron M. (1991) Production of transgenic rice (*Oryza sativa* L.) plants from agronomically important indica and japonica varieties via electric discharge particle acceleration of exogenous DNA into immature zygotic embryos. *Biotechnology* **9**, 957–62.

Chu G., Hayakawa H. & Berg P. (1987) Electroporation for the efficient transfection of mammalian cells with DNA. *Nucleic Acids Res.* **15**, 1311–26.

Chu, G. & Sharp P.A. (1981) SV40 DNA transfection of cells in suspension: analysis of the efficiency of transcription and translation of T-antigen. *Gene* **13**, 197–202.

Clackson T. & Winter G. (1989) 'Sticky feet' – directed mutagenesis and its application to swapping antibody domains. *Nucleic Acids Res.* **17**, 10163–70.

Clackson T., Hoogenboom H.R., Griffiths A.D. & Winter G. (1991) Making antibody fragments using phage display libraries. *Nature* **352**, 624–8.

Clare J.J., Rayment F.B., Ballantine S.P., Sreekrishna K. & Romanos M.A. (1991) High level expression of tetanus toxin fragment C in *Pichia pastoris* strains containing multiple tandem integrations of the gene. *Biotechnology* **9**, 455–60.

Clark W., Register J., Eichholtz D., Sanders P., Fraley R. & Beachy R. (1990) Tissue-specific expression of the TMV coat protein in transgenic tobacco plants affects the level of coat protein-mediated virus protection. *Virology* **179**, 640–7.

Clarke L. & Carbon J. (1980) Isolation of a yeast centromere and construction of functional small circular chromosomes. *Nature* **287**, 504–9.

Cockett M.I., Bebbington C.R. & Yarranton G.T. (1990) High level expression of tissue inhibitor of metalloproteinases in Chinese hamster ovary cells using glutamine synthetase gene amplification. *Biotechnology* **8**, 662–7.

Cohen J.D., Eccleshall T.R., Needleman R.B., Federoff H., Buchferer B.A. & Marmur J. (1980) Functional expression in yeast of the *Escherichia coli* plasmid gene coding for chloramphenicol acetyl-transferase. *Proc. Natl. Acad. Sci. USA* **77**, 1078–82.

Cohen S.N., Chang A.C.Y. & Hsu L. (1972) Nonchromosomal antibiotic resistance in bacteria: genetic transformation of *Escherichia coli* by R-factor DNA. *Proc. Natl. Acad. Sci. USA* **69**, 2110–4.

Colbère-Garapin F., Horodniceanu F., Kourilsky P. & Garapin A.C. (1981) A new dominant hybrid selective marker for higher eukaryotic cells. *J. Mol. Biol.* **150**, 1–14.

Collins F.S., Drumm M.L., Cole J.L., Lockwood W.K., Vande Woude G.F. & Iannuzzi

M.C. (1987) Construction of a general human chromosome jumping library, with application to cystic fibrosis. *Science* **235**, 1046–9.

Collins F.S. & Weissman S.M. (1984) Directional cloning of DNA fragments at a large distance from an initial probe: a circularization method. *Proc. Natl. Acad. Sci. USA* **81**, 6812–16.

Collins J. & Brüning H.J. (1978) Plasmids usable as gene-cloning vectors in an *in vitro* packaging by coliphage λ: 'cosmids'. *Gene* **4**, 85–107.

Collins J. & Hohn B. (1979) Cosmids: a type of plasmid gene-cloning vector that is packageable *in vitro* in bacteriophage λ heads. *Proc. Natl. Acad. Sci. USA* **75**, 4242–6.

Colman A. (1984) Translation of eukaryotic messenger RNA in *Xenopus* oocytes. In: *Transcription and Translation – A Practical Approach*, eds Hames BD & Higgens SJ, pp. 271–302. IRL Press, Oxford.

Colman A., Lane C., Craig R., Boulton A., Mohun T. & Morser J. (1981) The influence of topology and glycosylation on the fate of heterologous secretory proteins made in *Xenopus* oocytes. *Eur. J. Biochem.* **113**, 339–348.

Comai L., Facciotti D., Hiatt W.R., Thompson G., Rose R.T. & Stalker D.M. (1985) Expression in plants of a mutant *aroA* gene from *Salmonella typhimurium* confers tolerance to glyphosate. *Nature* **317**, 741–4.

Compton J. (1991) Nucleic acid sequence-based amplification. *Nature* **350**, 91–2.

Cone R.D. & Mulligan R.C. (1984) High-efficiency gene transfer into mammalian cells: generation of helper-free recombinant retrovirus with broad mammalian host range. *Proc. Natl. Acad. Sci. USA* **81**, 6349–53.

Conner B.J., Reyes, A.A., Morin C., Itakura K., Teplitz R. & Wallace R.B. (1983) Detection of sickle cell β^s-globin allele by hybridization with synthetic oligonucleotides. *Proc. Natl. Acad. Sci. USA* **80**, 278–82.

Contente S. & Dubnau D. (1979) Characterization of plasmid transformation in *Bacillus subtilis*: kinetic properties and the effect of DNA conformation. *Mol. Gen. Genetics* **167**, 251–8.

Cooley L., Berg C. & Spradling A. (1988) Controlling P element insertional mutagenesis. *Trends Genetics* **4**, 254–8.

Cooney E.L., Collier A.C., Greenberg P.D., Coombs R.W., Zarling J., Arditti D.E., Hoffman M.C., Hu S.-L. & Corey L. (1991) Safety and immunological response to a recombinant vaccinia virus vaccine expressing HIV envelope glycoprotein. *Lancet* **337**, 567–72.

Coppella S.J., Acheson C.M. & Dhurjati P. (1987) Measurement of copy number using HPLC. *Biotechnol. Bioeng.* **XXIX**, 646–7.

Cosloy S.D. & Oishi M. (1973) Genetic transformation in *Escherichia coli* K12. *Proc. Natl. Acad. Sci. USA* **70**, 84–7.

Courtney M., Jallat S., Terrier L.-H., Benavente A., Crystal R.G. & Lecocq J.-P. (1985) Synthesis in *E. coli* of alpha$_1$-antitrypsin variants of therapeutic potential for emphysema and thrombosis. *Nature* **313**, 149–51.

Cousens L.S., Shuster J.R., Gallegos C., Ku L., Stempien M.M., Urden M.S., Sanchez-Pescador R., Taylor A. & Tekamp-Olson P. (1987) High level of expression of pro-insulin in the yeast, *Saccharomyces cerevisiae*. *Gene* **61**, 265–75.

Cousens D.J., Wilson M.J. & Hinchliffe E. (1990) Construction of a regulated PGK expression vector. *Nucleic Acids Res.* **18**, 1308.

Cox D.W., Woo S.L.C. & Mansfield T. (1985) DNA restriction fragments associated with alpha-antitrypsin indicate a single origin for deficiency allele PIZ. *Nature* **316**, 79–81.

Crabeel M., Huygen R., Cunin R. & Glansdorff N. (1983) The promoter region of the *arg3* gene in *Saccharomyces cerevisiae*. Nucleotide sequence and regulation in an *arg3–lacZ* gene fusion. *EMBO J.* **2**, 205–12.

Crawley M.J., Hails R.S., Rees M., Kohn D. & Buxton J. (1993) Ecology of transgenic oilseed rape in natural habitats. *Nature* **363**, 620–3.

Crouse G.F., Simonsen C.C., McEwan R.N. & Schrinke R.T. (1982) Structure of amplified normal and variant dihydrofolate reductase genes in mouse sarcoma S180 cells. *J. Biol. Chem.* **257**, 7887–97.

Crouzet J., Levy-Schil S., Cauchois L. & Cameron B. (1992) Construction of a broad-host-range non-mobilizable stable vector carrying RP4 *par*-region. *Gene* **110**, 105–8.

Crowe J.S., Cooper H.J., Smith M.A., Sims M.J., Parker D. & Gewert D. (1990) Improved cloning efficiency of polymerase chain reaction (PCR) products after proteinase K digestion. *Nucleic Acids Res.* **19**, 184.

Crutz A.-M., Steinmeta M., Aymerich S., Richter R. & Lecoq D. (1990) Induction of levansucrase in *Bacillus subtilis*: an antitermination mechanism negatively controlled by the phosphotransferase system. *J. Bacteriol.* **172**, 1043–50.

Cryz S.J. (1992) Live attenuated vaccines for human use. *Curr. Opinion Biotechnol.* **3**, 298–302.

Cunningham B.C. & Wells J.A. (1989) High-resolution epitope mapping of hGH-receptor interactions by alanine-scanning mutagenesis. *Science* **244**, 1081–5.

Cunningham B.C., Jhurani P., Ng P. & Wells J.A. (1989) Receptor and antibody epitopes in human growth hormone identified by homolog-scanning mutagenesis. *Science* **243**, 1330–6.

Cunningham T.P., Montelaro R.C. & Rushlow K.E. (1993) Lentivirus envelope sequences and proviral genomes are stabilized in *Escherichia coli* when cloned in low-copy-number plasmid vectors. *Gene* **124**, 93–8.

Currier T.C. & Nester E.W. (1976) Evidence for diverse types of large plasmids in tumor-inducing strains of *Agrobacterium*. *J. Bacteriol.* **126**, 157–65.

Cwirla S.E., Peters E.A., Barrett R.W. & Dower W.J. (1990) Peptides on phage: a vast library of peptides for identifying ligands. *Proc. Natl. Acad. Sci. USA* **87**, 309–14.

Dalbadie-McFarland G., Cohen L.W., Riggs A.D., Morin C., Itakura K. & Richards J.H. (1982) Oligonucleotide-directed mutagenesis as a general and powerful method for studies of protein functions. *Proc. Natl. Acad. Sci. USA* **79**, 6409–13.

Dani G.M. & Zakian U.A. (1983) Mitotic and meiotic stability of linear plasmids in yeast. *Proc. Natl. Acad. Sci. USA* **80**, 3406–10.

Danos O. & Mulligan R.C. (1988) Safe and efficient generation of recombinant retroviruses with amphotropic and ecotropic host ranges. *Proc. Natl. Acad. Sci USA* **85**, 6460–4.

Datta S.K., Peterhaus A., Datta K. & Potrykus I. (1990) Genetically engineered fertile indica-rice recovered from protoplasts. *Biotechnology* **8**, 736–40.

Davidson B.L., Allen E.D., Kozarsky K.F., Wilson J.M. & Roessler B.J. (1993) A model system for *in vivo* gene transfer into the central nervous system using an adenoviral vector. *Nature Genet.* **3**, 219–23.

Davies J.W. & Stanley J. (1989) Geminivirus genes and vectors. *Trends Genetics* **5**, 77–81.

Davis B. & MacDonald R.J. (1988) Limited transcription of rat elastase I transgene repeats in transgenic mice. *Genes Devel.* **2**, 13–22.

Davison J., Brunel F. & Merchez M. (1979) A new host-vector system allowing selection for foreign DNA inserts in bacteriophage λ gt *WES*. *Gene* **8**, 69–80.

Davison J., Chevalier N. & Brunel F. (1989) Bacteriophage T7 RNA polymerase-controlled specific gene expression in *Pseudomonas*. *Gene* **83**, 371–5.

Day A.G., Bejarano E., Buck K.W., Burrell M. & Lichtenstein C. (1991) Expression of an antisense viral gene in transgenic tobacco confers resistance to the DNA virus tomato golden mosaic virus. *Proc. Natl. Acad. Sci. USA* **88**, 6721–5.

De Block M., Botterman J., Vandewiele M., Dockx J., Thoen C., Gossele V., Rao Movra N., Thompson C., Van Montagu M. & Leemans J. (1987) Engineering herbicide resistance in plants by expression of a detoxifying enzyme. *EMBO J.* **6**, 2513–18.

De Block M., Herrera-Estrella L., Van Montagu M., Schell J. & Zambryski P. (1984) Expression of foreign genes in regenerated plants and their progeny. *EMBO J.* **3**, 1681–9.

De Block M., Schell J. & Van Montagu M. (1985) Chloroplast transformation by *Agrobacterium tumefaciens*. *EMBO J.* **4**, 1367–72.

De Boer H.A. & Hui A.S. (1990) Sequences within ribosome binding site affecting messenger RNA translatability and method to direct ribosomes to single messenger RNA species. *Methods Enzymol.* **185**, 103–14.

de Boer H.A., Comstock L.J. & Vasser M. (1983a) The *tac* promoter: a functional hybrid derived from the *trp* and *lac* promoters. *Proc. Natl. Acad. Sci. USA* **80**, 21–5.

de Boer H.A., Hui A., Comstock L.J., Wong E. & Vasser M. (1983b) Portable Shine–Dalgarno regions: a system for a systematic study of defined alterations of nucleotide sequences within *E. coli* ribosome binding sites. *DNA* **2**, 231–41.

de Framond A.J., Barton K.A. & Chilton M.-D. (1983) Mini-Ti: a new vector strategy for plant genetic engineering. *Biotechnology* **1**, 262–9.

De Greef W., Delon R., De Block M., Leemans J. & Botterman J. (1989) Evaluation of herbicide resistance in transgenic crops under field conditions. *Biotechnology* 7, 61–4.

De Greve H., Decraemer H., Senrinck J., Van Montagu M. & Schell J. (1981) The functional organization of the octopine *Agrobacterium tumefaciens* plasmid pTi B653. *Plasmid* 6, 235–48.

De Greve H., Leemans J., Hernalsteens J.P., Thia-Toong L., De Benckeleer M., Willmitzer L., Olten L., Van Montagu M. & Schell J. (1982a) Regeneration of normal fertile plants that express octopine synthase from tobacco crown galls after deletion of tumor-controlling functions. *Nature* 300, 752–5.

De Greve H., Phaese P., Seurwick J., Lemmers M., Van Montagu M. & Schell J. (1982b) Nucleotide sequence and transcript map of the *Agrobacterium tumefaciens* Ti plasmid-encoded octopine synthase gene. *J. Mol. Appl. Genetics* 1, 499–511.

de Saint Vincent B.R., Delbruck S., Eckhart W., Meinkoth J., Vitto L. & Wahl G. (1981) The cloning and reintroduction into animal cells of a functional CAD gene, a dominant amplifiable genetic marker. *Cell* 27, 267–77.

de Vos W.M., Venema G., Canosi U. & Trautner T.A. (1981) Plasmid transformation in *Bacillus subtilis*: fate of plasmid DNA. *Mol. Gen. Genetics* 181, 424–33.

de Wet J.R., Wood K.V., De Luca M., Helinski D.R. & Subramani S. (1987) Firefly luciferase gene: structure and expression in mammalian cells. *Mol. Cell Biol.* 7, 725–37.

Dean D. (1981) A plasmid cloning vector for the direct selection of strains carrying recombinant plasmids. *Gene* 15, 99–102.

Deans R.J., Denis K.A., Taylor A. & Wall R. (1984) Expression of an immunoglobulin heavy chain gene transfected into lymphocytes. *Proc. Natl. Acad. Sci. USA* 81, 1292–6.

Debenham P.G. (1992) Probing identity: the changing face of DNA fingerprinting. *Trends Biotechnol.* 10, 96–102.

Delannay X., La Valle B.J., Proksch R.K., Fuchs R.L., Sims S.R., Greenplate J.T., Marrone P.G., Dodson R.B., Augustine J.J., Layton J.G. & Fischhoff D.A. (1989) Field performance of transgenic tomato plants expressing the *Bacillus thuringiensis* var. *kurstaki* insect control protein. *Biotechnology* 7, 1265–9.

Della-Cioppa G., Garger S.J., Sverlow G.G., Turpen T.J. & Grill L.K. (1990) Melanin production in *Escherichia coli* from a cloned tyrosinase gene. *Biotechnology* 8, 634–38.

Delseny M. & Hull R. (1983) Isolation and characterization of faithful and altered clones of the genomes of cauliflower mosaic virus isolates Cabb B–J1, CM4–184 and Bari 1. *Plasmid* 9, 31–41.

Demolder J., Fiers W. & Contreras R. (1992) Efficient synthesis of secreted murine interleukin-2 by *Saccharomyces cerevisiae*: influence of 3'-untranslated regions and codon usage. *Gene* 111, 207–13.

Deng G. & Wu R. (1981) An improved procedure for utilizing terminal transferase to add homopolymers to the 3' termini of DNA. *Nucleic Acids Res.* 9, 4173–88.

Deng W.P. & Nickeloff J.A. (1992) Site-directed mutagenesis of virtually any plasmid by eliminating a unique site. *Anal. Biochem.* 200, 81–8.

Denhardt D.T. (1966) A membrane-filter technique for the detection of complementary DNA. *Biochem. Biophys. Res. Commun.* 23, 641–6.

Denman J., Hayes M., O'Day C., Edmunds T., Bartlett C., Hirani S., Ebert K.M., Gordon K. & McPherson J.M. (1991) Transgenic expression of a variant of human tissue-type plasminogen activator in goat milk: purification and characterization of the recombinant enzyme. *Biotechnology* 9, 839–43.

Dente L., Cesareni G. & Cortese R. (1983) pEMBL: a new family of single stranded plasmids. *Nucleic Acids Res.* 11, 1645–55.

Depicker A., Stachel S., Dhaese P., Zambryski P. & Goodman H.M. (1982) Nopaline synthase: transcript mapping and DNA sequence. *J. Mol. Appl. Genetics* 1, 561–74.

Deretic V., Chandrasekharappa S., Gill J.F., Chaterjee D.K. & Chakrabarty A.M. (1987) A set of cassettes and improved vectors for genetic and biochemical characterization of *Pseudomonas* genes. *Gene* 57, 61–72.

Derynck R., Singh A. & Goeddel D. (1983) Expression of the human interferon-γ cDNA in yeast. *Nucleic Acids Res.* 11, 1819–37.

Deshayes A., Herrera-Estrella L. & Caboche M. (1985) Liposome-mediated transformation of tobacco mesophyll protoplasts by an *Escherichia coli* plasmid. *EMBO J.* 4, 2731–7.

Devlin J.J., Panganiban L.C. & Devlin P.E. (1990) Random peptide libraries: a source of specific protein binding molecules. *Science* **249**, 404–6.

Devlin P.E., Drummond R.J., Toy P., Mark D.F., Watt K.W. & Devlin J.J. (1988) Alteration of amino-terminal codons of human granulocyte-colony-stimulating factor increases expression levels and allows efficient processing by methionine aminopeptidase in *Escherichia coli*. *Gene* **65**, 13–22.

Devor E.J., Ivanovich A.K., Hickok J.M. & Todd R.D. (1988) A rapid method for confirming cell line identity: DNA fingerprinting with a minisatellite probe from M13 bacteriophage. *Biotechniques* **6**, 200–1.

Di Maio D., Triesman R. & Maniatis T. (1982) A bovine papillomavirus vector which propagates as an episome in both mouse and bacterial cells. *Proc. Natl. Acad. Sci. USA* **79**, 4030–4.

Diaz R. & Staudenbauer W.L. (1982) Replication of the broad host range plasmid RSF1010 in *Escherichia coli* K-12. *J. Bacteriol.* **134**, 1117–22.

Ditta G., Stanfield S., Corbin D. & Helinski D.R. (1980) Broad host range DNA cloning system for Gram-negative bacteria. Construction of a gene bank of *Rhizobium meliloti*. *Proc. Natl. Acad. Sci. USA* **77**, 7347–51.

Dobson M.J., Tuite M.F., Roberts N.A., Kingsman A.J., Kingsman S.M., Perkins R.E., Conroy S.C., Dunbar B. & Fothergill L.A. (1982) Conservation of high efficiency promoter sequences in *Saccharomyces cerevisiae*. *Nucleic Acids Res.* **10**, 2625–37.

Dobson M.J., Tuite M.F., Mellor J., Roberts N.A., King R.M., Burke D.C., Kingsman A.J. & Kingsman S.M. (1983) Expression in *Saccharomyces cerevisiae* of human interferon-alpha directed by the TRP 5' region. *Nucleic Acids Res.* **11**, 2287–302.

Donnelly D.F., Birkenhead K. & O'Gara F. (1987) Stability of IncQ and IncP-1 vector plasmids in *Rhizobium* spp. *FEMS Microbiol. Lett.* **42**, 141–5.

Doolittle R.F. (1986) *Of Urfs and Orfs: a primer on how to analyze derived amino acid sequences*. University Science Books, Mill Valley, California.

Dotto G.P., Enea V. & Zinder N.D. (1981) Functional analysis of bacteriophage f1 intergenic region. *Virology* **114**, 463–73.

Dotto G.P. & Horiuchi K. (1981) Replication of a plasmid containing two origins of bacteriophage f1. *J. Mol. Biol.* **153**, 169–76.

Dower W.J., Miller J.F. & Ragsdale C.W. (1988) High efficiency transformation of *E. coli* by high voltage electroporation. *Nucleic Acids Res.* **16**, 6127–45.

Drummond M.H. & Chilton M.-D. (1978) Tumor-inducing (Ti) plasmids of *Agrobacterium* share extensive regions of DNA homology. *J. Bacteriol.* **136**, 1178–83.

Dueschle U., Kammerer W., Genz R. & Bujard H. (1986) Promoters of *Escherichia coli*: a hierarchy of *in vivo* strength indicates alternate structure. *EMBO J.* **5** 2987–94.

Dugaiczyk A., Boyer H.W. & Goodman H.M. (1975) Ligation of EcoRI endonuclease-generated DNA fragments into linear and circular structures. *J. Mol. Biol.* **96**, 171–84.

Dunphy W.G. & Rothman J.E. (1985) Compartmental organisation of the Golgi stack. *Cell* **42**, 13–21.

Durand H., Clanet M. & Tiraby G. (1988) Genetic improvement of *Trichoderma reesei* for large scale cellulase production. *Enzyme Microbial Technol.* **10**, 341–5.

Durnam D.M. & Palmiter R.D. (1981) Transcriptional regulation of the mouse metallothionein-1 gene by heavy metals. *J. Biol. Chem.* **256**, 5712–16.

Duronio R.J., Jackson-Machelski E., Heuckeroth R.O., Olins P.O., Devine C.S., Yonemoto W., Slice L.W., Taylor S.S. & Gordon J.I. (1990) Protein N-myristoylation in *Escherichia coli*: reconstitution of a eukaryotic protein modification in bacteria. *Proc. Natl. Acad. Sci. USA* **87**, 1506–10.

Durfee T., Becherer K., Chen P.L., Yen S.H., Yang Y., Kilburn A.E., Lee W.H. & Elledge S.J. (1993) The retinoblastoma protein associates with the protein phosphatase type 1 catalytic subunit. *Genes Devel.* **7**, 555–69.

Dussoix D. & Arber W. (1962) Host specificity of DNA produced by *Escherichia coli*. II. Control over acceptance of DNA from infecting phage λ. *J. Mol. Biol.* **5**, 37–49.

Dycaico M.J., Grant S.G.O., Felts K., Nichols W.S., Geller S.A., Hager J.M., Pollard A.J., Kohler S.W., Short H.P., Jirik F.R., Hanahan D. & Sorge J.A. (1988) Neonatal hepatitis induced by α$_1$-antitrypsin: a transgenic mouse model. *Science* **242**, 1409–12.

Ebert K.M., Selgrath J.P., Ditullio P., Denman J., Smith T.E., Memon M.A., Schindler J.E., Monastersky G.M., Uitale J.A. & Gordon K. (1991) Transgenic production of a variant of human tissue-type plasminogen activator in goat milk: generation of transgenic goats and analysis of expression. *Biotechnology* **9**, 835–8.

Eckert K.A. & Kunkel T.A. (1990) High fidelity DNA synthesis by the *Thermus aquaticus* DNA polymerase. *Nucleic Acids Res.* **18**, 3739–44.

Eckert K.A. & Kunkel T.A. (1991) DNA polymerase fidelity and the polymerase chain reaction. *PCR: Methods Appl.* **1**, 17–24.

Efstratiadis A., Kafatos F.C., Maxam A.M. & Maniatis T. (1976) Enzymatic *in vitro* synthesis of globin genes. *Cell* **7**, 279–88.

Ehrlich H.A., Cohen S.N. & McDevitt H.O. (1978) A sensitive radioimmunoassay for detecting products translated from cloned DNA fragments. *Cell* **13**, 681–9.

Ehrlich S.D. (1977) Replication and expression of plasmids from *Staphylococcus aureus* in *Bacillus subtilis*. *Proc. Natl. Acad. Sci. USA* **74**, 1680–2.

Ehrlich S.D. (1978) DNA cloning in *Bacillus subtilis*. *Proc. Natl. Acad. Sci. USA* **75**, 1433–6.

Ehrlich S.D., Niaudet B. & Michel B. (1981) Use of plasmids from *Staphylococcus aureus* for cloning of DNA in *Bacillus subtilis*. *Curr. Top. Microbiol. Immunol.* **96**, 19–29.

Ehrlich S.D., Noirot P., Petit M.A., Janniere L., Michel B. & Te Riele H. (1986) Structural instability of *Bacillus subtilis* plasmids. In *Genetic Engineering*, eds Setlow J.K. & Hollaender A., Vol. 8, pp. 71–83. Plenum, New York.

Ellis D.D., McCabe D.E., McInnis S., Ramchandran R., Russell D.R., Wallace K.M., Martinell B.J., Roberts D.R., Raffa K.F. & McCown B.H. (1993) Stable transformation of *Picea glauca* by particle acceleration – a model system for conifer transformation. *Biotechnology* **11**, 84–9.

Eloit M., Gilardihebenstreit P., Toma B. & Perriccaudet M. (1990) Construction of a defective adenovirus vector expressing the pseudorabies virus glycoprotein gp50 and its use as a live vaccine. *J. Gen. Virol.* **71**, 2425–31.

Emerman M. & Temin H.M. (1986) Quantitative analysis of gene suppression in integrated retrovirus vectors. *Mol. Cell. Biol.* **6**, 792–800.

Enfors S.-O. (1992) Control of *in vivo* proteolysis in the production of recombinant proteins. *Trends Biotechnol.* **10**, 310–15.

Engebrecht J., Simon M. & Silverman M. (1985) Measuring gene expression with light. *Science* **227**, 1345–7.

Engleberg N.C., Cianciotto N., Smith J. & Eisenstein B.I. (1988) Transfer and maintenance of small, mobilizable plasmids with ColE1 replication origins: *Legionella pneumophila*. *Plasmid* **20**, 83–91.

Engler G., Depicker A., Maenhaut R., Villarroel R., Van Montagu M. & Schell J. (1981) Physical mapping of DNA base sequence homologies between an octopine and a nopaline Ti plasmid of *Agrobacterium tumefaciens*. *J. Mol. Biol.* **152**, 183–208.

Ensley B.D., Ratzkin B.J., Osslund T.D., Simon M.J., Wackett L.P. & Gibson D.T. (1983) Expression of naphthalene oxidation genes in *Escherichia coli* results in biosynthesis of indigo. *Science* **222**, 167–9.

Epp J.K., Huber M.L.B., Turner J.R., Goodson T. & Schoner B.E. (1989) Production of a hybrid macrolide antibiotic in *Streptomyces aureofaciens* and *Streptomyces lividans* by introduction of a cloned carbomycin biosynthetic gene from *Streptomyces thermotolerans*. *Gene* **85**, 293–301.

Erlich H.A. (ed.) (1989) *PCR Technology: Principles and Applications for DNA Amplification*. Stockton Press.

Erlich H.A., Gelfand D.H. & Sakai R.K. (1988) Specific DNA amplification. *Nature* **331**, 461–2.

Ernst J.F. (1988) Codon usage and gene expression. *Trends Biotechnol.* **6**, 196–9.

Errington J. (1993) *Bacillus subtilis* sporulation: regulation of gene expression and control of morphogenesis. *Microbiol. Rev.* **57**, 1–33.

Errington J. & Pughe N. (1987) Upper limit for DNA packaging by *Bacillus subtilis* bacteriophage φ105: isolation of phage deletion mutants by induction of oversized prophages. *Mol. Gen. Genetics* **210**, 347–51.

Eskin B. & Linn S. (1972) The deoxyribonucleic modification and restriction enzymes of *Escherichia coli* B. II. Purification, subunit structure, and catalytic properties of the restriction endonuclease. *J. Biol. Chem.* **247**, 6183–91.

Etkin L.D., Pearman B. & Ansah-Yiadom R. (1987) Replication of injected DNA templates in *Xenopus* embryos. *Exp. Cell Res.* **169**, 468–77.

Evans G.A., Lewis K. & Rothenberg B.E. (1989) High efficiency vectors for cosmid microcloning and genomic analysis. *Gene* **79**, 9–20.

Falkow S. (1975) *Infectious Multiple Drug Resistance*. London, Pion.

Faust P.L., Wall D.A., Perara E., Lingappa V.R. & Kornfield S. Expression of human cathepsin D in *Xenopus* oocytes: phosphorylation and intracellular targeting. *J. Cell Biol.* **105**, 1937–45.

Feldmann K.A. (1991) T-DNA insertion mutagenesis in *Arabidopsis*: mutational spectrum. *Plant J.* **1**, 71–82.

Felgner P.L., Gadek T.R., Holm M., Roman R., Chan H.W., Wenz M., Northrop J.P., Ringold G.M. & Danielsen M. (1987) Lipofection: a highly efficient lipid-mediated DNA-transfection procedure. *Proc. Natl. Acad. Sci. USA* **84**, 7413–17.

Ferretti L. & Sgaramella V. (1981) Temperature dependence of the joining by T4 DNA ligase of termini produced by type II restriction endonucleases. *Nucleic Acids Res* **9**, 85–93.

Field L.J., Veress A.T., Steinhelper M.E., Cochrane K., Sonnenberg H. (1991) Kidney function in ANF-transgenic mice: effect of blood volume expansion. *Am. J. Physiol.* **260**, R1–R5.

Fincham J.R.S. (1989) Transformation in fungi. *Microbiol. Rev.* **53**, 148–70.

Finer J.J. & McMullen M.D. (1990) Transformation of cotton (*Gossypium hirsutum* L.) via particle bombardment. *Plant Cell Reports* **8**, 586–9.

Fischetti V.A., Medaglini D., Oggioni M. & Pozzi G. (1993) Expression of foreign proteins on Gram-positive commensal bacteria for mucosal vaccine delivery. *Curr. Opinion Biotechnol.* **4**, 603–10.

Fischhoff D.A., Bowdish K.S., Perlak F.J., Marrone P.G., McCormick S.M., Niedermeyer J.G., Dean D.A., Kusano-Kretzmer K., Mayer E.J., Rochester D.E., Rogers S.G. & Fraley R.T. (1987) Insect tolerant transgenic tomato plants. *Biotechnology* **5**, 807–13.

Fitchen J.H. & Beachy R.N. (1993) Genetically engineered protection against viruses in transgenic plants. *Annu. Rev. Microbiol.* **47**, 739–63.

Fitzgerald-Hayes M., Buhler J.M., Cooper T.G. & Carbon J. (1982) Isolation and subcloning analysis of functional centromere DNA (cen11) from yeast chromosome XI. *Mol. Cell. Biol.* **2**, 82–7.

Flamand A., Coulon P., Lafay F., Kappeler A., Artois M., Aubert M., Blancou J. & Wandeler A. (1992) Eradication of rabies in Europe. *Nature* **360**, 115–16.

Fleer R., Yeh P., Amellal N., Maury I., Fournier A., Baccheta F., Baduel P., Jung G., L'Hôte H. & Becquart J. (1991) Stable multicopy vectors for high-level secretion of recombinant human serum albumin by *Kluyveromyces* yeasts. *Biotechnology* **9**, 968–75.

Flintoff W.F., Davidson S.V. & Siminovitch L. (1976) Isolation and partial characterization of three methotrexate-resistant phenotypes from Chinese hamster ovary cells. *Somat. Cell Genetics* **2**, 245–61.

Folger K.R., Wong E.A., Wahl G. & Capecchi M. (1982) Patterns of integration of DNA microinjected into cultured mammalian cells: evidence for homologous recombination between injected plasmid DNA molecules. *Mol. Cell. Biol.* **2**, 1372–87.

Forkmann G. (1993) Control of pigmentation in natural and transgenic plants. *Curr. Opinion Biotechnol.* **4**, 159–65.

Fraley R., Rogers S.G., Horsch R.B., Sanders P.R., Flick J.S., Adams S.P., Bittner M.L., Brand L.A., Fink C.L., Fry J.S., Galuppi G.R., Goldberg S.B., Hoffmann N.L. & Woo S.C. (1983) Expression of bacterial genes in plant cells. *Proc. Natl. Acad. Sci. USA* **80**, 4803–7.

Francisco J., Earhart C.F. & Georgiou G. (1992) Transport and anchoring of β-lactamase to the external surface of *Escherichia coli*. *Proc. Natl. Acad. Sci. USA* **89**, 2713–17.

Franck A., Guilley H., Jonard G., Richards K. & Hirth L. (1980) Nucleotide sequence of cauliflower mosaic virus DNA. *Cell* **21**, 285–94.

Franklin F.C.H., Bagdasarian M., Bagdasarian M.M. & Timmis K.N. (1981) Molecular and functional analysis of the TOL plasmid pWWO from *Pseudomonas putida* and cloning of genes for the entire regulated aromatic ring *meta* cleavage pathway. *Proc. Natl. Acad. Sci. USA* **78**, 7458–62.

Fray R.G. & Grierson D. (1993) Molecular genetics of tomato fruit ripening. *Trends Genetics* **9**, 438–43.

French R., Janda M. & Ahlquist P. (1986) Bacterial gene inserted in an engineered RNA virus: efficient expression in monocotyledonous plant cells. *Science* **231**, 1294–7.

Frey J., Bagdasarian M., Feiss D., Franklin C.H. & Deshusses J. (1983) Stable cosmid vectors that enable the introduction of cloned fragments into a wide range of Gram-negative bacteria. *Gene* **24**, 299–308.

Frey J., Bagdasarian M.M. & Bagdasarian M. (1992) Replication and copy number control of the broad-host-range plasmid RSF1010. *Gene* **113**, 101–6.

Friedmann T. (1987) HPRT-deficient mice: a useful new animal model for human disease. *Trends Biotechnol.* **5**, 157–8.

Frischauf A.-M., Lehrach H., Poustka A. & Murray N. (1983) Lambda replacement vectors carrying polylinker sequences. *J. Mol. Biol.* **170**, 827–42.

Frohman M.A., Dush M.K. & Martin G. (1988) Rapid production of full-length cDNAs from rare transcripts; amplification using a single gene-specific oligonucleotide primer. *Proc. Natl. Acad. Sci. USA* **85**, 8998–9002.

Frohman M.A. & Martin G.R. (1989) Rapid amplification of cDNA ends using nested primers. *Technique* **1**, 165–70.

Fuchs R.L. & Perlak F.J. (1992) Commercialization of genetically engineered plants. *Curr. Opinion Biotechnol.* **3**, 181–4.

Fuerst T.R., Niles E.G., Studier W.F. & Moss B. (1986) Eukaryotic transient-expression system based on recombinant vaccinia virus that synthesizes bacteriophage T7 RNA polymerase. *Proc. Natl. Acad. Sci. USA* **83**, 8122–6.

Fukunaga R., Sokawa Y. & Nagata S. (1984) Constitutive production of human interferons by mouse cells with bovine papilloma virus as vector. *Proc. Natl. Acad. Sci. USA* **81**, 5086–90.

Fulton A.M., Adams S.E., Mellor J., Kingsman S.M. & Kingsman A.J. (1987) The organisation and expression of the yeast transposon, Ty. *Microbiol. Sci.* **4**, 180–185.

Futcher A.B. & Carbon J. (1986) Toxic effects of excess cloned centromeres. *Mol. Cell. Biol.* **6**, 2213–22.

Gallie D.R., Gay P. & Kado C.I. (1988) Specialized vectors for members of Rhizobiaceae and other Gram-negative bacteria. In *Vectors. A Survey of Molecular Cloning Vectors and Their Uses*, eds Rodriguez R.L. & Denhardt D.T., pp. 333–42. Butterworths, London.

Gallwitz D. & Sures I. (1980) Structure of a split yeast gene. Complete nucleotide sequence of the actin gene in *Saccharomyces cerevisiae*. *Proc. Natl. Acad. Sci. USA* **77**, 2546–50.

Ganoza M.C., Kofoid E.C., Marliere P. & Louis B.G. (1987) Potential secondary structure at translation initiation sites. *Nucleic Acids Res.* **15**, 345–59.

Gardner R.C., Howarth A.J., Hahn P., Brown-Luedi M., Shepherd R.J. & Messing J. (1981) The complete nucleotide sequence of an infectious clone of cauliflower mosaic virus by M13 mp7 shotgun sequencing. *Nucleic Acids Res.* **9**, 2871–87.

Garfinkel D.J., Boeke J.D. & Fink G.R. (1985) Ty element transposition: reverse transcription and virus-like particles. *Cell* **42**, 507–17.

Garfinkel D.J. & Nester E.W. (1980) *Agrobacterium tumefaciens* mutants affected in crown gall tumorigenesis and octopine catabolism. *J. Bacteriol.* **144**, 732–43.

Gasson M.J. & Davies F.L. (1985) The genetics of dairy lactic-acid bacteria. In *Advances in the Microbiology and Biochemistry of Cheese and Fermented Milk*, eds Davies F.L. & Law B.A., pp. 99–126. Elsevier, London.

Gehrlach W., Llewellyn D. & Haseloff J. (1987) Construction of a plant disease resistance gene from the satellite RNA of tobacco ringspot virus. *Nature* **328**, 802–5.

Georgiou G., Poetschke H.L., Stathopoulos C. & Francisco J.A. (1993) Practical applications of engineering Gram-negative bacterial cell surfaces. *Trends Biotechnol.* **11**, 6–10.

Gerard R.D. & Gluzman Y. (1985) A new host cell system for regulated simian virus 40 DNA replication. *Mol. Cell. Biol.* **5**, 3231–40.

Gerbaud C., Fournier P., Blanc H., Aigle M., Heslot H. & Geurineau M. (1979) High frequency of yeast transformation by plasmids carrying part or entire 2 μm yeast plasmid. *Gene* **5**, 233–53.

Germino J. & Bastia D. (1984) Rapid purification of a cloned gene product by genetic fusion and site-specific proteolysis. *Proc. Natl. Acad. Sci. USA* **81**, 4692–6.

Gershoni J.M. & Palade G.E. (1982) Electrophoretic transfer of proteins from sodium dodecyl sulfate-polyacrylamide gels to a positively charged membrane filter. *Anal. Biochem.* **124**, 396–405.

Gething M.-J. & Sambrook J. (1981) Cell-surface expression of influenza haemogglutinin from a cloned DNA copy of the RNA gene. *Nature* **293**, 620–5.

Gheysen D., Iserentant D., Derom C. & Fiers W. (1982) Systematic alteration of the nucleotide sequence preceding the translation initiation codon and the effects on bacterial expression of the cloned SV-40 small-t antigen gene. *Gene* **17**, 55–63.

Gibbs C.S. & Zoller M.J. (1991) Rational scanning mutagenesis of a protein kinase identifies functional regions in catalysis and substrate interactions. *J. Biol. Chem.* **266**, 8923–31.

Gibson R.M. & Errington J. (1992) A novel *Bacillus subtilis* expression vector based on bacteriophage Φ105. *Gene* **121**, 137–42.

Gill P. & Werrett D.J. (1987) Exclusion of a man charged with murder by DNA finger-printing. *Forensic Sci. Int.* **35**, 145–8.

Gill P., Ivanov P.L., Kimpton C., Piercy R., Benson N., Tully G., Evett I., Hagelberg E. & Sullivan K. (1994) Identification of the remains of the Romanov family by DNA analysis. *Nature Genet.* **6**, 130–35.

Gill P., Jeffreys A.J. & Werrett D.J. (1985) Forensic application of DNA 'fingerprints'. *Nature* **318**, 577–9.

Gill P., Lygo J.E., Fowler S.J. & Werrett D.J. (1987) An evaluation of DNA fingerprinting for forensic purposes. *Electrophoresis* **8**, 38–44.

Gillam S., Astell C.R. & Smith M. (1980) Site-specific mutagenesis using oligodeoxyri-bonucleotides: isolation of a phenotypically silent ϕX174 mutant, with a specific nucleotide deletion, at very high efficiency. *Gene* **12**, 129–37.

Girgis S.I., Alevizaki M., Denny P., Ferrier G.J.M. & Legon S. (1988) Generation of DNA probes for peptides with highly degenerate codons using mixed primer PCR. *Nucleic Acids Res.* **16**, 10371.

Giri I. & Danos O. (1986) Papillomavirus genomes: from sequence data to biological properties. *Trends Genetics* **2**, 227–32.

Gluzman Y. (1981) SV40-transformed simian cells support the replication of early SV40 mutants. *Cell* **23**, 175–82.

Gocayne J., Robinson D.A., Fitzgerald M.G., Chung F.Z., Kerlavage A.R., Leutes K.U., Lai J., Wang C.D., Fraser C.M. & Venter J.C. (1987) Primary structure of rat cardiac beta-adrenergic and muscarinic cholinergic receptors obtained by automated DNA sequence analysis – further evidence for a multigene family. *Proc. Natl. Acad. Sci. USA* **84**, 8296–300.

Goeddel D.V. (1990) Systems for heterologous gene expression. *Methods. Enzymol.* **185**, 3–7.

Goff S.P. & Berg P. (1976) Construction of hybrid viruses containing SV40 and λ phage DNA segments and their propagation in cultured monkey cells. *Cell* **9**, 695–705.

Gold L. & Stormo G.D. (1990) High-level translation initiation. *Methods Enzymol.* **185**, 89–93.

Goldberg D.A., Posakony J.W. & Maniatis T. (1983) Correct developmental expression of a cloned alcohol dehydrogenase gene transduced into the *Drosophila* germ line. *Cell* **34**, 59–73.

Goldfarb M., Shimizu K., Perucho M. & Wigler M. (1982) Isolation and preliminary characterization of a human transforming gene from T24 bladder carcinoma cells. *Nature* **296**, 404–9.

Goldin A.L. (1991) Expression of ion channels in oocytes. In *Methods in Cell Biology*, Vol. 36. *Xenopus laevis: Practical Uses in Cell and Molecular Biology*, eds Kay KB & Peng HB, pp. 487–509. Academic Press, New York.

Gordon J.W. & Ruddle F.H. (1981) Integration and stable germ line transmission of genes injected into mouse pronuclei. *Science* **214**, 1244–6.

Gordon K., Lee E., Vitale J.A., Smith A.E., Westphal H. & Henninghausen L. (1987) Production of human tissue plasminogen activator in transgenic mouse milk. *Bio-technology* **5**, 1183–7.

Gordon M.P. (1980) In *Proteins and Nucleic Acids. The Biochemistry of Plants*, ed. Marcus A., Vol. 6, pp. 531–70. Academic Press, New York.

Gordon M.P., Farrand S.K., Sciaky D., Montoya A.L., Chilton M.-D., Merlo D.J. &

Nester E.W. (1979) In *Molecular Biology of Plants, Symposium*, University of Minnesota, ed. Rubenstein I. Academic Press, London.

Gordon-Kamm W.J., Spencer T.M., Maugano M.L., Adams T.R., Daines R.J., Start W.G., O'Brien J.V., Chambers S.A., Adams W.R. Jr., Willetts N.G., Rice T.B., Mackey C.J., Krueger R.W., Kausch A.P. & Lemaux P.G. (1990) Transformation of maize cells and regeneration of fertile transgenic plants. *Plant Cell* **2**, 603–18.

Gorman C.M., Merlino G.T., Willingham M.C., Pastan I. & Howard B. (1982a) The Rous sarcoma virus long terminal repeat is a strong promoter when introduced into a variety of eukaryotic cells by DNA-mediated transfection. *Proc. Natl. Acad. Sci. USA* **79**, 6777–81.

Gorman C.M., Moffat L.F. & Howard B.H. (1982b) Recombinant genome which expresses chloramphenicol acetyl transferase in mammalian cells. *Mol. Cell. Biol.* **2**, 1044–51.

Gorman C., Padmanabhan R. & Howard B. (1983) High efficiency DNA mediated transformation of primate cells. *Science* **221**, 551–3.

Gormley E.P. & Davies J. (1991) Transfer of plasmid RSF1010 by conjugation from *Escherichia coli* to *Streptomyces lividans* and *Mycobacterium smegmatis*. *J. Bacteriol.* **173**, 6705–8.

Gottesman S. (1990) Minimizing proteolysis in *Escherichia coli*: genetic solutions. *Methods Enzymol.* **185**, 119–29.

Gottesmann M.E. & Yarmolinsky M.D. (1968) The integration and excision of the bacteriophage lambda genome. *Cold Spring Harbor Symp. Quant. Biol.* **33**, 735–47.

Gould J., Devey M., Hasegawa O., Vlian E.C., Peterson G. & Smith R.H. (1991) Transformation of *Zea mays* L. using *Agrobacterium tumefaciens* and the shoot apex. *Plant Physiol.* **95**, 426–34.

Gourse R.L., De Boer H.A. & Nomura M. (1986) DNA determinants of RNA synthesis in *E. coli*: growth rate dependent regulation, feedback inhibition, upstream activation, anti-termination. *Cell* **44**, 197–205.

Gouy M. & Gautier C. (1982) Codon usage in bacteria. Correlation with gene expressivity. *Nucleic Acids Res.* **10**, 7055–74.

Goze A. & Ehrlich S.D. (1980) Replication of plasmids from *Staphylococcus aureus* in *Escherichia coli*. *Proc. Natl. Acad. Sci. USA* **77**, 7333–7.

Green S. (1993) Promiscuous liaisons. *Nature* **361**, 590–1.

Green M.R., Maniatis T. & Melton D.A. (1983) Human β-globin pre-mRNA synthesized *in vitro* is accurately spliced in *Xenopus* oocyte nuclei. *Cell* **32**, 681–94.

Greener A., Lehman S.M. & Helinski D.R. (1992) Promoters of the broad host range plasmid RK2: analysis of transcription (initiation) in five species of Gram-negative bacteria. *Genetics* **130**, 27–36.

Gregory R.J., Cheng S.H., Rich D.P., Marshall J., Paul S., Hehir K., Ostedgaard L., Klinger K.W., Welsh M.J. & Smith A.E. (1990) Expression and characterization of the cystic fibrosis transmembrane conductance regulator. *Nature* **347**, 382–6.

Gribskov M. & Burgess R.R. (1983) Overexpression and purification of the sigma subunit of *Escherichia coli* RNA polymerase. *Gene* **26**, 109–18.

Grimsley N., Hohn T., Davies J.W. & Hohn B. (1987) *Agrobacterium*-mediated delivery of infectious maize streak virus into maize plants. *Nature* **325**, 177–9.

Grinter N.J. (1983) A broad-host-range cloning vector transposable to various replicons. *Gene* **21**, 133–43.

Gronenborn B., Gardner R.C., Schaefer S. & Shepherd R.J. (1981) Propagation of foreign DNA in plants using cauliflower mosaic virus as vector. *Nature* **294**, 773–6.

Gronenborn B. & Messing J. (1978) Methylation of single-stranded DNA *in vitro* introduces new restriction endonuclease cleavage sites. *Nature* **272**, 375–7.

Grosjean J. & Fiers W. (1982) Preferential codon usage in prokaryotic genes. The optimal codon–anticodon interaction energy and the selective codon usage in efficiently expressed genes. *Gene* **18**, 199–209.

Grosveld F.G., Lund T., Murray E.J., Mellor A.L., Dahl H.H.M. & Flavell R.A. (1982) The construction of cosmid libraries which can be used to transform eukaryotic cells. *Nucleic Acids Res.* **10**, 6715–32.

Grosveld F., van Assendelft G.B., Greaves D.R. & Kollias G. (1987) Position independent, high level expression of the human β-globin gene in transgenic mice. *Cell* **51**, 975–85.

Grunstein M. & Hogness D.S. (1975) Colony hybridization: a method for the isolation of cloned DNAs that contain a specific gene. *Proc. Natl. Acad. Sci. USA* **72**, 3961–5.

Gruss A. & Ehrlich S.D. (1988) Insertion of foreign DNA into plasmids from Gram-positive bacteria induces formation of high-molecular-weight plasmid multimers. *J. Bacteriol.* **170**, 1183–90.

Gruss A. & Ehrlich S.D. (1989) The family of highly interrelated single-related deoxyribonucleic acid plasmids. *Microbiol. Rev.* **53**, 231–41.

Gruss P., Efstratiadis A., Karathanasis S., Konig M. & Khoury G. (1981a) Synthesis of stable unspliced mRNA from an intronless simian virus 40-rat preproinsulin gene recombinant. *Proc. Natl. Acad. Sci. USA* **78**, 6091–5.

Gruss P., Ellis R.W., Shih T.Y., Konig M., Scolnick E.M. & Khoury G. (1981b) SV40 recombinant molecules express the gene encoding p21 transforming protein of Harvey murine sarcoma virus. *Nature* **293**, 486–8.

Gruss P. & Khoury G. (1981) Expression of simian virus 40-rat preproinsulin recombinants in monkey kidney cells: use in preproinsulin RNA processing signals. *Proc. Natl. Acad. Sci. USA* **78**, 133–7.

Gryczan T.J., Contente S. & Dubnau D. (1980a) Molecular cloning of heterologous chromosomal DNA by recombination between a plasmid vector and a homologous resident plasmid in *Bacillus subtilis*. *Mol. Gen. Genetics* **177**, 459–67.

Guarente L. (1987) Regulatory proteins in yeast. *Annu. Rev. Genet.* **21**, 425–52.

Guarente L. & Ptashne M. (1981) Fusion of *Escherichia coli* lacZ to the cytochrome *c* gene of *Saccharomyces cerevisiae*. *Proc. Natl. Acad. Sci. USA* **78**, 2199–203.

Gubler U. & Hoffman B.J. (1983) A simple and very efficient method for generating cDNA libraries. *Gene* **26**, 263–9.

Gumport R.I. & Lehman I.R. (1971) Structure of the DNA ligase adenylate intermediate: lysine (ε-amino) linked AMP. *Proc. Natl. Acad. Sci. USA* **68**, 2559–63

Gurdon J.B, Lane C.D., Woodland H.R. & Marbaix G. (1971) Use of frog eggs and oocytes for the study of messenger RNA and its translation in living cells. *Nature* **233**, 177–182.

Gusella J.F., Keys C., Varsanyi-Breiner A., Kao F.-T., Jones C., Puck T.T. & Housman D. (1980) Isolation and localization of DNA segments from specific human chromosomes. *Proc. Natl. Acad. Sci. USA* **77**, 2829–33.

Haas R., Kahrs A.F., Facius D., Allmeier H., Schmitt R. & Meyer T.F. (1993) Tn*Max* – a versatile mini-transposon for the analysis of cloned genes and shuttle mutagenesis. *Gene* **130**, 23–31.

Hackett J. (1993) Use of *Salmonella* for heterologous gene expression and vaccine delivery systems. *Curr. Opinion Biotechnol.* **4**, 611–15.

Hadi S.M., Bachi B., Iida S. & Bickle T.A. (1982) DNA restriction-modification enzymes of phage P1 and plasmid p15B. Subunit functions and structural homologies. *J. Mol. Biol.* **165**, 19–34.

Haenlin M., Steller H., Pirotta V. & Momer E. (1985) A 43 kb cosmid P transposon rescues the fs(1)K10 morphogenetic locus and three adjacent *Drosophila* developmental mutants. *Cell* **40**, 827–37.

Hagan C.E. & Warren G.J. (1983) Viability of palindromic DNA is restored by deletions occurring at low but variable frequency in plasmids of *Escherichia coli*. *Gene* **24**, 317–26.

Hager P.W. & Rabinowitz J.C. (1985) Translational specificity in *Bacillus subtilis*, p1–29. In *The Molecular Biology of the Bacilli*, ed. Dubnau D. Academic Press, New York.

Hahn S., Hoar E.T. & Guarente L. (1985) Each of three 'TATA elements' specifies a subset of the transcription initiation sites at the CYC-1 promoter of *Saccharomyces cerevisiae*. *Proc. Natl. Acad. Sci. USA* **82**(24), 8562–6.

Haima P., Bron S. & Venema G. (1987) The effect of restriction on shotgun cloning and plasmid stability in *Bacillus subtilis* Marburg. *Mol. Gen. Genetics* **209**, 335–42.

Haima P., Bron S. & Venema G. (1988) A quantitative analysis of shotgun cloning in *Bacillus subtilis* protoplasts. *Mol. Gen. Genetics* **213**, 364–9.

Haima P., Bron S. & Venema G. (1990) Novel plasmid marker rescue transformation system for molecular cloning in *Bacillus subtilis* enabling direct selection of recombinants. *Mol. Gen. Genetics* **223**, 185–91.

Hain R., Stabel P., Czernilofsky A.P., Steinbiss H.H., Herrera-Estrella L. & Schell

J. (1985) Uptake, integration, expression and genetic transmission of a selectable chimaeric gene by plant protoplasts. *Mol. Gen. Genetics* **199**, 161–8.

Hain R., Reif H.-J., Krause E., Langebartels R., Kindl H., Vornam B., Wiese W., Schmelzer E., Schreier P.H., Stocker R.H. & Stenzel K. (1993) Disease resistance results from foreign phytoalexin expression in a novel plant. *Nature* **361**, 153–6.

Hall M.N., Hereford L. & Herskowitz I. (1984) Targeting of *E. coli* β-galactosidase to the nucleus in yeast. *Cell* **36**, 1057–65.

Hamer D.H. & Leder P. (1979a) Expression of the chromosomal mouse β maj-globin gene cloned in SV40. *Nature* **281**, 35–40.

Hamer D.H. & Leder P. (1979b) SV40 recombinants carrying a functional RNA splice junction and polyadenylation site from the chromosomal mouse β maj-globin gene. *Cell* **17**, 737–47.

Hamer D.H., Hu S., Magnuson V.L., Hu N. & Pattatucci A.M.L. (1993) A linkage between DNA markers on the X chromosome and male sexual orientation. *Science* **261**, 321–7.

Hamer J.E. & Timberlake W.E. (1987) Functional organisation of the *Aspergillus nidulans trp*C promoter. *Mol. Cell. Biol.* **7**, 2352–9.

Hammer R.E., Krumlauf R., Camper S.A., Brinster R.L. & Tilghman S.M. (1987) Diversity of alpha-fetoprotein gene expression in mice is generated by a combination of separate enhancer elements. *Science* **235**, 53–8.

Hamptman R.M., Vasil V., Ozias-Akins P., Tabaizadeh Z., Rogers S.G., Fraley R.T., Horsch R.B. & Vasil I.K. (1988) Evaluation of selectable marker for obtaining stable transformants in the Gramineae. *Plant Physiol.* **86**, 602–6.

Hanahan D. (1983) Studies on transformation of *Escherichia coli* with plasmids. *J. Mol. Biol.* **166**, 557–80.

Hanahan D., Lane D., Lipsich L., Wigler M. & Botchan M. (1980) Characteristics of an SV40-plasmid recombinant and its movement into and out of the genome of a murine cell. *Cell* **21**, 127–39.

Hanahan D. & Meselson M. (1980) Plasmid screening at high colony density. *Gene* **10**, 63–7.

Harbers K., Jahner D. & Jaenisch R. (1981) Microinjection of cloned retroviral genomes into mouse zygotes: integration and expression in the animal. *Nature* **293**, 540–3.

Hardies S.C. & Wells R.D. (1976) Preparative fractionation of DNA by reversed phase column chromatography. *Proc. Natl. Acad. Sci. USA* **73**, 3117–21.

Harley C.B. & Reynolds R.P. (1987) Analysis of *E. coli* promoter sequences. *Nucleic Acids Res.* **15**, 2343–61.

Harwood C.R. (1992) *Bacillus subtilis* and its relatives: molecular biological and industrial workhorses. *Trends Biotechnol.* **10**, 247–56.

Harwood C.R. & Cutting S.M. (1990) *Molecular Biology Methods for Bacillus*. John Wiley, Chichester. 580 pp.

Hashimoto-Gotoh T., Franklin F.C.H., Nordheim A. & Timmis K.N. (1981) Low copy number, temperature-sensitive, mobilization-defective pSC101-derived containment vectors. *Gene* **16**, 227–35.

Hashimoto-Gotoh T., Kume A., Masahashi W., Takeshita S. & Fukuda A. (1986) Improved vector, pHSG664, for direct streptomycin-resistance selection: cDNA cloning with G:C-tailing procedure and subcloning of double digest DNA fragments. *Gene* **41**, 125–8.

Hasty P., Ramirez-Solis R., Krumlauf R. & Bradley A. (1991) Introduction of a subtle mutation into the *Hox-2.6* locus in embryonic stem cells. *Nature* **350**, 243–6.

Hasty P., Rivera-Perez J. & Bradley A. (1991) The length of homology required for gene targeting in embryonic stem cells. *Mol. Cell. Biol.* **11**, 5586–91.

Hawley D.K. & McClure W.R. (1983) Compilation and analysis of *Escherichia coli* promoter DNA sequences. *Nucleic Acids Res.* **11**, 2237–55.

Hayes W. (1968) *The Genetics of Bacteria and their Viruses*, 2nd edn. Blackwell Scientific Publications, Oxford.

Heasman J., Holwill S. & Wylie C.C. (1991) Fertilization of cultured *Xenopus* oocytes and use in studies of maternally inherited molecules. *Methods Cell Biol.* **36**, 213–30.

Hedgepeth J., Goodman H.M. & Boyer H.W. (1972) DNA nucleotide sequence restricted by the RI endonuclease. *Proc. Natl. Acad. Sci. USA* **69**, 3448–52.

Heidecker G. & Messing J. (1983) Sequence analysis of Zein cDNAs by an efficient mRNA cloning method. *Nucleic Acids Res.* **11**, 4891–906.

Heinemann J.A. & Sprague G.F. (1989) Bacterial conjugative plasmids mobilize DNA transfer between bacteria and yeast. *Nature* **340**, 205–9.

Helfman D.M., Feramisco J.R., Fiddes J.C., Thomas G.P. & Hughes S.H. (1983) Identification of clones that encode chicken tropomyosin by direct immunological screening of a cDNA expression library. *Proc. Natl. Acad. Sci. USA* **80**, 31–5.

Henikoff S., Tatchell K., Hall B.D. & Nosmyth K.A. (1981) Isolation of a gene from *Drosophila* by complementation in yeast. *Nature* **289**, 33–7.

Hennecke H., Günther I. & Binder F. (1982) A novel cloning vector for the direct selection of recombinant DNA in *E. coli*. *Gene* **19**, 231–4.

Hermes J.D., Blacklow S.C. & Knowles J.R. (1990) Searching sequence space by definably random mutagenesis: improving the catalytic potency of an enzyme. *Proc. Natl. Acad. Sci. USA* **87**, 696–700.

Hermesz E., Olasz F., Dorgai L. & Orosz L. (1992) Stable incorporation of genetic material into the chromosome of *Rhizobium meliloti* 41: construction of an integrative vector system. *Gene* **119**, 9–15.

Herrera-Estrella L., Depicker A., Van Montagu M. & Schell J. (1983a) Expression of chimaeric genes transferred into plant cells using a Ti-plasmid-derived vector. *Nature* **303**, 209–13.

Herrera-Estrella L., DeBlock M., Messens E., Hernalsteens J.P., Van Montagu M. & Schell J. (1983b) Chimeric genes as dominant selectable markers in plant cells. *EMBO J.* **2**, 987–95.

Hershfield V., Boyer H.W., Yanofsky C., Lovett M.A. & Helinski D.R. (1974) Plasmid Col El as a molecular vehicle for cloning and amplification of DNA. *Proc. Natl. Acad. Sci. USA* **71**, 3455–9.

Herskowitz I. (1974) Control of gene expression in bacteriophage lambda. *Annu. Rev. Genet.* **7**, 289–324.

Hiatt A.H., Cafferkey R. & Bowdish K. (1989) Production of antibodies in transgenic plants. *Nature* **342**, 76–8.

Hicks J.B., Strathern J.N., Klar A.J.S. & Dellaporta S.L. (1982) Cloning by complementation in yeast. The mating type genes. In *Genetic Engineering*, eds Setlow J.K. & Hollaender A., pp. 219–48. Plenum Press, New York.

Higgins C.F., Peltz S.W. & Jacobson A. (1992) Turnover of mRNA in prokaryotes and lower eukaryotes. *Curr. Opinion Genetics Devel.* **2**, 739–47.

Higo K.E., Otaka E. & Osawa S. (1982) Purification and characterization of 30S ribosomal proteins from *Bacillus subtilis*: correlation to *Escherichia coli* 30S proteins. *Mol. Gen. Genetics* **185**, 239–44.

Higuchi R., Paddock G.V., Wall R. & Salser W. (1976) A general method for cloning eukaryotic structural gene sequences. *Proc. Natl. Acad. Sci. USA* **73**, 3146–50.

Higuchi R., Krummel B. & Saiki R.K. (1988) A general method of *in vitro* preparation and specific mutagenesis of DNA fragments: study of protein and DNA interactions. *Nucleic Acids Res.* **16**, 7351–67.

Hill A. & Bloom K. (1987) Genetic manipulation of centromere function. *Mol. Cell. Biol.* **7**, 2397–405.

Hill A.V.S. & Jeffreys A.J. (1985) Use of minisatellite DNA probes for determination of twin zygosity at birth. *Lancet* **2**, 1394–5.

Hinnen A., Hicks J.B. & Fink G.R. (1978) Transformation of yeast. *Proc. Natl. Acad. Sci. USA* **75**, 1929–33.

Hitzeman R.A., Hagie F.E., Levine H.L., Goeddel D.V., Ammerer G. & Hall B.D. (1981) Expression of a human gene for interferon in yeast. *Nature* **293**, 717–22.

Hitzeman R.A., Leung D.W., Perry L.J., Kohr W.J., Levine H.L. & Goeddel D.V. (1983) Secretion of human interferons by yeast. *Science* **219**, 620–5.

Hoekma A., Hirsch P.R., Hooykass P.J.J. & Schilperoort R.A. (1983) A binary plant vector strategy based on separation of *vir*- and T-regions of the *Agrobacterium tumefaciens* Ti-plasmid. *Nature* **303**, 179–83.

Hoekma A., Kastelein R.A., Vasser M. & de Boer H.A. (1987) Codon replacement in the PGK1 gene of *Saccharomyces cerevisiae*: experimental approach to study the role of biased codon usage in gene expression. *Mol. Cell. Biol.* **7**, 2914–24.

Hoekstra W.P.M., Bergmans J.E.N. & Zuidweg E.M. (1980) Role of *rec*BC nuclease in *Escherichia coli* transformation. *J. Bacteriol.* **143**, 1031–2.

Hogan B. & Lyons K. (1988) Gene targeting: getting nearer the mark. *Nature* **336**, 304–5.

Hohn B. (1975) DNA as substrate for packaging into bacteriophage lambda, *in vitro. J. Mol. Biol.* **98**, 93–106.

Hohn B. & Murray K. (1977) Packaging recombinant DNA molecules into bacteriophage particles *in vitro. Proc. Natl. Acad. Sci. USA* **74**, 3259–63.

Hohn T., Hohn B. & Pfeiffer P. (1985) Reverse transcription in CaMV. *Trends Biochem. Sci.* 205–9.

Holben W.E. & Tiedje (1988) Applications of nucleic acid hybridization in microbial ecology. *Ecology* **69**, 561–8.

Holmes D.L. & Stellwagen N.C. (1990) The electric field dependence of DNA mobilities in agarose gels: a reinvestigation. *Electrophoresis* **11**, 5–15.

Holsters M., Silva B., Van Vliet F., Genetello C., De Block M., Dhaese P., Depicker A., Inze D., Engler G., Villarroel R., Van Montagu M. & Schell J. (1980) The functional organization of the nopaline *A. tumefaciens* plasmid pTi C58. *Plasmid* **3**, 212–30.

Holton T.A. & Graham M.W. (1991) A simple and efficient method for direct cloning of PCR products using ddT-tailed vectors. *Nucleic Acids Res.* **19**, 1156.

Holwill S., Heasman J., Crawley C.R. & Wylie C.C. (1987) Axis germ line deficiencies caused by u.v. irradiation of *Xenopus* oocytes cultured *in vivo. Development* **100**, 735–43.

Holzman D.M., Li Y., Dearmond S.J., McKinley M.P., Gage F.H., Epstein C.J. & Mobley W.C. (1992) Mouse model of neurodegeneration: atrophy of basal forebrain cholinergic neurons in trisomy 16 transplants. *Proc. Natl. Acad. Sci. USA* **89**, 1383–7.

Hooykaas-Van Slogteren G.M.S., Hooykaas P.J.J. & Schilperoort R.A. (1984) Expression of Ti plasmid genes in monocotyledonous plants infected with *Agrobacterium tumefaciens. Nature* **311**, 763–4.

Hopkins A.S., Murray N.E. & Brammar W.J. (1976) Characterization of λtrp-transducing bacteriophages made *in vitro. J. Mol. Biol.* **107**, 549–69.

Hopwood D.A. (1993) Genetic engineering of *Streptomyces* to create hybrid antibiotics. *Curr. Opinion Biotechnol.* **4**, 531–7.

Hopwood D.A., Bibb M.J., Chater K.F. & Kieser T. (1987) Plasmid and phage vectors for gene cloning and analysis in *Streptomyces. Methods Enzymol.* **153**, 116–66.

Hopwood D.A., Malpartida F., Kieser H.M., Iheda H., Duncan J., Fujii I., Rudd B.A.M., Floss H.G. & Omura S. (1985) Production of 'hybrid' antibiotics by genetic engineering. *Nature* **314**, 642–4.

Horland P., Flick J., Johnston M. & Sclafani R.A. (1989) Galactose as a gratuitous inducer of *GAL* gene expression in yeasts growing on glucose. *Gene* **83**, 57–64.

Horsch R.B., Fraley R.T., Rogers S.G., Sanders P.R., Lloyd A. & Hoffmann N. (1984) Inheritance of functional genes in plants. *Science* **223**, 496–8.

Horsch R.B., Fry J.E., Hoffmann N.L., Eicholtz D., Rogers S.G. & Fraley R.T. (1985) A simple and general method for transferring genes into plants. *Science* **227**, 1229–31.

Howell S.H., Walker L.L. & Dudley R.K. (1980) Cloned cauliflower mosaic virus DNA infects turnips (*Brassica rapa*). *Science* **208**, 1265–7.

Hsiao C.L. & Carbon J. (1981) Characterization of a yeast replication origin (ars2) and construction of stable minichromosomes containing cloned yeast centromere DNA (CEN 3). *Gene* **15**, 157–66.

Hu S.-L., Kosowski S.P. & Dalrymple J.M. (1986) Expression of AIDS virus envelope gene by a recombinant vaccinia virus. *Nature* **320**, 537–40.

Hughes S. & Kosick E. (1984) Mutagenesis of the region between *env* and *src* of the SR-A strain of Rous sarcoma virus for the purpose of constructing helper-independent vectors. *Virology* **136**, 89–99.

Hui A., Hayflick J., Dinkelspiel K. & de Boer H.A. (1984) Mutagenesis of the three bases preceding the start codon of the β-galactosidase mRNA and its effect on translation in *Escherichia coli. EMBO J.* **3**, 623–9.

Hull R. & Al-Hakim A. (1988) Nucleic acid hybridization in plant virus diagnosis and characterization. *Trends Biotechnol.* **6**, 213–18.

Hull R. & Covey S.N. (1983a) Replication of cauliflower mosaic virus DNA. *Sci. Prog.* **68**, 403–22.

Hull R. & Covey S.N. (1983b) Does cauliflower mosaic virus replicate by reverse transcription? *Trends Biol. Sci.* **8**, 119–21.

Hultman T., Murby M., Stahl S., Hornes E. & Uhlén M. (1990) Solid phase *in vitro* mutagenesis using plasmid DNA template. *Nucleic Acids Res.* **18**, 5107–11.

Huston J.S., Levinson D., Mudgett-Hunter M., Tai M.-S., Novotny J., Margolies M.N., Ridge R.J., Bruccoleri R.E., Haber E., Crea R. & Opperman H. (1988) Protein engineering of antibody-binding sites: recovery of specific activity in an anti-digoxin single-chain Fv analogue produced in *Escherichia coli. Proc. Natl. Acad. Sci. USA* **85**, 5879–83.

Huttner K.M., Barbosa J.A., Scangos G.A., Pratchera D.D. & Ruddle F.H. (1981) DNA-mediated gene transfer without carrier DNA. *J. Cell. Biol.* **91**, 153–6.

Hwu P., Yannelli J., Kriegler M., Anderson W.F., Perez C., Chiand Y., Schwarz S., Cowherd R., Degado C. & Mule J. (1993) Functional and molecular characterization of tumor-infiltrating lymphocytes transduced with tumor necrosis factor-alpha cDNA for the gene therapy of cancer in humans. *J. Immunol.* **150**, 4101–15.

Iida S., Meyer J., Bachi B., Stalhammer-Carlemalm M., Schrickel S., Bickle T.A. & Arber W. (1982) DNA restriction-modification genes of phage P1 and plasmid p15B. Structure and *in vitro* transcription. *J. Mol. Biol.* **165**, 1–18.

Ikemura T. (1981a) Correlation between the abundance of *Escherichia coli* transfer RNAs and the occurrence of the respective codons in its protein genes. *J. Mol. Biol.* **146**, 1–21.

Ikemura T. (1981b) Correlation between the abundance of *Escherichia coli* transfer RNAs and the occurrence of the respective codons in its protein genes. A proposal for a synonymous codon choice that is optimal for the *E. coli* translational system. *J. Mol. Biol.* **151**, 389–409.

Imanaka T., Shibazaki M. & Takagi M. (1986) A new way of enhancing the thermostability of proteases. *Nature* **324**, 695–7.

Innis M.A., Gelfand D.H., Sninsky J.J. & White T.J. (eds) (1990) *PCR Protocols: A Guide to Methods and Applications*. Academic Press, New York.

Innis M.A., Myambo K.B., Gelfand D.H. & Brow M.A.D. (1988) DNA sequencing with *Thermus aquaticus* DNA polymerase and direct sequencing of polymerase chain reaction-amplified DNA. *Proc. Natl. Acad. Sci. USA* **85**, 9436–40.

Inouye M. & Inouye S. (1991) Retroelements in bacteria. *Trends Biochem. Sci.* **16**, 18–21.

Iserentant D. & Fiers W. (1980) Secondary structure of mRNA and efficiency of translation initiation. *Gene* **9**, 1–12.

Ish-Horowicz D. & Burke J.F. (1981) Rapid and efficient cosmid cloning. *Nucleic Acids Res.* **9**, 2989–98.

Israel D.I. & Kaufman R.J. (1989) Highly inducible expression from vectors containing multiple GREs in CHO cells overexpressing the glucocorticoid receptor. *Nucleic Acids Res.* **17**, 4589–604.

Itakura K., Hirose T., Crea R., Riggs A.D., Heyneker H.L., Bolivar F. & Boyer H.W. (1977) Expression in *Escherichia coli* of a chemically synthesized gene for the hormone somatostatin. *Science* **198**, 1056–63.

Ito K., Bassford P.J. & Beckwith J. (1981) Protein localization in *E. coli*: is there a common step in the secretion of periplasmic and outer-membrane proteins? *Cell* **24**, 707–17.

Jackson D.A., Symons R.H. & Berg P. (1972) Biochemical method for inserting new genetic information into DNA of Simian virus 40: circular SV40 DNA molecules containing lambda phage genes and the galactose operon of *Escherichia coli. Proc. Natl. Acad. Sci. USA* **69**, 2904–9.

Jacob A.E., Cresswell J.M., Hedges R.W., Coetzee J.N. & Beringer J.E. (1976) Properties of plasmids constructed by *in vitro* insertion of DNA from *Rhizobium leguminosarum* or *Proteus mirabilis* into RP4. *Mol. Gen. Genetics* **147**, 315–23.

Jacobs E., Dewerchin M. & Boeke J.D. (1988) Retrovirus-like vectors for *Saccharomyces cerevisiae*: integration of foreign genes controlled by efficient promoters into yeast chromosomal DNA. *Gene* **67**, 259–69.

Jaenisch R. (1988) Transgenic animals. *Science* **240**, 1468–74.

Jaenisch R. & Mintz B. (1974) Simian virus 40 DNA sequences in DNA of healthy adult mice derived from preimplantation blastocysts injected with viral DNA. *Proc. Natl. Acad. Sci. USA* **71**, 1250–4.

Jannière L., Bruand C. & Ehrlich S.D. (1990) Structurally stable *Bacillus subtilis* cloning vectors. *Gene* **81**, 53–61.

Jannière L. & Ehrlich S.D. (1987) Recombination between short repeated sequences is more frequent in plasmids than in the chromosome of *Bacillus subtilis*. *Mol. Gen. Genetics* **210**, 116–21.

Jefferson R.A., Burgess S.M. & Hirsh D. (1986) β-Glucuronidase from *Escherichia coli* as a gene-fusion marker. *Proc. Natl. Acad. Sci. USA* **83**, 8447–51.

Jefferson R.A., Kavanagh T.A. & Bevan M.W. (1987) GUS fusions: β-glucuronidase as a sensitive and versatile gene fusion marker in higher plants. *EMBO J.* **6**, 3901–7.

Jeffreys A.J., Brookfield J.F.Y. & Semenoff R. (1985b) Positive identification of an immigration test case using human DNA fingerprints. *Nature* **317**, 577–9.

Jeffreys A.J. & Morton D.B. (1987) DNA fingerprints of cats and dogs. *Animal Genet.* **18**, 1–15.

Jeffreys A.J., Wilson V. & Thein S.L. (1985a) Individual-specific 'fingerprints' of human DNA. *Nature* **316**, 76–9.

Jeffreys A.J., McLeod A., Tamaki K., Neil D.L. & Monckton D.G. (1991) Minisatellite repeat coding as a digital approach to DNA typing. *Nature* **354**, 204–9.

Jen G.C. & Chilton M.D. (1986) The right border region of pTiT37 T-DNA is intrinsically more active than the left border region in promoting T-DNA transformation. *Proc. Natl. Acad. Sci. USA* **83**, 3895–9.

Jensen J.S., Marcker K.A., Otten L. & Schell J. (1986) Nodule-specific expression of a chimaeric soybean leghaemoglobin gene in transgenic *Lotus corniculatus*. *Nature* **321**, 669–74.

Jimenez A. & Davies J. (1980) Expression of a transposable antibiotic resistance element in *Saccharomyces*. *Nature* **287**, 869–71.

Johnson K., Murphy C.K. & Beckwith J. (1992) Protein export in *Escherichia coli*. *Curr. Opinion Biotechnol.* **3**, 481–5.

Johnston M. (1987) A model fungal gene regulatory mechanism: the GAL genes of *Saccharomyces cerevisiae*. *Microbiol. Rev.* **51**, 458–76.

Johnston S.A., Anziano P.Q., Shark K., Sanford J.C. & Butow R.A. (1988) Mitochondrial transformation in yeast by bombardment with microprojectiles. *Science* **240**, 1538–41.

Jonas E.A., Snape A.M. & Sargent T.D. (1989) Transcriptional regulation of a *Xenopus* embryonic epidermal keratin gene. *Development* **106**, 399–405.

Jones E.W. (1990) Vacuolar proteases in yeast *Saccharomyces cerevisiae*. *Methods Enzymol.* **185**, 372–86.

Jones D. & Errington J. (1987) Construction of improved bacteriophage φ105 vectors for cloning by transfection in *Bacillus subtilis*. *J. Gen. Microbiol.* **133**, 483–92.

Jones I.M., Primrose S.B. & Ehrlich S.D. (1982) Recombination between short direct repeats in a *recA* host. *Mol. Gen. Genetics* **188**, 486–9.

Jones I.M., Primrose S.B., Robinson A. & Ellwood D.C. (1980) Maintenance of some Col E1-type plasmids in chemostat culture. *Mol. Gen. Genetics* **180**, 579–84.

Jones K. & Murray K. (1975) A procedure for detection of heterologous DNA sequences in lambdoid phage by *in situ* hybridization. *J. Mol. Biol.* **51**, 393–409.

Julius D.J., Blair L.C., Brake A.J., Sprague G.F. & Thurner J. (1983) Yeast alpha-factor is processed from a larger precursor polypeptide: the essential role of a membrane-bound dipeptidyl amino-peptidase. *Cell* **32**, 839–52.

Julius D., Scheckman R. & Thorner J. (1984) Glycosylation and processing of prepro-alpha-factor through the yeast secretory pathway. *Cell* **36**, 309–18.

Kan Y.W. & Dozy A.M. (1978) Polymorphisms of DNA sequence adjacent to human β-globin structural gene: relation to sickle mutation. *Proc. Natl. Acad. Sci. USA* **75**, 5631–5.

Kanalas J.J. & Suttle D.P. (1984) Amplification of the Ump synthase gene and enzyme overproduction in pyrazofurin-resistant rat hepatoma cells. *J. Biol. Chem.* **259**, 1848–53.

Kane S.E., Reinhard D.E., Fordis C.M., Pastau I. & Gottesman M.M. (1989) A new vector using human multidrug resistance as a selectable marker enables overexpression of foreign genes in eukaryotic cells. *Gene* **84**, 439–46.

Kane S.E., Troen B.R., Gal S., Ueda K., Pastan I. & Gottesman M.M. (1988) Use of a cloned multidrug resistance gene for amplification and overproduction of major excreted protein, a transformation regulated secreted acid protease. *Mol. Cell. Biol.* **8**, 3316–21.

Kaper J.B., Lockman H., Baldini M.M. & Levine M.M. (1984) A recombinant live oral cholera vaccine. *Biotechnology* **1**, 345–9.

Kareiva P. (1993) Transgenic plants on trial. *Nature* **363**, 580–1.

Karn J., Brenner S., Barnett L. & Cesareni G. (1980) Novel bacteriophage λ cloning vector. *Proc. Natl. Acad. Sci. USA* **77**, 5172–6.

Kaufman R.J. & Sharp P.A. (1982) Construction of a modular dihydrofolate reductase cDNA gene: analysis of signals utilized for efficient expression. *Mol. Cell. Biol.* **9**, 1304–19.

Kaufman R.J., Murtha P., Ingolia D.E., Yeung C.-Y. & Kellems R.E. (1986) Selection and amplification of heterologous genes encoding adenosine deaminase in mammalian cells. *Proc. Natl. Acad. Sci. USA* **83**, 3136–40.

Kaufman R.J., Wasley L.C., Spiliotes A.T., Gossels S.D., Latt S.A., Larsen G.R. & Kay R.M. (1985) Coamplification and coexpression of human tissue-type plasminogen activator and murine dihydrofolate reductase sequences in Chinese hamster ovary cells. *Mol. Cell. Biol.* **5**, 1730–59.

Kawasaki E.S. (1990) Amplification of RNA. In *PCR Protocols: A Guide to Methods and Applications*, eds Innis *et al.*, pp. 21–7. Academic Press, New York.

Kay B.K. & Peng H.B. (eds) (1991) *Xenopus laevis*: Practical uses in cell and molecular biology. *Methods Cell Biol.* **36**.

Keggins K.M., Lovett P.S. & Duvall E.J. (1978) Molecular cloning of genetically active fragments of *Bacillus* DNA in *Bacillus subtilis* and properties of the vector plasmid pUB110. *Proc Natl. Acad. Sci. USA* **75**, 1423–7.

Keilty S. & Rosenberg M. (1987) Constitutive function of a positively regulated promotor reveals new sequences essential for activity. *J. Biol. Chem.* **262**, 6389–95.

Kelley W.S., Chalmers K. & Murray N.E. (1977) Isolation and characterization of a λ*polA* transducing phage. *Proc. Natl. Acad. Sci. USA* **74**, 5632–6.

Kelly J.M. & Hynes M.J. (1987) Multiple copies of the *amdS* gene of *Aspergillus nidulans* cause titration of *trans*-activity regulatory proteins. *Curr. Genetics* **12**, 21–31.

Kelly T.J. & Smith H.O. (1970) A restriction enzyme from *Hemophilus influenzae*. II. Base sequence of the recognition site. *J. Mol. Biol.* **51**, 393–409.

Kenny B., Haigh R. & Holland I.B. (1991) Analysis of the haemolysin transport process through the secretion from *Escherichia coli* of PCM, CAT or β-galactosidase fused to the Hly C-terminal signal domain. *Mol. Microbiol.* **5**, 2557–68.

Keohavong P. & Thilly W.G. (1989) Fidelity of DNA polymerase in DNA amplification. *Proc. Natl. Acad. Sci. USA* **86**, 9253–7.

Kerem B.-S., Rommens J.M., Buchanan J.A., Markiewicz D., Cox T.K., Chakravarti A., Buchwald M. & Tsui L.-C. (1989) Identification of the cystic fibrosis gene: genetic analysis. *Science* **245**, 1073–80.

Kessler C., Neumaier P.S. & Wolf W. (1985) Recognition sequences of restriction endo-nucleases and methylases – a review. *Gene* **33**, 1–102.

Khoury G. & Gruss P. (1983) Enhancer elements. *Cell* **33**, 313–14.

Kimura S., Mullins J.J., Bunnemann B., Metzger R., Hilgenfeldt U., Zimmermann F., Jacob H., Fuxe K., Ganten D. & Kaling M. (1992) High blood pressure in transgenic mice carrying the rat angiotensinogen gene. *EMBO J.* **11**, 821–7.

Kinashi H., Shimaji M. & Sakai A. (1987) Giant linear plasmids in *Streptomyces* which code for antibiotic biosynthesis genes. *Nature* **328**, 454–6.

Kinashi H., Shimaji-Murayama M. & Hanafusa T. (1992) Integration of SCP1, a giant linear plasmid, into the *Streptomyces coelicolor* chromosome. *Gene* **115**, 35–41.

King L.A. & Possee R.D. (1992) *The Baculovirus Expression System: A Laboratory Guide.* Chapman & Hall, London.

Kingsman A.J., Clarke L., Mortimer R.K. & Carbon J. (1979) Replication in *Saccharomyces cerevisae* of plasmid pBR313 carrying DNA from the yeast *trp* 1 region. *Gene* **7**, 141–52.

Kishore G.M. & Somerville C.R. (1993) Genetic engineering of commercially useful biosynthetic pathways in transgenic plants. *Curr. Opinion Biotechnol.* **4**, 152–8.

Klauser T., Pohlner J. & Meyer T.F. (1992) Selective extracellular release of cholera toxin

B subunit by *Escherichia coli*: dissection of *Neisseria* Iga$_B$-mediated outer membrane transport. *EMBO J.* **11**, 2327–35.

Klee H., Montoya A., Horodyski F., Lichtenstein C. & Garfinkel D. (1984) Nucleotide sequence of the *tms* genes of the pTiA6NC octopine Ti plasmid: two gene products in plant tumorigenesis. *Proc. Natl. Acad. Sci. USA* **81**, 1728–32.

Klee H.J., Hayford M.B., Kretzmer K.A., Barry G.F. & Kishore G.M. (1991) Control of ethylene synthesis by expression of a bacterial enzyme in transgenic tomato plants. *Plant Cell* **3**, 1187–93.

Kleid D.G., Yansura D., Small B., Dowbenko D., Moore D.M., Grubman M.J., McKercher P.D., Morgan D.O., Robertson B.H. & Bachrach H.L. (1981) Cloned viral protein vaccine for foot-and-mouth disease: responses in cattle and swine. *Science* **214**, 1125–9.

Klein T.M. & Fitzpatrick-McElligott (1993) Particle bombardment: a universal approach for gene transfer to cells and tissues. *Curr. Opinion Biotechnol.* **4**, 583–90.

Klein T.M., Wolf E.D., Wu R. & Sanford J.C. (1987) High-velocity micro-projectiles for delivering nucleic acids into living cells. *Nature* **327**, 70–3.

Klein-Lankhorst R., Rietveld P., Machiels B., Verkerk R., Weide R., Gebharot C., Koorn-neef M. & Zabel P. (1991) RFLP markers linked to the root knot nematode resistance gene *Mi* in tomato. *Theor. Appl. Genetics* **81**, 661–7.

Klotsky R.-A. & Schwartz I. (1987) Measurement of *cat* expression from growth-rate-regulated promoters employing β-lactamase activity as an indicator of plasmid copy number. *Gene* **55**, 141–6.

Kniskern P.J., Hagopian A., Montgomery D.L., Burke P., Dunn N.R., Hofmann K.J., Miller W.J. & Ellis R.W. (1986) Unusually high-level expression of a foreign gene (hepatitis B virus core antigen) in *Saccharomyces cerevisiae*. *Gene* **46**, 135–41.

Knutzon D.S., Thompson G.A., Radke S.E., Johnson W.B., Knauf V.C. & Kridl J.C. (1992) Modification of *Brassica* seed oil by antisense expression of a stearoyl-acyl carrier protein desaturase gene. *Proc. Natl. Acad. Sci. USA* **89**, 2624–8.

Koekman B.P., Ooms G., Klapwijk P.M. & Schilperoort R.A. (1979) Genetic map of an octopine Ti plasmid. *Plasmid* **2**, 347–57.

Kogan S.C., Doherty M. & Gitschier J. (1987) An improved method for prenatal diagnosis of genetic diseases by analysis of amplified DNA sequences. *N. Engl. J. Med.* **317**, 985–90.

Kohler S.W., Provost G.S., Kretz P.L., Kieck A., Sorge J.A. & Short J.M. (1990) The use of transgenic mice for short-term, *in vivo* mutagenicity testing. *Genet. Anal. Tech. Appl. (USA)* **7**, 212–18.

Koizumi M., Yamaguchi-Shinozaki K., Tguji H. & Shinozaki K. (1993) Structure and expression of two genes that encode distinct drought-inducible cysteine proteinases in *Arabidopsis thaliana*. *Gene* **129**, 175–82.

Kolberg R. (1992) Gene-transfer virus contaminant linked to monkey's cancer. *J. NIH Res.* **4**, 43–4.

Kollias G., Wrighton N., Hurst J., Grosveld F. (1986) Regulated expression of human ^{A}γ-, β-, and hybrid γβ-globin genes in transgenic mice: manipulation of the developmental expression patterns. *Cell* **46**, 89–94.

Kondo J.K. & McKay L.L. (1985) Gene transfer systems and molecular cloning in group N streptococci: a review. *J Dairy Sci.* **68**, 2143–59.

Kondoh H., Yasuda K. & Okada T.S. (1983) Tissue-specific expression of a cloned chick δ-crystallin gene in mouse cells. *Nature* **301**, 440–2.

Koonin E.V., Bork P. & Sander C. (1994) Yeast chromosome III: new gene functions. *EMBO J.* **13**, 493–503.

Kooter J.M. & Mol J.N.M. (1993) *Trans*-inactivation of gene expression in plants. *Curr. Opinion Biotechnol.* **4**, 166–71.

Kornacker M.G. & Pugsley A.P. (1990) The normally periplasmic enzyme β-lactamase is specifically and efficiently translocated through the *Escherichia coli* outer membrane when it is fused to the cell surface enzyme pullulanase. *Mol. Microbiol.* **4**, 1101–9.

Korona R., Korona B. & Levin B.R. (1993) Sensitivity of naturally occurring coliphages to type I and type II restriction and modification. *J. Gen. Microbiol.* **139**, 1283–90.

Kotula L. & Curtis J. (1991) Evaluation of foreign gene codon optimisation in yeast: expression of a mouse Ig kappa chain. *Biotechnology* **9**, 1386–9.

Koukolikova-Nicola Z., Shillito R.D., Hohn B., Wang K., Van Montagu M. & Zambryski P. (1985) Involvement of circular intermediates in the transfer of T-DNA from *Agrobacterium tumefaciens* to plant cells. *Nature* **313**, 191–6.

Kozak M. (1984) Compilation and analysis of sequences upstream from the translation start site in eukaryotic messenger RNAs. *Nucleic Acids Res.* **12**, 857–72.

Kozak M. (1986) Point mutations define a sequence flanking the AUG initiator codon that modulates translation by eukaryotic ribosomes. *Cell* **44**, 283–92.

Kramer R.A., Cameron J.R. & Davis R.W. (1976) Isolation of bacteriophage λ containing yeast ribosomal RNA genes: screening by *in situ* RNA hybridization to plaques. *Cell* **8**, 227–32.

Kramer B., Kramer W. & Fritz H.-J. (1984a) Different base/base mismatches are corrected with different efficiencies by the methyl-directed DNA mismatch-repair system of *E. coli*. *Cell* **38**, 879–88.

Kramer W., Drutsa V., Jansen H.-W., Kramer B., Pflugfelder M. & Fritz H.-J. (1984b) The gapped duplex DNA approach to oligonucleotide-directed mutation construction. *Nucleic Acids Res.* **12**, 9441–56.

Kreft J. & Hughes C. (1981) Cloning vectors derived from plasmids and phage of *Bacillus*. *Curr. Top. Microbiol. Immunol.* **96**, 1–17.

Krieg P.A. & Melton D.A. (1985) Developmental regulation of a gastrula-specific gene injected into fertilized *Xenopus* eggs. *EMBO J.* **4**, 3463–71.

Krisch H.M. & Selzer G.B. (1981) Construction and properties of a recombinant plasmid containing gene 32 of bacteriophage T4D. *J. Mol. Biol.* **148**, 199–218.

Kruger D.H. & Bickle T.A. (1983) Bacteriophage survival: multiple mechanisms for avoiding deoxyribonucleic acid restriction systems of their hosts. *Microbiol. Rev* **47**, 345–60.

Kruger D.H., Schroeder C., Reuter M., Bogdarina I.G., Buryanov Y.I. & Bickle T.A. (1985) DNA methylation of bacterial viruses T3 and T7 by different DNA methylases in *Escherichia coli* K12 cells. *Eur. J. Biochem.* **150**, 323–30.

Kuchler K. (1993) Unusual routes of protein secretion: the easy way out. *Trends Cell Biol.* **3**, 421–6.

Kudla B. & Nicolas A. (1992) A multisite integrative cassette for the yeast *Saccharomyces cerevisiae*. *Gene* **119**, 49–56.

Kuehn M.R., Bradley A., Robertson E.J. & Evans M.J. (1987) A potential model for Lesch–Nyhan syndrome through introduction of HPRT mutations in mice. *Nature* **326**, 295–8.

Kues U. & Stahl U. (1989) Replication of plasmids in Gram-negative bacteria. *Microbiol. Rev.* **53**, 491–516.

Kukuruzinska M.A., Bergh M.L.E. & Jackson B.J. (1987) Protein glycosylation in yeast. *Ann. Rev. Biochem.* **56**, 915–44.

Kunkel T.A. (1985) Rapid and efficient site-specific mutagenesis without phenotypic selection. *Proc. Natl. Acad. Sci. USA* **82**, 488–92.

Kuo C.-L. & Campbell J.L. (1983) Cloning of *Saccharomyces cerevisiae* DNA replication genes: isolation of the CDC8 gene and two genes that compensate for the cdc8–1 mutation. *Mol. Cell. Biol.* **3**, 1730–7.

Kurjan J. & Herskowitz I. (1982) Structure of a yeast pheromone (MF alpha). A putative alpha factor precursor contains four tandem copies of mature alpha factor. *Cell* **30**, 933–43.

Kurland C.G. (1987) Strategies for efficiency and accuracy in gene expression. 1. The major codon preference: a growth optimization strategy. *Trends Biochem. Sci.* **12**, 126–8.

Kuroda K., Hauser C., Rott R., Klenk H.-D. & Doerfler W. (1986) Expression of the influenza virus haemagglutinin in insect cells by a baculovirus vector. *EMBO J.* **5**, 1359–65.

Kurtz D.T. & Nicodemus C.F. (1981) Cloning of $\alpha_{2\mu}$ globulin cDNA using a high efficiency technique for the cloning of trace messenger RNAs. *Gene* **13**, 145–152.

Kwoh D.Y., Davis G.R., Whitfield K.M., Chappelle H.L., Dimichele L.J. & Gingeras T.R. (1989) Transcription-based amplification system and detection of amplified human immunodeficiency virus type I with a bead-based sandwich hybridization format. *Proc. Natl. Acad. Sci. USA* **86**, 1173–7.

Labes M., Puhler A. & Simon R. (1990) A new family of RSF1010-derived expression and *lac*-fusion broad-host-range vectors for Gram-negative bacteria. *Gene* **89**, 37–46.

Lacks S.A., Lopez P., Greenberg B. & Espinosa M. (1986) Identification and analysis of genes for tetracycline resistance and replication functions in the broad-host-range plasmid pLS1. *J. Mol. Biol.* **192**, 753–5.

Lacy E., Roberts S., Evans E.P., Burtenshaw M.D. & Constantini F.D. (1983) A foreign β-globin gene in transgenic mice: integration at abnormal chromosomal positions and expression in inappropriate tissues. *Cell* **34**, 343–58.

Lamond A.I. & Travers A.A. (1983) Requirement for an upstream element for optimal transcription of a bacterial tRNA gene. *Nature* **304**, 248–50.

Land H., Grey M., Hanser H., Lindenmaier W. & Schutz G. (1981) 5'-Terminal sequences of eucaryotic mRNA can be cloned with a high efficiency. *Nucleic Acids Res.* **9**, 2251–66.

Lane C.D., Colman A., Mohun T., Morser J., Champion J., Kourides I., Craig R., Higgins S., James T.C., Appelbaum S.W., Ohlsson R.I., Pauch E., Houghton M., Matthews J. & Miflin B.J. (1980) The *Xenopus* oocyte as a surrogate secretory system. The specificity of protein export. *Eur. J. Biochem* **111**, 225–35.

Lane D., Prentki P. & Chandler M. (1992) Use of gel retardation to analyze protein–nucleic acid interactions. *Microbiol. Rev.* **56**, 509–28.

Langford C., Nellen W., Niessing J. & Gallwitz D. (1983) Yeast is unable to excise foreign intervening sequences from hybrid gene transcripts. *Proc. Natl. Acad. Sci. USA* **80**, 1496–500.

Lapeyre B. & Amalric F. (1985) A powerful method for the preparation of cDNA libraries: isolation of cDNA encoding a 100-kDa nucleolar protein. *Gene* **37**, 215–20.

LaSalle G.L., Robert J.J., Berrard S., Ridoux V., Stratford-Perricaudet L.D., Perricaudet M. & Mallet J. (1993) An adenovirus vector for gene transfer into neurons and glia in the brain. *Science* **259**, 988–90.

Laski F.A., Rio D.C. & Rubin G.M. (1986) Tissue specificity of *Drosophila* P element transposition is regulated at the level of mRNA splicing. *Cell* **44**, 7–19.

Lathe R. (1985) Synthetic oligonucleotide probes deduced from amino acid sequence data. *J. Mol. Biol.* **183**, 1–12.

Lau D., Kuzma G., Wei C.-M., Livingston D.J. & Hsiung N. (1987) A modified human tissue plasminogen activator with extended half-life *in vivo*. *Biotechnology* **5**, 953–8.

La Vallie E.R., Di Blasio E.A., Kovacic S., Grant K.L., Schendel P.F. & McCoy J.M. (1993) A thioredoxin gene fusion expression system that circumvents inclusion body formation in the *E. coli* cytoplasm. *Bio/technology* **11**, 187–93.

Law M.-F., Byrne J. & Hawley P.M. (1983) A stable bovine papillomavirus hybrid plasmid that expresses a dominant selective trait. *Mol. Cell. Biol.* **3**, 2110–15.

Lawyer F.C., Stoffel S., Saiki R., Myambo K., Drummond R. & Gelfand D.H. (1989) Isolation, characterization, and expression in *E. coli* of the DNA polymerase from *Thermus aquaticus*. *J. Biol. Chem.* **264**, 6427–37.

Lay Thein S. & Wallace R.B. (1986) In *Human Genetic Diseases. A Practical Approach*, ed. Davies I.E., pp. 33–50. IRL Press, Oxford.

Lazarus R.A., Seymour J.L., Stafford R.K., Dennis M.S., Lazarus M.G., Marks C.B. & Anderson S. (1990) A biocatalytic approach to Vitamin C production: metabolic pathway engineering of *Erwinia herbicola*. In *Biocatalysis*, ed. Abramowitz D., pp. 136–55. Van Nostrand Reinhold, New York.

Le Hegarat J.C. & Anagnostopoulos C. (1977) Detection and characterization of naturally occurring plasmids in *Bacillus subtilis*. *Mol. Gen. Genetics* **157**, 164–74.

Leach D.R.F. & Stahl F. (1983) Viability of lambda phages carrying a perfect palindrome in the absence of recombination nucleases. *Nature* **305**, 448–51.

Leder P., Tiemeier D. & Enquist L. (1977) EK2 derivatives of bacteriophage lambda useful in the cloning of DNA from higher organisms: the λgt WES system. *Science* **196**, 175–7.

Lederberg S. (1957) Suppression of the multiplication of heterologous bacteriophages in lysogenic bacteria. *Virology* **3**, 496–513.

Lederberg S. & Meselson M. (1964) Degradation of non-replicating bacteriophage DNA in non-accepting cells. *J. Mol. Biol.* **8**, 623–8.

Lee F., Mulligan R., Berg P. & Ringold G. (1981) Glucocorticoids regulate expression of dihydrofolate reductase cDNA in mouse mammary tumour virus chimaeric plasmids. *Nature* **294**, 228–32.

Lee C.C., Wu X., Gibbs R.A., Cook R.G., Muzuy D.M. & Caskey C.T. (1988) Generation of cDNA probes directed by amino acid sequence: cloning of urate oxidase. *Science* **239**, 1288–91.

Leemans J., Deblaere R., Willmitzer L., De Greve H., Hernalsteens J.P., Van Montagu M. & Schell J. (1982a) Genetic identification of functions of T_L–DNA transcripts in octopine crown galls. *EMBO J.* **1**, 147–52.

Leemans J., Langenakens J., De Greve H., Deblaere R., Van Montagu M. & Schell J. (1982b) Broad-host-range cloning vectors derived from the W-plasmid Sa. *Gene* **19**, 361–4.

Leenhouts K.J., Tolner B., Bron S., Kok J., Venema G. & Seegers J.F. (1991) Nucleotide sequence and characterization of the broad-host-range lactococcal plasmid pWV01. *Plasmid* **26**, 55–66.

Legrice S.F.J. (1990) Regulated promoter for high-level expression of heterologous genes in *Bacillus subtilis*. *Methods Enzymol.* **185**, 201–14.

Lemmers M., De Beuckeleer M., Holsters M., Zambryski P., Hernalsteens J.P., Van Montagu M. & Schell J. (1980) Internal organization, boundaries and integration of Ti plasmid DNA in nopaline crown gall tumours. *J. Mol. Biol.* **144**, 353–76.

Leonhardt H. & Alonso J.C. (1991) Parameters affecting plasmid stability in *Bacillus subtilis*. *Gene.* **103**, 107–11.

Lerner C.G., Kobayashi T. & Inouye M. (1990) Isolation of subtilisin pro-sequence mutations that affect formation of active protease by localized random polymerase chain reaction mutagenesis. *J. Biol. Chem.* **265**, 20085–6.

Lesley S.A., Brow M.A. & Burgess R.R. (1991) Use of *in vitro* protein synthesis from polymerase chain reaction-generated templates to study interaction of *Escherichia coli* transcription factors with core RNA polymerase and for epitope mapping of monoclonal antibodies. *J. Biol. Chem.* **266**, 2632–8.

Lewontin R.C. & Hartl D.L. (1991) Population genetics in forensic DNA typing. *Science* **254**, 1745–50.

Li H.H., Gyllensten U.B., Cui X.F., Sakai R.K., Erlich H.A. & Arnhein N. (1988) Amplification and analysis of DNA sequences in single human sperm and diploid cells. *Nature* **335**, 414–17.

Liao H., McKenzie T. & Hageman R. (1986) Isolation of a thermostable enzyme variant by cloning and selection in a thermophile. *Proc. Natl. Acad. Sci. USA* **83**, 576–80.

Lin Q., Chen Z., Antoniw J. & White R. (1991) Isolation and characterization of a cDNA clone encoding the anti-viral protein from *Phytolacca americana*. *Plant Mol. Biol.* **17**, 609–14.

Liu H. & Rashidbaigi A. (1990) Comparison of various competent cell preparation methods for high efficiency DNA transformation. *Biotechniques* **8**, 21–5.

Lloyd A., Walbot V. & Davis R. (1992) Dicot anthocyanin production activated by maize anthocyanin-specific regulators R and C1. *Science* **258**, 1773–5.

Lobban P.E. & Kaiser A.D. (1973) Enzymatic end-to-end joining of DNA molecules. *J. Mol. Biol.* **78**, 453–71.

Loenen W.A.M. & Brammar W.J. (1980) A bacteriophage lambda vector for cloning large DNA fragments made with several restriction enzymes. *Gene* **10**, 249–59.

Logan J.S. (1993) Transgenic animals: beyond 'funny milk'. *Curr. Opinion Biotechnol.* **4**, 591–5.

Lohnes D., Kastner A., Dierich M., Mark M., LeMeur M. & Chambon P. (1993) Function of retinoic acid receptor gamma in the mouse. *Cell* **73**, 643–58.

Lorz H., Baaker B. & Schell J. (1985) Gene transfer to cereal cells mediated by protoplast transformation. *Mol. Gen. Genetics* **199**, 178–182.

Lowman H.B., Bass S.H., Simpson N. & Wells J.A. (1991) Selecting high-affinity binding proteins by monovalent phage display. *Biochemistry* **30**, 10832–8.

Lowy I., Pellicer A., Jackson J.F., Sim G.K., Silverstein S. & Axel R. (1980) Isolation of transforming DNA: Cloning of the hamster *hprt* gene. *Cell* **22**, 817–23.

Lu A.L., Clark S. & Modrich P. (1983) Methyl-directed repair of DNA basepair mismatches *in vitro*. *Proc. Natl. Acad. Sci. USA* **80**, 4639–43.

Lubbert H., Hoffman B.J., Snutch T.P., Vandyke T., Levine A.J., Hartig P.R., Lester H.A. & Davidson N. (1987) cDNA cloning of a serotonin 5-HT1C receptor by electrophysiological assays of messenger RNA-injected *Xenopus* oocytes. *Proc. Natl. Acad. Sci. USA* **84**, 4332–6.

Lundberg K.S., Shoemaker D.D., Adams M.W., Short J.M., Sorge J.A. & Mathur E.J. (1991) High fidelity amplification using a thermostable DNA polymerase isolated from *Pyrococcus furiosus*. *Gene* **108**, 1–4.

Lundbergh U., Von Gabain A. & Melefors O. (1990) Cleavages in the 5' region of the *ompA* and *bla* mRNA control stability: studies with an *E. coli* mutant altering mRNA stability and a novel endoribonuclease. *EMBO J.* **9**, 2731–41.

Luo Z. & Wu R. (1988) A simple method for the transformation of rice via the pollen-tube pathway *Plant Mol. Biol. Rep.* **6**, 165–74.

Luria S.E. (1953) Host-induced modifications of viruses. *Cold Spr. Harb. Symp. Quant. Biol.* **18**, 237–44.

Luria S.E. & Human M.L. (1952) A nonhereditary host-induced variation of bacterial viruses. *J. Bacteriol.* **64**, 557–9.

Lusky M. & Botchan M. (1981) Inhibitory effect of specific pBR322 DNA sequences upon SV40 replication in simian cells. *Nature* **293**, 79–81.

Ma H., Kunes S., Schatz P.J. & Botstein D. (1987) Plasmid construction by homologous recombination in yeast. *Gene* **58**, 201–16.

Maeda S., Kawai T., Obinata M., Fujiwara H., Horiuchi T., Saeki Y., Sato Y. & Furusawa M. (1985) Production of human α-interferon in silkworm using a baculovirus vector. *Nature* **315**, 592–4.

Mahillon J. & Kleckner N. (1992) New IS10 transposition vectors based on a Gram-positive replication origin. *Gene* **116**, 69–74.

Mandel M. & Higa A. (1970) Calcium-dependent bacteriophage DNA infection. *J. Mol. Biol.* **53**, 159–62.

Manen D. & Caro L. (1991) The replication of plasmid pSC101. *Mol. Microbiol.* **5**, 233–7.

Manen D., Goebel T. & Caro L. (1990) The *par* region of pSC101 affects plasmid copy number as well as stability. *Mol. Microbiol.* **4**, 1839–46.

Maniatis T., Goodbourn S. & Fischer J.A. (1987) Regulation of inducible and tissue-specific gene expression. *Science* **236**, 1237–45.

Maniatis T., Hardison R.C., Lacy E., Lauer J., O'Connell C., Quon D., Sim G.K. & Efstratiadis A. (1978) The isolation of structural genes from libraries of eucaryotic DNA. *Cell* **15**, 687–701.

Mann R.S., Mulligan R.C. & Baltimore D. (1983) Construction of a retrovirus packaging mutant and its use to produce helper-free defective retrovirus. *Cell* **32**, 871–9.

Mansour S.L., Thomas K.R. & Capecchi M.R. (1988) Disruption of the proto-oncogene int-2 in mouse embryo-derived stem cells: a general strategy for targeting mutations to non-selectable genes. *Nature* **336**, 348–52.

Marchuk D., Drumm M., Saulino A. & Collins F.S. (1990) Construction of T-vectors, a rapid and general system for direct cloning of unmodified PCR products. *Nucleic Acids Res.* **19**, 1154.

Marini N., Hiiyanna K.T. & Benbow R.M. (1989) Differential replication of circular DNA molecules co-injected into early *Xenopus laevis* embryos. *Nucleic Acids Res.* **17**, 5793–808.

Marinus M.G., Carraway M., Frey A.Z., Brown L. & Arraj J.A. (1983) Insertion mutations in the *dam* gene of *Escherichia coli* K-12. *Mol. Gen. Genetics* **192**, 288–9.

Marotti K.R., Castle C.K., Boyle T.P., Lin A.H., Murray R.W. & Melchior G.W. (1993) Severe atherosclerosis in transgenic mice expressing simian cholesterol ester transfer protein. *Nature* **364**, 73–5.

Martin G.B., Ganal M.W. & Tanskley S.D. (1992) Construction of a yeast artificial chromosome library of tomato and identification of cloned segments linked to two disease resistance loci. *Mol. Gen. Genetics* **233**, 25–32.

Matsumoto K., Yoshimatsu T. & Oshima Y. (1983) Recessive mutations conferring resistance to carbon catabolite repression of galactokinase synthesis in *Saccharomyces cerevisiae*. *J. Bacteriol.* **153**, 1405–14.

Masu Y., Nakayama K., Tamaki H., Harada Y., Kuno M. & Nakanishi S. (1987) cDNA cloning of bovine substance-K receptor through oocyte expression system. *Nature* **329**, 836–8.

Masui Y., Mizuno T. & Inouye M. (1984) Novel high-level expression cloning vehicles: 10^4-fold amplification of *Escherichia coli* minor protein. *Biotechnology* **2**, 81–5.

Matthews J., Brown J. & Hall T. (1981) Phaseolin mRNA is translated to yield glycosylated polypeptides in *Xenopus* oocytes. *Nature* **294**, 175–176.

Mattila P., Korpela J., Tenkanen T. & Pitkanen K. (1991) Fidelity of DNA synthesis by the *Thermococcus litoralis* DNA polymerase – an extremely heat stable enzyme with proofreading activity. *Nucleic Acids Res.* **18**, 4967–73.

Matzke A.J.M. & Chilton M.-D. (1981) Site-specific insertion of genes into T-DNA of the *Agrobacterium* tumour-inducing plasmid: an approach to genetic engineering of higher plant cells. *J. Mol. Appl. Genetics* **1**, 39–49.

Mazodier P., Petter R. & Thompson C. (1989) Intergeneric conjugation between *Escherichia coli* and *Streptomyces* species. *J. Bacteriol.* **171**, 3583–5.

McAlpine J.B., Tuan J.S., Brown D.P., Grebner K.D., Whittern D.N., Buko A. & Katz L. (1987) New antibiotics from genetically engineered actinomycetes. I. 2-Norerythromycins, isolation and structural determinations. *J. Antibiotics* **40**, 1115–22.

McClarin J.A., Frederick C.A., Wang B.-C., Green P., Boyer H.W., Grable J. & Rosenberg J.M. (1986) Structure of the DNA-*Eco*RI endonuclease recognition complex at 3 Å resolution. *Science* **234**, 1526–41.

McClelland M., Kessler L.G. & Bittner M. (1984) Site-specific cleavage of DNA at 8- and 10-base-pair sequences. *Proc. Natl. Acad. Sci. USA* **81**, 983–7.

McLaughlin J.R., Murray C.L. & Rabinowitz J.C. (1981) Unique features in the ribosome binding site sequence of the Gram-positive *Staphylococcus aureus* β-lactamase gene. *J. Biol. Chem.* **256**, 11283–91.

McMahon A.P. & Bradley A. (1990) The *Wnt* 1 (*int* 1) proto-oncogene is required for development for a large region of the mouse brain. *Cell* **62**, 1073–85.

McNeall J., Sanchez A., Gray P.P., Chesterman C.N. & Sleigh M.J. (1989) Hyperinducible expression from a metallothionein promoter containing additional metal responsive elements. *Gene* **76**, 81–8.

McPherson D.T. (1988) Codon preference reflects mistranslational constructs: a proposal. *Nucleic Acids Res.* **16**, 4111–20.

Meade H., Gates L., Lacy E. & Lonberg N. (1990) Bovine alpha$_{S1}$-casein gene sequences direct high level expression of active human urokinase in mouse milk. *Biotechnology* **8**, 443–6.

Meagher R.B., Tait R.C., Betlach M. & Boyer H.W. (1977) Protein expression in *E. coli* minicells by recombinant plasmids. *Cell* **10**, 521–36.

Meeley R.B. & Walton J.D. (1991) Enzymatic detoxification of HC-toxin, the host selective peptide from *Cochliobolus carbonum*. *Plant Physiol.* **97**, 1080–6.

Mellon P., Parker V., Gluzman Y. & Maniatis T. (1981) Identification of DNA sequences required for transcription of the human α-globin gene in a new SV40 host–vector system. *Cell* **27**, 279–88.

Mellor J., Dobson M.J., Roberts M.A., Tuite M.F., Emtage J.S., White S., Lowe P.A., Patel T., Kingsman A.J. & Kingsman S.M. (1983) Efficient synthesis of enzymatically active calf chymosin in *Saccharomyces cerevisiae*. *Gene* **24**, 1–14.

Mellor J., Malim M., Gull K., Tuite M.F., McCready S., Sibbayawan T., Kingsman S.M. & Kingsman A.J. (1985) Reverse transcriptase activity and Ty RNA are associated with virus-like particles in yeast. *Nature* **318**, 583–6.

Mellor J., Dobson M., Kingsman A.J. & Kingsman S.M. (1987) A transcriptional activator is located in the coding region of the yeast PGK gene. *Nucleic Acids Res.* **15**, 6243–59.

Melnick L.M., Turner B.G., Puma P., Price-Tillotson B., Salvato K.A., Dumais D.R., Moir D.T., Broeze R.J. & Augerinos G.C. (1990) Characterisation of a nonglycosylated single chain urinary plasminogen activator secreted from yeast. *J. Biol. Chem.* **265**, 801–7.

Melton D.A. (1987) In *Methods in Enzymology*, eds Berger & Kimmel, Vol. 152, pp. 288–96. Academic Press, New York.

Mendel D., Ellman J.A., Chang Z., Veenstra D.L., Kollman P.A. & Schultz P.G. (1992) Probing protein stability with unnatural amino acids. *Science* **256**, 1798–802.

Mendel D., Ellman J. & Schultz P. (1991) Construction of a light-activated protein by unnatural amino acid mutagenesis. *J. Am. Chem. Soc.* **113**, 2758–60.

Mermod N., Ramos J.L., Lehrbach P.R. & Timmis K.N. (1986) Vector for regulated expression of cloned genes in a wide range of Gram-negative bacteria. *J. Bacteriol.* **167**, 447–54.

Mertz J.E. & Berg P. (1974) Defective simian virus 40 genomes. Isolation and growth of individual clones. *Virology* **62**, 112–24.

Meselson M. & Yuan R. (1968) DNA restriction enzyme from *E. coli. Nature* **217**, 1110–14.

Messeguer R., Ganal M., De Vincente M.C., Young N.D., Bolkan H. & Tanksley S.D. (1991) High resolution RFLP map around the root knot nematode resistance gene (Mi) in tomato. *Theor. Appl. Genetics* **82**, 529–36.

Messing J., Crea R. & Seeburg P.H. (1981) A system for shotgun DNA sequencing. *Nucleic Acids Res.* **9**, 309–21.

Messing J., Gronenborn B., Muller-Hill B. & Hofschneider P.H. (1977) Filamentous coliphage M13 as a cloning vehicle: insertion of a *Hind*II fragment of the *lac* regulatory region in M13 replicative form *in vitro. Proc. Natl. Acad. Sci. USA* **74**, 3642–6.

Messing J. & Vieira J. (1982) A new pair of M13 vectors for selecting either DNA strand of double-digest restriction fragments. *Gene* **19**, 269–76.

Metzger D., White J.H. & Chambon P. (1988) The human estrogen receptor functions in yeast. *Nature* **334**, 31–6.

Meyer P., Heidmann I., Forkmann G. & Saedler H. (1987) A new petunia flower colour generated by transformation of a mutant with a maize gene. *Nature* **330**, 677–8.

Michel B. & Ehrlich S.D. (1986) Illegitimate recombination at the replication origin of bacteriophage M13. *Proc. Natl. Acad. Sci. USA* **83**, 3386–90.

Michel B., Niaudet B. & Ehrlich S.D. (1982) Intramolecular recombination during plasmid transformation of *Bacillus subtilis* competent cells. *EMBO J.* **1**, 1565–71.

Michel B., Palla E., Niaudet B. & Ehrlich S.D. (1980) DNA cloning in *Bacillus subtilis*. III. Efficiency of random-segment cloning and insertional inactivation vectors. *Gene* **12**, 147–54.

Michel B., Niaudet B. & Ehrlich S.D. (1983) Intermolecular recombination during transformation of *Bacillus subtilis* competent cells by monomeric and dimeric plasmids. *Plasmid* **10**, 1–10.

Michel P., Palla E., Niaudet B. & Ehrlich S.D. (1980) DNA cloning in *Bacillus subtilis*. III. Efficiency of random segment cloning and insertional inactivation vectors. *Gene* **12**, 147–54.

Michelson A.M. & Orkin S.H. (1982) Characterization of the homopolymer tailing reaction catalyzed by terminal deoxynucleotidyl transferase. Implications for the cloning of cDNA. *J. Biol. Chem.* **256**, 1473–82.

Miller A.D. (1992) Human gene therapy comes of age. *Nature* **357**, 455–60.

Miller A.D., Law M.F. & Verma I.M. (1985) Generation of helper-free amphitropic retroviruses that transduce a dominant-acting, methotrexate-resistant dihydrofolate reductase gene. *Mol. Cell. Biol.* **5**, 431–7.

Miller C.A., Beaucage S.L. & Cohen S.N. (1990) Role of DNA superhelicity in partitioning of the pSC101 plasmid. *Cell* **62**, 127–33.

Miller C.G., Strauch K.L., Kukral A.M., Miller J.L., Wingfield P.T., Mazzei G.J., Werlen R.C., Graber P. & Movva N.R. (1987) *N*-terminal methionine-specific peptidase in *Salmonella typhimurium. Proc. Natl. Acad. Sci. USA* **84**, 2718–22.

Miller O.K. & Temin H.M. (1983) High efficiency ligation and recombination of DNA fragments by vertebrate cells. *Science* **200**, 606–9.

Miyamoto C., Smith G.E., Farrell-Towt J., Chizzonite R., Summers M.D. & Ju G. (1985) Production of human *c-myc* protein in insect cells infected with baculovirus expression vector. *Mol. Cell. Biol.* **5**, 2860–5.

Modell B. & Kuliev A. (1993) A scientific basis for cost-benefit analysis of genetics services. *Trends Genetics* **9**, 46–52.

Moir A. & Brammar W.J. (1976) Use of specialized transducing phages in amplification of enzyme production. *Mol. Gen. Genetics* **149**, 87–99.

Moir D.T. & Dumais D.R. (1987) Glycosylation and secretion of human alpha-1-antitrypsin of yeast. *Gene* **56**, 209–17.

Moks T., Abrahmsen L., Holmgren E. *et al.* (1987) Expression of human insulin-like growth factor I in bacteria: use of optimized gene fusion vectors to facilitate protein purification. *Biochemistry* **26**, 5239–44.

Montaya A.L., Chilton M.-D., Gordon M.P., Sciaky D. & Nester E.W. (1977) Octopine and nopaline metabolism in *Agrobacterium tumefaciens* and crown gall tumor cells: role of plasmid genes. *J. Bacteriol.* **129**, 101–7.

Moran C.P., Lang N., Le Grice S.F.J., Lee G., Stephens M., Sonenshein A.L., Pero J. & Losick R. (1982) Nucleotide sequences that signal the initiation of transcription and translation in *Bacillus subtilis*. *Mol. Gen. Genetics* **186**, 339–46.

Moriarty A.M., Hoyer B.H., Shih J.W.-K., Gerin J.L. & Hamer D.H. (1981) Expression of the hepatitis B virus surface antigen gene in cell culture by using a Simian virus 40 vector. *Proc. Natl. Acad. Sci USA* **78**, 2606–10.

Morris R.O. (1986) Genes specifying auxin and cytokinin biosynthesis in phytopathogens. *Annu. Rev. Plant Physiol.* **37**, 509–38.

Morrish F.M. & From M.E. (1992) Cereal transformation methods. *Curr. Opinion Biotechnol.* **3**, 141–6.

Moss B., Smith G.L., Gerin J.L. & Purcell R.H. (1984) Live recombinant vaccinia virus protects chimpanzees against hepatitis B. *Nature* **311**, 67–9.

Mottes M., Grandi G., Sgaramella V., Canosi U., Morelli G. & Trautner T.A. (1979) Different specific activities of the monomeric and oligomeric forms of plasmid DNA in transformation of *B. subtilis* and *E. coli*. *Mol. Gen. Genetics* **174**, 281–6.

Mudd E.A., Krisch H.M. & Higgins C.F. (1990) RNase E, an endoribonuclease, has a general role in the chemical decay of *Escherichia coli* mRNA: evidence that *rne* and *ams* are the same genetic locus. *Mol. Microbiol.* **4**, 2127–35.

Muesing M., Tamm J., Shepard H.M. & Polisky B. (1981) A single base pair alteration is responsible for the DNA overproduction phenotype of a plasmid copy-number mutant. *Cell* **24**, 235–42.

Muller-Rober B., Sonnewald U. & Willmitzer L. (1992) Inhibition of the ADP-glucose pyrophosphorylase in transgenic potatoes leads to sugar-storing tubers and influences tuber formation and expression of tuber storage proteins. *EMBO J.* **11**, 1229–38.

Mulligan R.C. & Berg P. (1980) Expression of a bacterial gene in mammalian cells. *Science* **209**, 1422–7.

Mulligan R.C. & Berg P. (1981a) Factors governing the expression of a bacterial gene in mammalian cells. *Mol. Cell. Biol.* **1**, 449–59.

Mulligan R.C. & Berg P. (1981b) Selection for animal cells that express the *Escherichia coli* gene coding for xanthine-guanine phosphoribosyl-transferase. *Proc. Natl. Acad. Sci. USA* **78**, 2072–6.

Mulligan R.C., Howard B.H. & Berg P. (1979) Synthesis of rabbit β-globin in cultured monkey kidney cells following infection with SV40 β-globin recombinant genome. *Nature* **277**, 108–114.

Mullis K.B. (1990) The unusual origin of the polymerase chain reaction. *Sci. Amer.* **262**, 56–65.

Mullis K.B. & Faloora F. (1987) Specific synthesis of DNA *in vitro* via a polymerase catalyzed chain reaction. *Methods Enzymol.* **155**, 335–50.

Murdock D., Ensley B.D., Serdar C. & Thalen M. (1993) Construction of metabolic operons catalyzing the *de novo* biosynthesis of indigo in *Escherichia coli*. *Biotechnology* **11**, 381–6.

Murphy D.J. (1992) Modifying oilseed crops for non-edible products. *Trends Biotechnol.* **10**, 84–7.

Murray A.W. & Szostak J.W. (1983a) Pedigree analysis of plasmid segregation in yeast. *Cell* **34**, 961–70.

Murray A.W. & Szostak J.W. (1983b) Construction of artificial chromosomes in yeast. *Nature* **305**, 189–93.

Murray J.A.H. (1987) Bending the rules: the 2μ plasmid of yeast. *Mol. Microbiol.* **1**, 1–14.

Murray K. & Murray N.E. (1975) Phage lambda receptor chromosomes for DNA fragments made with restriction endonuclease III of *Haemophilus influenzae* and restriction endonuclease I of *Escherichia coli*. *J. Mol. Biol.* **98**, 551–64.

Murray M.J., Shilo B.-Z., Chiaho S., Cowing D., Hsu H.W. & Weinberg R.A. (1981) Three different human tumor cell lines contain different oncogenes. *Cell* **25**, 355–61.

Murray N.E. (1983) Phage lambda and molecular cloning. In *The Bacteriophage Lambda*, eds Hendrix R.W., Roberts J.W., Stahl F.W. & Weisberg R.A., Lambda II (Monograph No. 13), Vol. 2. Cold Spring Harbor Laboratory, Cold Spring Harbor, New York.

Murray N.E. & Kelley W.S. (1979) Characterization of λpol A transducing phages. Effective expression of the *E. coli pol A* gene. *Mol. Gen. Genetics* **175**, 77–87.

Murray N.E., Manduca de Ritis P. & Foster L.A. (1973a) DNA targets for the *Escherichia coli* K restriction system analysed genetically in recombinants between phages phi-80 and Lambda. *Mol. Gen. Genetics* **120**, 261–81.

Murray N.E., Batten P.L. & Murray K. (1973b) Restriction of bacteriophage lambda by *Escherichia coli* K. *J. Mol. Biol.* **81**, 395–407.

Murray N.E., Bruce S.A. & Murray K. (1979) Molecular cloning of the DNA ligase gene from bacteriophage T4. II. Amplification and preparation of the gene product. *J. Mol. Biol.* **132**, 493–505.

Murray N.E., Daniel A.S., Cowan G.M. & Sharp P.M. (1993) Conservation of motifs within the unusually variable polypeptide sequences of type 1 restriction and modification enzymes. *Mol. Microbiol.* **9**, 133–43.

Myers R.M. & Tjian R. (1980) Construction and analysis of simian virus 40 origins defective in tumor antigen binding and DNA replication. *Proc. Natl. Acad. Sci. USA* **77**, 6491–5.

Nabel E.G., Plautz G. & Nabel G.J. (1990) Site-specific gene expression *in vivo* by direct gene transfer into the arterial wall. *Science* **249**, 1285–88.

Nakamura Y., Leppert M., O'Connell P., Holm T., Culver M., Martin C., Fujimoto E., Hoff M., Kumlin E. & White R. (1987) Variable number of tandem repeat (VNTR) markers for human gene mapping. *Science* **235**, 1616–21.

Napoli C., Lemieux C.H. & Jorgensen R. (1990) Introduction of a chimeric chalcone synthese gene into petunia results in reversible co-suppression of homologous genes *in trans*. *Plant Cell* **2**, 279–89.

Nasmyth K.A. & Reed S.I. (1980) Isolation of genes by complementation in yeast. Molecular cloning of a cell-cycle gene. *Proc. Natl. Acad. Sci. USA* **77**, 2119–23.

Nathans J. & Hogness D.S. (1983) Isolation, sequence analysis, and intron-exon arrangement of the gene encoding bovine rhodopsin. *Cell* **34**, 807–14.

Nedivi E., Hevroni D., Nato D., Israeli D. & Citri Y. (1993) Numerous candidate plasticity-related genes revealed by differential cDNA cloning. *Nature* **363**, 718–22.

Neumann E., Schaefer-Ridder W., Wang Y. & Hofschneider P.H. (1982) Gene transfer into mouse lymphoma cells by electroporation in high electric fields. *EMBO J.* **1**, 841–5.

Neve R.L. (1993) Adenovirus vectors enter the brain. *Trends Neurosci.* **16**, 251–3.

Newman A.J., Linn T.G. & Hayward R.S. (1979) Evidence for cotranscription of the RNA polymerase genes *rpo*BC with a ribosome protein of *Escherichia coli*. *Mol. Gen. Genetics* **169**, 195–204.

Newton S.M.C., Jacob C.O. & Stocker B.A.D. (1989) Immune response to cholera toxin epitope inserted in *Salmonella* flagellin. *Science* **244**, 70–2.

Ng R. & Abelson J. (1980) Isolation and sequence of the gene for actin in *Saccharomyces cerevisiae*. *Proc. Natl. Acad. Sci. USA* **77**, 3912–16.

Niaudet B., Goze A. & Ehrlich S.D. (1982) Insertional mutagenesis in *Bacillus subtilis*. Mechanism and use in gene cloning. *Gene* **19**, 277–84.

Niaudet B., Jannière L. & Ehrlich S.D. (1984) Recombination between repeated DNA sequences occurs most often in plasmids than in the chromosome of *Bacillus subtilis*. *Mol. Gen. Genetics* **197**, 46–54.

Nilsson B., Holmgren E., Josephson S., Gatenbeck S., Philipson L. & Uhlen M. (1985) Efficient secretion and purification of human insulin-like growth factor I with a gene fusion vector in staphylococci. *Nucleic Acids Res.* **13**, 1151–62.

Noiret P., Petit M.A. & Ehrlich S.D. (1987) Plasmid replications stimulates DNA recombination in *Bacillus subtilis*. *J. Mol. Biol.* **196**, 39–48.

Noma Y., Sideras P., Natto T., Bergstedtlindguist S., Azuma C., Severinson E., Tanabe T., Kinashi T., Matsuda F., Yaoita Y. & Honjo T. (1986) Cloning of cDNA encoding the murine IgG1 induction factor by a novel strategy using SP6 promoter. *Nature* **319**, 640–46.

Nordstrom K. & Uhlin B.E. (1992) Runaway-replication plasmids as tools to produce large quantities of proteins from cloned genes in bacteria. *Biotechnology* **10**, 661–6.

Noren C.J., Anthony-Cahill S.J., Griffith M.C. & Schultz P.G. (1989) A general method for site-specific incorporation of unnatural amino acids into proteins. *Science* **244**, 182–8.

Normanly J., Kleina L.G., Masson J.-M., Abelson J. & Miller J.H. (1990) Construction of *Escherichia coli* amber suppressor tRNA genes. III: determination of tRNA specificity. *J. Mol. Biol.* **213**, 719–26.

Norrander J., Kempe T. & Messing J. (1983) Construction of improved M13 vectors using oligodeoxynucleotide-directed mutagenesis. *Gene* **27**, 101–6.

Novick R.P., Clowes R.C., Cohen S.N., Curtiss R., Datta N. & Falkow S. (1976) Uniform nomenclature for bacterial plasmids: a proposal. *Bact. Rev.* **40**, 168–89.

Nugent M.E., Primrose S.B. & Tacon W.C.A. (1983) The stability of recombinant DNA. *Dev. Ind. Microbiol.* **24**, 271–85.

Nunberg J.H., Wright D.K., Cole G.E., Petrovskis E.A., Post L.E., Compton T. & Gilbert J.H. (1989) Identification of the thymidine kinase gene of feline herpesvirus: use of degenerate oligonucleotides in the polymerase chain reaction to isolate herpesvirus gene homologs. *J. Virol.* **63**, 3240–9.

Nyyssonen E., Penttila M., Harkki A., Saloheimo A., Knowles J.K.C. & Keranen S. (1993) Efficient production of antibody fragments by the filamentous fungus *Trichoderma reesei*. *Biotechnology* **11**, 591–5.

Ochman H., Gerber S.A. & Hartl D.L. (1988) Genetic applications of an inverse polymerase chain reaction. *Genetics* **120**, 621–5.

Oeller P.W., Miw-Wong L., Taylor L.P., Pike D.A. & Theologis A. (1991) Reversible inhibition of tomato fruit senescence by antisense RNA. *Science* **254**, 437–9.

O'Hare K., Benoist C. & Breathnach R. (1981) Transformation of mouse fibroblasts to methotrexate resistance. *Proc. Natl. Acad. Sci. USA* **78**, 1527–31.

O'Hare K. & Rubin G.M. (1983) Structures of P transposable elements and their sites of insertion and excision in the *Drosophila melanogaster* genome. *Cell* **34**, 25–35.

Okamoto T., Fujita Y. & Irie R. (1985) Interspecific protoplast fusion between *Streptococcus cremoris* and *Streptococcus lactis*. *Agric. Biol. Chem.* **49**, 1371–6.

O'Kane C.J. & Gehring W.J. (1987) Detection *in situ* of genetic regulatory elements in *Drosophila*. *Proc. Natl. Acad. Sci. USA* **84**, 9123–7.

O'Kane C.J. & Moffat K.G. (1992) Selective cell ablation and genetic surgery. *Curr. Opinion Genetics Devel.* **2**, 602–7.

Okayama H. & Berg P. (1982) High-efficiency cloning of full-length cDNA. *Mol. Cell. Biol.* **2**, 161–70.

Old R., Murray K. & Roizes G. (1975) Recognition sequence of restriction endonuclease III from *Haemophilus influenzae*. *J. Mol. Biol.* **92**, 331–9.

Old J.M., Ward R.H.T., Petrov M., Karagozlu F., Modell B. & Weatherall D.J. (1982a) First trimester diagnosis for haemoglobinopathies: a report of 3 cases. *Lancet* **2**, 1413–16.

Oliver S.G. & 146 others (1992) The complete DNA sequence of yeast chromosome III. *Nature* **357**, 38–46.

Olivera B.M., Hall Z.W. & Lehman I.R. (1968) Enzymatic joining of polynucleotides. V. A DNA adenylate intermediate in the polynucleotide joining reaction. *Proc. Natl. Acad. Sci. USA* **61**, 237–44.

Olsen D.B. & Eckstein F. (1990) High-efficiency oligonucleotide-directed plasmid mutagenesis. *Proc. Natl. Acad. Sci. USA* **87**, 1451–5.

Olsen R.H., DeBusccher G. & McCombie W.R. (1982) Development of broad host-range vectors and gene banks. Self-cloning of the *Pseudomonas aeruginosa* PAO chromosome. *J. Bacteriol.* **150**, 60–9.

Olson M.V. (1981) Applications of molecular cloning to *Saccharomyces*. In *Genetic Engineering*, eds Setlow J.K. & Hollaender A. Plenum Press, New York.

Olszewska E. & Jones K. (1988) Vacuum blotting enhances nucleic acid transfer. *Trends Genetics* **4**, 92–4.

Ooms G., Hooykaas P.J.J., Moolenaar G. & Schilperoort R.A. (1981) Crown gall plant tumours of abnormal morphology induced by *Agrobacterim tumefaciens* carrying mutated octopine Ti plasmids: analysis of T-DNA functions. *Gene* **14**, 33–50.

Ooms G., Klapwijk P.M., Poulis J.A. & Schilperoort R.A. (1980) Characterization of

Tn*904* insertion in octopine Ti plasmid mutants of *Agrobacterium tumefaciens*. *J. Bacteriol*. **144**, 82–91.

O'Reilly D.R., Miller L.K. & Luckow V.A. (1992) *Baculovirus Expression Vectors: A Laboratory Manual*. W.H. Freeman, San Francisco.

Orkin S.H. (1982) Genetic diagnosis of the foetus. *Nature* **296**, 202–3.

Orkin S.H., Little P.F.R., Kazazian H.H. & Boehm C. (1982) Improved detection of the sickle mutation by DNA analysis. *N. Engl. J. Med*. **307**, 32–6.

Orkin S.H., Daddona P.E., Shewach D.S., Markham A.F., Bruns G.A., Goff S.C. & Kelley W.N. (1983) Molecular cloning of human adenosine deaminase gene sequences. *J. Biol. Chem*. **258**, 2753–6.

Orr-Weaver T.L., Szostak J.W. & Rothstein R.L. (1981) Yeast transformation: a model system for the study of recombination. *Proc. Natl. Acad. Sci. USA* **78**, 6354–8.

Osburne M.S., Craig R.J. & Rothstein D.M. (1984) Thermoinducible transcription system for *Bacillus subtilis* that utilizes control elements from temperate phage Φ105. *J. Bacteriol*. **163**, 1101–8.

Ostrowski M.C., Richard-Foy H., Wolford R.G., Berard D.S. & Hager G.L. (1983) Glucocorticoid regulation of transcription at an amplified episomal promoter. *Mol. Cell. Biol*. **3**, 2045–57.

Ou C.Y., Kwok S., Mitchell S.W., Mack D.H., Sninsky J.J., Krebs J.W., Feorino P., Warfield D. & Schochetman G. (1988) DNA amplification for direct detection of HIV-1 in DNA of peripheral blood mononuclear cells. *Science* **239**, 295–7.

Ow D.W., Wood K.V., DeLuca M., de Wet J.R., Helinski D.R. & Howell S.H. (1986) Transient and stable expression of the firefly luciferase gene in plant cells and transgenic plants. *Science* **234**, 856–9.

Palmiter R.D. & Brinster R.L. (1986) Germ-line transformation of mice. *Annu. Rev. Genetics* **20**, 465–99.

Palmiter R.D., Brinster R.L., Hammer R.E., Trumbauer M.E., Rosenfeld M.G., Birnberg N.C. & Evans R.M. (1982a) Dramatic growth of mice that develop from eggs micro-injected with metallothionein-growth hormone fusion genes. *Nature* **300**, 611–15.

Palmiter R.D., Chen H.Y. & Brinster R.L. (1982b) Differential regulation of metallothionein-thymidine kinase fusion genes in transgenic mice and their offspring. *Cell* **29**, 701–10.

Palmiter R.D., Chen H.Y., Messing A. & Brinster R.L. (1985) SV40 enhancer and large T-antigen are instrumental in development of choroid plexus tumors in transgenic mice. *Nature* **316**, 457–60.

Palmiter R.D., Norstedt G., Gelinas R.E., Hammer R.E. & Brinster (1983) Metallothionein-human GH fusion genes stimulate growth of mice. *Science* **222**, 809–14.

Panasenko S.M., Cameron J.R., Davis R.W. & Lehman I.R. (1977) Five hundredfold overproduction of DNA ligase after induction of a hybrid lambda lysogen contructed *in vitro*. *Science* **196**, 188–9.

Parmley S.E. & Smith P.G. (1988) Antibody-selectable filamentous fd phage vectors: affinity purification of target genes. *Gene* **73**, 305–18.

Paszkowski J., Shillito R.D., Saul M., Mandak V., Hohn T., Hohn B. & Potrykus I. (1984) Direct gene transfer to plants. *EMBO J*. **3**, 2717–22.

Paterson B.M., Roberts B.E. & Kuff E.L. (1977) Structural gene identification and mapping by DNA. mRNA hybrid-arrested cell-free translation. *Proc. Natl. Acad. Sci. USA* **74**, 4370–4.

Patzer E.J., Nakamura G.R., Hershberg R.D., Gregory T.J., Crowley C., Levinson A.D. & Eichberg J.W. (1986) Cell culture derived recombinant HBsAg is highly immunogenic and protects chimpanzees from infection with hepatitis B virus. *Biotechnology* **4**, 630–6.

Peden K.W.C. (1983) Revised sequence of the tetracycline-resistance gene of pBR322. *Gene* **22**, 277–80.

Peebles C.L., Ogden R.C., Knapp G. & Abelson J. (1979) Splicing of yeast transfer RNA precursors: a two-stage reaction. *Cell* **18**, 27–36.

Peeters B.P.H., De Boer J.H., Bron S. & Venema G. (1988) Structural plasmid instability in *Bacillus subtilis*: effect of direct and inverted repeats. *Mol. Gen. Genetics* **212**, 450–8.

Peijnenburg A.A.C.M., Bron S. & Venema G. (1987) Structural plasmid instability in recombination- and repair-deficient strains of *Bacillus subtilis*. *Plasmid* **17**, 167–70.

Pen J., Verwoerd T.C., Van Paridow P.A., Beudeker R.F., Van den Elzen P.J.M., Geerse K., Van der Klis J.D., Versteegh H.A.J., Van Ooyen A.J.J. & Hoekema A. (1993) Phytase-containing transgenic seeds as a novel feed additive for improved phosphorus utilization. *Biotechnology* **11**, 811–14.

Pennica D., Holmes W.E., Kohr W.J. *et al.* (1983) Cloning and expression of human tissue-type plasminogen activator cDNA in *E. coli. Nature* **301**, 214–21.

Penttila M.E., Andre L., Lehtovaara P., Bailey M., Teeri T.T. & Knowles J.K.C. (1988) Efficient secretion of two fungal cellobiohydrolases by *Saccharomyces cerevisiae*. *Gene* **63**, 103–12.

Peralta E.G., Hellmiss R. & Ream W. (1986) *Overdrive*, a T-DNA transmission enhancer on the *A. tumefaciens* tumour-inducing plasmid. *EMBO J.* **5**, 1137–42.

Perlak F.J., Deaton R.W., Armstrong T.A., Fuchs R.L., Sims S.R., Greenplate J.T. & Fischhoff D.A. (1990) Insect resistant cotton plants. *Biotechnology* **8**, 939–43.

Perlak F.J., Fuchs R.L., Dean D.A., McPherson S.L. & Fischhoff D.A. (1991) Modification of the coding sequence enhances plant expression of insect control protein genes. *Proc. Natl. Acad. Sci. USA* **88**, 3324–8.

Perry L.J. & Wetzel R. (1984) Disulfide bond engineered into T4 lysozyme: stabilization of the protein toward thermal inactivation. *Science* **226**, 555–7.

Perucho M., Goldfarb M., Shimizu K., Lama C., Fogh J. & Wigler M. (1981) Human-tumor-derived cell lines contain common and different transforming genes. *Cell* **27**, 467–76.

Perucho M., Hanahan D., Lipsich L. & Wigler M. (1980a) Isolation of the chicken thymidine kinase gene by plasmid rescue. *Nature* **285**, 207–10.

Perucho M., Hanahan D. & Wigler M. (1980b) Genetic and physical linkage of exogenous sequences in transformed cells. *Cell* **22**, 309–17.

Petrusyte M., Bitinaite J., Menkevicius S., Klimasauskas S., Butkus V. & Janulaitis A. (1988) Restriction endonucleases of a new type. *Gene* **74**, 89–91.

Pfeiffer P. & Hohn T. (1983) Involvement of reverse transcription in the replication of cauliflower mosaic virus. A detailed model and test of some aspects. *Cell* **33**, 781–9.

Pierce J.C., Sauer B. & Sternberg N. (1992) A positive selection vector for cloning high molecular weight DNA by the bacteriophage P1 system: improved cloning efficiency. *Proc. Natl. Acad. Sci. USA* **89**, 2056–60.

Pirrotta V., Hadfield C. & Pretorius G.H.J. (1983) Microdissection and cloning of the *white* locus and the 3B1–3C2 region of the *Drosophila* X chromosome. *EMBO J.* **2**, 927–34.

Pittius C.W., Hennighausen L., Lee E., Westphal H., Nicols E., Vitale J. & Gordon K. (1988) A milk protein gene promoter directs expression of human tissue plasminogen activator cDNA to the mammary gland in transgenic mice. *Proc. Natl. Acad. Sci. USA* **85**, 5874–8.

Plaskon R.R. & Wartell R.M. (1987) Sequence distributions associated with DNA curvature are found upstream of strong *E. coli* promoters. *Nucleic Acids Res.* **15**, 785–96.

Poirier Y.P., Dennis D.E., Klomparens K. & Somerville C.R. (1992) Production of polyhydroxybutyrate, a biodegradable thermoplastic, in higher plants. *Science* **256**, 520–3.

Potrykus I., Paszkowski J., Saul M.W., Petruska J. & Shillito R.D. (1985a) Molecular and general genetics of a hybrid foreign gene introduced into tobacco by direct gene transfer. *Mol. Gen. Genetics* **199**, 169–77.

Potrykus I., Saul M.W., Petruska J., Paszkowski J. & Shillito R. (1985b) Direct gene transfer to cells of a graminaceous monocot. *Mol. Gen. Genetics* **199**, 183–8.

Potter H. (1988) Electroporation in biology: methods applications and instrumentation. *Anal. Biochem.* **174**, 361–73.

Potter H., Weir L. & Leder P. (1984) Enhancer-dependent expression of human K immunoglobulin genes introduced into mouse pre-B lymphocytes by electroporation. *Proc. Natl. Acad. Sci. USA* **81**, 7161–5.

Poustka A. & Lehrach H. (1986) Jumping libraries and linking libraries: the next generation of molecular tools in mammalian genetics. *Trends Genetics* **2**, 174–9.

Poustka A., Pohl T.M., Barlow D.P., Frischauf A.M. & Lehrach H. (1987) Construction and use of human chromosome jumping libraries from *Not*I-digested DNA. *Nature* **325**, 353–5.

Powell-Abel P., Nelson R.S., De B., Hoffmann N., Rogers S.G., Fraley R.T. & Beachy R.N. (1986) Delay of disease development in transgenic plants that express the tobacco mosaic virus coat protein gene. *Science* **232**, 738–43.

Pratt J.M., Boulnois G.J., Darby V., Orr E., Wahle E. & Holland I.B. (1981) Identification of gene products programmed by restriction endonuclease DNA fragments using an *E. coli in vitro* system. *Nucleic Acids Res.* **9**, 4459–74.

Primrose S.B., Derbyshire P., Jones I.M., Nugent M.E. & Tacon W.C.A. (1983) Hereditary instability of recombinant DNA molecules. In *Bioactive Microbial Products 2: Development and Production*, eds Nisbet L.J. & Winstanley D.J., pp. 63–77. Academic Press, London.

Primrose S.B. & Ehrlich S.D. (1981) Isolation of plasmid deletion mutants and a study of their instability. *Plasmid* **6**, 193–201.

Prober J.M., Trainor G.L., Dam R.J., Hobbs F.W., Robertson C.W., Zagursky R.J., Cocuzza A.J., Jensen M.A. & Baumeister K. (1987) A system for rapid DNA sequencing with fluorescent chain-terminating dideoxynucleotides. *Science* **238**, 336–41.

Ptashne M. (1967a) Isolation of the λ phage repressor. *Proc. Natl. Acad. Sci. USA* **57**, 306–13.

Ptashne M. (1967b) Specific binding of the λ phage repressor to λDNA. *Nature* **214**, 232–4.

Ptashne M. (1986) *A Genetic Switch. Gene Control and Phage λ*. Cell Press and Blackwell Scientific Publications, Oxford.

Pugsley A.P. (1993) The complete general secretory pathway in Gram-negative bacteria. *Microbiol. Rev.* **57**, 50–108.

Pursel V.G., Pinkert C.A., Miller K.F., Bolt D.J., Campbell R.G., Palmiter R.D., Brinster R.L. & Hammer R.E. (1989) Genetic engineering of livestock. *Science* **244**, 1281–8.

Putney S.D., Herlihy W.C. & Schimmel P. (1983) A new troponin T and cDNA clones for 13 different muscle proteins, found by shotgun sequencing. *Nature* **302**, 718–21.

Puyet A., Sandoval M., López P., Aguilar A., Martin J.F. & Espinosa M. (1987) A simple medium for rapid regeneration of *Bacillus subtilis* protoplasts transformed with plasmid DNA. *FEBS Microbiol. Lett.* **40**, 1–5.

Quon D., Catalano W.R., Scardina J.M., Murakami K. & Cordell B. (1991) Formation of β-amyloid protein deposits in the brains of transgenic mice. *Nature* **352**, 239–41.

Radloff R., Bauer W. & Vinograd J. (1967) A dye-bouyant-density method for the detection and isolation of closed circular duplex DNA: the closed circular DNA in HeLa cells. *Proc. Natl. Acad. Sci. USA* **57**, 1514–21.

Radman M. & Wagner R. (1984) Effects of DNA methylation on mismatch repair, mutagenesis, and recombination of *Escherichia coli*. *Curr. Top. Microbiol. Immunol.* **108**, 23–28.

Rafalski J.A. & Tingey S.V. (1993) Genetic diagnostics in plant breeding: RAPDs, microsatellites and machines. *Trends Genetics* **9**, 275–80.

Ragot T., Vincent N., Chafey P., Vigne E., Gilgenkvantz H., Couton D., Cartand J., Briand P., Kaplan K.C., Perricaudet M. & Kahn A. (1993) Efficient adenovirus-mediated transfer of a human minidystrophin gene to skeletal muscle of *mdx* mice. *Nature* **361**, 647–50.

Raleigh E.A., Murray N.E., Revel H., Blumenthal R.M., Westaway D., Reith A.D., Rigby P.W.J., Elhai J. & Hanahan D. (1988) Mcr A and Mcr B restriction phenotypes of some *E. coli* strains and implications for gene cloning. *Nucleic Acids Res.* **16**, 1563–75.

Raleigh E.A. & Wilson G. (1986) *Escherichia coli* K-12 restricts DNA containing 5-methylcytosine. *Proc. Natl. Acad. Sci. USA* **83**, 9070–4.

Rassoulzadegan M., Binetruy B. & Cuzin F. (1982) High frequency of gene transfer after fusion between bacteria and eukaryotic cells. *Nature* **295**, 257–9.

Ratzkin B. & Carbon J. (1977) Functional expression of cloned yeast DNA in *Escherichia coli*. *Proc. Natl. Acad. Sci. USA* **74**, 487–91.

Razin A. & Riggs A.D. (1980) DNA methylation and gene function. *Science* **210**, 604–10.

Reidhaar-Olson J.F. & Sauer R.T. (1988) Combinatorial cassette mutagenesis as a probe of the informational content of protein sequences. *Science* **241**, 53–7.

Renart J. & Sandoval I.V. (1984) Western blots. *Methods Enzymol.* **104**, 455–60.

Rennell D., Bouvier S., Hardy L. & Poteete A. (1991) Systematic mutation of bacteriophage T4 lysozyme. *J. Mol. Biol.* **222**, 67–87.

Revel H.R. (1967) Restriction of non-glycosylated T-even bacteriophages: properties of permissive mutants of *Escherichia coli* B and K-12. *Virology* **31**, 688–701.

Rice C.M., Fuchs R., Higgins D.G., Stoehr P.J. & Cameron G.N. (1993) The EMBL data library. *Nucleic Acids Res.* **21**, 2967–71.

Richards J.E., Gilliam T.C., Cole J.L., Drumm M.L., Wasmuth J.J., Gusella J.F. & Collins F.S. (1988) Chromosome jumping from D4S10 (G8) toward the Huntington disease gene. *Proc. Natl. Acad. Sci. USA* **85**, 6437–41.

Richards K.E., Guilley H. & Jonard G. (1981) Further characterization of the discontinuities in cauliflower mosaic virus DNA. *FEBS Lett.* **134**, 67–70.

Rinas U., Tsai L.B., Lyons D., Fox G.M., Stearns G., Fieschko J., Fenton D. & Bailey J.E. (1992) Cysteine to serine substitutions in basic fibroblast growth factor: effect on inclusion body formation and proteolytic susceptibility during *in vitro* refolding. *Biotechnology* **10**, 435–40.

Rio D.C., Clark S.G. & Tjian R. (1985) A mammalian host–vector system that regulates expression and amplification of transfected genes by temperature induction. *Science* **227**, 23–8.

Riordan J.R., Rommens J.M., Kerem B.-S., Alon N., Rozmahel R., Zbyszko G., Zielenski J., Lok S., Plavsic N., Chou J.-L., Drumm M.L., Iannuzzi M.C., Collins F.S. & Tsui L.-C. (1989) Identification of the cystic fibrosis gene: cloning and characterization of complementary DNA. *Science* **245**, 1066–73.

Robbins D.M., Ripley S., Henderson A. & Axel R. (1981) Transforming DNA integrates into the host chromosome. *Cell* **23**, 29–39.

Roberts R.J. (1978) Restriction endonucleases: a new role *in vivo*? *Nature* **271**, 502.

Roberts R.J. (1990) Restriction enzymes and their isoschizomers. *Nucleic Acids Res.* **18**, 2331–65 (Suppl.).

Roberts L. (1991) Fight erupts over DNA fingerprinting. *Science* **254**, 1721–3.

Roberts L. (1992) DNA fingerprinting: Academy reports. *Science* **256**, 300–1.

Roberts M.W. & Rabinowitz J.C. (1989) The effect of *Escherichia coli* ribosomal protein S1 on the translational specificity of bacterial ribosomes. *J. Biol. Chem.* **264**, 2228–35.

Roberts R.C. & Helinski D.R. (1992) Definition of a minimal plasmid stabilization system from the broad-host-range plasmid RK2. *J. Bacteriol.* **174**, 8119–32.

Roberts T.M., Kacich R. & Ptashne M. (1979) A general method for maximizing the expression of a cloned gene. *Proc. Natl. Acad. Sci. USA* **76**, 760–4.

Roberts T.M., Swanberg S.L., Poteete A., Riedal G. & Backman K. (1980) A plasmid cloning vehicle allowing a positive selection for inserted fragments. *Gene* **12**, 123–7.

Robinson M., Lilley R., Little S., Emtage J.S., Yarranton G., Stephens P., Millican A., Eaton M. & Humphreys G. (1984) Codon usage can affect efficiency of translation of genes in *Escherichia coli*. *Nucleic Acids Res.* **12**, 6663–71.

Rodriguez R.L., West R.W., Heyneker H.L., Bolivar P. & Boyer H.W. (1979) Characterizing wild-type and mutant promoters of the tetracycline resistance gene in pBR313. *Nucleic Acids Res.* **6**, 3267–87.

Rogers D.T. & Hiller E. (1990) Genetic engineering of carbon fluxes in *Saccharomyces cerevisiae*. *Proceedings of the Sixth International Symposium on Genetics of Industrial Microorganisms*, eds Heslot H., *et al.* Société Française de Microbiologie, Strasbourg, pp. 533–44.

Romanos M.A., Clare J.J., Beesley K.M., Rayment F.B., Ballantine S.P., Makoff A.J., Dougan G., Fairweather N.F. & Charles I.G. (1991) Recombinant *Bordetella pertussis* pertactin (P69) from the yeast *Pichia pastoris*: high level production and immunological properties. *Vaccine* **9**, 901–6.

Romanos M.A., Scorer C.A. & Clare J.J. (1992) Foreign gene expression in yeast: a review. *Yeast* **8**, 423–88.

Rommens J.M., Iannuzzi M.C., Kerem B.-S., Drumm M.L., Melmer G., Dean M., Rozmahel R., Cole J.L., Kennedy D., Hidaka N., Zsiga M., Buchwald M., Riordan J.R., Tsui L.-C. & Collins F.S. (1989) Identification of the cystic fibrosis gene: chromosome walking and jumping. *Science* **245**, 1059–65.

Rood J.I., Sneddon M.K. & Morrison J.F. (1980) Instability in *tyrR* strains of plasmid carrying the tyrosine operon: isolation and characterization of plasmid derivatives with insertions or deletions. *J. Bacteriol.* **144**, 552–9.

Rosamond J., Endlich B. & Linn S. (1979) Electron microscopic studies of the mechanism of action of the restriction endonuclease of *Escherichia coli* B. *J. Mol. Biol.* **129**, 619–35.

Rose M., Casadaban M.J. & Botstein D. (1981) Yeast genes fused to β-galactosidase in *Escherichia coli* can be expressed normally in yeast. *Proc. Natl. Acad. Sci. USA* **78**, 2460–4.

Rose M.D., Novick P., Thomas J.H., Botstein D. & Fink G.R. (1987) A *Saccharomyces cerevisiae* genomic plasmid bank based on a centromere-containing shuttle vector. *Gene* **60**, 237–43.

Rosenberg A.H., Goldman E., Dunn J.J., Studier F.W. & Zubay G. (1993) Effects of consecutive AGG codons on translation in *Escherichia coli*, demonstrated with a versatile codon test system. *J. Bacteriol.* **175**, 716–22.

Rosenberg A.H., Lade B.N., Chui D.-S., Lin S.-W., Dunn J.J. & Studier F.W. (1987) Vectors for selective expression of cloned DNAs by T7 RNA polymerase. *Gene* **56**, 125–35.

Rosenberg S. plus 14 others (1990) Gene transfer into humans – immunotherapy of patients with advanced melonoma, using tumor-infiltrating lymphocytes modified by retroviral gene transduction. *N. Engl. J. Med.* **323**, 570–8.

Rosenfeld M.A., Yoshimura K., Trapnell B.C., Yoneyama K., Rosenthal E.R., Dalemans W., Fukayama M., Bargon J., Stier L.E., Stratford-Perricandet L., Perricandet M., Guggino W.B., Pavirani A., Lecocq J.P. & Crystal R.G. (1992) *In vivo* transfer of the human, cystic fibrosis transmembrane conductance regulator gene to the airway epithelium. *Cell* **68**, 143–55.

Rotbart H.A. (1991) Nucleic acid detection systems for enteroviruses. *Clin. Microbiol. Rev.* **4**, 156–68.

Rothman J.E. & Orci L. (1992) Molecular dissection of the secretory pathway. *Nature* **355**, 409–15.

Rothstein R.J., Lau L.F., Bahl C.P., Narang S.A. & Wu R. (1979) Synthetic adaptors for cloning DNA. *Methods Enzymol.* **68**, 98–109.

Rubin E.M., Wilson G.A. & Young F.E. (1980) Expression of thymidylate synthetase activity in *Bacillus subtilis* upon integration of a cloned gene from *Escherichia coli*. *Gene* **10**, 227–35.

Rubin G.M., Kidwell M.G. & Bingham P.M. (1982) The molecular basis of P–M hybrid dysgenesis: the nature of induced mutations. *Cell* **29**, 987–94.

Rubin G.M. & Spradling A.C. (1982) Genetic transformation of *Drosophila* with transposable element vectors. *Science* **218**, 348–53.

Rubin G.M. & Spradling A. (1983) Vectors for P element-mediated gene transfer in *Drosophila*. *Nucleic Acids Res.* **11**, 6341–51.

Rubin J.S., Joyner A.L., Bernstein A. & Whitmore G.F. (1983) Molecular identification of a human DNA repair gene following DNA-mediated gene transfer. *Nature* **306**, 206–8.

Ruby S.W. & Abelson J. (1991) Pre-mRNA splicing in yeast. *Trends Genetics* **7**, 79–85.

Rudnicki M.A., Braun B., Hinuma S. & Jaenisch, R. (1992) Inactivation of *myoD* in mice leads to up-regulation of the myogenic HLH gene *myf-5* and results in apparently normal muscle development. *Cell* **71**, 383–90.

Rudolph H. & Hinnen A. (1987) The yeast PHO5 promoter: phosphate control elements and sequences mediating mRNA start-site selection. *Proc. Natl. Acad. Sci. USA* **84**, 1340–4.

Rudolph H.K., Antebi A., Fink G.R., Buckley C.M., Dorman T.E., Levitre J., Davidow L.S., Mao J. & Moir D.T. (1989) The yeast secretory pathway is perturbed by mutations in PMR1, a member of a Ca ATPase family. *Cell* **58**, 133–45.

Rusconi S. & Schaffner W. (1981) Transformation of frog embryos with a rabbit β-globin gene. *Proc. Natl. Acad. Sci. USA* **78**, 5051–5.

Russell D.R. & Bennett G.N. (1982) Construction and analysis of *in vivo* activity of *E. coli* promoter hybrids and promoter mutants that alter the −35 to −10 spacing. *Gene* **20**, 231–43.

St. Germain D.L., Dittrich W., Morganelli C.M. & Cryns V. (1990) Molecular cloning by hybrid arrest of translation in *Xenopus laevis* oocytes: identification of a cDNA encoding the type-I iodothyronine 5′-deiodinase from rat liver. *J. Biol. Chem.* **265**, 20087–90.

Saloheimo M. & Niku-Paavola M.-J. (1991) Heterologous production of a ligninolytic enzyme: expression of the *Phlebia radiata* laccase gene in *Trichoderma reesei*. *Biotechnology* **9**, 987–990.

Saloman F., Deblaere R., Leemans J., Hernalsteens J.P., Van Montagu M. & Schell J. (1984) Genetic identification of functions of TR-DNA transcripts in octopine crown galls. *EMBO J.* **3**, 141–6.

Salter D.W. & Crittenden L.B. (1988) Gene insertion into the avian germ line. *Occasional Publications of the British Society of Animal Production* **12**, 32–57.

Sambrook J., Rodgers L., White J. & Getling M.J. (1985) Lines of BPV-transformed murine cells that constitutively express influenza virus haemagglutinin. *EMBO J.* **4**, 91–103.

Sancar A., Hack A.M. & Rupp W.D. (1979) Simple method for identification of plasmid-coded proteins. *J. Bacteriol.* **137**, 692–3.

Sanford J.C., Devit M.J., Russell J.A., Smith F.D., Harpending P.R., Roy M.K. & Johnston S.A. (1991) An improved, helium-driven biolistic device. *Technique* **3**, 3–16.

Sanger F., Nicklen S. & Coulson A.R. (1977) DNA sequencing with chain-terminating inhibitors. *Proc. Natl. Acad. Sci. USA* **74**, 5463–7.

Sanger F., Coulson A.R., Hong G.-F., Hill D.F. & Petersen G.B. (1982) Nucleotide sequence of bacteriophage lambda DNA. *J. Mol. Biol.* **162**, 729–73.

Sarkar G. & Sommer S.S. (1990) The 'megaprimer' method of site-directed mutagenesis. *Biotechniques* **8**, 404–7.

Sankar P., Hutton M.E., Van Bogelen R.A., Clark R.L. & Neidhardt F.C. (1993) Expression analysis of cloned chromosomal segments of *Escherichia coli*. *J. Bacteriol.* **175**, 5145–52.

Sarver N., Gruss P., Law M.-F., Khoury G. & Howley P.M. (1981a) Bovine papilloma virus deoxyribonucleic acid: a novel eucaryotic cloning vector. *Mol. Cell. Biol.* **1**, 486–96.

Sarver N., Gruss P., Law M.-F., Khoury G. & Howley P.M. (1981b) Rat insulin gene covalently linked to bovine papilloma virus DNA is expressed in transformed mouse cells. In *Development Biology Using Purified Genes*, eds Brown D. & Fox C.R., ICN-UCLA Symposia on Molecular and Cellular Biology, Vol. 23. Academic Press, New York.

Sassenfeld H.M. (1990) Engineering proteins for purification. *Trends Biotechnol.* **9**, 59–63.

Sayers J.R. & Eckstein F. (1991) A single-strand specific endonuclease activity copurifies with overexpressed T5 D15 exonuclease. *Nucleic Acids Res.* **19**, 4127–32.

Scahill S.J., Devos R., Van der Heyden J. & Fiers W. (1983) Expression and characterisation of the product of a human immune interferon cDNA gene in Chinese hamster ovary cells. *Proc. Natl. Acad. Sci. USA* **80**, 4654–8.

Scalenghe F., Turco E., Edstrom J.E., Pirotta V. & Melli M. (1981) Micro-dissection and cloning of DNA from a specific region of *Drosophila melanogaster* polytene chromosomes. *Chromosoma* **82**, 205–16.

Scangos G.A., Huttner K.M., Juricek D.K. & Ruddle F.H. (1981) DNA-mediated gene transfer in mammalian cells: molecular analysis of unstable transformants and their progression to stability. *Mol. Cell. Biol.* **1**, 111–20.

Schafer A., Kalinowski J., Simon R., Seep-Feldhaus A.-H. & Puhler A. (1990) High-frequency conjugal plasmid transfer from Gram-negative *Escherichia coli* to various Gram-positive coryneform bacteria. *J. Bacteriol.* **172**, 1663–6.

Schafer W., Gorz A. & Kahl G. (1987) T-DNA integration and expression in a monocot crop plant after induction of *Agrobacterium*. *Nature* **327**, 529–31.

Schaffner W. (1980) Direct transfer of cloned genes from bacteria to mammalian cells. *Proc. Natl. Acad. Sci. USA* **77**, 2163–7.

Scharf S.J., Horn G.T. & Erlich H.A. (1986) Direct cloning and sequence analysis of enzymatically amplified genomic sequences. *Science* **233**, 1076–8.

Scheidereit C., Greisse S., Westphal H.M. & Beato M. (1983) The glucocorticoid receptor binds to defined nucleotide sequences near the promoter of mouse mammary tumour virus. *Nature* **304**, 749–52.

Schein C.H. (1991) Optimizing protein folding to the native state in bacteria. *Curr. Opinion Biotechnol.* **2**, 746–50.

Schein C.H. & Noteborn M.H.M. (1988) Formation of soluble recombinant proteins in *Escherichia coli* is favored by lower growth temperature. *Biotechnology* **6**, 291–4.

Schell J. & Van Montagu M. (1977) The Ti-plasmid of *Agrobacterium tumefaciens*, a natural vector for the introduction of fix genes in plants. In *Genetic Engineering for Nitrogen Fixation*, ed. Hollaender A., pp. 159–79. Plenum Press, New York.

Scheller R.H., Dickerson R.E., Boyer H.W., Riggs A.D. & Itakura K. (1977) Chemical synthesis of restriction enzyme recognition sites useful for cloning. *Science* **196**, 177–80.

Schena M. & Yamamoto K.R. (1988) Mammalian glucocorticoid receptor derivatives enhance transcription in yeast. *Science* **241**, 965–7.

Scherzinger E., Bagdasarian M.M., Scholz P., Lurz R., Ruchert B. & Bagdasarian M. (1984) Replication of the broad host range plasmid RSF1010: requirement for three plasmid-encoded proteins. *Proc. Natl. Acad. Sci. USA* **81**, 654–8.

Schiestl R.H. & Petes T.D. (1991) Integration of DNA fragments by illegitimate recombination in *Saccharomyces cerevisiae*. *Proc. Natl. Acad. Sci. USA* **88**, 7585–9.

Schimke R.T. (1984) Gene amplification in cultured animal cells. *Cell* **37**, 705–13.

Schimke R.T., Kaufman R.J., Alt F.W. & Kellems R.F. (1978) Gene amplification and drug resistance in cultured murine cells. *Science* **202**, 1051–5.

Schmidhauser T.J., Ditta G. & Helinski D.R. (1988) Broad-host-range plasmid cloning vectors for Gram-negative bacteria. In *Vectors. A Survey of Molecular Cloning Vectors and Their Uses*, eds Rodriguez R.L. & Denhardt D.T., pp. 287–332. Butterworths, London.

Schmidhauser T.J., Filutowicz M. & Helinski D.R. (1983) Replication of derivatives of the broad host range plasmid RK2 in two distantly related bacteria. *Plasmid* **9**, 325–30.

Schoepfer R. (1993) The pRSET family of T7 promoter expression vectors for *Escherichia coli*. *Gene* **124**, 83–5.

Scholz P., Haring V., Wittman-Liebold B., Ashman K., Bagdasarian M. & Scherzinger E. (1989) Complete nucleotide sequence and gene organization of the broad-host-range plasmid RSF1010. *Gene* **75**, 271–88.

Schroder G., Waffenschidt S., Weiler E.W. & Schroder J. (1984) The T-region of Ti plasmids codes for an enzyme synthesizing indole-3-acetic acid. *Eur. J. Biochem.* **138**, 387–91.

Schultz L.D., Hofmann K.J., Mylin L.M., Montgomery D.L., Ellis R.W. & Hopper J.E. (1987) Regulated over-production of the *GAL4* gene product greatly increases expression from galactose-inducible promoters on multi-copy expression vectors in yeast. *Gene* **61**, 123–33.

Schwartz D.C. & Cantor C.R. (1984) Separation of yeast chromosomal-sized DNAs by pulsed field gradient gel electrophoresis. *Cell* **37**, 67–75.

Schwartz D.C. & Koval M. (1989) Conformational dynamics of individual DNA molecules during gel electrophoresis. *Nature* **338**, 520–2.

Sciaky D., Montoya A.L. & Chilton M.D. (1978) Fingerprints of *Agrobacterium* Ti plasmids. *Plasmid* **1**, 238–54.

Scott J.K. & Smith G.P. (1990) Searching for peptide ligands with an epitope library. *Science* **249**, 386–90.

Seed B. (1983) Purification of genome sequences from bacteriophage libraries by recombination and selection *in vivo*. *Nucleic Acids Res.*, **8**, 2427–45.

Seehaus T., Breitling F., Dubel S., Klewinghaus I. & Little M. (1992) A vector for the removal of deletion mutants from antibody libraries. *Gene* **114**, 235–7.

Seelke R., Kline B., Aleff R., Porter R.D. & Shields M.S. (1987) Mutations in the *recD* gene of *Escherichia coli* that raise the copy number of certain plasmids. *J. Bacteriol.* **169**, 4841–4.

Segal G., Sarfatti M., Schaffer M.A., Ori N., Zamir D. & Fluhr R. (1992) Correlation of genetic and physical structure in the region surrounding the I_2 *Fusarium oxysporum* resistance locus in tomato. *Mol. Gen. Genetics* **231**, 179–85.

Seki M., Shigemoto N., Komeda Y., Imamura J., Yamada Y. & Morikawa H. (1991) Transgenic *Arabidopsis thaliana* plants obtained by particle-bombardment-mediated transformation. *Applied Microbiol. Biotechnol.* **36**, 228–30.

Serres R.A., McCowan B.H., McCabe D.E., Stang E., Russell D. & Martinelli B. (1990) Stable transformation of cranberry using electric discharge particle acceleration. *Hort. Sci.* **25**, 1130.

Sgaramella V. (1972) Enzymatic oligomerization of bacteriophage P22 DNA and of linear simian virus 40 DNA. *Proc. Natl. Acad. Sci. USA* **69**, 3389–93.

Shah D.M., Horsch R.B., Klee H.J., Kishare G.M., Winter J.A., Turner N.E., Hironaka C.M., Sanders P.R., Gasser C.S., Aykent S., Siegel N.R., Rogers S.G. & Fraley R.T. (1986) Engineering herbicide tolerance in transgenic plants. *Science* **233**, 478–81.

Sharp P.M. (1986) Molecular evolution of bacteriophage: evidence of selection against the recognition sites of host restriction enzymes. *Mol. Biol. Evol.* **3**, 75–83.

Sharp P.M. & Bulmer M. (1988) Selective differences among translation termination codons. *Gene* **63**, 141–5.

Sharp P.M. & Cowe E. (1991) Synonymous codon usage in *Saccharomyces cerevisiae. Yeast* **7**, 657–78.

Shaw C.H., Leemans J., Shaw C.H., Van Montague M. & Schell J. (1983) A general method for the transfer of cloned genes to plant cells. *Gene* **23**, 315–30.

Shaw C.H., Watson M.D., Carter G.H. & Shaw C.H. (1984) The right hand copy of the nopaline Ti-plasmid 25 bp repeat is required for tumour formation. *Nucleic Acids Res.* **12**, 6031–41.

Shaw G. & Kamen R. (1986) A conserved AU sequence from the 3'-untranslated region of GM-CSF messenger RNA mediates selective messenger RNA degradation. *Cell* **46**, 659–67.

Sheehy R.E., Kramer M. & Hiatt W.R. (1988) Reduction of polygalacturonase in tomato fruit by antisense RNA. *Proc. Natl. Acad. Sci. USA* **85**, 8805–9.

Shepard H.M., Yelverton E. & Goeddel D.V. (1982) Increased synthesis in *E. coli* of fibroblast and leukocyte interferons through alterations in ribosome binding sites. *DNA* **1**, 125–31.

Shimada N., Toyoda-Yamamoto A., Nagamine J. *et al.* (1990) Control of expression of *Agrobacterium vir* genes by synergistic actions of phenolic signal molecules and monosaccharides. *Proc. Natl. Acad. Sci. USA* **87**, 6684–88.

Shimamoto K., Tereda R., Izawa T. & Fujimoto H. (1989) Fertile transgenic rice plants regenerated from transformed protoplasts. *Nature* **338**, 274–6.

Shimotohno K. & Temin H.M. (1981) Formation of infectious progeny virus after insertion of herpes simplex thymidine kinase gene into DNA of an avian retrovirus. *Cell* **26**, 67–77.

Shimotohno K. & Temin H.M. (1982) Loss of intervening sequences in genomic mouse α-globin DNA inserted in an infectious retrovirus vector. *Nature* **299**, 265–8.

Shine J. & Dalgarno L. (1975) Determinant of cistron specificity in bacterial ribosomes. *Nature* **254**, 34–8.

Shine J., Fettes I., Lan N.C.Y., Roberts J.L. & Baxter J.D. (1980) Expression of cloned β-endorphin gene sequences by *Escherichia coli. Nature* **285**, 456–61.

Shorenstein R.G. & Losick R. (1973) Comparative size and properties of the sigma subunits of ribonucleic acid polymerase from *Bacillus subtilis* and *Escherichia coli. J. Biol. Chem.* **248**, 6170–3.

Short J.M., Fernandez J.M., Sorge J.A. & Huse W.D. (1988) λZAP: a bacteriophage lambda expression vector with *in vivo* excision properties. *Nucleic Acids Res.* **16**, 7583–600.

Sikorski R.S., Michaud W., Levin H.L., Boeke J.D. & Hieter P. (1990) *Trans*-kingdom promiscuity. *Nature* **345**, 581–2.

Simmler M.O., Johnsson C., Petit C., Rouyer F., Vergnaud G. & Weissenbach J. (1987) Two highly polymorphic minisatellites from the pseudoautosomal region of the human sex chromosomes. *EMBO J.* **6**, 963–9.

Simonen M. & Palva I. (1993) Protein secretion in *Bacillus* species. *Microbiol. Rev.* **57**, 109–37.

Simpson R., O'Hara P., Lichtenstein C., Montoya A.L., Kwok K., Gordon M.P. & Nester E.W. (1982) The DNA from A6S/Z tumor contains scrambled Ti plasmid sequence near its junction with plant DNA. *Cell* **29**, 1005–14.

Singer-Sam J., Simmer R.L., Keith D.H., Shively L., Teplitz M., Itakura K., Gartler S.M. & Riggs A.D. (1983) Isolation of a cDNA clone for human X-linked 3-phosphoglycerate kinase by use of a mixture of synthetic oligodeoxyribonucleotides as a detection probe. *Proc. Natl. Acad. Sci. USA* **80**, 802–6.

Sinn E., Muller W., Pattengale P., Tepler I., Wallace R. & Leder P. (1987) Coexpression of MMTV/v-Ha-*ras* and MMTV/c-*myc* genes in transgenic mice: synergistic action of oncogenes *in vivo. Cell* **49**, 465–75.

Sive H.L. & StJohn T. (1988) A simple subtractive hybridization technique employing photoactivable biotin and phenol extraction. *Nucleic Acids Res.* **16**, 10937.

Skalka A. & Shapiro L. (1976) *In situ* immunoassays for gene translation products in phage plaques and bacterial colonies. *Gene* **1**, 65–79.

Skogman G., Nilsson J. & Gustafsson P. (1983) The use of a partition locus to increase stability of tryptophan–operon-bearing plasmids in *Escherichia coli. Gene* **23**, 105–15.

Slauch J.M. & Silhavy T.J. (1991) Genetic fusions as experimental tools. *Methods Enzymol.* **204**, 213–48.

Sleep D., Belfield G.P., Ballance D.J., Steven J., Jones S., Evans L.R., Moir P.D. & Goodey A.R. (1991) *Saccharomyces cerevisiae* strains that overexpress heterologous proteins. *Biotechnology* **9**, 183–7.

Smith C., Watson C., Ray J., Bird C., Morris P., Schuch W. & Grierson D. (1988) Antisense RNA inhibition of polygalacturonase gene expression in transgenic tomatoes. *Nature* **334**, 724–6.

Smith D.F., Searle P.F. & Williams J.G. (1979) Characterization of bacterial clones containing DNA sequences derived from *Xenopus laevis. Nucleic Acids Res.* **6**, 487–506.

Smith D.B. & Johnson K.S. (1988) Single-step purification of polypeptides expressed in *Escherichia coli* as fusions with glutathione S-transferase. *Gene* **67**, 31–40.

Smith D.P., Mason C.S., Jones E. & Old R. (1994) Expression of a dominant negative retinoic acid receptor gamma in *Xenopus* embryos leads to partial resistance to retinoic acid. *Roux's Archive Dev. Biol.* In press.

Smith E.F. & Townsend C.O. (1907) A plant-tumor of bacterial origin. *Science* **25**, 671–3.

Smith G.E., Summers M.D. & Fraser M.J. (1983a) Production of human β-interferon in insect cells infected with a baculovirus expression vector. *Mol. Cell. Biol.* **3**, 2156–65.

Smith G.L., Mackett M. & Moss B. (1983b) Infectious vaccinia virus recombinants that express hepatitis B virus surface antigen. *Nature* **302**, 490–5.

Smith G.L., Murphy B.R. & Moss B. (1983c) Construction and characterization of an infectious vaccinia virus recombinant that expresses the influenza hemagglutinin gene and induces resistance to influenza virus infection in hamsters. *Proc. Natl. Acad. Sci. USA* **80**, 7155–9.

Smith H., Bron S., Ven Ee J. & Venema G. (1987) Construction and use of signal sequence selection vectors in *Escherichia coli* and *Bacillus subtilis. J. Bacteriol.* **169**, 3321–8.

Smith H.B., Larimer F.W. & Hartman F.C. (1990) An engineered change in substrate specificity of ribulosebisphosphate carboxylase/oxygenase. *J. Biol. Chem.* **265**, 1243–5.

Smith H.O. & Nathans D. (1973) A suggested nomenclature for bacterial host modification and restriction systems and their enzymes. *J. Mol. Biol.* **81**, 419–23.

Smith H.O. & Wilcox K.W. (1970) A restriction enzyme from *Hemophilus influenzae*. I. Purification and general properties. *J. Mol. Biol.* **51**, 379–91.

Smith J.C., Derbyshire R.B., Cook E., Dunthorne L., Viney J., Brewer S.J., Sassenfeld H.M. & Bell L.D. (1984) Chemical synthesis and cloning of a poly(arginine)-coding gene fragment designed to aid polypeptide purification. *Gene* **32**, 321–7.

Smith L.M., Sanders J.Z., Kaiser R.J., Hughes P., Dodd C., Connell C.R., Heiner C., Kent S.B.H. & Hood L.E. (1986) Fluorescence detection in automated DNA sequence analysis. *Nature* **321**, 674–9.

Smith S.B., Aldridge P.K. & Callis J.B. (1989) Observation of individual DNA molecules undergoing gel electrophoresis. *Science* **243**, 203–6.

Smithies O., Gregg R.G., Boggs S.S., Koralewski M.A. & Kucherlapati R. (1985) Insertion of DNA sequences into the human β-globin locus by homologous recombination. *Nature* **317**, 230–4.

Sorge J., Wright D., Erdman V.D. & Cutting A.E. (1984) Amphitropic retrovirus vector system for human cell gene transfer. *Mol. Cell. Biol.* **4**, 1730–7.

Southern E.M. (1975) Detection of specific sequences among DNA fragments separated by gel electrophoresis. *J. Mol. Biol.* **98**, 503–17.

Southern E.M. (1979) Gel electrophoresis of restriction fragments. *Methods Enzymol.* **68**, 152–76.

Southern P.J. & Berg P. (1982) Transformation of mammalian cells to antibiotic resistance with a bacterial gene under the control of the SV40 early region promoter. *J. Mol. Appl. Genet.* **1**, 327–41.

Spoerel N., Herlich P. & Bickle T.A. (1979) A novel bacteriophage defence mechanism: the anti-restriction protein. *Nature* **278**, 30–4.

Spradling A.C. & Rubin G.M. (1982) Transposition of cloned P elements into *Drosophila* germ line chromosomes. *Science* **218**, 341–7.

Sprague K.V., Faulds D.H. & Smith G.R. (1978) A single base-pair change creates a *chi* recombinational hotspot in bacteriophage λ. *Proc. Natl. Acad. Sci. USA* **75**, 6182–6.

Stacey G.N., Bolton B.J. & Doyle E. (1992) DNA fingerprinting transforms the art of cell authentication. *Nature* **357**, 261–2.

Stachel S.E., Messens E., Van Montagu M. & Zambryski P. (1985) Identification of the signal molecules produced by wounded plant cells that activate T-DNA transfer in *Agrobacterium tumefaciens*. *Nature* **318**, 624–9.

Stachel S.E., Timmerman B. & Zambryski P. (1986) Generation of single stranded T-DNA molecules during the initial stages of T-DNA transfer from *Agrobacterium tumefaciens* to plant cells. *Nature* **322**, 706–11.

Stachel S.E. & Zambryski P. (1986) *VirA* and *virG* control the plant-induced activation of the T-DNA transfer process in *A. tumefaciens*. *Cell* **46**, 325–33.

Stahl D.A., Flesher B., Mansfield H.R. & Montgomery L. (1988) Use of phylogenetically based hybridization probes for studies of ruminal microbial ecology. *Appl. Env. Microbiol.* **54**, 1079–84.

Stallcup M.R., Sharrock W.J. & Rabinowitz J.C. (1974) Ribosome and messenger specificity in protein synthesis by bacteria. *Biochem. Biophys. Res Commun.* **58**, 92–8.

Stark D.M., Timmermann K.P., Barry G.F., Preiss J. & Kishore G.M. (1992) Regulation of the amount of starch in plant tissues by ADP glucose phosphorylase. *Science* **258**, 287–92.

Stark M.J.R. (1987) Multicopy expression vectors carrying the *lac* repressor gene for regulated high-level expression of genes in *Escherichia coli*. *Gene* **51**, 255–67.

Stassi D.L. & Lacks S.A. (1982) Effect of strong promoters on the cloning in *Escherichia coli* of DNA fragments from *Streptococcus pneumoniae*. *Gene* **18**, 319–28.

Staub J. & Maliga P. (1992) Long regions of homologous DNA are incorporated into the tobacco plastid genome by transformation. *Plant Cell* **4**, 39–45.

Staudt L.M., Clerc R.G., Singh H., LeBowitz J.H., Sharp P.A. & Baltimore D. (1988) Cloning of a lymphoid-specific cDNA encoding a protein binding the regulatory octamer DNA motif. *Science* **241**, 577–9.

Stearns T., Ma H. & Botstein D. (1990) Manipulating yeast genome using plasmid vectors. *Methods Enzymol.* **185**, 280–97.

Steller H. & Pirotta V. (1985) A transposable P vector that confers selectable G418 resistance of *Drosophila* larvae. *EMBO J.* **4**, 167–71.

Sternberg N. (1990) Bacteriophage P1 cloning system for the isolation, amplification and recovery of DNA fragments as large as 100 kilobase pairs. *Proc. Natl. Acad. Sci. USA* **87**, 103–7.

Stewart G., Smith T. & Denyer S. (1990) Genetic engineering for bioluminescent bacteria. *Food Sci. Technol. Today* **3**, 19–22.

Stewart T.A., Pattengale P.K. & Leder P. (1984) Spontaneous mammary adenocarcinomas in transgenic mice that carry and express MTV/myc fusion genes. *Cell* **38**, 627–37.

Stinchcomb D.T., Struhl K. & Davis R.W. (1979) Isolation and characterization of a yeast chromosomal replicator. *Nature* **282**, 39–43.

Stinchcomb D.T., Mann C. & Davis R.W. (1982) Centromeric DNA from *Saccharomyces cerevisiae*. *J. Mol. Biol.* **158**, 157–79.

Stoker N.G., Fairweather N.F. & Spratt B.G. (1982) Versatile low-copy-number plasmid vectors for cloning in *Escherichia coli*. *Gene* **18**, 335–41.

Storb U., O'Brien R.L., McMullen M.D., Gollahon K.A. & Brinster R.L. (1984) High expression of cloned immunoglobulin kappa gene in transgenic mice is restricted to B-lymphocytes. *Nature* **310**, 238–48.

Stormo G.D., Schneider T.D. & Gold L.M. (1982) Characterization of translational initiation sites in *E. coli*. *Nucleic Acids Res.* **10**, 2971–96.

Storms R.K., McNeil J.B., Khandekar P.S., An G., Parker J. & Friesen J.D. (1979) Chimaeric plasmids for cloning of deoxyribonucleic acid sequences in *Saccharomyces cerevisiae. J. Bacteriol.* **140**, 73–82.

Stover C.K. & 13 others (1991) New use of BCG for recombinant vaccines. *Nature* **351**, 456–60.

Strivastava S.K., Cannistraro V.J. & Kennel D. (1992) Broad specificity endoribonucleases and mRNA degradation in *Escherichia coli. J. Bacteriol.* **174**, 56–62.

Struhl K. & Davis R.D. (1980) A physical, genetic and transcriptional map of the cloned *his3* gene region of *Saccharomyces cerevisiae. J. Mol. Biol.* **136**, 309–32.

Struhl K., Cameron J.R. & Davis R.W. (1976) Functional genetic expression of eukaryotic DNA in *Escherichia coli. Proc. Natl. Acad. Sci. USA* **73**, 1471–5.

Struhl K., Stinchcomb D.T., Scherer S. & Davis R.W. (1979) High-frequency transformation of yeast: autonomous replication of hybrid DNA molecules. *Proc. Natl. Acad. Sci. USA* **76**, 1035–9.

Studier F.W., Rosenberg A.H., Dunn J.J. & Dubendorff J.W. (1990) Use of T7 RNA polymerase to direct expression of cloned genes. *Methods Enzymol.* **185**, 60–89.

Subramani S., Mulligan R. & Berg P. (1981) Expression of mouse dihydrofolate reductase complementary deoxyribonucleic acid in simian virus 40 vectors. *Mol. Cell. Biol.* **1**, 854–64.

Suggs S.V., Wallace R.B., Hirose T., Kawashima E.H. & Itakura K. (1981) Use of synthetic oligonucleotides as hybridization probes. III. Isolation of cloned cDNA sequences for human beta-2-microglobulin. *Proc. Natl. Acad. Sci. USA* **78**, 6613–17.

Suihko M.-L., Blomqvist K., Penttila M., Gisler R. & Knowles J. (1990) Recombinant brewer's yeast strains suitable for accelerated brewing. *J. Biotechnol.* **14**, 285–300.

Sulston J. & 19 others (1992) The *C. elegans* genome sequencing project: a beginning. *Nature* **356**, 37–41.

Sumikawa K., Houghton M., Emtage J., Richards B. & Barnard E. (1981) Active multi-subunit ACh receptor assembled by translation of heterologous mRNA in *Xenopus* oocytes. *Nature* **292**, 862.

Summers D.K. & Sherratt D.J. (1984) Multimerization of high copy number plasmids causes instability: Col E1 encodes a determinant essential for plasmid monomerization and stability. *Cell* **36**, 1097–103.

Sussman D.J. & Milman G. (1984) Short-term, high-efficiency expression of transfected DNA. *Mol. Cell. Biol.* **4**, 1641–3.

Sutcliffe J.G. (1979) Complete nucleotide sequence of the *Escherichia coli* plasmid pBR322. *Cold Spring Harbor Symp. Quant. Biol.* **43** (1), 77–90.

Sutherland E., Coe L. & Raleigh E.A. (1992) McrBC: a multisubunit GTP-dependent restriction endonuclease. *J. Mol. Biol.* **225**, 327–48.

Svab Z., Hajdukiewcz P. & Maliga P. (1990) Stable transformation of plastids in higher plants. *Proc. Natl. Acad. Sci. USA* **87**, 8526–30.

Sveda M.M. & Lai C.-J. (1981) Functional expression in primate cells of cloned DNA coding for the hemagglutinin surface glycoprotein of influenza virus. *Proc. Natl. Acad. Sci. USA* **78**, 5488–92.

Swick A.G., Janicot M., Chenevalkastelic T., McLenithan J.C. & Lane M.D. (1992) Promoter cDNA-directed heterologous protein expression in *Xenopus laevis* oocytes. *Proc. Natl. Acad. Sci. USA* **89**, 1812–16.

Swift G.H., Hammer R.E., MacDonald R.J. & Brinster R.L. (1984) Tissue specific expression of the rat pancreatic elastase 1 gene in transgenic mice. *Cell* **38**, 639–46.

Swinfield T.J., Jannière L., Ehrlich S.D. & Minton N.P. (1991) Characterization of a region of the *Enterococcus faecalis* plasmid pAMβ1 which enhances the segregational stability of pAMβ1-derived cloning vectors in *Bacillus subtilis. Plasmid* **26**, 209–21.

Swyryd E.A., Seaver S. & Stark G.R. (1974) N-(phosphonacetyl)-L-aspartate, a potent transition state analog inhibitor of aspartate transcarbamylase, blocks proliferation of mammalian cells in culture. *J. Biol. Chem.* **249**, 6945–50.

Szittner R. & Meighen E. (1990) Nucleotide sequence expression and properties of luciferase coded by *lux* genes from a terrestrial bacterium. *J. Biol. Chem.* **265**, 16581–7.

Szostak J.W. & Blackburn E.H. (1982) Cloning yeast telomeres on linear plasmid vectors. *Cell* **29**, 245–55.

Tabin C.J., Hoffman J.W., Groff S.P. & Weinberg R.A. (1982) Adaptation of a retrovirus as a eucaryotic vector transmitting the herpes simplex virus thymidine kinase gene. *Mol. Cell. Biol.* **2**, 426–36.

Tachibana K., Watanabe T., Sekizawa Y. & Takematsu T. (1986) Accumulation of ammonia in plants treated with bialaphos. *J. Pesticide Sci.* **11**, 33–7.

Tacon W.C.A., Bonass W.A., Jenkins B. & Emtage J.S. (1983) Plasmid expression vectors containing tryptophan promoter transcriptional regulons lacking the attenuator. *Gene* **23**, 255–65.

Tacon W., Cary N. & Emtage S. (1980) The construction and characterization of plasmid vectors suitable for the expression of all DNA phases under the control of the *E. coli* tryptophan promoter. *Mol. Gen. Genetics* **177**, 427–38.

Tait R.C., Close T.J., Lundquist R.C., Hagiya M., Rodriguez R.L. & Kado C.I. (1983) Construction and characterization of a versatile broad host range DNA cloning system for Gram-negative bacteria. *Biotechnology* **1**, 269–75.

Tait R.C., Lundquist R.C. & Kado C.I. (1982) Genetic map of the crown gall suppressive IncW plasmid pSa. *Mol. Gen. Genetics* **186**, 10–15.

Takagi H., Morinaga Y., Tsuchiya M., Ikemura H. & Inouye M. (1988) Control of folding of proteins secreted by a high expression secretion vector, pIN-111-ompA: 16-fold increase in production of active subtilisin E in *Escherichia coli*. *Biotechnology* **6**, 948–50.

Takamatsu N., Ishikawa M., Meshi T. & Okada Y. (1987) Expression of bacterial chloramphenicol acetyltransferase gene in tobacco plants mediated by TMV-RNA. *EMBO J.* **6**, 307–11.

Talmadge K. & Gilbert W. (1980) Construction of plasmid vectors with unique *Pst*I cloning sites in a signal sequence coding region. *Gene* **21**, 235–41.

Talmadge K. & Gilbert W. (1982) Cellular location affects protein stability in *Escherichia coli*. *Proc. Natl. Acad Sci. USA* **79**, 1830–3.

Taniguchi T., Guarente L., Roberts T.M., Kimelman D., Douhan J. & Ptashne M. (1980) Expression of the human fibroblast interferon gene in *Escherichia coli*. *Proc. Natl. Acad. Sci. USA* **77**, 5230–3.

Tartaglia J. & Paoletti E. (1988) Recombinant vaccinia virus vaccines. *Trends Biotechnol.* **6**, 43–6.

Taylor J.W., Schmidt W., Cosstick R., Okruszek A. & Eckstein F. (1985) The use of phosphorothioate-modified DNA in restriction enzyme reactions to prepare nicked DNA. *Nucleic Acids Res.* **13**, 8749–64.

Tenover F.C. (1988) Diagnostic deoxyribonucleic acid probes for infectious diseases. *Clin. Microbiol. Rev.* **1**, 82–101.

Tepfer D. (1984) Transformation of several species of higher plants by *Agrobacterium rhizogenes*: sexual transmission of the transformed genotype and phenotype. *Cell* **37**, 959–67.

Thomas C.M., Stalker D.M., Guiney D.G. & Helinski D.R. (1979) Essential regions for the replication and conjugal transfer of the broad host range plasmid RK2. In *Plasmids of Medical, Environmental and Commercial Importance*, eds Timmis K.N. & Puhler A., pp. 375–85. Elsevier/North-Holland Biomedical Press, Amsterdam.

Thomas C.M., Stalker D.M. & Helinski D.R. (1981) Replication and incompatibility properties of segments of the origin region of replication of the broad host range plasmid RK2. *Mol. Gen. Genetics* **81**, 1–7.

Thomas K.R. & Cappechi M.R. (1986) Introduction of homologous DNA sequences into mammalian cells induces mutations in the cognate gene. *Nature* **324**, 34–8.

Thomas K.R., Folger K.R. & Cappechi M.R. (1986) High frequency targeting of genes to specific sites in the mammalian genome. *Cell* **44**, 419–28.

Thomas M., Cameron J.R. & Davis R.W. (1974) Viable molecular hybrids of bacteriophage lambda and eukaryotic DNA. *Proc. Natl. Acad. Sci. USA* **71**, 4579–83.

Thomas P.S. (1980) Hybridization of denatured RNA and small DNA fragments transferred to nitrocellulose. *Proc. Natl. Acad. Sci. USA* **77**, 5201–5.

Thomashow M.F., Nutter R., Montoya A.L., Gordon M.P. & Nester E.W. (1980) Integration and organization of Ti plasmid sequences in crown gall tumours. *Cell* **19**, 729–39.

Thompson C.J., Movva N.R., Tizard R., Crameri R., Davies J.E., Lauwereys M. & Botterman J. (1987) Characterization of the herbicide-resistance gene *bar* from *Streptomyces hygroscopicus*. *EMBO J.* **6**, 2519–23.

Tikchonenko T.I., Karamov E.V., Zavizion B.A. & Naroditsky B.S. (1978) *Eco*RI* activity:

enzyme modification or activation of accompanying endonuclease? *Gene* **4**, 195–212.

Tilghman S.M., Tiemeier D.C., Polsky F., Edgell M.H., Seidman J.G., Leder A., Enquist L.W., Norman B. & Leder P. (1977) Cloning specific segments of the mammalian genome: bacteriophage λ containing mouse globin and surrounding gene sequences. *Proc. Natl. Acad. Sci. USA* **74**, 4406–10.

Timberlake W.E. & Marshall M.A. (1988) Genetic regulation of development in *Aspergillus nidulans*. *Trends Genetics* **4**, 162–9.

Tinland B., Koukolikova-Nicola Z., Hall M.N. & Hohn B. (1992) The T-DNA-linked *vir*D2 protein contains two distinct functional nuclear localization signals. *Proc. Natl. Acad. Sci. USA* **89**, 7442–6.

Tomizawa J.-I. & Itoh T. (1981) Plasmid ColE1 incompatibility determined by interaction of RNA I with primer transcript. *Proc. Natl. Acad. Sci. USA* **78**, 6096–100.

Tomizawa J.-I. & Itoh T. (1982) The importance of RNA secondary structure in ColE1 primer formation. *Cell* **31**, 575–83.

Toneguzzo F., Hayday A.C. & Keating A. (1986) Electric field-mediated gene transfer: transient and stable gene expression in human and mouse lymphoid cells. *Mol. Cell. Biol.* **6**, 703–6.

Toole J.J., Knopf J.L., Wozney J.M., Sultzman L.A., Buecker J.L., Pittman D.D., Kaufman R.J., Brown E., Shoemaker C., Orr E.C., Amphlett G.W., Foster W.B., Coe M.L., Knutson G.J., Fass D.N. & Hewick R.M. (1984) Molecular cloning of a cDNA encoding human antihaemophilic factor. *Nature* **312**, 342–7.

Towbin H., Staehelin T. & Gordon J. (1979) Electrophoretic transfer of proteins from polyacrylamide gels to nitrocellulose sheets: procedure and some applications. *Proc. Natl. Acad. Sci. USA* **76**, 4350–4.

Traboni C., Cortese R., Cilibert G. & Cesarini G. (1983) A general method to select M13 clones carrying base pair substitution mutants constructed *in vitro*. *Nucleic Acids Res.* **11**, 4229–39.

Trieu-Cuot P., Carlier C. & Courvalin P. (1988) Conjugative plasmid transfer from *Enterococcus faecalis* to *Escherichia coli*. *J. Bacteriol.* **170**, 4388–91.

Trieu-Cuot P., Carlier C., Martin P. & Courvalin P. (1987) Plasmid transfer by conjugation from *Escherichia coli* to Gram positive bacteria. *FEMS Microbiol. Lett.* **48**, 289–94.

Triglia T., Peterson M.G. & Kemp D.J. (1988) A procedure for *in vitro* amplification of DNA segments that lie outside the boundaries of known sequences. *Nucleic Acids Res.* **16**, 8186.

Tschumper G. & Carbon J. (1983) Copy number control by a yeast centromere. *Gene* **23**, 221–32.

Tschumper G. & Carbon J. (1987) *Saccharomyces cerevisiae* mutants that tolerate centromere plasmids at high copy number. *Proc. Natl. Acad. Sci. USA* **84**, 7203–7.

Tuite M.F., Dobson M.J., Roberts N.A., King R.M., Burke D.C., Kingsman S.M. & Kingsman A.J. (1982) Regulated high efficiency expression of human interferon-alpha in *Saccharomyces cerevisiae*. *EMBO J.* **1**, 603–8.

Turgeon R., Wood H.N. & Braun A.C. (1976) Studies on the recovery of crown gall tumor cells. *Proc. Natl. Acad. Sci. USA* **73**, 3562–4.

Twigg A.J. & Sherratt D. (1980) *Trans*-complementable copy-number mutants of plasmid ColE1. *Nature* **283**, 216–18.

Uchimaya H., Fushimi T., Hashimoto H., Harada H., Syono K. & Sugawara Y. (1986) Expression of a foreign gene in callus derived from DNA-treated protoplasts of rice (*Oryza sativa*). *Mol. Gen. Genetics* **204**, 204–7.

Uchimiya H., Iwata M., Nojiri C., Samarajeewa P.K., Takamatsu S., Ooba S., Anzai H., Christensen A.H., Quail P.H. & Toki S. (1993) Bialaphos treatment of transgenic rice plants expressing a *bar* gene prevents infection by the sheath blight pathogen (*Rhizoctonia solani*). *Biotechnology* **11**, 835–6.

Uhlin B.E., Molin S., Gustafsson P. & Nordstrom K. (1979) Plasmids with temperature-dependent copy number for amplification of cloned genes and their products. *Gene* **6**, 91–106.

Ullrich A., Bell J.R., Chen E.Y., Herrera R., Petruzzelli L.M., Dull T.J., Gray A., Coussens L., Liao Y.C., Tsubokawa M., Mason A., Seeburg P.H., Grunfield C., Rosen O.M. & Ramachandran J. (1985) Human insulin receptor and its relationship to the tyrosine kinase family of oncogenes. *Nature* **313**, 756–61.

Upshall A. (1986) Filamentous fungi in biotechnology. *Biotechniques* **4**, 158–66.

Vaeck M., Reynaerts A., Hofte H., Jansens S., De Beuckeleer M., Dean C., Zabeau M., van Montagu M. & Leemans J. (1987) Transgenic plants protected from insect attack. *Nature* **328**, 33–7.

Valenzuela P., Medina A., Rutter W.J., Ammerer G. & Hall B.D. (1982) Synthesis and assembly of hepatitis B virus surface antigen particles in yeast. *Nature* **298**, 347–50.

Van der Krol A.R., Lenting P.E., Veenstra J., Van der Meer I.M., Koes R.E., Gerats A.G.M., Mol J.N.M. & Stuitje A.R. (1988) An antisense chalcone synthase gene in transgenic plants inhibits flower pigmentation. *Nature* **333**, 866–9.

Van Dyk T.K., Gatenby A.A. & LaRossa R.A. (1989) Demonstration by genetic suppression of interaction of GroE products with many proteins. *Nature* **342**, 451–3.

Van Larbeke N., Engler G., Holsters M., van den Elsacker S., Zaenen I., Schilperoort R.A. & Schell J. (1974) Large plasmid in *Agrobacterium tumefaciens* essential for crown gall-inducing ability. *Nature* **252**, 169–70.

van Randen J. & Venema G. (1984) Direct plasmid transfer from replica-plated *E. coli* colonies to competent *B. subtilis* cells. Identification of an *E. coli* clone carrying the *his*H and *tyr*A genes of *B. subtilis*. *Mol. Gen. Genetics* **195**, 57–61.

Van Sluys M.A., Tempe J. & Fedoroff N. (1987) Studies on the introduction and mobility of the maize *Activator* element in *Arabidopsis thaliana* and *Daucus carota*. *EMBO J.* **6**, 3881–9.

Varadarajan R., Szabo A. & Boxer S.G. (1985) Cloning, expression in *Escherichia coli*, and reconstitution of human myoglobin. *Proc. Natl. Acad. Sci. USA* **82**, 5681–4.

Vasil V., Castillo A., Fromm M. & Vasil I. (1992) Herbicide resistant fertile transgenic wheat plants obtained by microprojectile bombardment of regenerable embryogenic callus. *Biotechnology* **10**, 667–74.

Vassart G., Georges M., Monsieur R., Brocas H., Lequarre A.S. & Christophe D. (1987) A sequence in M13 phage detects hypervariable minisatellites in human and animal DNA. *Science* **235**, 683–4.

Veale R.A., Guiseppin M.L.F., Van Eijk H.M.J., Sudberry P.E. & Verrips C.T. (1992) Development of a strain of *Hansenula polymorpha* for the efficient expression of guar α-galactosidase. *Yeast* **8**, 361–72.

Vellanoweth R.L. (1993) Translation and its regulation. *Bacillus subtilis* and other Gram-positive bacteria. In *Biochemistry, Physiology and Molecular Genetics*, eds Sonenshein A.L., Hoch J.A., Losick B.R. ASM, Washington DC.

Velten J., Fukada K. & Abelson J. (1976) *In vitro* construction of bacteriophage λ and plasmid DNA molecules containing DNA fragments from bacteriophage T4. *Gene* **1**, 93–106.

Venema G. (1979) Bacterial transformation. *Adv. Microbiol. Physiol.* **19**, 245–331.

Vieira J. & Messing J. (1982) The pUC plasmids, an M13mp7-derived system for insertion mutagenesis and sequencing with synthetic universal primers. *Gene* **19**, 259–68.

Vieira J. & Messing J. (1987) Production of single-stranded plasmid DNA. *Method Enzymol.* **153**, 3–11.

Vierling E. & Kimpel J.A. (1992) Plant responses to environmental stress. *Curr. Opinion Biotechnol.* **3**, 164–70.

Vilette D., Uzest M., Ehrlich S.D. & Michel B. (1992) DNA transcription and repressor binding affect deletion formation in *Escherichia coli* plasmids. *EMBO J.* **11**, 3629–34.

Villafane R., Bechhofer D.H., Narayanan C.S. & Dubnau E. (1987) Replication control genes of plasmid pE194. *J. Bacteriol.* **169**, 4822–9.

Villa-Komaroff L., Efstratiadas A., Broome S., Lomedico P., Tizard R., Naber S.P., Chick W.L. & Gilbert W. (1978) A bacterial clone synthesizing proinsulin. *Proc. Natl. Acad. Sci. USA* **75**, 3727–31.

Vinson C.R., LaMarco K.L., Johnson P.F., Landschulz W.H. & McKnight S.L. (1987) *In situ* detection of sequence-specific DNA binding activity specified by a recombinant bacteriophage. *Genes Devel.* **2**, 801–6.

Visser R.G.F., Somhorst I., Kuipers G.F.J., Ruys N.J., Feenstra W.J. & Jacobsen E. (1991) Inhibition of the expression of the gene for granule-bound synthase in potato by antisense constructs. *Mol. Gen. Genetics* **225**, 289–6.

Vize P.D. & Melton D.A. (1991) Assays for gene function in developing *Xenopus* embryos.

Methods Cell Biol. **36**, 367–87.

Voelker T.A., Worrell A.C., Anderson L., Bleibaum J., Fanc C., Hawkins D.J., Radke S.E. & Davies H.M. (1992) Fatty acid biosynthesis redirected to medium chains in transgenic oilseed plants. *Science* **257**, 72–4.

Vogel J., Hinrichs S.H., Reynolds R.K., Luciw P.A. & Jay G. (1988) The HIV *tat* gene induces dermal lesions resembling Kaposi's sarcoma in transgenic mice. *Nature* **335**, 606–11.

Wahl G.M., de Saint Vincent B.R. & DeRose M.L. (1984) Effect of chromosomal position on amplification of transfected genes in animal cells. *Nature* **307**, 516–20.

Wahl G.M., Lewis K.A., Ruiz J.C., Rothenberg B., Zhao J. & Evans G.A. (1987) Cosmid vectors for rapid genomic walking, restriction mapping, and gene transfer. *Proc. Natl. Acad. Sci. USA* **84**, 2160–4.

Walden R., Hayashi H. & Schell J. (1991) T-DNA as a gene tag. *Plant J.* **1**, 281–8.

Walker M.D., Edlund T., Boulet A.M. & Rutter W.J. (1983) Cell-specific expression controlled by the 5'-flanking region of insulin and chymotrypsin genes. *Nature* **306**, 557–61.

Wallace R.B., Johnson P.F., Tanaka S., Schold M., Itakura K. & Abelson J. (1980) Directed deletion of a yeast transfer RNA intervening sequence. *Science* **209**, 1396–400.

Wallace R.B., Schold M., Johnson M.J., Dembek P. & Itakura K. (1981) Oligonucleotide directed mutagenesis of the human β-globin gene: a general method for producing specific point mutations in cloned DNA. *Nucleic Acids Res.* **9**, 3647–56.

Wandersman C. (1992) Secretion across the bacterial outer membrane. *Trends Genetics* **8**, 317–22.

Ward A., Etessami P. & Stanley J. (1988) Expression of a bacterial gene in plants mediated by infectious geminivirus DNA. *EMBO J.* **7**, 1583–7.

Ward P.P., Lo J.-Y., Duke M., May G.S., Headon D.R. & Conneely O.M. (1992) Production of biologically active recombinant human lactoferrin in *Aspergillus oryzae*. *Biotechnology* **10**, 784–80.

Wasylyk B. (1988) Enhancers and transcription factors in the control of gene expression. *Biochim. Biophys. Acta* **951**, 17–35.

Watson B., Currier T.C., Gordon M.P., Chilton M.-D. & Nester E.W. (1975) Plasmid requirement for virulence of *Agrobacterium tumefaciens*. *J. Bacteriol.* **123**, 255–64.

Watson J.D. (1972) Origin of concatameric T7 DNA. *Nature New Biol.* **239**, 197–201.

Weatherall D.J. & Clegg J.B. (1982) Thalassemia revisited. *Cell* **29**, 7–9.

Weatherall D.J. & Old J.M. (1983) Antenatal diagnosis of the haemoglobin disorders by analysis of foetal DNA. *Mol. Biol. Med.* **1**, 151–5.

Weatherall D.J. (1991) *The New Genetics and Clinical Practice*, 3rd edn. Oxford University Press, Oxford.

Webb N.R. & Summers D.K. (1990) Expression of proteins using recombinant baculoviruses. *Technique* **2**, 173–88.

Weber J.M., Leung J.O., Swanson S.J., Idler K.B. & McAlpine J.B. (1991) An erythromycin derivative produced by targeted gene disruption in *Saccharopolyspora erythraea*. *Science* **252**, 114–17.

Wei C.-M., Gibson P., Spear P.G. & Scolnick E.M. (1981) Construction and isolation of a transmissible retrovirus containing the *src* gene of Harvey Sarcoma Virus and the thymidine kinase gene of herpes simplex virus type 1. *J. Virol.* **39**, 935–44.

Weiler E.W. & Schroder J. (1987) Hormone genes and crown gall disease. *TIBS* **12**, 271–5.

Weinstein M., Roberts R.C. & Helinski D.R. (1992) A region of the broad-host-range plasmid RK2 causes stable in plant inheritance of plasmids in *Rhizobium meliloti* cells isolated from alfalfa root nodules. *J. Bacteriol.* **174**, 7486–9.

Weinstock G.M., Ap Rhys C., Berman M.L., Hampar B., Jackson D. & Silhavy T.J. (1983) Open reading frame expression vectors: a general method for antigen production in *Escherichia coli* using protein fusions to β-galactosidase. *Proc. Natl. Acad. Sci. USA* **80**, 4432–6.

Weiss R., Teich N., Varmus H. & Coffin J. (1985) *RNA Tumour Viruses*, 2nd edn. Cold Spring Harbor Laboratory, Cold Spring Harbor, New York.

Wells J.A. & Estell D.A. (1988) Subtilisin – an enzyme designed to be engineered. *Trends Biochem.* **13**, 291–7.

Wells J.A. & Lowman H.B. (1992) Rapid evolution of peptide and protein binding properties *in vitro*. *Curr. Opinion Biotechnol.* **3**, 355–62.

Wells J.A., Vasser M. & Powers D.B. (1985) Cassette mutagenesis: an efficient method for generation of multiple mutations at defined sites. *Gene* **34**, 315–23.

Wensink P.C., Finnegan D.J., Donelson J.E. & Hogness D.S. (1974) A system for mapping DNA sequences in the chromosomes of *Drosophila melanogaster*. *Cell* **3**, 315–25.

Wetzel R., Perry L.J., Baase W.A. & Becktel W.J. (1988) Disulphide bonds and thermal stability in T4 lysozyme. *Proc. Natl. Acad. Sci. USA* **85**, 401–5.

White F.F. & Nester E.W. (1980) Relationship of plasmids responsible for hairy root and crown gall tumorigenicity. *J. Bacteriol.* **144**, 710–20.

Widera G., Gautier F., Lindenmaier W. & Collins J. (1978) The expression of tetracycline resistance after insertion of foreign DNA fragments between the *Eco*RI and *Hind*III sites of the plasmid cloning vector pBR322. *Mol. Gen. Genetics* **163**, 301–5.

Wigler M., Perucho M., Kurtz D., Dana S., Pellicer A., Axel R. & Silverstein S. (1980) Transformation of mammalian cells with an amplifiable dominant acting gene. *Proc. Natl. Acad. Sci. USA* **77**, 3567–70.

Wigler M., Silverstein S., Lee L.S., Pellicer A., Cheng Y.C. & Axel R. (1977) Transfer of purified herpes virus thymidine kinase gene to cultured mouse cells. *Cell* **11**, 223–32.

Wigler M., Sweet R., Sim G.K., Wold B., Pellicer A., Lacy E., Maniatis T., Silverstein S. & Axel R. (1979) Transformation of mammalian cells with genes from procaryotes and eucaryotes. *Cell* **16**, 777–85.

Wilkinson K.D. (1990) Detection and inhibition of ubiquitin-dependent proteolysis. *Methods Enzymol.* **185**, 387–97.

Wilkinson A.J., Fersht A.R., Blow D.M., Carter P. & Winter G. (1984) A large increase in enzyme–substrate affinity by protein engineering. *Nature* **307**, 187–8.

Wilks A.F. (1989) Two putative protein-tyrosine kinases identified by application of the polymerase chain reaction. *Proc. Natl. Acad. Sci. USA* **86**, 1063–7.

Williams D.C., Van Frank R.M., Muth W.L. & Burnett J.P. (1982) Cytoplasmic inclusion bodies in *Escherichia coli* producing biosynthetic human insulin proteins. *Science* **215**, 687–8.

Williams R.S., Johnston S.A., Riedy M., Devit M.J., McElligott S.G. & Sanford J.C. (1991) Introduction of foreign genes into tissues of living mice by DNA-coated microprojectiles. *Proc. Natl. Acad. Sci. USA* **88**,, 2726–30.

Williamson R., Eskdale J., Coleman D.V., Niazi M., Loeffler F.E. & Modell B. (1981) Direct gene analysis of chorionic villi: a possible technique for first trimester diagnosis of haemoglobinopathies. *Lancet* **2**, 1127.

Willmitzer L., Dhaese P., Schreier P.H., Schmalenbach W., Van Montagu M. & Schell J. (1983) Size, location & polarity of transferred DNA encoded transcripts in nopaline crown gall tumours: common transcripts in octopine and nopaline tumours. *Cell* **32**, 1045–6.

Willmitzer L., Simons G. & Schell J. (1982) The Ti DNA in octopine crown gall tumours codes for seven well-defined polyadenylated transcripts. *EMBO J.* **1**, 139–46.

Wilson C., Cross G.S. & Woodland H.R. (1986) Tissue-specific expression of actin genes injected into *Xenopus* embryos. *Cell* **47**, 589–99.

Wilson G.G. & Muway N.E. (1991) Restriction and modification systems. *Annu. Rev. Genetics* **25**, 585–627.

Wilson R. & 54 others (1994) 2.2 Mb of contiguous nucleotide sequence from chromosome III of *C. elegans*. *Nature* **368**, 32–8.

Windass J.D., Worsey M.J., Pioli E.M., Pioli D., Barth P.T., Atherton K.T., Dart E.C., Byrom D., Powell K. & Senior P.J. (1980) Improved conversion of methanol to single-cell protein by *Methylophilus methylotrophus*. *Nature* **287**, 396–401.

Winter G., Fersht A.R., Wilkinson A.J., Zoller M. & Smith M. (1982) Redesigning enzyme structure by site-directed mutagenesis: tyrosyl tRNA synthetase and ATP binding. *Nature* **299**, 756–8.

Winter J.A., Wright R.L. & Gurley W.B. (1984) Map locations of five transcripts homologous to TR-DNA in tobacco and sunflower crown gall tumours. *Nucleic Acids Res.* **12**, 2391–2406.

Wirth R., Frieseneger A. & Fielder S.F. (1989) Transformation of various species of Gram-negative bacteria belonging to 11 different genera by electroporation. *Mol. Gen. Genetics* **216**, 175–7.

Wong C., Dowling C.E., Sakai R.K., Higuchi R.G., Erlich H.A. & Kazazian H.H. (1987) Characterization of β-thalassaemia mutations using direct genomic sequencing of amplified single copy DNA. *Nature* **330**, 384–6.

Wong C.-H., Chen S.-T., Hennen W.J., Bibbs J.A., Wang Y.-F., Liu J.L.-C., Pantoliano M.W., Whitlow M. & Bryan P.N. (1990) Enzymes in organic synthesis: use of sub-tilisin and a highly stable mutant derived from multiple site-specific mutation. *J. Am. Chem. Soc.* **112**, 945–53.

Wong S.-L. (1989) Development of an inducible and enhanceable expression and secre-tion system in *Bacillus subtilis*. *Gene* **83**, 215–23.

Wong Z., Wilson V., Jeffreys A.J. & Thein S.L. (1986) Cloning a selected fragment from a human DNA 'fingerprint': isolation of an extremely polymorphic minisatellite. *Nucleic Acids Res.* **14**, 4605–16.

Wood K.V. (1991) *The Origin of Beetle Luciferase, Bioluminescence & Chemiluminescence Current Status*, eds Stanley P. & Kricka L. John Wiley, Chichester.

Wood C.R., Boss M.A., Kenlen J.H., Calvert J.E., Roberts N.A. & Emtage J.S. (1988) The synthesis and *in vivo* assembly of functional antibodies in yeast. *Nature* **314**, 446–9.

Wood K.V. & DeLuca M. (1987) Photographic detection of luminescence in *Escherichia coli* containing the gene for firefly luciferase. *Analytical Biochem.* **161**, 501–7.

Wood K.V., Lam Y.A., Seliger H.H. & McElroy W.D. (1989) Complementary DNA coding click beetle luciferases can elicit bioluminescence of different colors. *Science* **244**, 700–2.

Wood W.I., Capon D.J., Simonsen C.C., Eaton D.L., Gitschier J., Keyt B., Seeburg P.H., Smith D.H., Hollingshead P., Wion K.L., Delwart E., Tuddenham E.G.D., Vehar G.A. & Lawn R.M. (1984) Expression of active human factor VIII from recombinant DNA clones. *Nature* **312**, 330–7.

Woodcock D.M., Crowther P.J., Doherty J., Jefferson S., DeCruz E., Noyer-Weidner M., Smith S.S., Michael M.Z. & Graham M.W. (1989) Quantitative evaluation of *Escherichia coli* host strains for tolerance to cytosine methylation in plasmid and phage recom-binants. *Nucleic Acids Res.* **17**, 3469–78.

Woolcott M. (1992) Advances in nucleic acid-based detection methods. *Clin. Microbiol. Rev.* **5**, 370–86.

Woolford J.L. Jr. & Peebles C.L. (1992) RNA splicing in lower eukaryotes. *Curr. Opinion Genetics Devel.* **2**, 712–19.

Woolston C.J., Covey S.N., Penswick J.R. & Davies J.W. (1983) Aphid transmission and a polypeptide are specified by a defined region of the cauliflower mosaic virus genome. *Gene* **23**, 15–23.

Wright G., Carver A., Cottom D., Reeves D., Scott A., Simons P., Wilmut I., Garner I. & Colman A. (1991) High-level expression of active human alpha-1-antitrypsin in the milk of transgenic sheep. *Biotechnology* **9**, 830–4.

Wu X.-C., Lee W., Tran L. & Wong S.-L. (1991) Engineering a *Bacillus subtilis* expression–secretion system with a strain deficient in six extracellular proteases. *J. Bacteriol.* **173**, 4952–8.

Wullems G.J., Molendijk L., Ooms G. & Schilperoort R. (1981a) Differential expression of crown gall tumor markers in transformants obtained after *in vitro Agrobacterium tume-faciens* induced transformation of cell wall regenerating protoplasts derived from *Nicotiana tabacum*. *Proc. Natl. Acad. Sci. USA* **78**, 4344–8.

Wullems G.J., Molendijk L., Ooms G. & Schilperoort R.A. (1981b) Retention of tumor markers in F1 progeny plants formed from *in vitro* induced octopine and nopaline tumor tissues. *Cell* **24**, 719–28.

Wychowiski C., Emerson S.U., Silver J. & Feinstone S.M. (1990) Construction of recom-binant DNA molecules by the use of a single stranded DNA generated by the polymerase chain reaction: its application to chimeric hepatitis A virus/poliovirus subgenomic cDNA. *Nucleic Acids Res.* **18**, 913–18.

Wyman A.R., Wolfe L.B. & Botstein D. (1985) Propagation of some human DNA sequences in bacteriophage lambda vectors requires mutant *Escherichia coli* hosts. *Proc. Natl. Acad. Sci. USA* **82**, 2880–4.

Yadav N.S., Vanderleyden J., Bennet D., Barnes W.M. & Chilton M.-D. (1982) Short direct repeats flank the T-DNA on a nopaline Ti plasmid. *Proc. Natl. Acad. Sci. USA* **79**, 6322–6.

Yagi Y., McLellan T., Frez W. & Clewell D. (1978) Characterization of a small plasmid determining resistance to erythromycin, lincomycin and vernamycin B_α in a strain of *Streptococcus sanguis* isolated from dental plaque. *Antimicrob. Agents Chemother.* **13**, 884–7.

Yang N.-S., Burkholder J., Roberts B., Martinelli B. & McCabe D.E. (1990) *In vivo* and *in vitro* gene transfer to mammalian somatic cells by particle bombardment. *Proc. Natl. Acad. Sci. USA* **87**, 9568–72.

Yang X.C., Karschin A., Labarca C., Elroystein O., Moss B., Davidson N. & Lester H.A. (1991) Expression of ion channels and receptors in *Xenopus* oocytes using vaccinia virus. *FASEB J* **5**, 2209–16.

Yanisch-Perron C., Vieira J. & Messing J. (1985) Improved M13 phage cloning vectors and host strains: nucleotide sequences of the M13 mp18 and pUC19 vectors. *Gene* **33**, 103–19.

Yanofsky C. & Kolter R. (1982) Attenuation in amino acid biosynthetic operons. *Annu. Rev. Genetics* **16**, 113–34.

Yansura D.G. & Henner D.J. (1984) Use of the *Escherichia coli lac* repressor and operator to control gene expression in *Bacillus subtilis*. *Proc. Natl. Acad. Sci. USA* **81**, 439–43.

Yarranton G.T., Wright E., Robinson M.K. & Humphreys G.O. (1984) Dual origin plasmid vectors whose origin of replication is controlled by the coliphage lambda promoter P_L. *Gene* **28**, 293–300.

Ymer S., Schofield P.R., Draguhn A., Werner P., Kohler M. & Seeburg P.H. (1989) GABA receptor beta-subunit heterogeneity: functional expression of cloned cDNAs. *EMBO J* **6**, 1665–70.

Yoshizumi H. & Ashikari T. (1987) Expression, glycosylation and secretion of fungal hydrolases in yeast. *Trends Biotechnol.* **5**, 277–81.

Young B.D., Birnie G.D. & Paul J. (1976) Complexity and specificity of polysomal poly $(A)^+$ RNA in mouse tissues. *Biochemistry* **15**, 2823–8.

Young M. & Ehrlich S.D. (1989) Stability of reiterated sequences in the *Bacillus subtilis* chromosome. *J. Bacteriol.* **171**, 2653–6.

Young R.A. & Davis R.W. (1983) Efficient isolation of genes by using antibody probes. *Proc. Natl. Acad. Sci. USA* **80**, 1194–8.

Youngman P. (1990) Use of transposons and integrational vectors for mutagenesis and construction of gene fusions in *Bacillus* species. In *Molecular Biological Methods for Bacillus*, eds Harwood C.R. & Cutting S.M., 580 pp. John Wiley, Chichester.

Yuan R., Hamilton D.L. & Burckhardt J. (1980) DNA translocation by the restriction enzyme from *E. coli* K. *Cell* **20**, 237–44.

Yuan R., Bickle T.A., Ebbers W. & Brack C. (1975) Multiple steps in DNA recognition by restriction endonuclease from *E. coli* K. *Nature* **256**, 556–60.

Zaballos A., Salas M. & Mellado R.P. (1987) A set of expression plasmids for the synthesis of fused and unfused polypeptides in *Escherichia coli*. *Gene* **58**, 67–76.

Zaenen I., Van Larbeke N., Teuchy H., Van Montagu M. & Schell J. (1974) Super-coiled circular DNA in crown-gall inducing *Agrobacterium* strains. *J. Mol. Biol.* **86**, 109–27.

Zambryski P., Holsters M., Kruger K., Depicker A., Schell J., Van Montagu M. & Goodman H. (1980) Tumor DNA structure in plant cells transformed by *A. tumefaciens*. *Science* **209**, 1385–91.

Zambryski P., Depicker A., Kruger H. & Goodman H. (1982) Tumor induction by *Agrobacterium tumefaciens*: analysis of the boundaries of T-DNA. *J. Mol. Appl. Genetics* **1**, 361–70.

Zambryski P., Joos H., Genetello C., Leemans J., Van Montagu M. & Schell J. (1983) Ti plasmid vector for the introduction of DNA into plant cells without alteration of their normal regeneration capacity. *EMBO J.* **2**, 2143–50.

Zhong Z., Liu J.L.-C., Dinterman L.M., Finkelman M.A., Mueller W.T., Rollence M.L., Whitlow M. & Wong C.-H. (1991) Engineering subtilisin for reaction in dimethylformamide. *J. Am. Chem. Soc.* **113**, 683–4.

Zimmer A. & Gruss P. (1989) Production for chimaeric mice containing embryonic stem cell (ES) cells carrying a homeobox *Hox1.1* allele mutated by homologous recombination. *Nature* **338**, 150–3.

Zimmerman S.B. & Pheiffer B. (1983) Macromolecular crowding allows blunt end ligation by DNA ligases from rat liver or *Escherichia coli*. *Proc. Natl. Acad. Sci. USA* **80**, 5852–6.

Zimmerman U. & Vienken J. (1983) Electric field induced cell to cell fusion. *J. Membr. Biol.* **67**, 165–82.

Zinder N.D. & Boeke J.D. (1982) The filamentous phage (Ff) as vectors for recombinant DNA – a review. *Gene* **19**, 1–10.

Zoller M.J. & Smith M. (1983) Oligonucleotide-directed mutagenesis of DNA fragments cloned into M13 vectors. *Methods Enzymol.* **100**, 468–500.

Zubay G., Lederman M. & DeVries J.K. (1967) DNA-directed peptide synthesis and inhibition of repressor by inducer in a cell-free system. *Proc. Natl. Acad. Sci. USA* **58**, 1669–75.

Additional references

Burke D.T., Carle G.F. & Olson M.V. (1987) Cloning of large segments of exogenous DNA into yeast by means of artificial chromosome vectors. *Science* **236**, 806–13.

Burke T. & Bruford M.W. (1987) DNA fingerprinting in birds. *Nature* **327**, 149–52.

Clarke L. & Carbon J. (1976) A colony bank containing synthetic Col E1 hybrid plasmids representative of the entire *E. coli* genome. *Cell* **9**, 91–9.

Debatisse M., Hyrien O., Petit-Koskas E., de Saint Vincent B.R. & Buttin G. (1980) Secretion and rearrangement of coamplified genes in different lineages of mutant cells that over produce adenylate deaminase. *Mol. Cell. Biol.* **6**, 1776–81.

Frech G.C., Vandongen A.M.J., Schuster G., Brown A.M. & Joho R.H. (1989) A novel potassium channel with delayed rectifier properties isolated from rat brain by expression cloning. *Nature* **340**, 642–5.

Hediger M.A., Coady M.J., Ikeda T.S. & Wright E.M. (1987) Expression cloning and cDNA sequencing of the Na^+/glucose co-transporter. *Nature* **330**, 379–81.

Hollmann M., Osheagreenfield A., Rogers S.W. & Heinemann S. (1989) Cloning by functional expression of a member of the glutamate receptor family. *Nature* **342**, 643–8.

Honda Z., Nakamura M., Miki I., Minami M., Watanabe T., Seyama Y., Okado H., Toh H., Ito K., Miyamoto T. & Shimizu T. (1991) Cloning by functional expression of platelet activating factor receptor from guinea-pig lung. *Nature* **349**, 342–6.

Kaufman R.J. (1990) Strategies for obtaining high level expression in mammalian cells. *Technique* **2**, 221–36.

Paulmichl M., Li Y., Wickman K., Ackerman M., Peralta E. & Clapham D. (1992) New mammalian chloride channel identified by expression cloning. *Nature* **356**, 238–41.

Straub R.E., Frech G.C., Joho R.H. & Gershengorn M.C. (1990) Expression cloning of a cDNA encoding the mouse pituitary thyrotropin-releasing hormone receptor. *Proc. Natl. Acad. Sci. USA* **87**, 9514–18.

Tate S.S., Yan N. & Udenfriend S. (1992) Expression cloning of a Na^+-independent neutral amino acid transporter from rat kidney. *Proc. Natl. Acad. Sci. USA* **89**, 1–5.

Index

Page numbers in italic refer to figures, those in bold refer to tables.